Innovative Strategies and Approaches for End–User Computing Advancements

Ashish Dwivedi
Hull University, UK

Steve Clarke
Hull University, UK

Managing Director:	Lindsay Johnston
Editorial Director:	Joel Gamon
Book Production Manager:	Jennifer Romanchak
Publishing Systems Analyst:	Adrienne Freeland
Assistant Acquisitions Editor:	Kayla Wolfe
Typesetter:	Henry Ulrich
Cover Design:	Nick Newcomer

Published in the United States of America by
    Information Science Reference (an imprint of IGI Global)
    701 E. Chocolate Avenue
    Hershey PA 17033
    Tel: 717-533-8845
    Fax: 717-533-8661
    E-mail: cust@igi-global.com
    Web site: http://www.igi-global.com

Library of Congress Cataloging-in-Publication Data

Innovative strategies and approaches for end-user computing advancements / Ashish Dwivedi and Steve Clarke, editors.
    p. cm.
 Includes bibliographical references and index.
 Summary: "This book presents comprehensive research on the implementation of organizational and end user computing initiatives to further understand this discipline and its related fields"--Provided by publisher.
 ISBN 978-1-4666-2059-9 (hardcover) -- ISBN 978-1-4666-2060-5 (ebook) -- ISBN 978-1-4666-2061-2 (print & perpetual access) 1. End-user computing--
Technological innovations. I. Dwivedi, Ashish N. II. Clarke, Steve, 1950-
 QA76.9.E53I56 2013
 004--dc23
             2012013062

British Cataloguing in Publication Data
A Cataloguing in Publication record for this book is available from the British Library.

The views expressed in this book are those of the authors, but not necessarily of the publisher.

Table of Contents

Detailed Table of Contents

Yongmei Bentley, University of Bedfordshire Business School, UK
Steve Clarke, University of Hull Business School, UK

Information strategy is often relegated to an information technology element of corporate strategy, or worse, ignored in favour of IT operational planning. This research, conducted over a five-year period, stresses the correct framing of an information strategy and its implementation. The authors propose a framework that assists in the evaluation of such strategies, primarily those at higher education institutions, but also in a wider range of organisations seeking to improve the understanding and implementation of their information strategy.

Chad Anderson, Georgia State University, USA
Said S. Al-Gahtani, King Khalid University, Saudi Arabia
Geoffrey S. Hubona, Virginia Commonwealth University, USA

Theoretical models are often conceived and tested in western countries. However, culture influences theoretical models, and the importance of evaluating models in non-western cultures has grown with the accelerating pace of globalization. The technology acceptance model (TAM) is no exception, and more TAM research is being conducted in non-western countries. TAM constructs of perceived usefulness and perceived ease of use are difficult to act on, which has led to several studies that identified valid antecedents to these constructs that make the model more practically actionable. These antecedents were conceived and tested in a western country but have yet to be evaluated in the context of a non-western country. In this paper, the authors evaluate these antecedents in Saudi Arabia and find that they function in the specific context of general computer use by Saudi knowledge workers.

Chapter 3

Yinglei Wang, Acadia University, Canada

Darren B. Meister, The University of Western Ontario, Canada

Peter H. Gray, University of Virginia, USA

The way individuals use internal and external knowledge sources influences organizational knowledge integration, an important source of competitive advantage. Drawing on research into knowledge sourcing and consumer switching behavior, the authors develop an integrated model to understand individuals' choices between internal and external knowledge sources in contemporary work settings, where information technology has made both easily accessible. A test of the model using survey data collected from an international consulting firm yields an important new insight: satisfied individuals in knowledge reuse friendly environments are likely to use internal knowledge sources while they may also be tempted by easily accessible external knowledge sources. The implications for researchers and practitioners are also discussed.

Chapter 4

Sandra Barker, University of South Australia, Australia

Brenton Fiedler, University of South Australia, Australia

The acceleration of technology in business since the 1980s suggests that traditional management techniques, systems, and strategies employed in a business environment should be challenged. As a consequence of this acceleration, end-user computing (EUC) and end-user development (EUD) have also grown. Definitions of EUC developed in the 1980s continue to be used by contemporary researchers without regard to the changing technological environment, user experience, and user needs. Therefore, the authors challenge traditional definitions of EUC developed and used by researchers to ascertain whether they meet the needs of management for the 21st century. There is a conflict among traditional definitions that has not been addressed since the early 1990s (Downey & Bartczak, 2005). In this regard, the authors proffer that the management strategies for end-user (EU) systems development in the 21st century should suggest a different and proactive role for users. This paper summarises key traditional definitions from the literature and evaluates their consonance with the technology and business system environment. The impetus for researchers to rethink the traditional definition of EUC is provided through a real world management project involving the development of a university staff workload database that investigated the role of end-users in system enhancement and development.

Chapter 5

Brian Bishop, Dundalk Institute of Technology, Ireland

Kevin McDaid, Dundalk Institute of Technology, Ireland

The reliability of end-user developed spreadsheets is poor. Research studies find that 94% of 'real-world' spreadsheets contain errors. Although some research has been conducted in the area of spreadsheet testing, little is known about the behaviour or processes of individuals during the debugging task. In this paper, the authors investigate the performance and behaviour of expert and novice end-users in the debugging of an experimental spreadsheet. To achieve this aim, a spreadsheet debugging experiment was conducted, with professional and student participants requested to debug a spreadsheet seeded with errors. The work utilises a novel approach for acquiring experimental data through the unobtrusive recording of participants' actions using a custom built VBA tool. Based on findings from the experiment, a debugging tool is developed, and its effects on debugging performance are investigated.

Chapter 6
Exploring the Dimensions and Effects of Computer Software Similarities in Computer Skills

Yuan Li, Columbia College, USA
Kuo-Chung Chang, Yuan Ze University, Taiwan

Computer software similarities play important roles in users' skills transfer from one application to another. Despite common software attributes recognized in extant literature, a systematic understanding of the components and structure of software similarities has not been fully developed. To address the issue, a Delphi study was conducted to explore the underlying dimensions of software similarities. Inputs gathered from 20 experienced Information Systems instructors show that Computer Software Similarity is a multi-dimensional construct made up of interface similarity, function similarity, and syntax similarity. Each dimension consists of software attributes that users perceive to be transferable in learning new applications. A field study was carried out to test the impact of the construct. Results from a survey on students' learning two software applications confirm the expectation that Computer Software Similarity facilitates the students' skills transfer between the applications. These studies provide a basis to better design training programs for improved training performance.

Chapter 7
Preparing IS Students for Real-World Interaction with End Users Through Service Learning:

Laura L. Hall, University of Texas at El Paso, USA
Roy D. Johnson, University of Pretoria, South Africa

Although teaching the technical skills required of Information Systems (IS) graduates is a straightforward\ process, it is far more difficult to prepare students in the classroom environment for the challenges they will face interacting with end users in the real world. The ability to establish a successful relationship with end users is a critical success factor for any IS project. One way to prepare students for interaction with end users is through the implementation of service learning projects. Service learning projects provide a rich environment for students to experience real world interactions with users. This paper presents an organizational model to guide the implementation of service learning projects in IS curriculums. Service learning projects better prepare students to assume important management positions by giving them experience in applying the system development life cycle to an IS project and working with people. This organizational model uses the system development life cycle approach to integrate typical curriculum and service learning models. The organizational model is grounded in anecdotal evidence from prior experiences with IS students in service learning environments.

Chapter 8
Construct Validity Assessment in IS Research: Methods and Case Example of User Satisfaction

Dewi Rooslani Tojib, Monash University, Australia
Ly-Fie Sugianto, Monash University, Australia

Valid and reliable measures are critical to theory development as they facilitate theory testing in empirical research. Efforts in scale development have been put on ensuring aspects of validity. In this paper, the authors address a specific topic of construct validity assessment in scale development. Using data from the five leading IS journals between 1989-2008, in this paper, the authors determine if and how the field has advanced in construct validity assessment. Findings suggest that the proportion of studies reporting construct validity had increased and Confirmatory Factor Analysis (CFA), Exploratory Factor Analysis (EFA), and Multi-Trait Multi-Method (MTMM) were the three most common methods of construct validity assessment. The authors also apply a popular method from psychology and exemplify how the correlation analysis technique can be used to measure construct validity.

Chapter 9

Lixuan Zhang, Augusta State University, USA

Mary C. Jones, University of North Texas, USA

Attracting and retaining information technology (IT) professionals is a current concern for companies. Although research has been conducted about the job behavior and attitudes of IT professionals over the past three decades, little research has explored the effect of IT professionals' social capital. The primary research question that this study addresses is how social capital affects IT professionals' work attitude and behavior, including job satisfaction and job performance. Data were collected from 128 IT professionals from a range of jobs, organizations and industries. Results indicate that the strength of the ties an IT professional has in his or her organization is positively related to job satisfaction. The number of ties that an IT professional has outside the organization is also positively related to job performance. Several implications for research and practice are offered based on these findings.

Chapter 10

Richard D. Johnson, University at Albany – SUNY, USA

Although previous research has suggested that women may be at a learning disadvantage in e-learning environments, this study examines communication differences between women and men, arguing that women's communication patterns may provide them with a learning advantage. Using data from 303 males and 252 females, this paper discusses gender differences in course communication processes and course outcomes in a web-based introductory information systems course. Results indicate that women communicated more, perceived the environment to have greater social presence, were more satisfied with the course, found the course to be of greater value, and had marginally better performance than men. Despite the challenges facing women in e-learning environments, the results of this study suggest that e-learning environments that allow for peer to peer communication and connectedness can help females overcome some of these disadvantages. Implications for research and practice are also discussed.

Chapter 11

Katherine M. Chudoba, Utah State University, USA

Mary Beth Watson-Manheim, University of Illinois, Chicago, USA

Kevin Crowston, Syracuse University, USA

Chei Sian Lee, Nanyang Technological University, Singapore

Meetings are a common occurrence in contemporary organizations. The authors' exploratory study at Intel, an innovative global technology company, suggests that meetings are evolving beyond their familiar definition as the pervasive use of information and communication technologies (ICTs) changes work practices associated with meetings. Drawing on data gathered from interviews prompted by entries in the employees' electronic calendar system, the authors examine the multiple ways in which meetings build and reflect work in the organization and derive propositions to guide future research. Specifically, the authors identify four aspects of meetings that reflect work in the 21st century: meetings are integral to work in team-centered organizations, tension between group and personal objectives, discontinuities, and ICT support for fragmented work environment.

Chapter 12

This paper examines the role of fairness and how it shapes a user's view in IT-enabled change. Drawing from several fairness theories, components of fairness are identified and examined in two studies. The first study examines the role of fairness through user interviews and finds that all five components of fairness are considered by users in enterprise system implementations. The second study operationalizes and analyzes the components of fairness through a questionnaire distributed to users. This second study finds that fairness is comprised of all five components that were proposed and a significant relationship exists with user dissatisfaction. The two studies lead to a new theoretical perspective and provide practical implications regarding the role of fairness in IT-enabled change and their strategic implications.

Chapter 13

The success of engineering work depends on the ability of individuals to improvise in response to emerging challenges and opportunities (Kappel & Rubenstein, 1999). Building on experiential learning theory (Eisenhardt & Tabrizi 1995; Kolb, 1984) and improvisation theory (Miner, Bassoff, & Moorman, 2001), this authors argue that information systems facilitate the generation of new product and process design ideas by providing richer feedback, creating shorter learning cycles, and enabling engineers to try a variety of new ideas more easily. An empirical research model of the antecedents of improvisation in IT-enabled engineering work is proposed. This model is examined using a sample of 208 individuals engaged in computer-intensive engineering design work. The multiple regression results suggest that software capability, autonomy, problem solving/decision support usage, system use for work planning, and length of use explain the extent of new product and process ideas that are generated. The practical and theoretical implications of these findings are discussed.

Chapter 14

Computer competence is poorly conceptualized and inconsistently measured. This study clarifies computer competence and examines its relationship with anxiety, affect, and pessimism, along with self-efficacy and previous experience. Using a survey of 610 end users, the strengths of anxiety, affect (positive), pessimism, self-efficacy, and previous experience were compared for nine different competency measures in seven different domains, including word processing, email applications, spreadsheets, graphic programs, databases, web design, and overall computing. Results suggest that for most domains, affect and anxiety are significant predictors, as are self-efficacy and previous experience, but pessimism is not. In addition, competence in a domain was found to mediate the relationship between competence and its antecedents. These results suggest that organizations focus not only on skills training, but on ways to enhance computing attitudes during the training process.

Chapter 15

Scientists and engineers are increasingly developing software to enable them to do their work. A number of characteristics differentiate the software development environment in which a scientist or engineer works from the development environment in which a more traditional business/IT software developer works. This paper describes a case study, specifically about the development of a mesh-generation code. The goal of this case study was to understand the process for developing the code and identify some lessons learned that can be of use to other similar teams. Specifically, the paper reports on lessons learned concerning: requirements evolution, programming language choice, methods of communication among teammates, and code structure.

In this paper, the authors present the results of a qualitative case-study seeking to characterize data discovery needs and barriers of principal investigators and research support staff in clinical translational science. Several implications for designing and implementing translational research systems have emerged through the authors' analysis. The results also illustrate the benefits of forming early partnerships with scientists to better understand their workflow processes and end-user computing practices in accessing data for research. The authors use this user-centered, iterative development approach to guide the implementation and extension of i2b2, a system they have adapted to support cross-institutional aggregate anonymized clinical data querying. With ongoing evaluation, the goal is to maximize the utility and extension of this system and develop an interface that appropriately fits the swiftly evolving needs of clinical translational scientists.

Chapter 17

Cloud computing services, which allow users to lease time on remote computer systems, must be particularly attractive to smaller engineering organizations that use engineering simulation software. Such organizations have occasional need for substantial computing power but may lack the budget and in-house expertise to purchase and maintain such resources locally. The case study presented in this paper examines the potential benefits and practical challenges that a medium-sized manufacturing firm faced when attempting to leverage computing resources in a cloud computing environment to do model-based simulation. Results show substantial reductions in execution time for the problem of interest, but several socio-technical barriers exist that may hinder more widespread adoption of cloud computing within engineering.

When software supports the complex and poorly understood application domain of cutting-edge science, effective engagement between its users/customers and developers is crucial. Drawing on recent literature, the authors examine barriers to such engagement. Significant among these barriers is the effects of the experience that many research scientists have of local scientific end-user development. Through a case study, the authors demonstrate that involving such scientists in a team developing software for a widely distributed group of scientists can have a positive impact on establishing requirements and promoting adoption of the software. However, barriers to effective engagement exist, which scientific end-user developers can do little to address. Such barriers stem from the essential nature of scientific practice.

The development of scientific software is usually carried out by a scientist who has little professional training as a software developer. Concerns exist that such development produces low-quality products, leading to low-quality science. These concerns have led to recommendations and the imposition of software engineering development processes and standards on the scientists. This paper utilizes different frameworks to investigate and map characteristics of the scientific software development environment to the assumptions made in plan-driven software development methods and agile software development methods. This mapping exposes a mismatch between the needs and goals of scientific software development and the assumptions and goals of well-known software engineering development processes.

Preface

Technology continues to strongly influences the way that work is carried out in modern day organisations, and creates new opportunities and new demands for a range of different approaches for modern day organisations (Feldman and Gainey, 1997). Technology-led change opens up opportunities for new working methods in three main ways: allowing existing activities to be carried out more rapidly, with more consistency and at a lower cost than could previously be achieved. The explosive growth of the internet has promoted investment in information and communication devices, which is further accelerating this trend (Kazman and Westerheim, 1999). IT expenditure is approximately around the 7-10% mark in other information-intensive industries (Moran, 1998). Managers are under pressure to examine the costs associated with delivering IT solutions, and to discover approaches that would help to carry out their activities better, faster and cheaper (Davis and Klein, 2000; Latamore, 1999). Latamore (1999) argues that workflow and associated internet technologies are seen as an instrument to cut costs as specifically-designed IT implementations (such as workflow tools) are used to reduce administrative expenses. For example - Data has become a major asset for organisations and recent innovations in Information Technology (IT) have transformed the way that most organisations function. Concepts such as data warehousing and data mining have exponentially increased the amount of information to which a organisation has access – this has created the problem of "information explosion".

THE PROBLEM AND ITS CONTEXT: INFORMATION OVERLOAD

One of the major challenges that face managers is how to make effective decisions based on the data at hand. It is acknowledged that the selection of a particular direction is both constrained and influenced by the availability of data, the ability to transform data into information and then to make recognition of it by deriving knowledge from information. The fact that modern day organisations are facing a deluge of data and information (Sieloff, 1999) whilst simultaneously lacking knowledge is very well documented (Dwivedi, Bali, James, & Naguib, 2001; Dwivedi & Clarke, 2012; Dwivedi, Wickramasinghe, Bali, & Naguib, 2008). The 20th century has also witnessed a business environment where revolutionary technologies have resulted in new products and reduced product lifecycles. Thus the twin forces of economic and technological revolutions have put organizations under pressure to adopt innovative information management practices so as to be in a position to adapt swiftly to the new business environment (Sieloff, 1999). Technological innovations relating to workflow and groupware systems in conjunction with the growth of the WWW has brought about a radical transformation in the way organizations can interact internally and externally. These new ways of collaboration have resulted in organizations deluged with information to an unprecedented degree resulting in data/information overload (Sieloff, 1999).

The widespread use of Internet applications using client server architecture and web browsers has further increased exponentially our ability to draw on information in ways not previously possible. It is now possible to allow every computer access to the entire organizational knowledgebase. Moreover, technologies such as Data Integration, Document and Content Management (support applications enabling users to have personalised access to the organizational knowledgebase) continue to grow at an exponential rate. This has implications for decision makers across, as they have to deal with large amounts of data.

As a solution to the massive problem of the information explosion, Sieloff (1999) has recommended that organisations try to use KM to manage the flow of information and knowledge as it deals intrinsically with the learning capacity of human resources of a company. Consequently, KM has become an important focus area for organisations (Earl & Scott, 1999). According to Covin and Stivers (1997), the following factors have brought focus on for contemporary KM solutions (almost all KM solutions involve significant IS and End-User Involvement)

1. **Organisational Restructuring:** The reshaping of the organisational landscape has led to a situation where rapid changes in technology and business climates (i.e. the external business environment) have forced organisations to invent, discover, and transfer knowledge.
2. **Investments in Information Technology:** Organisations have made massive investments in organisational infrastructures which, in-turn, have provided a means to create and capture knowledge.
3. **Customer Contact:** The amount of contact that organisational employees have with their customers has increased exponentially, resulting in the requirement for a mechanism that allows organisations to capture the tacit knowledge of the relationship.
4. **Globalisation:** The emergence of a globalised world for business has diminished geographic barriers. This has accentuated the need for systems that effectively allow organisations to gain knowledge from different sources (i.e. subsidiaries, joint ventures, markets collaborators etc) all over the world.
5. **Collaborative Knowledge work:** Highly specialised work has resulted in the emergence of workflow and collaborative tools. These tools have formed the substrate for the development of tools that support the codification of information present in a system. These tools have raised the challenge for organisations to have a mechanism that supports contextual and effective dissemination of information in an organisation.
6. **The need for increased speed and reduced cycle time:** Organisations function in a highly competitive environment. As such, they are under constant pressure to use their resources in a manner which allows access and adaptation of information present in the organisational database, rather then be forced to repeatedly reinvent the wheel.

As disciplines, Knowledge Management and Information and Communication Technology do not have commonly accepted or de-facto definitions. However, we would argue that some common ground has been established which covers the following points:

KM is a multi-disciplinary paradigm, which uses technologies in order to support knowledge acquisition, generation, codification, and transfer in the context of specific organisational processes.

ICT refers to the recent advances in applications of communication technologies which have enabled access to large amounts of data and information, when seeking to identify problems or solutions to specific issues.

It has been argued that the main aim behind any KM strategy is to ensure that knowledge workers have access to the right knowledge, at the right place and at the right time (Dove, 1999). According to some KM experts, the above outlook forms the crucible of any organisational attempts to build a personalised KM system. KM can be construed as having origins in both of the ancient disciplines of the sciences and the arts.

KM BUILDING BLOCKS: DATA, INFORMATION, KNOWLEDGE AND WISDOM

Bellinger, Castro, and Mills (1996) while explaining the interrelationship between data, information, knowledge and wisdom, submits that data on its own has no value. Data becomes information when a relationship is established between different elements of data, i.e. information is data *with* meaning. It is contented that the main difference between knowledge and information is that, in knowledge, collated information has a value greater then the individual parts that compose it. This has resulted in what we call the *knowledge paradox*: knowledge for one individual could be data for another individual. Bellinger, Castro, and Mills (1996) further state that wisdom is an "extrapolative and non-deterministic, non-probabilistic process", whilst understanding is an interpolative and probabilistic process and that knowledge is a deterministic process. While interpreting the above-mentioned definition of wisdom, it should be noted that emphasis has been laid on the use of *distinction* (i.e. wisdom is said to arise when knowledge is used as a process that allows for a distinction to be made between different aspects of known knowledge). The concept of wisdom and knowledge is acceptable as it supports the notion that wisdom requires interpretation from a human (i.e. sociological) perspective, whilst knowledge in certain cases can be acquired through a technological process.

A number of leading management theorists have affirmed that the Hungarian chemist, economist and philosopher Michael Polanyi was among the earliest theorists to have popularised the concept of characterising knowledge as tacit or explicit. This is now recognised as the de-facto knowledge categorisation approach (Beijerse, 1999; Gupta, Iyer and Aronson, 2000; Hansen, Nohria, and Tierney, 1999; Nonaka, 1988; Nonaka and Konno, 1998; Zack, 1999). Explicit knowledge typically takes the form of company documents and is easily available whilst Tacit knowledge is subjective and cognitive. The cornerstone of any KM project is to transform tacit knowledge into explicit knowledge so as to allow its effective dissemination (Gupta, Iyer and Aronson 2000).

Nonaka and Konno (1998) have also substantiated Polanyi's contention that there are two kinds of knowledge: explicit knowledge and tacit knowledge. They add that there are significant differences between the way these two kinds of knowledge are viewed in the western and the eastern-based management entities. Western management organisations lay more emphasis on explicit knowledge as it is more recognised and scientifically organised. Eastern management entities and in particular Japanese firms "view knowledge as being primarily tacit, something not easily visible and expressible" (Nonaka and Konno, 1998, pp.42). They also mention that whilst there are two dimensions of tacit knowledge: (1) technical, (2) cognitive, it is the cognitive dimension that moulds the manner in which we perceive the information available to us. This supports the notion we raised earlier that creation of knowledge requires human insight.

Zack (1999) also affirms that tacit knowledge is not easily expressed, as we comprehend tacit knowledge at a subconscious level. Explicit knowledge, conversely, is more precisely and formally articulated, although removed from the original context, creation or use. This definition of explicit knowledge brings to light an important feature of explicit knowledge: that the knowledge passed on is not original. An analogy can be drawn between explicit knowledge and secondary data. Explicit knowledge can never be like primary data: (i.e. based upon a new concept, which has not been validated). Explicit knowledge has to be based upon real life experiences, which have already taken place. Another distinguishing feature of tacit knowledge is that it is often based upon first hand experiences that others have gone through. The transfer component of tacit knowledge comes into effect when it is disseminated. When dissemination is carried out in written format (manuals etc), tacit knowledge becomes explicit knowledge. Personal knowledge is the outcome of an intellectual process which includes an individual's personal experiences, values and sagacity. Organisational knowledge refers to the codified knowledge that is present in organisational storage (document repositories etc) and communication mediums (organisational practices and norms). We argue that the key to success is an effective integration of technology with the human decision-making process, and that it is therefore important to consider frameworks that encompass technological, organisational and managerial perspectives. Any such framework would involve bringing in various aspects of Organizational and End-User Computing.

In-order to provide Information Technology educators, researchers, and practitioners with a concise summary of the most recent advances in theory and practice of organizational and end user computing, as editors, we have structured current publications from the Journal of Organizational and End User Computing in this book. As detailed below – we have selected 3 book chapters (Chapters 1 to 3) which present new models on how to increase organizational and end user productivity and performance, 4 book chapters (Chapters 4 to 7) which discuss new insights into how current IT practice is enhancing organizational and end user productivity and performance, 7 book chapters (Chapters 8 to 14) which highlight new theoretical perspectives on various issues surrounding organizational and end user productivity themes, and 5 book chapters (Chapters 15 to 19) which put forward new insights on how innovations in computing practices and products are enabling the creation, development and enhancement of IT applications that result in better organizational and end user productivity and performance.

Organisation of this Book

Chapter 1, "Evaluation of Information Strategy Implementation: A Critical Approach" by Bentley and Clarke notes that Information strategy is often relegated to an information technology element of corporate strategy, or worse, ignored in favour of IT operational planning. The authors propose a framework that assists in the evaluation of such strategies. Chapter 2, "The Value of TAM Antecedents in Global IS Development and Research" by Anderson, Al-Gahtani and Hubona notes that theoretical models are often conceived and tested in western countries. However, culture influences theoretical models, and the importance of evaluating models in non-western cultures has grown with the accelerating pace of globalization. They present an evaluation of the antecedents of the technology acceptance model (TAM) model in an in non-western culture.

Chapter 3, "In or Out: An Integrated Model of Individual Knowledge Source Choice" by Wang, Meister and Gray looks into the way that the use by individuals of internal and external knowledge sources influences organizational knowledge integration, an important source of competitive advantage. Drawing on research into knowledge sourcing and consumer switching behavior; the authors develop an integrated model to understand individuals' choices between internal and external knowledge sources in contemporary work settings, where information technology has made both easily accessible. Chapter 4, "Developers, Decision Makers, Strategists or Just End-Users? Redefining End-User Computing for the 21st Century: A Case Study" by Barker and Fiedler notes that the acceleration of technology in baseness since the 1980s suggests that traditional management techniques, systems, and strategies employed in a business environment should be challenged. In this regard, the authors proffer that the management strategies for end-user (EU) systems development in the 21st century should suggest a different and proactive role for users.

Chapter 5, "Expert and Novice End-User Spreadsheet Debugging: A Comparative Study of Performance and Behaviour" by Bishop and McDaid notes that the reliability of end-user developed spreadsheets is poor and research studies find that 94% of spreadsheets contain errors. The work utilises a novel approach for acquiring experimental data through the unobtrusive recording of participants' actions using a custom built IBA tool. Based on findings from the experiment, a debugging tool is developed. and its effects on debugging performance are investigated. Chapter 6, "Exploring the Dimensions and Effects of Computer Software Similarities in Computer Skills Transfer" by Li and Chang asserts that computer software similarities play important roles in users' skills transfer from one application to another. Despite common software attributes recognized in extant literature, a systematic understanding of the components and structure of software similarities has not been fully developed. To address the issue, the results of a study conducted to explore the underlying dimensions of software similarities are presented.

Chapter 7, "Preparing IS Students for Real-World Interaction with End Users Through Service Learning: A Proposed Organizational Model" by Hall and Johnson states that although teaching the technical skills required of Information Systems (IS) graduates is a straightforward process, it is far more difficult to prepare students in the classroom environment for the challenges they will face interacting with end users in the real world. They present an organizational model to guide the implementation of service learning projects in IS curriculums. Service learning projects better prepare students to assume important management positions by giving them experience in applying the system development life cycle to an IS project and working with people. This organizational model uses the system development life cycle approach to integrate typical curriculum and service learning models. The organizational model is grounded in anecdotal evidence from prior experiences with IS students in service learning environments.

Chapter 8, "Construct Validity Assessment in IS Research: Methods and Case Example of User Satisfaction Scale" by Tojib and Sugianto notes that valid and reliable measures are critical to theory development as they facilitate theory testing in empirical research. The authors also apply a popular method from psychology and exemplify how the correlation analysis technique can be used to measure construct validity. Chapter 9, "A Social Capital Perspective on IT Professionals' Work Behavior and Attitude" by Zhang and Jones argues that though attracting and retaining information technology (IT) professionals is a current concern for companies, research has been conducted about the job behavior and attitudes of IT professionals over the past three decades. They add that little research has explored the effect of IT professionals' social capital. They present the results of study on how social capital affects IT professionals' work attitude and behavior, including job satisfaction and job performance.

Chapter 10, "Gender Differences in E-Learning: Communication, Social Presence, and Learning Outcomes" by Johnson maintains that although previous research has suggested that women may be at a learning disadvantage in e-learning environments, their research has examined communication differences between women and men, arguing that women's communication patterns may provide them with a learning advantage. The results of this study suggest that e-learning environments that allow for peer to peer communication and connectedness can help females overcome some of these disadvantages. Implications for research and practice are also discussed. Chapter 11, "Participation in ICT-Enabled Meetings" by Chudoba, Watson-Manheim, Crowston and Lee contends that though meetings are a common occurrence in contemporary organizations, their research notes that meetings are evolving beyond their familiar definition as the pervasive use of information and communication technologies (ICTs) changes work practices associated with meetings. Specifically the authors identify four aspects of meetings that reflect work in the 21st century: meetings are integral to work in team-centered organizations, tension between group and personal objectives, discontinuities, and ICT support for, fragmented work environment. Chapter 12, "Understanding User Dissatisfaction: Exploring the Role of Fairness in IT-Enabled Change" by Klaus examines the role of fairness and how it shapes a user's view in IT-enabled change. Their research leads to a new theoretical perspective and provides practical implications regarding the role of fairness in IT-enabled change and their strategic implications.

Chapter 13, "Antecedents of Improvisation in IT-Enabled Engineering Work" by Doll and Deng argues that the success of engineering work depends on the ability of individuals to improvise in response to emerging challenges and opportunities. Building on experiential learning theory and improvisation theory, the authors argue that information systems facilitate the generation of new product and process design ideas by providing richer feedback, creating shorter learning cycles, and enabling engineers to try a variety of new ideas more easily. Chapter 14, "The Role of Computer Attitudes in Enhancing Computer Competence in Training" by Downey and Smith clarifies the impact of computer competence and examines its relationship with anxiety affect, and pessimism, along with self-efficacy and previous experience. These results suggest that organizations focus not only on skills training, but on ways to enhance computing attitudes during the training process.

Chapter 15, "Development of a Mesh Generation Code with a Graphical Front-End: A Case Study" by Carver clarifies that although scientists and engineers are increasingly developing software to enable them to do their work, a number of characteristics differentiate the software development environment in which a scientist or engineer works from the development environment in which a more traditional business/IT software developer works. Specifically, this chapter reports on lessons learned concerning: requirements evolution, programming language choice, methods of communication among teammates, and code structure.

Chapter 16, "Characterizing Data Discovery and End-User Computing Needs in Clinical Translational Science" by Chilana, Fishman, Geraghty, Tarczy-Hornoch, Wolf and Anderson presents the result of a qualitative case-study seeking to characterize data discovery needs and barriers of principal investigators and research support staff in clinical translational science. Several implications for designing and implementing translational research systems have emerged through the authors' analysis. The results also illustrate the benefits of forming early partnerships with scientists to better understand their workflow processes and end-user computing practices in accessing data for research. Chapter 17, "Computational Engineering in the Cloud: Benefits and Challenges" by Hochstein, Schott and Graybill notes that cloud computing services, which allow users to lease time on remote computer systems, must be particularly attractive to smaller engineering organizations that use engineering simulation software. The case study presented in this chapter examines the potential benefits and practical challenges that a medium-sized manufacturing firm faced when attempting to leverage computing resources in a cloud computing environment to do model-based simulation.

Chapter 18, "Scientific End-User Developers and Barriers to User/Customer Engagement" by Segal and Morris argues that although Scientific End-User Developers recognise that effective engagement between its users/customers and developers is crucial, barriers to effective engagement exist, which scientific end-user developers can do little to address, and that such barriers stem from the essential nature of scientific practice.

Chapter 19, "An Analysis of Process Characteristics for Developing Scientific Software" by Kelly reports that the development of scientific software is usually carried out by a scientist who has little professional training as a software developer, and this raises concerns that such development produces low-quality products, leading to low-quality science. This chapter utilizes different frameworks to investigate and map characteristics of the scientific software development environment to the assumptions made in plan-driven software development methods and agile software development methods.

We hope you will enjoy reading these chapters, as much as we have, and that they will be of value to you in your respective roles as scholars, practitioners, managers, and consumers and end-users of modern-day IT products.

Ashish N Dwivedi
Hull University, UK

Steve Clarke
Hull University, UK

June 2012

REFERENCES

Beijerse, P. u. R. (1999). Questions in knowledge management: defining and conceptualising a phenomenon. *Journal of Knowledge Management, 3*(2), 94–109. doi:10.1108/13673279910275512

Bellinger, G., Castro, D., & Mills, A. (1996, March 7 2004). Data Information Knowledge and Wisdom. from http://www.systems-thinking.org/dikw/dikw.htm

Covin, T. J., & Stivers, B. P. (1997). Knowledge management in focus in UK and Canadian firms. *Creativity and Innovation Management, 6*(3), 140–150. doi:10.1111/1467-8691.00062

Davis, M., & Klein, J. (2000). Net holds breakthrough solutions. *Modern Healthcare, Feb 7*(2000), 14.

Dove, R. (1999). Knowledge management response ability and the agile enterprise. *Journal of Knowledge Management, 3*(1), 18–35. doi:10.1108/13673279910259367

Dwivedi, A., Bali, R. K., James, A. E., & Naguib, R. N. G. (2001, 25-28 October). *Telehealth Systems: Considering Knowledge Management and ICT Issues.* Paper presented at the 23rd Annual International Conference of the IEEE - Engineering in Medicine and Biology Society (EMBS), Istanbul, Turkey.

Dwivedi, A., & Clarke, S. (Eds.). (2012). *End-User Computing, Development and Software Engineering: New Challenges.* Hershey, PA, USA: IGI Global. doi:10.4018/978-1-4666-0140-6

Dwivedi, A. N., Wickramasinghe, N., Bali, R. K., & Naguib, R. N. G. (2008). Designing intelligent healthcare organizations with KM and ICT. *International Journal of Knowledge Management Studies, 2*(2), 198–213. doi:10.1504/IJKMS.2008.018321

Earl, M., & Scott, I. (1999). What is a chief knowledge officer? *Sloan Management Review*, *40*(2), 29–38.

Feldman, D., & Gainey, T. (1997). Patterns of telecommuting and their consequences: framing the research agenda. *Human Resource Management Review*, *7*(4), 369–388. doi:10.1016/S1053-4822(97)90025-5

Gupta, B., Iyer, L. S., & Aronson, J. E. (2000). Knowledge management: practices and challenges. *Industrial Management & Data Systems*, *100*(1), 17–21. doi:10.1108/02635570010273018

Hansen, M. T., Nohria, N., & Tierney, T. (1999). What's your strategy for managing knowledge? *Harvard Business Review*, *77*(2), 106–116.

Kazman, W., & Westerheim, A. A. (1999). Telemedicine leverages power of clinical information. *Health Management Technology*, *20*(9), 8–10.

Latamore, G. B. (1999). Workflow tools cut costs for high quality care. *Health Management Technology*, *20*(4), 32–33.

Moran, D. W. (1998). Health information policy: on preparing for the next war. *Health Affairs*, *17*(6), 9–22. doi:10.1377/hlthaff.17.6.9

Nonaka, I. (1988). *The Knowledge-Creating Company. Harvard Business Review on Knowledge Management* (pp. 21–45). Boston, MA: Harvard Business School Press.

Nonaka, I., & Konno, N. (1998). The concept of "ba":Building a foundation for knowledge creation. *California Management Review, 40*(3), pp - 40-54.

Sieloff, C. (1999). If only HP knew what HP knows: The roots of knowledge management at Hewlett-Packard. *Journal of Knowledge Management*, *3*(1), 47–53. doi:10.1108/13673279910259385

Zack, M. H. (1999). Managing codified knowledge. *Sloan Management Review*, *40*(4), 45–58.

Chapter 1
Evaluation of Information Strategy Implementation:
A Critical Approach

Yongmei Bentley
University of Bedfordshire Business School, UK

Steve Clarke
University of Hull Business School, UK

ABSTRACT

Information strategy is often relegated to an information technology element of corporate strategy, or worse, ignored in favour of IT operational planning. This research, conducted over a five-year period, stresses the correct framing of an information strategy and its implementation. The authors propose a framework that assists in the evaluation of such strategies, primarily those at higher education institutions, but also in a wider range of organisations seeking to improve the understanding and implementation of their information strategy.

INTRODUCTION

This research presents guidance for developing and evaluating information strategies of organisations, and in particular those of higher education institutions (HEIs). The guidance reflects a critical systems approach, and draws on a broad range of theory and empirical experience to provide analytical support for the evaluation. The process works by identifying specific elements of an information strategy, and for each element directing the evaluator to the relevant theoretical perspectives and empirical evidence that illuminate the issues raised. The primary aim is to identify underlying causes for problems and difficulties exhibited by an information strategy either under development or as implemented, and to suggest ways to resolve these.

DOI: 10.4018/978-1-4666-2059-9.ch001

Abai (2007) defines an information strategy as "an organization's unified blueprint for capturing, integrating, processing, delivery, and presentation of information in a clean, consistent, and timely manner. All information in an organization should meet a certain standard for quality." This definition emphasises the pragmatic aspects of such a strategy where the underlying aim is to make effective use of the organisation's vital information resources.

This research, by examining some of the problems encountered in information strategy development and implementation from the point of view of the end users, aims to advance the understanding of aspects of organizational and end-user computing (OEUC). In particular this work reflects ideas set out for a new direction for OEUC (Clarke, 2004) which emphasises the need for adopting critical systems approaches so as to overcome the limitations of using exclusively either technologically-based or human-centred approaches.

RESEARCH CONTEXT

The research context for this paper was a desire to improve understanding of the development process of information strategies in UK HEIs. Any information strategy has, at its heart, a need to provide information systems that facilitate the work of the organisation. Within the UK Higher Education sector, as part of the wider strategic planning that has become a requirement in these HEIs, there has been increasing emphasis on the development of information strategies. An important driver for this was the increasing competition faced by UK universities and colleges in the expanding global education market, and increased expectations of students and the wider society.

In this context, the Joint Information Systems Committee (JISC) of the Higher Education Funding Council in England has been charged since the 1990s with encouraging the development of information strategies within HEIs. To facilitate this development, JISC selected six UK HEIs as pilot sites, and guided these through the process of information strategy development. This process highlighted the requirement on UK HEIs to develop information strategies so as to ensure value-for-money from their IT, including exploiting technological advances, for coping with the rapidly increasing number of students, and most importantly, for attempting to bring about a change in attitude towards the ownership and accessibility of information within their institutions.

The experience with the pilot sites led to the publication by JISC of the *Guidelines for Developing an Information Strategy* (JISC, 1998a) and *Case Study Reports* drawn from the individual sites (JISC, 1998b). At the same time, JISC selected a further nine universities to be 'exemplars' to represent HEIs at different stages of information strategy development. Subsequently most HEIs in the UK have gone on to develop their own institution-wide information strategies.

A good definition of such a strategy, as it applies to an HEI, is that provided by the University of York (2005). This says that an information strategy embodies the following principles:

1. **Knowledge-based organisation:** A University is a knowledge-based organisation *par excellence*. Information is critical to a University's success and needs to be managed as a strategic resource.
2. **Quality:** Information should be fit for purpose – relevant, up-to-date, accurate, secure, and compliant with legislation and University policies. Information should normally be shared and duplication minimised.
3. **Ownership:** Each area of information or element of data should have a *custodian* who will be responsible for ensuring the quality of the data and for implementing the access policy.
4. **Users of Information:** All users should be fully aware of their rights and responsibilities in the handling of information.

5. **Information Infrastructure:** The University will provide an *information infrastructure* to facilitate information-handling processes and procedures across the University and to ensure that they are coherent and coordinated.

6. **Communications:** The University will provide a University-wide system for the rapid communication of information between staff, students and external stakeholders.

7. **Governance:** The critical importance of information to the activities of the University, together with the increasing requirements for regulatory compliance, requires the development of an effective information governance framework.

However, though attempted by JISC, there has been no *formal* approach put forward so far for assessing the problems and difficulties encountered in developing and implementing such strategies. It has been the aim of the research reported here to generate such a framework.

INFORMATION STRATEGY EVALUATION: CONCEPTUAL BASIS

There are many issues that bear on the success or otherwise of the development of an information strategy. They range from direct matters, such as correctly defining information needs, and ensuring adequate resources, through to deeper notions including the choice of paradigm for the development process, and the degree of control vs. emancipation of the users of the information. For this reason, a wide range of literature was surveyed - within the focus of critical systems thinking - in order to inform the information strategy evaluation process reported here. A full review of this literature is given in Bentley (2005), and a summary of some of the topics covered is given below. This starts by looking at the so-called hard vs. soft debate.

The 'Hard-Soft' Debate

There is abundant evidence (e.g. Checkland, 1999; Clarke, 2007) that information systems (IS) is a domain which primarily adheres to a functional engineering model, taking a structured, problem-solving approach to IS development and management. Human complexity in the system is seen as something which can be analysed, and toward which a specification can be written. However, the limitations of this purely technological approach gave rise in the 1960s and 1970s to the so-called *soft* or human-centred methods. Traditional 'hard' or 'engineering' approaches were premised on a world composed of determinable and rule-based systems. *Soft* methods, by contrast, took a more human-centred stance, where issues were seen as primarily determinable from the viewpoints of the human participants (Flood & Jackson, 1991). Soft systems thinking (Checkland, 1999) thus began to be more widely applied, with its argument that the study of organisations should be based on subjective meaning and interpretation.

In spite of this improved understanding, by the 1990s Beath and Orlykowski (1994) were still offering a convincing critique of the reality of the interaction between users and systems professionals in IS, concluding that the apparent commitment to 'user participation' in IS design was often more ideological than actual, with users frequently shown to be passive rather than active participants in the process. They saw the various systems development methodologies as often containing 'incompatible assumptions about the role of users and IS personnel during systems development' (Beath & Orlykowski, 1994, p. 358).

Ehn (1988) and Nygaard (1986) had offered a way forward through 'socio-technical' initiatives, seeking to link the social and the technical from a participatory standpoint. Other approaches to the inclusion of both technical and human perspectives into the design of IS include interactive planning (Ackoff, 1981) and soft systems methodology (Checkland & Haynes, 1994). These approaches essentially combined the social and technical in a more holistic framework. To understand an information system, the technology, organisation, and human activity all needed to be addressed interdependently, within a "whole systems" envelope.

In IS, these developments gave rise to a range of methodologies designed to address structural, technical and social issues, often seeking to mix different methods, and match them to relevant problem contexts. Three of the most widely used approaches to mixing methods (see Clarke & Lehaney, 1998; Clarke, 2001) have been Multiview (Avison & Wood-Harper, 1990) which sees IS development as a hybrid process involving computer specialists who build the system, and users for whom the system is being built; Client-led design (Stowell & West, 1994); and ETHICS (Effective Technical and Human Implementation of Computer-based Systems) by Mumford (1995).

It was into this melting pot of ideas that, during the late 1990s, a number of IS professionals and academics began to promote the concept of critical systems as a natural way of combining technical and social perspectives within the structural minefield of organisations (see Alvesson, 1985; Hirschheim, 1986; Laughlin, 1987; Lyytinen, 1992; Probert, 1993; Clarke, 2000). Moreover, Clarke (2007) argues extensively that this 'softer' view of IS, in which due consideration is given to the needs of end-users, should be applied to the wider area of information strategy, not just to information systems.

The broad underpinning to such 'critical systems thinking' is discussed next.

Critical Systems Thinking and Critical Social Theory

Soft systems thinking (SST) had been criticised for being unable to help practitioners address the problem of *coercion*, where participants in an organisation suffered varying degrees of control limiting their ability to participate to the best of their abilities in the planning and implementation of an information system. SST was also criticised for its inability to combine multiple methods.

Critical systems thinking (CST) was developed to address these shortcomings.

CST accepts the place of both 'hard' and 'soft' systems thinking, but also emphasizes the 'oppressing and inequitable' nature of social systems. CST is characterized by three commitments: to *critical awareness*, to *emancipation* of those involved, and to *pluralism* in approach (Jackson, 2000, 2003; Mingers & Gill, 1997).

Critical awareness and emancipation, in turn, require an understanding of critical social theory (CSoT). CSoT has been applied extensively in information management (Hirschheim & Klein, 1989; Lyytinen & Hirschheim, 1989; Clarke & Lehaney, 2002). CSoT sees our understanding of the world as too often determined by *a priori* conditions which are uncritically accepted. Critical theory seeks to expose these, and thereby release people from this 'false consciousness' to a position in which their full potential can be attained. In this context, the literature reviewed looked in particular at the works of two influential critical social theorists, Habermas (1972) and Foucault (1983).

Habermas (1972) was concerned with truth and rationality, and maintained that humans seek to achieve three interests - *technical, practical,* and *emancipatory*. Foucault (1983) saw power as necessary for the production of truth, explaining that power is not possessed, but exercised; and that power works to constrain people. Such philosophical perspectives see the world as socially constructed, and solutions to human issues being best achieved from such a viewpoint. The relevance here, of course, is to ask to what extent the information strategy being examined has been influenced by such deeper, and often hidden, factors.

By drawing on these developments as they have been applied to organisations in general and to IS in particular, this research has sought to provide a more holistic, inclusive approach to information strategy, in which the users are the prime concern. This is reflected in the framework for evaluating information strategy presented later in this paper.

'Planned' vs. 'Incremental' Strategy Development

For developing an organisation's strategy there are a wide variety of approaches. From these, two extremes may be distilled: strategy as something planned, or strategy as something that surfaces as the result of organisational activity. Mintzberg (1987) characterised this debate as the distinction between a 'plan' and a 'pattern', whilst other authors (e.g. Quinn, 1980; Johnson et al., 2007) refer to planned vs. 'emergent/incremental' strategies.

The 'planning' approach to strategy, developed from the so called 'design school' (Ansoff, 1964), can be traced to scientific reductionism. Ansoff (1964) referred to such an approach as a succession of different reduction steps: whereby a set of objectives was identified for the firm, the current situation with respect to the objectives was diagnosed, and the difference between these was determined. Strategy is concerned with finding those 'operators' which are best able to close this gap. The design school thus 'places primary emphasis on the appraisals of the external and internal situations' (Mintzberg, 1990, p. 179). Whilst it may consider organisational and managerial values and social responsibility, in Mintzberg's view these are almost always given secondary attention.

In an analysis of the pitfalls of strategic planning, Mintzberg (1994) identified two main categories: the lack of senior management support, and a climate in the organisation which was not congenial to planning. These pitfalls have in common the fact that they are attributable to people involved in the strategic planning other than the planners themselves. Hence the problems associated with planning were 'seldom technical deficiencies with the planning process or the analytical approaches' (Abell & Hammond, quoted in Mintzberg, 1994), rather they were to do with the people involved or organisational issues within the business: in either case the root of the problems is 'the nature of human beings' (Ansoff, 1965). Overall, the review of the literature in this area showed that the traditional 'planned' top-down approach to strategy development had been challenged more recently by an 'emergent', bottom-up approach (Bentley & Clarke, 2006a).

Insights from Practice: The Empirical Literature

In terms of reviewing the literature, of equal importance in developing an approach to information strategy evaluation within this research has been an understanding of practical issues. For this reason, an extensive range of empirical information on the development and implementation of information strategy at UK HEIs was examined. This included reports from the Higher Education Funding Council for England (HEFCE, 1998); from JISC (1998a, 1998b); and from the wider group of HEIs themselves. For information strategy development at HEIs, JISC recommended a six-step model - *preparing, planning, developing, implementing, monitoring and reviewing*. This model was widely adopted in UK HEIs and was reported to work well in practice, although adjustment was necessary in specific situations.

By examining the practice of information development and implementation at UK HEIs as described in these documents, the study reported here was able to link the conceptual ideas in the theoretical literature to actual information strategy development experience. Two findings stood out. The first was that the procedures adopted in practice had incorporated little in the way of theoretical grounding. The second was that although many HEIs had developed information strategies, so far there had been few attempts to evaluate these. JISC had had the original intention of doing so, but the difficulties in conducting such evaluations caused this effort to be abandoned.

This paper aims, therefore, to assist such evaluations by presenting a generalised framework under which these may be conducted. This framework is presented next.

GENERALISED FRAMEWORK FOR EVALUATING AN INFORMATION STRATEGY

Based on the concepts discussed above, a framework is proposed as Figure 1 for evaluating information strategies at higher education institutions.

The figure indicates in diagrammatic form the areas covered by the framework, and has at its centre those constituents, in terms of people and systems, which are at the core of an HEI's information strategy. These constituents need to be considered individually in any comprehensive evaluation of an information strategy.

The framework centres on three inter-related primary areas of an HEI's information strategy. First, the general *control structures* required for preparing, planning, developing, implementing, monitoring, and reviewing a strategy - this is based on JISC's model for information strategy development mentioned earlier. Second, the degree that the strategy takes account of the HEI's *external environment*, including its 'competitive advantage'. Third, the degree that the strategy takes account of aspects of the HEI's own *internal environment*. This covers many factors, and includes organisational culture and structure, resource management, information needs analysis, strategic alignment and management of strategic change.

The use of the framework is as follows: The strategy evaluator, whether examining an information strategy still in its development process or one that has already been implemented, is encouraged to look in turn at each of the framework elements, and where problems with the strategy have been identified, to raise analysis and questions based on the theoretical and practical information indicated. In particular, the framework is intended to encourage a more human-centred and critical perspective to this process, and thereby takes end-users as its primary focus. The framework's aim here is to assist the evaluator uncover people's *attitudes*, *perceptions* and *motivations* about the situation; to examine not only what *was* done, but also what *ought to* have been done; to recognise situations of empowerment or coercion; and to propose one or a number of methodologies from the literature to improve those aspects of the information strategy determined as problematic.

It is recognised that Figure 1 illustrates information strategy evaluation primarily as it relates to higher education institutions, but as the general nature of the literature reviewed above indicates, many of the principles covered apply to information strategy development across many types of organisations. These include the control structures adopted, the external environment that must be faced, and specific internal factors of the strategy such as correct information needs analysis, strategic alignment across the organisation, and human-centred issues associated with the organisation's structure and culture.

Table 1, tying into the framework categories, sets out in a condensed form specific concepts covered in the literature review that relate to elements of the information strategy on which an evaluator might wish to focus. The more detailed literature review from which this table is drawn is given in Bentley (2005). As can be seen, the table distinguishes between theoretical perspectives and empirical evidence, where the latter is drawn from JISC (1998a, 1998b, 2004) and other data sources, and also from information gathered in the course of the research carried out for this study, as described in later sections of this paper.

Table 2 adds to Table 1 by setting out, again in a very condensed form, those aspects of *critical systems theory* that need to be 'embedded' in the framework in the sense of specifically drawing the evaluator's attention to the need for methods of inquiry, analysis and intervention that incorporate the principles of *critical awareness*, methodological and theoretical *pluralism*, and attention to *emancipation*. The latter aims to allow an organisation to maximise the potential of each of its individuals, in part by understanding the limits to this potential as set by the organisation's existing structures of power and coercion.

Figure 1. Proposed framework for information strategy evaluation at HEIs

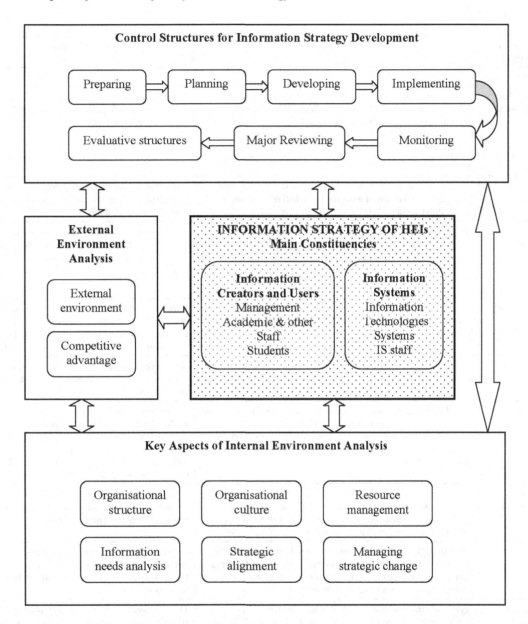

EMPIRICAL INVESTIGATIONS

The above framework has been applied to three empirical investigations into aspects of information strategies of UK HEIs: a student records system at a UK HEI; the process of information strategy development at the same HEI; and the experience of successful information strategy development and implementation at another HEI which had been one of JISC's 'pilot' sites.

Note that in discussing an information strategy, it is often not fully clear whether a problem of implementation should be seen as a strategic matter, or more operationally as poor performance of a specific information system. This dichotomy will become apparent in some of the examples

Table 1. Summary of theoretical perspectives and empirical evidence relevant to information strategy evaluation in HEIs

Framework Element	Theoretical Perspectives	Empirical Evidence
Control Structures		
Prepare, plan, develop, implement, monitor, and review	Strategic planning (Galliers, 1991); Planned and incremental strategies (Mintzberg, 1987; Quinn, 1980); 'Is' and 'ought' (Ulrich, 1983); SoSM (Jackson & Keys, 1984); Jackson, 1987); Strategic implementation (Johnson et al., 2007); Mixing Methods (Clarke, 2001)	Six-step model in the *Guidelines* (JISC, 1998a); Lessons to learn from the experiences of JISC pilot sites and UK HEIs
External environment		
Environment analysis	Five forces (Porter, 1980); Survey of the environment (Johnson et al., 2007); Environment analysis (Campbell & Stonehouse, 2002); PEST model, SWOT analysis	Internal and external environments are important. (JISC, 1998a); the empirical research of this study has supported this.
Competitive advantage	Five forces (Porter, 1980); The strategic advantage to be gained from information (Porter, 1990)	The investigation of the information strategy development for this study showed that ignoring this had put the university in a less competitive position.
Internal environment		
Organisational structure	Basic organisational structures (Johnson et al., 2007); 'Mechanistic to adhocratic' (Mintzberg, Quinn & Ghoshal, 1998); Metaphors (Morgan, 1986)	Experience of UK HEIs showed that a changed organisational structure in correspondence to the changed environments had facilitated its information strategy implementation (JISC, 2004).
Organisational culture	Cultural web (Johnson et al., 2007); General discussions on organizational culture (Schein, 1996; Wit & Meyer, 2005).	JISC's pilot sites' experience implied there was a need to move to more inclusive and participative culture (JISC, 1998b).
Resources management	IT and information – different resources (King, 1988); TSI (Flood & Jackson, 1991); Resource planning (Johnson et al., 2007); Information systems (Avison & Fitzgerald, 2006); IS functions (Savage & Mingers, 1996); Resource analysis framework (Campbell & Stonehouse, 2002)	The implementation of an information strategy is often found to be resource constrained. Some HEIs considered reduction of resources as one of the reasons for having an information strategy. This research showed there should be a balance between the two (JISC, 2004).
Information needs analysis	Three interests (Habermas, 1972); SSM (Checkland, 1999); System movement (Checkland, 1983); 'Is' and 'ought' (Ulrich, 1983); Management strategies for IT (Earl, 1989); Supply and demand of information (Smits, van der Poel & Ribbers, 1997)	Pilot sites case studies showed that incorporation of 'user' views in information needs analysis was weak; JISC had models for this.
Strategic alignment	Align strategies (Earl, 1989); Integrating various strategies (Galliers, 1993); Strategic alignment model (Henderson & Venkatraman, 1993); Linkage between information strategy and business strategy (Smits, van der Poel & Ribbers, 1997)	Alignment of different strategies in JISC pilot sits and other UK HEIs was poorly done; the terms IT, IS and information should not be used interchangeably.
Managing strategic change	Styles and other issues in change management (Johnson et al., 2007); Mechanistic approach (Mintzberg & Westley, 1992); Information systems strategic management (Clarke, 2007)	This was weak in most HEIs, with over-emphasis on operations, lack of long-term vision, and human issues poorly addressed. Lessons can be drawn from the 'negative' examples uncovered in this study.
Evaluative structures	Strategic choice (Johnson et al., 2007); information systems strategic management (Clarke, 2007)	JISC's evaluative structures can be helpful, but there should be monitoring and reviewing mechanisms for the overall evaluation.

discussed below. In general, while an information system problem is often seen as specific to that system, there have been underlying strategic decisions that have led to the problem arising.

These might include lack of recognition within the organisation's strategy of the value of accurate and timely information; lack of focus on rules for information ownership and generation; or lack of

Table 2. Aspects of critical systems theory to apply to information strategy evaluation

Aspect	Key Points	Means
Critical Awareness	To understand people's perceptions.	Observation, interview, discussion, action research, ethnography, case study.
	To understand limitations of inquiry methods.	Identification of boundary conditions, Asking *is* and *ought* questions.
Pluralism	To understand three types of knowledge - technical, practical, and emancipatory.	Empirical-analytic inquiry.
	To use of multiple social theories.	Critical social theories, organisational theories.
	To use appropriate intervention methodologies.	Soft system methodology; system of systems methodologies; critical systems heuristics; methodology selection frameworks.
Emancipation	To answer the questions: Who has the power? Is this deciding the truth?	Critique of ideology; assumption analysis; Need for radical change?

resource allocation from the organisation's overall time and management budget. It has been an aim in the investigation described below to try and tease out both sides of the picture: identifying which have been systems development problems and which deeper information strategy problems.

The student record system (SRS) was examined because in any HEI its SRS is a major component of the HEI's information system, and hence a key focus in terms of that HEI's wider information strategy. In particular, at the university in question the SRS had been exhibiting chronic failures typical of many such IS components. The question here, therefore, was whether the evaluative framework proposed above could shed useful light on, and offer solutions to, the problems that were being encountered. The second study, that of the information strategy development process underway at the same university, was of interest to see if lessons from examining the development process at this university could be related back in a meaningful way to issues highlighted within the framework. The final piece of empirical research had as its motivation the fact that it related to a completed and implemented information strategy process, and one where the resulting strategy was seen as achieving significant benefits. The question here, therefore, was whether the proposed framework sufficiently covered those activities that

had clearly been done correctly in implementing this particular information strategy. Each of these empirical investigations is described in turn below.

STUDENT RECORDS SYSTEM INVESTIGATION

The student records system (SRS) had played an important role in the management of information in the university in question. But it had shown a variety of problems and had generated a wide range of complaints from system users on issues such as inefficiency in reporting, and frequent errors in the information it provided. Indeed, the system had often failed to meet the basic needs of both internal and external information provision. As a result the SRS had become a focus of strategic management attention at the university. Thus the immediate objectives of this part of the research were to understand the main issues involved in the management of the SRS, identify problems with the system, trace both the immediate and deeper causes for these problems, and suggest solutions for improving the system.

The research was conducted over a period of eighteen months, allowing comprehensive observation of what people were actually doing, as well as what they said they were doing, and a proper understanding to be gained of the people involved, the organisation, and the broader context within which they worked.

The researcher first spent two weeks gaining 'hands-on' experience of data entry for the system, interacting under normal working conditions with other system users and understanding their problems. Then thirty-two semi-structured interviews were carried out with a wide range of people involved with the system, including senior management using the system, operational managers, data entry staff, and academic staff who used the output reports from the SRS. These interviews were based on a questionnaire designed to elicit not only the correspondents' experiences and views of the system, but also to help uncover deeper feelings about how human-centred issues connected with the system were being addressed. A transcript was made of the key issues discussed at each of the interviews, and a formal data analysis procedure – content analysis (Neuendorf, 2002) adopted. In addition, a range of more general conversations about the system's strength and weaknesses were held with a variety of colleagues.

The following is a summary of the key findings from the data analysis.

Most of the respondents reported a range of failings, including incorrect, incomplete, and ambiguous information, and a failure to provide information for key academic activities and external returns. Analysis of these problems with the SRS identified a number of proximate causes for these failings, including lack of system specification, poor data quality and system management, inflexibility, and lack of communication about the system. But the investigation also showed that deeper factors were at work, partly reflected in the development of a 'culture of blame' about the system. The SRS therefore appeared typical of many other failed information systems, where there was agreement about the system's aims, but dissatisfaction with the reality: confusion about what was wrong; little attention to users'

needs; and poor motivation of those involved. Some of the issues raised with the SRS seemed to be operational, and could be fixed relatively easily. But many were more strategic in nature, and required a more over-arching plan if they were to be addressed. Overall, from the study it became clear that the information needs analysis had been poorly done, in that key information was missing or unreliable. Some system users said that the problems with the system had been serious enough to be classed as a 'disaster' for the university.

This identification of a requirement for a comprehensive information needs analysis (Bentley & Clarke, 2006b) illustrated the use of the relevant section of the evaluative framework given above. In particular, the framework links back to JISC documents which detail a variety of methods for information needs analysis, thus placing an evaluator of an information strategy in a position to recommend one or more of these approaches as appropriate.

Again in terms of the framework, analysis of the SRS had also shown that the evaluative structures in place were inadequate, as the failures had become chronic but were not being picked up and addressed in any systematic way. It became evident that aspects of organisational culture associated with the SRS needed change, as people were encountering SRS problems on a daily basis but were either not motivated to, or by their level of authority were prevented from, getting these addressed. Here the evaluative framework is also potentially of use, as it draws attention to the theoretical literature on human behaviour and motivation that can be insightful in such situations.

Finally, strategic change as listed in the framework is another area where it appeared that well-established approaches were not being employed. Changes to the SRS were being introduced without sufficient discussion or training. Importantly also, the research indicated that strategy related to the SRS was being dictated from the top, with little attempt being made to allow for 'emergent' strategy formation.

Overall, in terms of the critical systems theory issues as summarised in Table 2, there appeared to be a need for more human-centred processes to be employed, encouraging greater openness in response to questions, and in finding ways for both planners and participants to consider more complex social and cultural issues, such as motivation, information ownership, and the implications of hierarchy and power.

Some of the insights from these discussions were reported back to the University's Information Strategy Steering Group (ISSG) which had the mission of the institutional information strategy development, and were subsequently adopted in the development of the university's information strategy. This researcher therefore not only collected primary data, but also helped improve the outcomes.

Information Strategy Development at a UK HEI

As mentioned above, the second piece of empirical research involved participation in the development process then underway of the information strategy at the same university. The University had set up the ISSG described above with the duty of overseeing this strategy's development. As part of the research described here, this researcher was given access to, and was involved in, all the ISSG meetings.

To gain a broader view of the university's information needs during the period of information strategy development, in addition to attending these ISSG meetings, interviews and discussions on the general topic of information needs and information management were conducted with senior managers and members on the ISSG, a range of other staff across the university who would use the information covered by the strategy on a regular basis, and with students' representatives. These interviews and discussions were mainly used to

elicit, in a relaxed atmosphere, the 'real' issues relating to the university's information needs and information management. Over forty people were interviewed, with each interview taking about 25 minutes on average.

Responses were categorised and tabulated, with key comments summarised. The following are some of the main findings from this phase of the study.

At the university, an information strategy was generally considered to be 'a good thing', but there was uncertainty over what such a strategy should comprise, and how it might be achieved. Moreover, the strategy development process tended to be 'top-down' and directed, despite attempts made by some of the members at the ISSG meetings to introduce more participative (and sometimes "critical") ideas. In terms of the issues covered in the evaluative framework proposed above, this finding relates to the key distinction between 'strategy as a *plan*' and 'strategy as an *emergent* process'. It was clear that the ISSG group was more comfortable with the former than the latter.

There was some evidence of unequal power in decision making within the ISSG meetings. On occasion people with more technical views of the process left the meetings early, feeling that their views were not getting a fair hearing. At other times, the views of the more senior members of the strategy group carried the day simply for reasons of hierarchy. In terms of the framework, this goes to the heart of issues addressed by CST; in particular, to Habermas' and Foucault's cautions about power being a threat to '*emancipation*', and hence a potential hindrance to good decision making.

It was noted that a key requirement of an information strategy was for the information users to understand the importance of information, including an awareness of the often-hidden costs associated with lack of information, unnecessary duplication, inaccuracy, and incompatibilities among systems. It was certainly clear that the moves towards a strategy so far had not properly addressed this issue of getting a sufficiently widespread acceptance of this concept of the importance of information to the effective functioning of the institution. In terms of the framework, this relates to the management fields of organisational culture, and of managing strategic change. Many organisations have needed to encourage new ways of thinking across their institutions, and the management literature covers a range of practical advice for helping management and staff work together adopt new paradigms.

The interviews found that a large knowledge base about the university's information systems was held in the heads of key staff members around the university. If the university were to lose these staff, this knowledge would be lost. Here the framework points to JISC experience with information strategy development, and indeed highlights one of the fundamental reasons why an organisation may need a formal information strategy in the first place.

In addition, it was found that the people included in the information strategy development process were mostly senior management staff, and administrators. There were few academic representatives or other systems users, and no student representatives. The focus appeared to be far too narrow to give any real insight into the university's information strategy needs. Two aspects of the framework address this issue: the theoretical perspective from CST for a critical awareness of the situation under examination, and the practical advice set out in the JISC *Guidelines* on the formalism to be adopted when carrying out the strategy's information needs analysis - in this case, in correctly identifying the groups that need to be involved in this process.

Information Strategy Implementation at a UK HEI

The final piece of empirical research carried out to support the evaluative framework development was a case study which looked in detail at the information strategy implementation process at another university. This had been one of JISC's original pilot sites for information strategy development. Twelve relevant staff members were interviewed at this university and they were very informative and co-operative. Data analysis from these interviews revealed useful findings which included the following:

•   This university was considered to have been very successful in its information strategy implementation. It had followed JISC's *Guidelines*, but also, importantly, given a high level of commitment to the process, and allocated an appropriated level of resources. The university maintained a senior committee with responsibility for implementing and monitoring the information strategy. This committee received regular reports from its project teams. Considerable planning and effort had been committed by the university in getting people to accept, understand and take responsibility for the information strategy, both in its development and its operation.

•   An important decision was taken to devolve responsibility for information quality away from the centre, passing this to departments and individuals. This improved information quality, and in 'soft-system' terms 'closed the human loop' by giving people ownership of information and a sense of partnership in the information provision process across the university.

•   It was found in the study that the effort expended in bringing staff 'on-board' with the information strategy implementation process was large, but so were the benefits. These were not only reflected in the information strategy process itself, but were also translated into improving management expertise within the university's general management functions.

- Many interesting specific issues emerged from the implementation of the information strategy at this university, for example, the question of restricted access versus shared ownership and responsibility; the move from paper reports to electronic display; reduction in data duplication; value-for-money gains; improvement in information quality; the use of 'quick-win' approaches; recommending 'experienced users' as a resource; and identifying the specific cases where the rules on non-duplication data should be broken.

The above findings all tie back to specific elements covered in the evaluation framework, in particular to the management of strategic change, organisational culture, data 'ownership', and issues of emancipation. The findings also point to the level of detailed planning and consequent commitment of resources necessary to successfully implement a wide-ranging information strategy. However, while it was clear that this university had found extensive benefits from the implementation of its information strategy, it had not yet had time to put in place a full range of monitoring and reviewing procedures.

STRENGTHS AND WEAKNESSES OF METHODOLOGY AND OUTCOMES

Like any empirical research, this study had a number of strengths and weaknesses. On the side of strengths, several well-established research methods were employed, including action research, questionnaire, and semi-structured interviews (for the investigation of the SRS); 'ethnography' (for investigating the development of an information strategy at a UK HEI); and case study (for the studying implementation of an IS at another UK HEI). In addition, the investigations were fairly extensive: 32 semi-structured interviews (plus many more *ad hoc* conversations) for the SRS; 40 or so interviews during the IS development process; and 12 interviews with key participants for the IS implementation. Considerable care was taken in these processes, making transcripts of all responses, and then tabulating and adopting formal content analysis of the tables so produced. Triangulation of findings was also carried out. In the case of the SRS, this was done against detailed findings from a second researcher, and in the IS development and implementation studies, against the experiences of JISC workshop participants of HEIs at similar stages of IS development across the UK.

Limitations of the studies included the following.

Questionnaires, used for all the three investigations, are good for collecting data but are sometimes difficult to interpret, and were at times superficial in the sense of providing facts about attitudes and behaviour but few explanations. That was why, in the SRS study, the questionnaires were sent about a week before the face-to-face interviews to give respondents time to reflect on the issues, and for questions during the interview to develop flexibly according to the conversation.

Action research (AR), used for the investigation of the SRS, involves a researcher influencing and changing the research situation. However, as the researcher for this project was acting as an ordinary system user, it was not always easy to gain support and co-operation of others who were also working under pressure. In addition, it was found that some people working in that environment, especially line managers, to some extent, exhibited an anti-research attitude, and did not want an 'outsider' to influence or change the situation.

Ethnography was used for the investigation of the information strategy development process. The profound strength of such research method is that "it is the most 'intensive' research method possible", but the main disadvantage is that "it takes a lot longer than most other kinds of research" (Mayers, 1999, pp. 5-6). For the two investigations (the SRS and the IS development), the research took eighteen months.

Perhaps the greatest limitation is the fact that there has been no opportunity to independently test the framework. This is included under 'Further Research'.

IMPLICATIONS FOR RESEARCHERS AND PRACTITIONERS, AND SCOPE FOR FURTHER RESEARCH

The main implication for researchers and practitioners in this field is to consider this framework when designing, implementing or evaluating an information strategy, and see where it can help orient the investigation. Detailed suggestions are given above in the section setting out the framework and its uses (see 'Generalised Framework for Evaluating an Information Strategy'). Note that while it may seem that the ideas presented are fairly broad, the problems encountered with information strategies usually have fairly explicit roots, even if it may take some investigation to uncover these.

The main primary need for further research, therefore, is to conduct an independent test of the information strategy framework proposed in this paper, either at a higher education institution, or perhaps more widely in any organisation which has developed such a strategy. The aim would be to use the framework as part of a human-centred enquiry to investigate if aspects of the strategy are operating below expectation, and if so to reflect on the problems encountered in light of the theories and practices set out in Figure 1 and Table 1, and then, on the basis of the literature cited, propose methodologies and techniques to address the issues encountered.

CONCLUSION

In this research, guidance in the form of a generalised framework has been proposed to assist the evaluation of information strategies implemented at HEIs. The framework reflects a critical approach, and assembles in one place a broad range of relevant theory and empirical experience to provide analytical support to the information strategy evaluation process. The knowledge presented includes critical social theory, critical systems concepts, and general theories of management, as well as management experience. Also included is experience drawn from empirical literature relating to information strategy development within the UK HEIs, drawn from JISC, from other HEFCE publications, from practice at a range of UK HEIs, as well as from findings from three investigations described here that were carried out specifically for this research. Although the empirical evidence is drawn from UK experience it is felt that much would apply to HEIs at other locations implementing information strategies, and indeed, to strategies implemented within other types of organisation.

Overall, the framework is designed to help an evaluator identify the underlying causes for the success or failure of aspects of an information strategy, and suggest ways in which problems uncovered might be rectified. In particular, the framework aims to point to the human dimensions of problems encountered, and to encompass and understand the broader issues involved, including situational complexity. In terms of methodologies, the framework identifies concepts from social theory, analysis techniques, systems methodologies, methodology selection frameworks, and other tools that can assist in the information strategy evaluation process. These approaches are not exclusive. Each has its strengths and weaknesses, and an evaluator has to determine, which, used singly or in combination, best fits the problem situation under review. As Flood and Jackson (1991, back cover) point out: "In the modern world organizations are faced with innumerable and multifaceted issues which cannot be captured in the minds of a few experts and solved with the aid of some super-method. We need a range of problem-solving methodologies".

In drawing attention to social theories that address the various human-centred issues likely to be encountered in evaluating information strategies, it is recognised that awareness of a social theory may not lead directly to a solution. But theory can provide insights into human attitudes and behaviour that allow situations to be tackled from a better-informed viewpoint. It is the contention of this research that such a theory-based approach leads to better outcomes for the situations under investigation.

REFERENCES

Abai, M. (2007). *Your organization needs an information strategy.* Retrieved July 8, 2009, from http://www.cioupdate.com/trends/article.php/3667451/Your-Organization-Needs-an-Information-Strategy

Ackoff, R. L. (1981). *Creating the corporate future.* New York: Wiley.

Alvesson, M. (1985). A critical framework for organizational analysis. *Organization Studies, 6*(2), 117–138. doi:10.1177/017084068500600202

Ansoff, H. I. (1964). A quasi-analytical approach of the business strategy problem. *Management Technology, IV,* 67–77.

Ansoff, H. I. (1965). *Corporate strategy: An analytic approach to business policy for growth and expansion.* New York: McGraw Hill.

Avison, D., & Fitzgerald, G. (2006). *Information systems development methodologies, techniques and tools* (4th ed.). London: McGraw-Hill International.

Avison, D. E., & Wood-Harper, A. T. (1990). *Multiview: An exploration in information systems development.* London: McGraw-Hill.

Beath, C. M., & Orlikowski, W. J. (1994). The contradictory structure of systems development methodologies: Deconstructing the IS-user relationship in information engineering. *Information Systems Research, 5*(4), 350–377. doi:10.1287/isre.5.4.350

Bentley, Y. (2005). *A critical approach to the development of a framework to support the evaluation of information strategies in UK higher education institutions.* Unpublished doctoral dissertation.

Bentley, Y., & Clarke, S. A. (2006a). Using Ethnographic research to investigate the development process of a university's information strategy. In *Proceedings of ECRM 2006: The 5th European Conference on Research Methods in Business and Management Studies,* Dublin, Ireland.

Bentley, Y., & Clarke, S. A. (2006b). Reflections from an information management project development. In *Proceedings of EMCIS 2006: The Third European and Mediterranean Conference on Information Systems,* Alicante, Spain.

Campbell, D., & Stonehouse, G. (2002). *Business strategy: An introduction* (2nd ed.). London: Bill Houston.

Checkland, P. B. (1983). OR and the System Movement: Mappings and Conflicts. *The Journal of the Operational Research Society, 34*(8), 661–675.

Checkland, P. B. (1999). *Systems Thinking, Systems Practice: includes a 30-year retrospective.* Chichester, UK: John Wiley and Sons.

Checkland, P. B., & Haynes, M. G. (1994). Varieties of systems thinking: the case of soft systems methodology. *System Dynamics, 10*(2-3), 189–197. doi:10.1002/sdr.4260100207

Clarke, S. A. (2000). From socio-technical to critical complementarist: A new direction for information systems development. In Coakes, E., Lloyd-Jones, R., & Willis, D. (Eds.), *The New Socio Tech: Graffiti on the Long Wall* (pp. 61–72). London: Springer.

Clarke, S. A. (2001). Mixing methods for organisational intervention: Background and current status. In M. G. Nicholls, S. Clarke, & B. Lehaney (Eds.), *Mixed-mode modelling: Mixing methodologies for organisational intervention* (1-18). Dordrecht, The Netherlands: Kluwer Academic.

Clarke, S. A. (2004). A new direction for organizational end-user computing. *Journal of Organizational and End User Computing, 16*(2), i–viii.

Clarke, S. A. (2007). *Information systems strategic management: an integrated approach* (2nd ed.). London: Routledge.

Clarke, S. A., & Lehaney, B. (1998). A theoretical framework for facilitating methodological choice. *Systemic Practice and Action Research, 11*(3), 295–318. doi:10.1023/A:1022952114289

Clarke, S. A., & Lehaney, B. (2002). Human-centered methods in information systems: Boundary setting and methodological choice. In Szewczak, E., & Snodgrass, C. (Eds.), *Human Factors in Information Systems* (pp. 20–30). Hershey, PA: IRM Press.

Earl, M. J. (1989). *Management strategies for information technology*. London: Prentice Hall.

Ehn, P. (1988). *Work-oriented design of computer artifacts*. Falköping, Sweden: Pelle Ehn and Arbetslivscentrum.

Flood, R. L., & Jackson, M. C. (1991). *Creative problem solving: Total systems intervention*. Chichester, UK: Wiley.

Foucault, M. (1983). The subject and power. In Dreyfus, H. L., & Rabinow, P. (Eds.), *Michel Foucault: Beyond structuralism and hermeneutics* (2nd ed.). Chicago: University of Chicago Press.

Galliers, R. D. (1991). Strategic information systems planning, myths and reality. *European Journal of Information Systems, 3*, 199–213.

Galliers, R. D. (1993). Towards a flexible information architecture: Integrating business strategies, information systems strategies and business process redesign. *Journal of Information Systems, 3*(3), 199–213. doi:10.1111/j.1365-2575.1993.tb00125.x

Habermas, J. (1972). *Knowledge and human interests*. London: Heinemann.

HEFCE. (1998). Information systems and technology management: Value for money study. In *Management Review Guide*. London: VFM Steering Group.

Henderson, J. C., & Venkatraman, N. (1993). Strategic alignment: Leveraging information technology for transforming organisations. *IBM Systems Journal, 32*(1), 4–16. doi:10.1147/sj.382.0472

Hirschheim, R. A. (1986). Understanding the office: A social- analytic perspective. *ACM Transactions on Office Information Systems, 4*(4), 331–344. doi:10.1145/9760.9763

Hirschheim, R. A., & Klein, H. K. (1989). Four paradigms of information systems development. *Communications of the ACM, 32*(10), 1199–1216. doi:10.1145/67933.67937

Jackson, M. C. (1987). Present positions and future prospects in management science. *Omega, 15*, 455. doi:10.1016/0305-0483(87)90003-X

Jackson, M. C. (2000). *Systems approaches to management*. New York: Kluwer Academic/Plenum Publishers.

Jackson, M. C. (2003). *Systems thinking: Creative holism for managers*. Chichester, UK: Wiley.

Jackson, M. C., & Keys, P. (1984). Towards a system of systems methodologies. *The Journal of the Operational Research Society, 35*, 473–486.

JISC. (1998a). *Guidelines for developing an information strategy*. London: Author.

JISC. (1998b). *Case studies on developing information strategies*. London: Author.

JISC. (2004). Renewal and growth: Using technology to support learning, teaching, management and research. *Inform (Silver Spring, Md.)*, 5.

Johnson, G., Scholes, K., & Whittington, R. (2007). *Exploring corporate strategy*. London: Prentice Hall.

King, W. R. (1988). How effective is your information systems planning? *Long Range Planning*, *21*(5), 103–112. doi:10.1016/0024-6301(88)90111-2

Laughlin, R. C. (1987). Accounting systems in organizational contexts: A case for critical theory. *Accounting, Organizations and Society*, *12*(5), 479–502. doi:10.1016/0361-3682(87)90032-8

Lyytinen, K. (1992). Information systems and critical theory. In Alvesson, M., & Willmott, H. (Eds.), *Critical Management Studies* (pp. 159–180). London: Sage.

Lyytinen, K., & Hirschheim, R. (1989). Information systems and emancipation: Promise or threat? In Klein, H. K., & Kumar, K. (Eds.), *Systems Development for Human Progress* (pp. 115–139). Amsterdam, The Netherlands: North Holland.

Mayers, M. D. (1999). Investigating information systems with ethnographic research. *Communications of the Association for Information Systems*, *2*, 23.

Mingers, J., & Gill, A. (1997). Commentary. In Mingers, J., & Gill, A. (Eds.), *Multimethodology: The theory and practice of combining management science methodologies*. Chichester, UK: Wiley.

Mintzberg, H. (1987). Crafting strategy. *Harvard Business Review*, *65*(4), 66–75.

Mintzberg, H. (1990). The design school: Reconsidering the basic premises of strategic management. *Strategic Management Journal*, *11*(3), 171–195. doi:10.1002/smj.4250110302

Mintzberg, H. (1994). Rethinking strategic planning. Part I: Pitfalls and fallacies. *Long Range Planning*, *27*(3), 12–21. doi:10.1016/0024-6301(94)90185-6

Mintzberg, H., Quinn, J. B., & Ghoshal, S. (1998). The strategy process (Revised European edition). Hemel Hempstead, UK: Prentice Hall.

Mintzberg, H., & Westley, F. (1992). Cycles of organizational change. *Strategic Management Journal*.

Morgan, G. (1986). *Images of Organisation*. Beverley Hills, CA: Sage.

Mumford, E. (1995). *Effective requirements analysis and systems design: The ETHICS Method*. Basingstoke, UK: Macmillan.

Neuendorf, K. A. (2002). *The content analysis guidebook*. Beverley Hills, CA: Sage.

Nygaard, K. (1986). *Program development as social activity in information processing*. Amsterdam, The Netherlands: Elsevier Science Publishers.

Porter, M. E. (1980). *Competitive strategy: Techniques for analyzing industries and competitors*. New York: Free Press.

Porter, M. E. (1990). *The competitive advantage of nations*. London: Macmillan.

Probert, S. K. (1993). Interpretive analytics and critical information systems: A framework for analysis. In *Systems Science*. Addressing Global Issues.

Quinn, J. B. (1980). Formulating strategy one step at a time. *The Journal of Business Strategy*, 42–63.

Savage, A., & Mingers, J. (1996). A framework for linking soft systems methodology (SSM) and Jackson system development (JSD). *Information Systems Journal, 6,* 109–129. doi:10.1111/j.1365-2575.1996.tb00008.x

Schein, E. H. (1996). Three cultures of management: The key to organizational learning. *Sloan Management Review.*

Smits, M. T., van der Poel, K. G., & Ribbers, P. M. A. (1997). Assessment of information strategies in insurance companies in the Netherlands. *The Journal of Strategic Information Systems, 6,* 129–148. doi:10.1016/S0963-8687(97)00004-8

Stowell, F., & West, D. (1994). *Client-led design – A systemic approach to information systems definition.* Maidenhead, UK: McGraw-Hill.

Ulrich, W. (1983). *Critical heuristics of social planning: A new approach to practical philosophy.* Chichester, UK: Berne, Haupt, and J. Wiley.

University of York. (2005). *Information strategy 2004-9.* Retrieved August 8, 2009, from http://www.york.ac.uk/admin/po/infostrat/informationstrategy

Wit, B., & Meyer, R. (2005). *Strategy synthesis: Resolving strategy paradoxes to create competitive advantage* (2nd ed.). London: International Thomson Business.

This work was previously published in the Journal of Organizational and End User Computing, Volume 23, Issue 1, edited by Mo Adam Mahmood, pp. 1-17, copyright 2011 by IGI Publishing (an imprint of IGI Global).

Chapter 2
The Value of TAM Antecedents in Global IS Development and Research

Chad Anderson
Georgia State University, USA

Said S. Al-Gahtani
King Khalid University, Saudi Arabia

Geoffrey S. Hubona
Virginia Commonwealth University, USA

ABSTRACT

Theoretical models are often conceived and tested in western countries. However, culture influences theoretical models, and the importance of evaluating models in non-western cultures has grown with the accelerating pace of globalization. The technology acceptance model (TAM) is no exception, and more TAM research is being conducted in non-western countries. TAM constructs of perceived usefulness and perceived ease of use are difficult to act on, which has led to several studies that identified valid antecedents to these constructs that make the model more practically actionable. These antecedents were conceived and tested in a western country but have yet to be evaluated in the context of a non-western country. In this paper, the authors evaluate these antecedents in Saudi Arabia and find that they function in the specific context of general computer use by Saudi knowledge workers.

INTRODUCTION

Does culture have an influence on the perceptions and acceptance of technology? In one of the earliest studies to address this question, Straub (1994) found that the cultural differences between Japan and the United States did have an effect on the perceptions and selective use of email and fax technologies. In a study of Arab cultures, Straub, Loch, and Hill (2001) gathered data from five Arab nations and determined that Arab cultural beliefs were powerful predictors of resistance to information systems technologies. Broader studies have also found similar patterns of cultural

DOI: 10.4018/978-1-4666-2059-9.ch002

influence on technology acceptance. Parboteeah, Parboteeah, Cullen, and Basu (2005) conducted a 24-nation study and found that cultural factors were influential in the perceived usefulness of information technology. The results of these studies would seem to provide clear evidence that culture does, in fact, have an influence on the perceptions and acceptance of technology.

So why is it important to study the influences of culture on technology acceptance? Information systems are expensive to implement and often involve considerable sunk costs before they are ready for use by end users. Since users can potentially undermine the intended purposes of the technology (Orlikowski & Robey, 1991), businesses need to have a good idea up front how these systems will be received and accepted. Increasing globalization is resulting in the need for multinational and trans-cultural organizations to utilize information technology to achieve economies of scale, coordinate global operations, and facilitate collaborative work across distributed locations and diverse cultures (Ford, Connelly, & Meister, 2003; Shin, Ishman, & Sanders, 2007). Understanding the influences of different sets of cultural values on the acceptance and use of information systems can help managers and system developers design systems for use by multinational organizations to improve their chances of being accepted and used.

How does this paper contribute to the study of the cultural influences on technology acceptance? The study of technology acceptance over the last two decades has predominantly centered on the technology acceptance model (TAM) (Davis, 1989). The extent to which it has been studied and applied over the last two decades is evident in recent meta-analyses (King & He, 2006; Schepers & Wetzels, 2007) and reviews (Lee, Kozar, & Larsen, 2003; Legris, Ingham, & Collerette, 2003) of TAM research. A part of the model's popularity and appeal to researchers can be explained by its parsimonious structure which has only four constructs, with perceived ease of use (PEOU) and perceived usefulness (PUSE) predicting behavioral

intention which leads to actual use. However, one of the criticisms of TAM is the difficulty in acting on the results of the model (Gefen & Keil, 1998; Venkatesh & Davis, 2000). One solution would be the identification of antecedents to perceived usefulness and perceived ease of use which could make TAM more practically actionable. In fact, Davis had considered antecedents to TAM in his initial research, but found that they were fully mediated by perceived usefulness and perceived ease of use (Davis, 1989). A decade later two studies re-examined potential TAM antecedents to produce a more actionable model of technology acceptance (Venkatesh, 2000; Venkatesh & Davis, 2000). These studies identified a number of antecedents to TAM that are effective predictors of perceived usefulness and perceived ease of use; however, both studies were conducted in North American organizations. Will those antecedents be valid in a non-western culture? Straub, Keil, and Brenner (1997) found that the TAM model held in the United States and Switzerland but not in Japan. The fact that culture has been shown to affect theoretical models that predict technology acceptance would seem to support the need for the evaluation of these models in multiple cultures. Since the validity of these antecedents has not been examined outside of a western country, the goal of this research is to evaluate the predictive validity of select TAM antecedents within an Arab country to determine whether they are valid for organizations operating within that context.

LITERATURE REVIEW AND HYPOTHESIS DEVELOPMENT

Information systems research in the context of non-western cultures has been increasing in recent years with TAM studies providing a good example of this trend. In the ten years following Davis's (1989) seminal article on the technology acceptance model, there were only eight TAM related articles published that either included culture as

Table 1. Growth in culture-related TAM research

Period	Culture-Related TAM Studies
1990 - 1999	De Vreede, Jones, & Mgaya, 1998; Ghorab, 1997; Igbaria, Iivari, & Maragahh, 1995; Igbaria, Zinatelli, Cragg, & Cavaye, 1997; Phillips, Calantone, & Lee, 1994; Rose & Straub, 1998; Straub, 1994; Straub et al., 1997
2000 - 2008	Chau & Hu, 2002; Elbeltagi, McBride, & Hardaker, 2005; Henderson & Divett, 2003; Koeszegi, Vetschera, & Kersten, 2004; Lee, Ahn, & Han, 2007; Liaw, Chang, Hung, & Huang, 2006; Liaw & Huang, 2003; Lin & Lu, 2000; Lu, Liu, Yu, & Yao, 2003; Mao & Palvia, 2006; Mao, Srite, Thatcher, & Yaprak, 2005; McCoy, Everard, & Jones, 2005; McCoy, Galletta, & King, 2007; Parboteeah et al., 2005; Selim, 2003; Stylianou, Robbins, & Jackson, 2003; Suh & Han, 2002; Sukkar & Hasan, 2005; Sun & Zhang, 2006; Teo, Chan, Wei, & Zhang, 2003; van Raaij & Schepers, 2008; Veiga, Floyd, & Dechant, 2001; Vetschera, Kersten, & Koeszegi, 2006

a component of the study or were conducted in a non-western country (see Table 1). In contrast, from 2000 to 2008, there have been more than 24 TAM related studies published that fell into this category (see Table 1).

There are probably a number of reasons for this trend, but the increasing globalization of business has certainly been a major contributing factor. Ford, Connelly, and Meister (2003) noted that globalization is particularly relevant for information systems practitioners and researchers because globalization has been largely facilitated by new information systems technologies. Since research has shown that cultural factors can make the difference between success and failure in the adoption and implementation of information systems (Png, Tan, & Khai-Ling, 2001; Straub et al., 1997), the importance of understanding the effectiveness of information systems that span nations and cultures has increased and led more researchers to study their impacts (Niederman, Boggs, & Kundu, 2002; Shin et al., 2007).

This increase in IS research in non-western cultures has included a number of studies focused on Arab countries and the impact the Arab culture has on the adoption and implementation of information systems. Ghorab (1997) was the first to use TAM in an Arab culture when he studied automation decisions in United Arab Emirate banks. Rose & Straub's (1998) TAM study that spanned five Arab countries is particularly well known in cultural IS studies. The authors chose to maintain a broad focus on generic IT use because

of the limited adoption of even basic technologies in these countries and found that in this context TAM was robust in its prediction of IT use. Al-Khaldi and Al-Jabri (1998) studied attitudes and computer utilization in a Saudi Arabian business school. Hill, Loch, Straub, and El-Sheshai (1998) conducted a field study across five Arab countries and found that socio-cultural factors were influential in the transfer of information technology. Straub, Loch, and Hill (2001) extended this study and suggest that instead of viewing culture as a barrier to information technology transfer, technology design and implementation should be adapted to the culture in which it will be used. Al-Gahtani (2003) tested whether the innovation diffusion model would function in a less-technologically developed country, specifically in Saudi Arabia, and found that it did hold in that context.

To make valid comparisons, IT user acceptance models operationalizing common constructs such as perceived usefulness and perceived ease of use in cross-cultural research should be robust across cultures. Otherwise, it will be difficult to judge whether the results obtained from the application of these models are, indeed, comparable in different nations or cultures. Understanding whether similar models are comparable across cultures is a critical first step needed to: (1) enhance our understanding of how culture affects IT acceptance and diffusion; and (2) improve the organizational management of IT in a global context. Since research has yet to be published that evaluates antecedents to perceived usefulness and perceived ease of use

Table 2. Hofstede's four cultural dimensions

Dimensions	Descriptions	United States	Arab World
Power-Distance (PDI)	*Degree of inequality among people which the population of a culture considers normal*	40	80
Uncertainty Avoidance (UAI)	*Degree to which people in a culture feel uncomfortable with uncertainty and ambiguity*	46	68
Individualism (IDV)	*Degree to which people in a culture prefer to act as individuals rather than as members of groups*	91	38
Masculinity (MAS)	*Degree to which values like assertiveness, performance, success, and competition prevail among people of a culture over gentler values like the quality of life, maintaining warm personal relationships, service, care for the weak, etc.*	62	52

in non-western countries, the contribution of this research will be to determine the predictive capability of these antecedents for organizations operating in Saudi Arabia, a non-western country. It is intended that the findings of this research will facilitate the development of policies and practices by organizations in non-western countries that will be more effective in influencing technology acceptance and adoption.

In order to theorize about the influences of cultural differences between western and Arab countries, it is necessary to have comparable cultural measures for those countries. Straub (1994) chose to use Hofstede's cultural dimensions (Hofstede, 1980) in his study of email and fax use in the US and Japan. This decision set a standard for cultural studies in information systems research with many subsequent studies looking to Hofstede's country scores for cultural comparison data (Anandarajan, Igbaria, & Anakwe, 2002; Igbaria et al., 1995; Mao & Palvia, 2006; Rose & Straub, 1998; Veiga et al., 2001).

However, several recent studies have argued that Hofstede's country scores may be inappropriate for IS research, particularly in the study of TAM (McCoy, Galletta, & King, 2005; Srite & Karahanna, 2006). McCoy et al. (2005) primarily argue that Hofstede's scores, generated over 30 years ago, may no longer be accurate in today's global environment. However, Hofstede's original scores have actually been updated based on a

number of more recent studies (Hofstede, 2001). In fact, Arab countries were not even included in Hofstede's original set of published scores, but were added in a later edition. Rather than individual country scores, Hofstede provides a single set of scores for the Arab world which includes; Egypt, Iraq, Kuwait, Lebanon, Libya, Saudi Arabia, and the United Arab Emirates. Bjerke and Al-Meer (1993) collected their own data using Hofstede's questionnaire to formulate scores for Saudi Arabia before Hofstede published his updated scores. Their results for Saudi Arabia were similar to Hofstede's subsequent scores for the combined Arab world. It is Hofstede's scores for the Arab world and those of the United States that are used in this study as a point of cultural comparison. Hofstede's four primary cultural dimensions and the standardized scores for the United States and the Arab world are shown in Table 2.

Srite and Karahanna (2006) have raised a more serious issue regarding the use of Hofstede's scores in IS studies. They argue that a level of analysis conflict exists because Hofstede's scores are aggregate values for an entire country, while models like TAM are theorized at the individual level. Srite and Karahanna (2006) argue that it would be more appropriate to collect and use individual level cultural scores when testing the effects of culture on these models. They go on to propose and test an instrument to measure individually-espoused national culture. While Srite and Karahanna's (2006) arguments are sound and the study is a step in the right direction, the results would suggest that the instrument needs further refinement for researchers to have confidence in its use. Therefore, until such time as a more robust set of individual level measures is constructed and authenticated we would argue that Hofstede's scores are still the most appropriate means of comparing cultural differences between individuals from different countries.

Hofstede's country scores on these four cultural dimensions (which reflect work related values) indicate some pertinent cultural differences between the Arab world and the United States. Specifically, there are large differences on the power-distance and individualism dimensions. We will theorize about the potential significance of these differences later in the paper. In addition, although the differences between the scores for uncertainty avoidance and masculinity are not as dramatic, they may still have an effect on the relative applicability of TAM antecedents in an Arab culture.

This study to evaluate the potential influence of Arab culture on TAM antecedents is based on the work of Venkatesh and Davis (2000) and Venkatesh (2000). In their study of TAM antecedents, Venkatesh and Davis (2000) focused specifically on the antecedents to perceived usefulness in developing a model they called TAM2. Their findings were based on four longitudinal field studies of North American organizations and the results showed that subjective norm, image, job relevance, output quality, and result demonstrability were all significant predictors of perceived usefulness. Venkatesh (2000) investigated the antecedents to perceived ease of use through three longitudinal field studies, also conducted in North America. He found that computer self-efficacy, perception of external control, computer anxiety, and computer playfulness were all significant anchors in predicting perceived ease of use, with perceived enjoyment and objective usability functioning as adjustment variables.

In developing our theoretical model we selected subjective norm, image, and result demonstrability from Venkatesh and Davis (2000) as viable antecedents to perceived usefulness in the context of Arab culture. We also selected computer self-efficacy, computer anxiety, and perceived enjoyment from Venkatesh (2000) as viable antecedents to perceived ease of use. The predicted variables of perceived usefulness, perceived ease of use, and intention to use were drawn directly from Davis (1989). Using the same model structure proposed by Venkatesh (2000), we did not include usage behavior as a part of the core TAM variables. The proposed model is shown in Figure 1. The model is segmented into three subsections, based on the underpinning studies which derive each set of research variables.

Because technology adoption occurs within a social environment, social influences are often included in theoretical models of acceptance or rejection of technology (Srite & Karahanna, 2006). Venkatesh and Davis (2000) included "subjective norm", adopted from the Theory of Reasoned Action (TRA), as one mechanism for capturing these social influences on perceived usefulness in their development of TAM2. Fishbein and Ajzen (1975), who developed TRA, defined subjective norm as a "person's perception that most people who are important to him think he should or should not perform the behavior in question" (p. 302). The relatively high power-distance in Arab cultures should lead subordinates to be more conscious of the opinions and expectations of their superiors resulting in a stronger influence of subjective norm on user perceptions and actions.

Venkatesh and Davis (2000) specifically theorized that subjective norm would have a direct effect on perceived usefulness through internalization and identification. Internalization means that, "when one perceives that an important referent thinks one should use a system, one incorporates the referent's belief into one's own belief structure", while identification refers to the "influence to accept information from another

Figure 1. Theoretical model

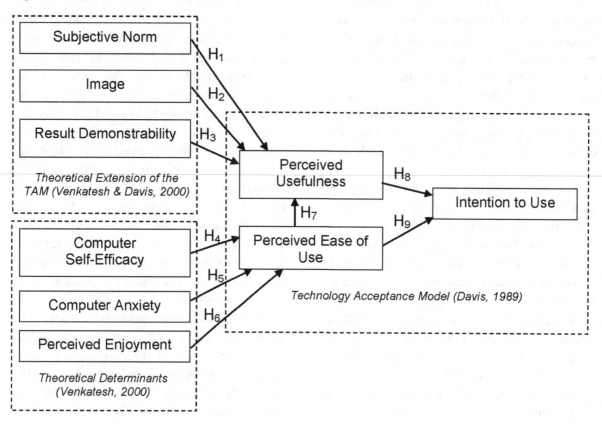

as evidence about reality" (Venkatesh & Davis, 2000, p. 189). Because decision making is more hierarchical in organizations within high power-distance cultures (e.g. technology is prescribed in a top-down approach), subordinates should have a clearer perception of their superiors beliefs about the usefulness of a particular technology. Combined with the relatively low individualism typical in Arab cultures, this environment is theorized to increase the prevalence of internalization and identification and thus enhance the direct effects of subjective norm on perceived usefulness.

Venkatesh and Davis (2000) adopted "image" from Moore and Benbasat (1991) as a second mechanism of social influence on perceived usefulness. Moore and Benbasat (1991) defined image as, "the degree to which use of an innovation is perceived to enhance one's … status in one's social system" (p. 195). In this case, the

cultural characteristics of high power-distance and low individualism typical of Arab cultures may have opposing effects. Low individualism will tend to weaken the effect of image on perceived usefulness because users will be less focused on standing out. Consequently, users in a collective culture may not consider a technology useful simply because its use will improve their social status in the organization. On the other hand, high power-distance will lead subordinates to seek approval and recognition from their superiors. Because technology is typically prescribed in a top-down approach in these cultures, subordinates will conclude that their superiors consider a prescribed technology useful and will therefore also find it useful in order to gain this approval and recognition.

Venkatesh and Davis (2000) also adopted "result demonstrability" from Moore and Benbasat (1991) who defined it as "tangibility of the results of using the innovation" (p. 203). While high power-distance and low individualism are expected to increase the influence of subjective norm on perceived usefulness, these influences should not drown out the need for a technology to produce tangible results. It is therefore hypothesized that result demonstrability will span cultural boundaries and continue to have a positive impact on perceived usefulness regardless of cultural differences. These beliefs about the antecedents of perceived usefulness are operationalized in the first three research hypotheses.

H₁: Subjective norm will have a positive effect on perceived usefulness.

H₂: Image will have a positive effect on perceived usefulness.

H₃: Result demonstrability will have a positive effect on perceived usefulness.

We expect that culture will also have some impact on the influence of antecedents for perceived ease of use. However, there is another important factor at play in this context that is also expected to affect the overall influence of perceived ease of use relative to perceived usefulness. Specifically, the less developed nature of the Saudi economy will mean that the typical Saudi knowledge worker will have had less exposure to computers and technology then would be typical in more developed countries. This should result in a greater emphasis on ease of use over usefulness.

Venkatesh (2000) theorized that there were a number of mechanisms that might influence perceived ease of use and function as valid antecedents to it. We first consider computer self-efficacy which Venkatesh (2000) defines as an internal control, "that represents one's belief about her/his ability to perform a specific task/job using a computer" (p. 347). Computer self-efficacy is linked to experience with computers and the likely wider range of experience levels for users in a developing country like Saudi Arabia

should make this factor a stronger predictor of perceived ease of use.

Computer anxiety typically has a negative correlation with computer self-efficacy and is therefore the one antecedent that is expected to produce a negative relationship in this model. Venkatesh (2000) draws on (Simonson, Maurer, Montag-Torardi, & Whitaker, 1987) for a definition of computer anxiety which is, "an individual's apprehension, or even fear, when she/he is faced with the possibility of using computers" (p. 349). Individuals with low computer self-efficacy will generally have a higher level of computer anxiety than individuals with high computer self-efficacy. Therefore, if computer self-efficacy is a good predictor of perceived ease of use then computer anxiety should also be a good predictor, but in the opposite direction of influence. Al-Gahtani (2004) studied employees in Saudi Arabia who used computers in their work routines and found that the relationship between computer anxiety and computer usage and satisfaction was highly negative. In contrast, Al-Khaldi and Al-Jabri (1998) found that computer anxiety was not a significant predictor of computer utilization by students at a Saudi Arabian university. In the context of this study we expect to find a negative relationship between computer anxiety and perceived ease of use.

The final antecedent to perceived ease of use in this model is perceived enjoyment. Venkatesh (2000) taps (Davis, Bagozzi, & Warshaw, 1992) for a definition of perceived enjoyment as, "the extent to which the activity of using a specific system is perceived to be enjoyable in its own right, aside from any performance consequences resulting from system use" (p. 351). The anticipated effects of these antecedents of perceived ease of use are operationalized in the next three hypotheses.

H₄: Computer self-efficacy will have a positive effect on perceived ease of use.

H₅: Computer anxiety will have a negative effect on perceived ease of use.

H₆: Perceived enjoyment will have a positive effect on perceived ease of use.

A number of studies have already shown that the core TAM model does hold in the context of non-western cultures. Specifically, Rose and Straub (1998) found that TAM did apply to the use of generic IT by knowledge workers in Jordan, Egypt, Saudi Arabia, Lebanon, and the Sudan. Sukkar and Hasan (2005) also successfully applied the core TAM model in a study of Internet banking in Jordan. Therefore, following the traditional TAM structure proposed by Davis, we complete our proposed model with the final three hypotheses.

H₇: Perceived ease of use will have a positive effect on perceived usefulness.

H₈: Perceived usefulness will have a positive effect on intention to use.

H₉: Perceived ease of use will have a positive effect on intention to use.

Each of these nine hypotheses is mapped to specific individual paths in the research model depicted as Figure 1.

RESEARCH METHOD

This research is based on a larger study conducted in Saudi Arabia and financed by the Saudi government to elicit perceptions and intentions for the general use of computer systems by Saudi knowledge workers. Major companies from all four of Saudi Arabia's main provinces were invited to participate in this study. The organizations that accepted this invitation included representatives from the petroleum, manufacturing, banking, and merchandising industries. The study utilized a written survey to elicit responses and included

question sets from the original technology acceptance model, TAM2 and UTAUT, as well as additional question sets for new constructs. The volume and scope of the data that resulted from this survey was intended to facilitate multiple points of investigation into aspects of technology acceptance and use in the context of an Arab culture. To this end, several papers have already been published based on this comprehensive data set.

Specifically, Al-Gahtani, Hubona, & Wang (2007) investigated a modified version of the unified theory of acceptance and use of technology (UTAUT). "Drawing on the theory of cultural dimensions, [they] hypothesized and tested the similarities and differences between North American and Saudi validations of UTAUT in terms of cultural differences that affected the organizational acceptance of IT in the two societies (Al-Gahtani et al., 2007, p. 681)." Next, Al-Gahtani (2008) investigated the applicability of the original TAM model in Saudi Arabia and found that it effectively predicted general computer technology adoption and use in the Saudi culture. The paper also extended the original TAM model by incorporating gender, age and educational level as moderators of the model's core relationships. In addition to these papers, groundwork was laid for this study by Al-Gahtani (2001) in which the applicability of TAM was tested in the United Kingdom as a precursor to its assessment in the context of Saudi Arabian culture.

Given Al-Gahtani's (2008) success in showing that the original TAM model holds well in a Saudi setting, the authors of this study determined to investigate the value of TAM antecedents as a supplement to enhance the picture of IT adoption and use in the Saudi culture. In this paper we are focusing specifically on the antecedents of perceived usefulness: image, result demonstrability, and subjective norm, proposed by Venkatesh and Davis (2000); and the antecedents of perceived ease of use: computer self-efficacy, computer anxiety, and perceived enjoyment, proposed by

Table 3. Predictor latent construct items for the antecedents of PUSE

Subjective Norm (SN)
SN1: Most people who are important to me think I should use computers.
SN2: Most people who are important to me would want me to use computers.
SN3: People whose opinions I value would prefer me to use computers.
Image (IMG)
IMG1: People in my organization who use computers have a high profile.
IMG2: Using computers is a status symbol in my organization.
IMG3: People in my organization who use computers have more prestige than those who do not.
Result Demonstrability (RD)
RD1: I would have no difficulty telling others about the results of using a computer.
RD2: I believe I could communicate to others the consequences of using a computer.
RD3: The results of using a computer are apparent to me.

Venkatesh (2000). This study of the antecedents to TAM was previously presented at two conferences for preliminary feedback to enhance the quality and clarity of the ideas presented here (Anderson, Al-Gahtani, & Hubona, 2007, 2008).

The study survey solicited responses from professional knowledge workers engaged in the use of desktop computers for the purpose of their work. Through this procedure, a total of 1,190 usable survey responses were collected. Of the 1,190 usable surveys, 102 responders indicated that they were foreign Nationals (e.g. non-Saudis) who were working in Saudi Arabia. Because we wanted our sample to exclusively represent Saudi workers in order to more effectively represent the influence of non-western culture, we excluded these 102 non-Saudi respondents, leaving a sample size of 1,088 survey responses. Due to the predominately male workforce in Saudi Arabia the sample was composed of 849 (78%) male and 239 (22%) female. Subjects were primarily in their 30's and 40's with an age distribution of: under 20 (.5%), 20-30 (39.3%), 31-40 (43.4%), 41-50 (15.3%), and above 50 (1.5%).

All survey items, originally published in English, were adapted for this study in Arabic using Brislin's (1986) back translation method.

The items were translated back and forth between English and Arabic by several bilingual professors. The process was repeated until both versions converged.

Table 3 presents the items used to estimate the predictor latent constructs for the antecedents of perceived usefulness. A seven point Likert scale with anchors of strongly disagree and strongly agree was used to measure each item. These constructs utilized items equivalent to those used by Venkatesh and Davis (2000). We measure and report on the reliabilities and discriminant validities of these constructs in the results section of this paper.

Table 4 presents the items used to estimate the predictor latent constructs for the antecedents of perceived ease of use. A seven point Likert scale with anchors of strongly disagree and strongly agree was also used to measure each of these items. These constructs utilized items equivalent to those used by Venkatesh (2000). We measure and report on the reliabilities and discriminant validities of these constructs in the results section of this paper.

Table 5 presents the items used to estimate the predicted latent constructs for perceived usefulness, perceived ease of use, and behavioral inten-

Table 4. Predictor latent construct items for the antecedents of PEOU

Computer Self-Efficacy (EFF)
EFF1: If I wanted to, I could easily operate the computer and related equipment on my own.
EFF2: I would be able to use the computer even if there is no one to show me how to use it.
EFF3: I would be able to use new computer systems on my own.
Computer Anxiety (ANX)
ANX1: I feel apprehensive about using a computer terminal.
ANX2: If given the opportunity to use a computer, I am afraid that I might damage it in some way.
ANX3: I have avoided computers because they are unfamiliar to me.
ANX4: I hesitate to use a computer for fear of making mistakes that I cannot correct.
Perceived Enjoyment (ENJ)
ENJ1: I believe using computers is enjoyable.
ENJ2: The actual process of using computers is pleasant.

tion. A seven point Likert scale with anchors of strongly disagree and strongly agree was used to measure each item. Items used to measure the PUSE, PEOU, and BI constructs are consistent with the essence of the generally accepted meanings of those constructs. We measure and report on the reliabilities and discriminant validities of these constructs in the results section below.

RESULTS

The research model depicted in Figure 1 was analyzed using SmartPLS (version 2.0), a Partial Least Squares (PLS) Structural Equation Modeling (SEM) tool (Ringle, Wende, & Will, 2005). SmartPLS assesses the psychometric properties of the *measurement model* (i.e. the reliability and validity

Table 5. Predicted latent construct items

Perceived Usefulness (PUSE)
PU1: I find computers useful in my job.
PU2: Using computers in my job enables me to accomplish tasks more quickly.
PU3: Using computers in my job increases my productivity.
PU4: Using computers enhances my effectiveness on the job.
Perceived Ease of Use (PEOU)
PE1: My interactions with computers are clear and understandable.
PE2: It is easy for me to become skillful using computers.
PE3: I find computers easy to use.
PE4: Learning to use computers is easy for me.
Behavioral Intention to Use (BI)
BI1: I would use computers rather than any other means available.
BI2: My intention would be to use computers rather than any other means available.
BI3: To do my work, I would use computers rather than any other means available.

Table 6. Assessment of the measurement model

Variable Constructs	The Composite Reliability (Internal Consistency Reliability)	Average Variance Extracted/Explained
Perceived Usefulness	.91	.71
Perceived Ease of Use	.90	.68
Behavioral Intention	.89	.73
Result Demonstrability	.90	.75
Subjective Norm	.95	.87
Image	.92	.80
Efficacy	.86	.68
Anxiety	.90	.69
Enjoyment	.93	.87

of the scales used to measure each variable), and estimates the parameters of the *structural model* (i.e. the strength of the path relationships among the model variables).

The Measurement Model

Reliability results from testing the measurement model are reported in Table 6. The data indicates that the measures are robust in terms of their internal consistency reliabilities as indexed by the composite reliability. The composite reliabilities of the different measures in the model range from .86 to .95, which exceed the recommended threshold value of 0.70 (Nunnally, 1978). In ad-

dition, consistent with guidelines promulgated by Fornell and Larcker (1981), the average variance extracted (AVE) for each measure well exceeds 0.50 (the minimum AVE is 0.68). Table 7 reports the results of testing the discriminant validity of the measure scales. The bolded elements in the matrix diagonals, representing the square roots of the AVEs, are greater in all cases than the off-diagonal elements in their corresponding row and column, supporting the discriminant validity of our scales.

We tested convergent validity using SmartPLS by extracting the factor loadings (and cross loadings) of all indicator items to their respective latent constructs. These results, presented in Table

Table 7. Discriminant validity (intercorrelations) of variable constructs

Latent Variables	1	2	3	4	5	6	7	8	9
1. Perceived Usefulness	**.84**								
2. Perceived Ease of Use	.47	**.83**							
3. Behavioral Intention	.47	.57	**.86**						
4. Result Demonstrability	.42	.57	.44	**.87**					
5. Subjective Norm	.31	.27	.34	.29	**.93**				
6. Image	.27	.21	.35	.23	.48	**.89**			
7. Efficacy	.32	.53	.35	.62	.16	.13	**.82**		
8. Anxiety	-.17	-.34	-.21	-.28	-.03	.05	-.30	**.83**	
9. Enjoyment	.36	.45	.41	.39	.32	.30	.29	-.16	**.93**

8, indicate that all items loaded: (1) on their respective construct (i.e. the bolded factor loadings) from a lower bound of 0.76 to an upper bound of 0.95; and (2) more highly on their respective construct than on any other construct (i.e. the non-bolded factor loadings in any one row). A common rule of thumb to indicate convergent validity is that all items should load greater than 0.7 on their own construct (Yoo & Alavi, 2001), and should load more highly on their respective construct than on the other constructs. Furthermore, each item's factor loading on its respective construct was highly significant ($p < 0.0001$) as indicated by the T-statistics of the outer model loadings in the SmartPLS output. These T-statistic values ranged from a low of 32.94 to a value of 165.14. The constructs' items' loadings and cross loadings presented in Table 8, and the highly significant T-statistic for each individual item loading, serve to confirm the convergent validity of these indicators as representing distinct latent constructs in the research model.

The Structural Model

Figure 2 presents the results of the structural model. The beta values of the path coefficients, indicating the direct influences of the predictor upon the predicted latent constructs, are presented. Result demonstrability exhibited a significant positive influence (beta = 0.19, $p < 0.001$) on perceived usefulness. Subjective norm also exhibited a significant positive influence (beta = 0.13, $p < 0.001$) on perceived usefulness, as did image (beta = 0.10, $p < 0.01$). Computer self-efficacy exhibited a significant positive influence (beta = 0.39, $p < 0.001$) on perceived ease of use. Computer anxiety exhibited a significant *negative* influence (beta = - 0.17, $p < 0.001$) on perceived ease of use. Perceived enjoyment also exhibited a significant positive influence (beta = 0.30, $p < 0.001$) on perceived ease of use. Finally, perceived ease of use exhibited significant positive influences on both perceived usefulness (beta = 0.31, $p < 0.001$) and on behavioral intention to use (beta = 0.44, $p < 0.001$). Lastly, perceived usefulness had a significant positive influence on behavioral intention to use (beta = 0.26, $p < 0.001$).

The direct influences of result demonstrability, subjective norm, image and perceived ease of use account for approximately 29% of the variance in perceived usefulness ($R^2 = .292$). The direct influences of computer self-efficacy, computer anxiety and perceived enjoyment account for approximately 40% of the variance in perceived ease of use ($R^2 = .400$). Lastly, the direct influences of perceived ease of use and perceived usefulness account for approximately 38% of the variance in behavioral intention ($R^2 = .377$) to use computers.

DISCUSSION

Table 9 presents the results of this study with respect to the nine hypotheses. Each hypothesis was supported. We compare our findings to those of: (1) Venkatesh and Davis (2000) with respect to the antecedents of perceived usefulness (hypotheses H_1, H_2 and H_3); (2) Venkatesh (2000) with respect to the antecedents of perceived ease of use (hypotheses H_4, H_5 and H_6); and (3) Davis (1989) with respect to the TAM variables (hypotheses H_7, H_8 and H_9). Particularly, we compare the magnitude and significance of the corresponding path coefficients from our findings to those of the corresponding predecessor studies. The focus of our discussion is on hypotheses H_1 through H_6, those pertaining to the antecedents of perceived usefulness and perceived ease of use. We are less interested in comparing at length the findings relative to the TAM variables (hypotheses H_7, H_8 and H_9).

We should note that a direct comparison of our findings to those of Venkatesh & Davis (2000) and Venkatesh (2000) must be made with caution. The predecessor studies from which we abstracted our antecedent variables differ from this study in at least three respects. First, the data from the current study was cross-sectional using a common

Table 8. Factoring loadings (bolded) and cross loadings

	PUSE	PEOU	BI	RD	SN	IMG	EFF	ANX	ENJ
PUSE1	**0.85**	0.38	0.37	0.35	0.25	0.20	0.28	-0.13	0.31
PUSE2	**0.87**	0.40	0.42	0.35	0.28	0.22	0.24	-0.15	0.28
PUSE3	**0.88**	0.40	0.41	0.38	0.30	0.23	0.28	-0.15	0.31
PUSE4	**0.76**	0.40	0.39	0.36	0.24	0.25	0.27	-0.15	0.30
PEOU1	0.38	**0.82**	0.45	0.47	0.17	0.16	0.43	-0.28	0.33
PEOU2	0.41	**0.87**	0.49	0.52	0.25	0.20	0.47	-0.32	0.39
PEOU3	0.41	**0.84**	0.48	0.44	0.24	0.22	0.41	-0.24	0.40
PEOU4	0.35	**0.76**	0.45	0.45	0.21	0.11	0.45	-0.27	0.35
BI1	0.45	0.52	**0.84**	0.42	0.28	0.24	0.32	-0.22	0.41
BI2	0.30	0.46	**0.85**	0.35	0.29	0.34	0.28	-0.11	0.29
BI3	0.44	0.48	**0.88**	0.37	0.31	0.33	0.30	-0.19	0.34
RD1	0.33	0.47	0.38	**0.84**	0.24	0.19	0.55	-0.21	0.29
RD2	0.37	0.49	0.39	**0.90**	0.25	0.21	0.55	-0.21	0.34
RD3	0.40	0.52	0.39	**0.87**	0.26	0.20	0.52	-0.29	0.37
SN1	0.29	0.27	0.33	0.27	**0.94**	0.44	0.15	-0.03	0.28
SN2	0.29	0.24	0.30	0.29	**0.95**	0.44	0.18	-0.03	0.27
SN3	0.30	0.24	0.33	0.26	**0.91**	0.45	0.11	-0.03	0.34
IMG1	0.26	0.21	0.36	0.22	0.43	**0.89**	0.14	0.03	0.27
IMG2	0.24	0.18	0.29	0.20	0.41	**0.90**	0.11	0.04	0.26
IMG3	0.22	0.17	0.29	0.20	0.44	**0.90**	0.09	0.07	0.26
EFF1	0.33	0.52	0.34	0.51	0.20	0.13	**0.83**	-0.28	0.31
EFF2	0.23	0.38	0.28	0.50	0.10	0.10	**0.85**	-0.22	0.21
EFF3	0.20	0.38	0.24	0.52	0.06	0.08	**0.79**	-0.23	0.19
ANX1	-0.11	-0.23	-0.15	-0.20	0.01	0.07	-0.22	**0.78**	-0.14
ANX2	-0.14	-0.30	-0.18	-0.24	-0.02	0.04	-0.28	**0.85**	-0.11
ANX3	-0.16	-0.26	-0.21	-0.21	-0.07	0.01	-0.20	**0.82**	-0.18
ANX4	-0.16	-0.31	-0.15	-0.25	-0.02	0.05	-0.27	**0.87**	-0.11
ENJ1	0.33	0.44	0.41	0.37	0.30	0.26	0.27	-0.17	**0.94**
ENJ2	0.33	0.39	0.35	0.35	0.30	0.29	0.27	-0.13	**0.93**

survey instrument for all participants at a single point in time. In contrast, the data collected in both predecessor studies was longitudinal with collection occurring across multiple time periods. Second, this study assessed the general use of desktop computers by knowledge workers engaged in a variety of job-related tasks. This is in contrast to both predecessor studies which focused on specific systems. Finally, the current study mixed voluntary and involuntary usage scenarios, whereas the predecessor studies segregated the involuntary from the voluntary usage scenarios. In spite of these material differences between our study and the predecessor studies, we justify the comparison on the basis that our research is preliminary and exploratory in nature with the primary purpose of validating the applicability of select antecedents to perceived usefulness and

Figure 2. Structural model results

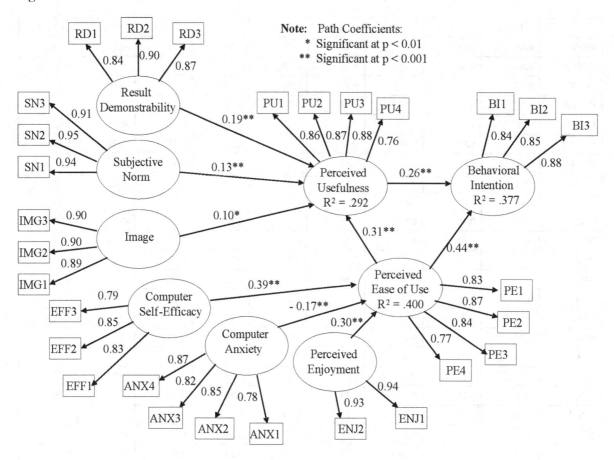

perceived ease of use in a non-western culture. This intention is in contrast to the more targeted purposes of the predecessor studies to materially refine an already mature research stream within a well-tested cultural context.

With respect to hypotheses H_1, H_2 and H_3, we found the influence of the antecedent variables on perceived usefulness to be significant and in the anticipated direction (e.g. all positive). With respect to the influence of subjective norm on perceived usefulness (H_1), we found a positive beta coefficient of 0.13, at a significance level of $p < 0.001$ (see Figure 2 and Table 9). Venkatesh and Davis (2000) reported that subjective norm had a more powerful positive influence on perceived usefulness (as high as beta = 0.50 at $p < 0.001$) early in a system implementation, but that this influence attenuated considerably to non-significant

levels three months after the implementation of the target system. The fact that subjective norm remains a significant predictor of perceived usefulness within the context of the ongoing use of desktop computers suggests a more sustained influence of subjective norm in this setting. We interpret this to be a manifestation of our earlier prediction that the higher power distance and lower individualism in Arab culture, relative to western cultures, would make subjective norm a more influential antecedent to perceived usefulness. The fact that this influence is potentially one of duration rather than magnitude would suggest that organizations functioning within this culture could sustain system usage behaviors through regular communication of management's expectations for system use.

Table 9. Hypotheses conclusions

Hypothesis	Conclusion/Result
H1: Subjective norm will have a positive effect on perceived usefulness.	Yes: (beta = 0.13, $p < 0.001$)
H2: Image will have a positive effect on perceived usefulness.	Yes: (beta = 0.10, $p < 0.01$)
H3: Result demonstrability will have a positive effect on perceived usefulness.	Yes: (beta = 0.19, $p < 0.001$)
H4: Computer self-efficacy will have a positive effect on perceived ease of use.	Yes: (beta = 0.39, $p < 0.001$)
H5: Computer anxiety will have a negative effect on perceived ease of use.	Yes: (beta = - 0.17, $p < 0.001$)
H6: Perceived enjoyment will have a positive effect on perceived ease of use.	Yes: (beta = 0.30, $p < 0.001$)
H7: Perceived ease of use will have a positive effect on perceived usefulness.	Yes: (beta = 0.31, $p < 0.001$)
H8: Perceived usefulness will have a positive effect on intention to use.	Yes: (beta = 0.26, $p < 0.001$)
H9: Perceived ease of use will have a positive effect on intention to use.	Yes: (beta = 0.44, $p < 0.001$)

With respect to the influence of image on perceived usefulness (H_2), we found a positive beta coefficient of 0.10, at a significance level of $p < 0.01$ (see Figure 2 and Table 9). Venkatesh and Davis (2000) reported higher beta coefficient magnitudes, ranging from 0.17 to 0.36 at a significance level of at least $p < 0.05$. This finding provides support for our predictions regarding the opposing influences that Arab culture will have on image. Specifically, high power-distance will tend to increase image, because the top-down approach of prescribing technology in these cultures will lead subordinates to conclude that their superiors consider a prescribed technology useful and will therefore also find it useful in order to gain approval and recognition. In opposition to the effect of high power-distance, low individualism will tend to weaken the effect of image on perceived usefulness because users will be less focused on standing out. Consequently, users in a collective culture may not consider a technology useful simply because its use will improve their social status in the organization. The reduced influence of image in this study, relative to the predecessor studies, would seem to indicate that low individualism has a stronger pull on image then does high power-distance. This would suggest that, although image may continue to be a significant predictor of perceived usefulness, organizations in this culture cannot rely on it to influence usage behavior to the extent they might in a western culture.

Finally, with respect to the influence of result demonstrability on perceived usefulness (H_3), we found a positive beta coefficient of 0.19, at a significance level of $p < 0.001$ (see Figure 2 and Table 9). Venkatesh and Davis (2000) again reported higher beta coefficient magnitudes, ranging from 0.22 to 0.34, at a significance level of $p < 0.01$. We have suggested that the need for a technology to produce tangible results would likely supersede the relative influences of cultural differences. The predecessor studies were focused on specific systems while this study assessed the more general use of desktop computers. We surmise that the slightly lower influence of result demonstrability in this study may be due to this broader focus on the general use of computers making the tangibility of results less distinct. In any case, the significance of result demonstrability in this study would suggest that the need for technology to produce tangible results is not limited to western cultures and is likely a significant antecedent for perceived usefulness in all cultures.

With respect to hypotheses H_4, H_5 and H_6, we found the influence of the antecedent variables on perceived ease of use to be significant and in the anticipated direction (e.g. positive for computer self-efficacy and perceived enjoyment, but *negative* for computer anxiety). We found that computer self-efficacy positively impacted (beta = 0.39, p < 0.001) the perceived ease of Saudi knowledge workers using desktop computers to perform job related tasks. Venkatesh (2000) reported similar findings, with computer self-efficacy positively impacting perceived ease of use, with beta coefficients ranging from 0.30 to 0.42 (at p < 0.001). Moreover, like Venkatesh (2000), we found that computer anxiety *negatively* impacted the perceived ease of using desktop computers. We found a beta coefficient of -0.17 (p < 0.001), whereas Venkatesh (2000) reported beta coefficients ranging from -0.25 to -0.30 (at p < 0.01). Finally, similar to our findings of 0.30 (p < 0.001), Venkatesh (2000) also reports a positive impact of perceived enjoyment on perceived ease of use, with beta coefficients ranging from 0.19 to 0.24 at p < 0.05. These results would suggest that, regardless of culture, organizations need to provide sufficient system training and other opportunities for users to develop computer self-efficacy and reduce computer anxiety, if they hope to establish and maintain appropriate system usage behavior.

Our findings also validate the applicability of the basic TAM structure (constructs and relationships) in Saudi Arabia (hypotheses H_7, H_8 and H_9). From the results of our survey, perceived ease of use has a positive impact on perceived usefulness (beta = 0.31, p < 0.001) and on behavioral intention (beta = 0.44, p < 0.001). Moreover, perceived usefulness also has a positive impact on behavioral intention (beta = 0.26, p < 0.001).

One additional approach to assess the relative 'goodness of fit' of the research model is to examine the relative proportions of the amounts of variance explained (e.g. R^2) for the predicted variables: perceived usefulness, perceived ease of use, and behavioral intention. The combined influences of subjective norm, image, and results demonstrability explain 29.2% of the variance of perceived usefulness. The combined influences of computer self-efficacy, computer anxiety, and perceived enjoyment explain 40% of the variance in perceived ease of use. Finally, the combined influences of perceived ease of use and perceived usefulness (which mediate the influences of all of the antecedent variables) explain 37.7% of the variance of intentions to use desktop computers amongst these Saudi knowledge workers. The relative amounts of the variance explained for these three TAM variables are consistent with other user acceptance studies conducted in western nations.

Through this study we have shown that antecedents to perceived usefulness and perceived ease of use do function within a non-western context. This would indicate that these antecedents have relevance for the development and implementation of information systems in a more global context then has been previously demonstrated. These findings also add to our knowledge base for research on information systems in non-western cultures.

LIMITATIONS

A limitation of this research was that data collection was cross sectional in nature and therefore the effects of time on the antecedents of perceived usefulness and perceived ease of use were not able to be evaluated. Also, in order to more meaningfully compare our findings with those of Venkatesh and Davis (2000) and Venkatesh (2000), it would be useful to examine the moderating influences of voluntariness of use upon the effects of these antecedent variables.

IMPLICATIONS AND FUTURE RESEARCH

The implications of our findings suggest that the antecedents of perceived usefulness and perceived ease of use do function outside of a western culture and therefore have potential as actionable constructs in the development and study of information technology in a global context. Based on these findings it would be useful to refine, and build upon, this current study in a number of ways. As previously mentioned, longitudinal studies would yield additional insights into the relative influences of these antecedent variables on the perceived usefulness and perceived ease of use of adopting new technologies in non-western countries. Moreover, the examination of specific target applications would further illuminate this process. With the increasing globalization of organizations, it would be instructive and productive to better understand those factors that influence the acceptance and adoption of new technologies in different cultural contexts.

CONCLUSION

This research investigated the influence of antecedents on perceived usefulness and perceived ease of use in a non-western culture. The study's findings indicate that the selected TAM antecedents: subjective norm, image, and result demonstrability as antecedents of perceived usefulness and self-efficacy, computer anxiety and perceived enjoyment as antecedents of perceived ease of use do function in the specific context of general computer use by knowledge workers in Saudi Arabia. These antecedents have been proposed and validated as a step toward making TAM more practically actionable by managers and system designers. It is hoped that the findings of this study will result in the development of better methods that businesses can implement to improve employee acceptance and use of new systems.

REFERENCES

Al-Gahtani, S. S. (2001). The applicability of TAM outside North America: An empirical test in the United Kingdom. *Information Resources Management Journal, 14*(3), 37.

Al-Gahtani, S. S. (2003). Computer technology adoption in Saudi Arabia: Correlates of perceived innovation attributes. *Information Technology for Development, 10*(1), 57–69. doi:10.1002/itdj.1590100106

Al-Gahtani, S. S. (2004). Computer technology acceptance success factors in Saudi Arabia: An exploratory study. *Journal of Global Information Technology Management, 7*(1), 5–29.

Al-Gahtani, S. S. (2008). Testing for the Applicability of the TAM Model in the Arabic Context: Exploring an Extended TAM with Three Moderating Factors. *Information Resources Management Journal, 21*(4), 1.

Al-Gahtani, S. S., Hubona, G. S., & Wang, J. (2007). Information technology (IT) in Saudi Arabia: Culture and the acceptance and use of IT. *Information & Management, 44*(8), 681–691. doi:10.1016/j.im.2007.09.002

Al-Khaldi, M. A., & Al-Jabri, I. M. (1998). The relationship of attitudes to computer utilization: New evidence from a developing nation. *Computers in Human Behavior, 14*(1), 23–42. doi:10.1016/S0747-5632(97)00030-7

Anandarajan, M., Igbaria, M., & Anakwe, U. P. (2002). IT acceptance in a less-developed country: A motivational factor perspective. *International Journal of Information Management, 22*(1), 47–65. doi:10.1016/S0268-4012(01)00040-8

Anderson, C., Al-Gahtani, S. S., & Hubona, G. S. (2007). *Evaluating the Antecedents of the Technology Acceptance Model in Saudi Arabia.* Paper presented at the Sixth Annual Workshop on HCI Research in MIS.

Anderson, C., Al-Gahtani, S. S., & Hubona, G. S. (2008). *Evaluating TAM Antecedents in Saudi Arabia*. Paper presented at the Southern Association of Information Systems Conference.

Bjerke, B., & Al-Meer, A. (1993). Culture's consequences: Management in Saudi Arabia. *Leadership and Organization Development Journal, 14*(2), 30–35. doi:10.1108/01437739310032700

Brislin, R. W. (1986). *The Wording and Translation of Research Instrument*. Beverly Hills, CA: Sage.

Chau, P. Y. K., & Hu, P. J.-H. (2002). Investigating healthcare professionals' decisions to accept telemedicine technology: An empirical test of competing theories. *Information & Management, 39*(4), 297–311. doi:10.1016/S0378-7206(01)00098-2

Davis, F. D. (1989). Perceived usefulness, perceived ease of use, and user acceptance of information technology. *Management Information Systems Quarterly, 13*(3), 318–340. doi:10.2307/249008

Davis, F. D., Bagozzi, R. P., & Warshaw, P. R. (1992). Extrinsic and intrinsic motivation to use computers in the workplace. *Journal of Applied Social Psychology, 22*(14), 1111–1132. doi:10.1111/j.1559-1816.1992.tb00945.x

De Vreede, G.-J., Jones, N., & Mgaya, R. J. (1998). Exploring the application and acceptance of group support systems in Africa. *Journal of Management Information Systems, 15*(3), 197–234.

Elbeltagi, I., McBride, N., & Hardaker, G. (2005). Evaluating the factors affecting DSS usage by senior managers in local authorities in Egypt. *Journal of Global Information Management, 13*(2), 42.

Fishbein, M., & Ajzen, I. (1975). *Belief, Attitude, Intention, and Behavior: An Introduction to Theory and Research*. Reading, MA: Addison-Wesley.

Ford, D. P., Connelly, C. E., & Meister, D. B. (2003). Information systems research and Hofstede's culture's consequences: An uneasy and incomplete partnership. *IEEE Transactions on Engineering Management, 50*(1), 8–25. doi:10.1109/TEM.2002.808265

Fornell, C., & Larcker, D. (1981). Evaluating structural equation models with unobservable variables and measurement error. *JMR, Journal of Marketing Research, 18*, 39–50. doi:10.2307/3151312

Gefen, D., & Keil, M. (1998). The impact of developer responsiveness on perceptions of usefulness and ease of use: An extension of the technology acceptance model. *The Data Base for Advances in Information Systems, 29*(2), 35–49.

Ghorab, K. E. (1997). The impact of technology acceptance considerations on system usage, and adopted level of technological sophistication: An empirical investigation. *International Journal of Information Management, 17*(4), 249–259. doi:10.1016/S0268-4012(97)00003-0

Henderson, R., & Divett, M. J. (2003). Perceived usefulness, ease of use and electronic supermarket use. *International Journal of Human-Computer Studies, 59*(3), 383–395. doi:10.1016/S1071-5819(03)00079-X

Hill, C. E., Loch, K. D., Straub, D. W., & El-Sheshai, K. (1998). A qualitative assessment of Arab culture and information technology transfer. *Journal of Global Information Technology Management*, 29–38.

Hofstede, G. (1980). *Culture's Consequences: International Differences in Work-Related Values*. Beverly Hills, CA: Sage.

Hofstede, G. (2001). *Culture's Consequences: Comparing Values, Behaviors, Institutions and Organizations Across Nations*. Newbury Park, CA: Sage.

Igbaria, M., Iivari, J., & Maragahh, H. (1995). Why do individuals use computer technology? A Finnish case study. *Information & Management, 29*(5), 227–238. doi:10.1016/0378-7206(95)00031-0

Igbaria, M., Zinatelli, N., Cragg, P., & Cavaye, A. L. M. (1997). Personal computing acceptance factors in small firms: A structural equation model. *Management Information Systems Quarterly, 21*(3), 279–305. doi:10.2307/249498

King, W. R., & He, J. (2006). A meta-analysis of the technology acceptance model. *Information & Management, 43*(6), 740–755. doi:10.1016/j.im.2006.05.003

Koeszegi, S., Vetschera, R., & Kersten, G. (2004). National cultural differences in the use and perception of Internet-based NSS: Does high or low context matter? *International Negotiation, 9*(1), 79. doi:10.1163/1571806041262070

Lee, H.-Y., Ahn, H., & Han, I. (2007). VCR: Virtual community recommender using the technology acceptance model and the user's needs type. *Expert Systems with Applications, 33*(4), 984–995. doi:10.1016/j.eswa.2006.07.012

Lee, Y., Kozar, K. A., & Larsen, K. R. T. (2003). The technology acceptance model: Past, present, and future. *Communications of the Association for Information Systems, 12*(50), 752–780.

Legris, P., Ingham, J., & Collerette, P. (2003). Why do people use information technology? A critical review of the technology acceptance model. *Information & Management, 40*(3), 191–204. doi:10.1016/S0378-7206(01)00143-4

Liaw, S.-S., Chang, W.-C., Hung, W.-H., & Huang, H.-M. (2006). Attitudes toward search engines as a learning assisted tool: Approach of Liaw and Huang's research model. *Computers in Human Behavior, 22*(2), 177–190. doi:10.1016/j.chb.2004.09.003

Liaw, S.-S., & Huang, H.-M. (2003). An investigation of user attitudes toward search engines as an information retrieval tool. *Computers in Human Behavior, 19*(6), 751–765. doi:10.1016/S0747-5632(03)00009-8

Lin, J. C.-C., & Lu, H. (2000). Towards an understanding of the behavioural intention to use a web site. *International Journal of Information Management, 20*(3), 197–208. doi:10.1016/S0268-4012(00)00005-0

Lu, J., Liu, C., Yu, C.-S., & Yao, J. E. (2003). Exploring factors associated with wireless Internet via mobile technology acceptance in mainland China. *Communications of the International Information Management Association, 3*(1), 101–120.

Mao, E., & Palvia, P. (2006). Testing an extended model of IT acceptance in the Chinese cultural context. *The Data Base for Advances in Information Systems, 37*(2-3), 20–32.

Mao, E., Srite, M., Thatcher, J. B., & Yaprak, O. (2005). A research model for mobile phone service behaviors: Empirical validation in the U.S. and Turkey. *Journal of Global Information Technology Management, 8*(4), 7.

McCoy, S., Everard, A., & Jones, B. M. (2005). An examination of the technology acceptance model in Uruguay and the US: A focus on culture. *Journal of Global Information Technology Management*, 27–45.

McCoy, S., Galletta, D. F., & King, W. R. (2005). Integrating national culture into IS research: The need for current individual level measures. *Communications of the Association for Information Systems, 15*, 1.

McCoy, S., Galletta, D. F., & King, W. R. (2007). Applying TAM across cultures: the need for caution. *European Journal of Information Systems, 16*(1), 81. doi:10.1057/palgrave.ejis.3000659

Moore, G. C., & Benbasat, I. (1991). Development of an instrument to measure the perceptions of adopting an information technology innovation. *Information Systems Research, 2*, 192–222. doi:10.1287/isre.2.3.192

Niederman, F., Boggs, D. J., & Kundu, S. (2002). International business and global information managment research: Toward a cumulative tradition. *Journal of Global Information Management, 10*(1), 33–47.

Nunnally, J. C. (1978). *Psychometric Theory.* New York: McGraw Hill.

Orlikowski, W., & Robey, D. (1991). Information technology and the structuring of organizations. *Information Systems Research, 2*(2), 143–169. doi:10.1287/isre.2.2.143

Parboteeah, D. V., Parboteeah, K. P., Cullen, J. B., & Basu, C. (2005). Perceived Usefulness Of Information Technology: A Cross-National Model. *Journal of Global Information Technology Management, 8*(4), 29.

Phillips, L. A., Calantone, R., & Lee, M.-T. (1994). International technology adoption: Behavior structure, demand certainty and culture. *Journal of Business and Industrial Marketing, 9*(2), 16. doi:10.1108/08858629410059762

Png, I. P. L., Tan, B. C. Y., & Khai-Ling, W. (2001). Dimensions of national culture and corporate adoption of IT infrastructure. *IEEE Transactions on Engineering Management, 48*(1), 36–45. doi:10.1109/17.913164

Ringle, C. M., Wende, S., & Will, A. (2005). *SmartPLS 2.0 (beta),* www.smartpls.de. Hamburg, Germany: University of Hamburg.

Rose, G., & Straub, D. W. (1998). Predicting general IT use: Applying TAM to the Arab world. *Journal of Global Information Management, 6*(3), 39–46.

Schepers, J., & Wetzels, M. (2007). A meta-analysis of the technology acceptance model: Investigating subjective norm and moderation effects. *Information & Management, 44*(1), 90–103. doi:10.1016/j.im.2006.10.007

Selim, H. M. (2003). An empirical investigation of student acceptance of course websites. *Computers & Education, 40*(4), 343–360. doi:10.1016/S0360-1315(02)00142-2

Shin, S. K., Ishman, M., & Sanders, G. L. (2007). An empirical investigation of socio-cultural factors of information sharing in China. *Information & Management, 44*(2), 165–174. doi:10.1016/j.im.2006.11.004

Simonson, M. R., Maurer, M., Montag-Torardi, M., & Whitaker, M. (1987). Development of a standardized test of computer literacy and a computer anxiety index. *Journal of Educational Computing Research, 3*(2), 231–247. doi:10.2190/7CHY-5CM0-4D00-6JCG

Srite, M., & Karahanna, E. (2006). The role of espoused national cultural values in technology acceptance. *Management Information Systems Quarterly, 30*(3), 679–704.

Straub, D. W. (1994). The effect of culture on IT diffusion: E-mail and fax in Japan and the United States. *Information Systems Research, 5*(1), 23–47. doi:10.1287/isre.5.1.23

Straub, D. W., Keil, M., & Brenner, W. (1997). Testing the technology acceptance model across cultures: A three country study. *Information & Management, 33*(1), 1–11. doi:10.1016/S0378-7206(97)00026-8

Straub, D. W., Loch, K. D., & Hill, C. (2001). Transfer of information technology to the Arab world: A test of cultural influence modeling. *Journal of Global Information Management, 9*, 6–28.

Stylianou, A. C., Robbins, S. S., & Jackson, P. (2003). Perceptions and attitudes about eCommerce development in China: An exploratory study. *Journal of Global Information Management, 11*(2), 31.

Suh, B., & Han, I. (2002). Effect of trust on customer acceptance of Internet banking. *Electronic Commerce Research and Applications, 1*(3-4), 247–263. doi:10.1016/S1567-4223(02)00017-0

Sukkar, A. A., & Hasan, H. (2005). Toward a model for the acceptance of Internet banking in developing countries. *Information Technology for Development, 11*(4), 381–398. doi:10.1002/itdj.20026

Sun, H., & Zhang, P. (2006). The role of moderating factors in user technology acceptance. *International Journal of Human-Computer Studies, 64*(2), 53–78. doi:10.1016/j.ijhcs.2005.04.013

Teo, H.-H., Chan, H.-C., Wei, K.-K., & Zhang, Z. (2003). Evaluating information accessibility and community adaptivity features for sustaining virtual learning communities. *International Journal of Human-Computer Studies, 59*(5), 671–697. doi:10.1016/S1071-5819(03)00087-9

van Raaij, E. M., & Schepers, J. J. L. (2008). The acceptance and use of a virtual learning environment in China. *Computers & Education, 50*(3), 838–852. doi:10.1016/j.compedu.2006.09.001

Veiga, J. F., Floyd, S., & Dechant, K. (2001). Towards modelling the effects of national culture on IT implementation and acceptance. *Journal of Information Technology, 16*(3), 145–158. doi:10.1080/02683960110063654

Venkatesh, V. (2000). Determinants of perceived ease of use: Integrating control, intrinsic motivation, and emotion into the technology acceptance model. *Information Systems Research, 11*(4), 342. doi:10.1287/isre.11.4.342.11872

Venkatesh, V., & Davis, F. D. (2000). A theoretical extension of the technology acceptance model: Four longitudinal field studies. *Management Science, 46*(2), 186–204. doi:10.1287/mnsc.46.2.186.11926

Vetschera, R., Kersten, G., & Koeszegi, S. (2006). User assessment of Internet-based negotiation support systems: An exploratory study. *Journal of Organizational Computing and Electronic Commerce, 16*(2), 123. doi:10.1207/s15327744joce1602_3

Yoo, Y., & Alavi, M. (2001). Media and group cohesion: Relative influences on social presence, task participation, and group consensus. *Management Information Systems Quarterly, 25*, 371–390. doi:10.2307/3250922

This work was previously published in the Journal of Organizational and End User Computing, Volume 23, Issue 1, edited by Mo Adam Mahmood, pp. 18-37, copyright 2011 by IGI Publishing (an imprint of IGI Global).

Chapter 3

In or Out:
An Integrated Model of Individual Knowledge Source Choice

Yinglei Wang
Acadia University, Canada

Darren B. Meister
The University of Western Ontario, Canada

Peter H. Gray
University of Virginia, USA

ABSTRACT

The way individuals use internal and external knowledge sources influences organizational knowledge integration, an important source of competitive advantage. Drawing on research into knowledge sourcing and consumer switching behavior, the authors develop an integrated model to understand individuals' choices between internal and external knowledge sources in contemporary work settings, where information technology has made both easily accessible. A test of the model using survey data collected from an international consulting firm yields an important new insight: satisfied individuals in knowledge reuse friendly environments are likely to use internal knowledge sources while they may also be tempted by easily accessible external knowledge sources. The implications for researchers and practitioners are also discussed.

INTRODUCTION

Organizations that engage in knowledge management (KM) often invest in two different kinds of initiatives to enhance employee performance (Alavi & Leidner, 2001). Internally-focused KM initiatives focus on capturing and storing employees' experiences and knowledge in information technology (IT) enabled knowledge repositories, thereby fostering knowledge sharing and reuse among employees (Markus, 2001). Externally-focused KM initiatives typically provide individuals with access to repositories of knowledge

DOI: 10.4018/978-1-4666-2059-9.ch003

produced by third parties, and convenient means of communication and knowledge sharing across organizational boundaries (Teigland & Wasko, 2003). Although such KM initiatives do not always succeed (Gilmour, 2003), many organizations have used them to create valuable pools of knowledge that employees can draw on when facing challenging problems.

However, improving the supply of knowledge that is available to employees is only the first step towards improving their performance. For organizations to benefit from KM initiatives, individuals as end users must use the resultant resources and tools to seek out others' knowledge and apply it when solving work-related problems (Gray & Meister, 2004). A recent research stream focuses on this demand side of KM and the impact of knowledge sourcing behaviors on individual performance, and has provided evidence for the positive impact of knowledge sourcing on individual performance and learning outcomes (Gray & Meister, 2006; Lin, Kuo, Kuo, Ho, & Kuo, 2007). By sourcing knowledge that is made available through KM initiatives, individuals are exposed to others' experiences and insights, which help them better understand the challenges they face, develop new skills, and improve performance (Gray & Meister, 2004, 2006).

Knowledge sourcing is a discretionary behavior. While individuals may have various options for obtaining knowledge, human limits on cognitive capacity often prevent them from consulting all available resources (Hansen & Haas, 2001); they must therefore make choices about which sources to tap. An important first decision is whether to use internal knowledge sources or external knowledge sources (Menon & Pfeffer, 2003). This decision is key as it may influence organizational knowledge integration, the fusion of knowledge from the outside and local expertise and understandings, which is a major source of performance and competitive advantage (Grant, 1996). For instance, if everyone follows the preference for outsiders suggested by Menon

and Pfeffer (2003) and only seeks knowledge from external sources, internal knowledge is likely to be left out, resulting in knowledge polarization rather than integration.

While research has revealed the importance of integrating internal knowledge with external knowledge at the group and firm levels (Mitchell, 2006), there is limited research about what drives individuals to choose either internal or external sources. This shortfall in knowledge constrains organizations' ability to achieve knowledge integration, as they would not be able to mobilize a crucial element, individuals themselves, effectively in this process. Therefore, the purpose of this paper is to shed light on the antecedents of individuals' choices between internal and external knowledge sources in volitional contexts, so that organizations can apply proper interventions to facilitate knowledge integration.

To develop our research model, we draw on marketing research that models consumers' choices between alternative services and products, which helps us identify potential antecedents of knowledge source choice. From a psychological perspective, individuals go through similar psychological processes when faced with choices; essentially, choice alternatives present a certain amount of conflict that urges individuals to evaluate salient elements in the situation and consequently adjust attitudes and behaviors to reduce the conflict (Einhorn & Hogarth, 1981). Prior research suggests that individuals only focus on limited elements in a conflict situation (Miller, 1956); and when faced with the conflicts of the same nature, they are likely to recall similar elements (Hansen, 1972, 1976). In this case, research on choice in one context is likely to provide meaningful references for identifying relevant elements in research on choice in similar contexts.

Indeed, there are strong parallels between an individual's choice of knowledge source and a consumer's choice between alternative services and products. Information technology has made more knowledge available to individuals by creating knowledge repositories, providing access to third party databases, and facilitating communication with experts (Alavi & Leidner, 2001; Markus, 2001). These various knowledge sources form a knowledge market in which individuals may switch between competing sources (Hansen & Haas, 2001), which resembles consumer markets where consumers switch between competing services/products (Kim, Shin, & Lee, 2006; van Birgelen, de Jong, & de Ruyter, 2006). Furthermore, in both cases, choice alternatives arouse the same conflict – the conflict between the different capacities of alternatives in satisfying individuals' functional and social needs (Menon & Pfeffer, 2003; Thaler, 1985). Thus, insights from the marketing literature may be helpful in shedding light on how individuals choose knowledge sources.

Integrating research on consumer switching behavior with research on knowledge source choice, we propose and test a model to explore the factors that influence individuals' intentions to use internal and external knowledge sources in work settings where these two kinds of sources compete for attention and use. Our research contributes to the literature by providing a valuable new perspective to understand internal/external knowledge use as well as tangible factors for potential managerial interventions.

In the following sections, we discuss our theory development, methodology, results and implications to both academics and practitioners.

THEORETICAL FRAMEWORK AND HYPOTHESES

Knowledge that is held by an organization and its members is often termed "internal" for employees, and that which is held by outsiders is termed "external" (Menon & Pfeffer, 2003). While internal and external knowledge may at times complement each other, following Menon and Pfeffer (2003), we pursue the possibility that individuals may also see them as comparable alternatives, especially at

the task level. In some (and perhaps many) cases, the proliferation of knowledge supply both within and outside the organization makes it likely that individuals can find ideas and task-related solutions on both sides. But whether through simple expediency or because of limited by cognitive capacity, individuals are likely to optimize their efforts by comparing sources and choosing to prioritize one over others (Hansen & Haas, 2001). Indeed, because one strong motivation for knowledge sourcing in organizational settings is time saving through knowledge reuse (Gray & Meister, 2006), it is unlikely that an individual would spend twice as much time and effort as necessary seeking the needed knowledge from both internal and external sources. Therefore, we situate our research in such contexts where individuals are able to access knowledge volitionally and emphasize the competition between internal and external knowledge, assuming a negative correlation between the use of internal and external knowledge sources.

A major finding in the literature regarding the dichotomy between internal and external knowledge sources at the individual level is that people often prefer to obtain information and knowledge from outsiders (Constant, Sproull, & Kiesler, 1996; Menon & Pfeffer, 2003). Nevertheless, this tendency runs counter to widespread organizational initiatives that promote the sharing of best practices and reuse of employees' knowledge (Markus, 2001). Over time, it may even interfere with knowledge integration and impair competitive advantage as internal knowledge may be overlooked or even intentionally rejected (Menon & Pfeffer, 2003).

Two distinct explanations for individuals' preference for external knowledge have been theorized. First, external knowledge is often perceived as being more difficult to access than internal knowledge; this lack of accessibility makes individuals less aware of its deficiencies, which in turn makes it appear more valuable and more novel than internal knowledge (Menon & Pfeffer, 2003). Second, re-using internal knowledge does not provide individuals with any competitive advantage over their co-workers (who have access to the same knowledge), and may even harm their ability to advance (Tesser, Millar, & Moore, 1988). Obtaining knowledge from another part or member of the firm may also impair an individual's status within the firm because it makes the individual look incompetent (Menon et al., 2006). As a result, individuals may avoid using internal knowledge sources because external knowledge seems more valuable and less threatening.

Although existing research provides a high-level set of explanations for why individuals may prefer external to internal knowledge, little effort has been made to produce a full nomological model that includes a broader range of antecedents. Prior studies focus principally on individuals' behavioral beliefs to explain their choice of knowledge sources, but few studies have tapped into the determinants, such as object-based beliefs and attitudes (individuals' perceptions and attitudes toward objects – for example, perceived quality and perceived accessibility of knowledge sources) and extrinsic factors, of these behavioral beliefs in organizational settings. A better understanding of these determinants is crucial for guiding managerial interventions, because behavioral beliefs are internal traits of individuals and cannot be directly changed (Lewis, Agarwal, & Sambamurthy, 2003). Organizations can only influence individuals' behaviors by changing extrinsic factors (e.g., rewards) and object attributes (e.g., knowledge accessibility) in order to achieve desired outcomes. Therefore, we set out to show how object-based beliefs and attitudes and extrinsic factors influence individual knowledge source choice, and in doing so, help managers tailor strategies and actions to facilitate knowledge integration.

As discussed in the previous section, marketing research on consumers' choices between alternative services and products provides meaningful references for identifying relevant elements in knowledge source choice, because of the simi-larities between these two decisions and the fact that individuals undergo similar psychological processes when faced with choices (Einhorn & Hogarth, 1981; Hansen, 1972, 1976; Hansen & Haas, 2001). This body of literature suggests that satisfaction and quality of alternatives are important factors in understanding individual choices (Kim et al., 2006; Li, Browne, & Chau, 2006; Li, Browne, & Wetherbe, 2007). Individuals are more likely to stay with a product/service when they feel satisfied with it (Bhattacherjee, 2001; Devaraj, Fan, & Kohli, 2002; McQuitty, Finn, & Wiley, 2000), and report being less likely to stay when there are high-quality alternatives available (Kim et al., 2006; Li et al., 2006; Rusbult & Farrell, 1983).

We adapt marketing theory of the effects of satisfaction and quality of alternatives on consumer product/service choice to the context of knowledge source choice, as depicted in Figure 1. We seek to explain and predict the extent to which individuals intend to use internal knowledge sources or external knowledge sources when they seek knowledge. In the following sections, we explain this model in detail.

Perceived Relative Value of Internal Knowledge

According to expectancy theory, the decision to engage in a behavior is a direct function of an evaluation of the benefits that are likely to result from that behavior (Kanfer, 1990; Morrison & Vancouver, 2000). However, such choices are rarely made in a vacuum, and often involve multiple alternatives. When comparing alternatives, relative value or relative advantage is often used to describe the benefits of making a choice (Moore & Benbasat, 1991; Rogers, 2003). We define perceived relative value of internal knowledge as the belief that using internal knowledge will be more valuable in improving performance and developing ability than using external knowledge. When individuals perceive a higher relative value

Figure 1. An integrated model of knowledge source choice

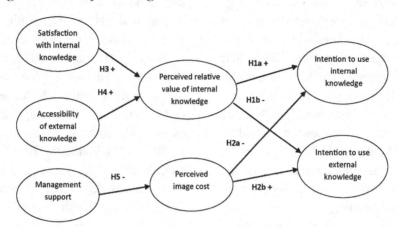

of knowledge sought from a certain source, they are more likely to adopt it (Menon & Pfeffer, 2003). In modeling individuals' choices between internal and external knowledge sources, perceived relative value of internal knowledge is thus a key indicator of the benefits that set apart the two. Therefore, we hypothesize that:

H1a: Perceived relative value of internal knowledge is positively related to individuals' intentions to use internal knowledge sources.

H1b: Perceived relative value of internal knowledge is negatively related to individuals' intentions to use external knowledge sources.

Perceived Image Cost

The loss of social status that results from admitting ignorance on a given topic is a significant cost of seeking knowledge within organizations (Borgatti & Cross, 2003; Menon et al., 2006). Consistent with previous conceptualizations of this cost (Morrison, 1993; Morrison & Vancouver, 2000; Tan & Zhao, 2003), we define perceived image cost of internal knowledge sourcing as the anticipated loss of desired social image or identity resulting from using internal knowledge sources. Previous studies have shown that perceived image cost has a negative effect on seeking knowledge

within organizations because individuals do not want to lose face (be humiliated or discredited) in front of others (Borgatti & Cross, 2003; Menon et al., 2006). Thus, we hypothesize that:

H2a: Perceived image cost of internal knowledge sourcing is negatively related to individuals' intentions to use internal knowledge sources.

H2b: Perceived image cost of internal knowledge sourcing is positively related to individuals' intentions to use external knowledge sources.

Satisfaction with Internal Knowledge

In the marketing literature, satisfaction with a product/service is considered as a key construct that determines whether an individual would switch to other products/services or stay with the current one (Kim et al., 2006; Li et al., 2006, 2007). In the context of knowledge source choice, we focus on individuals' satisfaction with the internal knowledge currently made available by their organizations. Prior research suggests that satisfaction may influence behavioral intention through its effect on behavioral beliefs (e.g., value judgments) (Ajzen & Fishbein, 2005). More specifically, as an object-based attitude, satisfaction affects individuals' beliefs regarding the value of relevant behaviors and the ease of conducting them, which in turn affect behavioral intention (Wixom & Todd, 2005). Thus, individuals may perceive internal knowledge to be less valuable if previous interactions with internal knowledge sources have failed to satisfy their knowledge needs, which also reduces its relative value when compared to seeking external knowledge (Cronin, Brady, & Hult, 2000; Menon & Pfeffer, 2003). This leads to our next hypothesis:

H3: Satisfaction with internal knowledge is positively related to perceived relative value of internal knowledge.

Accessibility of External Knowledge

One assumption made in existing research is that internal knowledge is easier to obtain than external knowledge given the ease of communication within organizations (Darr, Argote, & Epple, 1995; Menon & Pfeffer, 2003). However, with the proliferation of the Internet and public databases, this assumption does not necessarily hold true. Contemporary IT has made it easier for individuals to communicate and access resources beyond the boundaries of their organizations (Teigland & Wasko, 2003). As connections to external sources improve, internal knowledge sources no longer hold privileged supply positions.

We model such improvements in access to external knowledge sources as having a positive effect on improved perceptions of the relative value of internal sources. Research suggests that increased accessibility of external knowledge leads to devaluation of external knowledge in comparison with internal knowledge, for two reasons (Menon & Pfeffer, 2003). First, easy access to external knowledge leads to greater scrutiny and criticism, which makes external knowledge appear more flawed. Second, easy access to external knowledge reduces the attraction of external knowledge as a scarce resource. Therefore, increased accessibility makes using external knowledge less appealing when compared to using internal knowledge. Accordingly, we predict:

H4: Accessibility of external knowledge is positively related to perceived relative value of internal knowledge.

Management Support

KM research argues that management support has tremendous influence on the way that individuals seek knowledge (Argote, 1999; Tan & Zhao, 2003). We therefore include management support in our model, and define it as the degree to which individuals believe management encourages and supports internal knowledge sourcing behaviors. In highly competitive environments, individuals tend to avoid consulting internal knowledge sources because doing so may impair their social status (Menon et al., 2006). Management support mitigates this concern by providing incentives and building a healthy competitive environment (Sharma & Yetton, 2003). As such, it reduces the image cost associated with using internal knowledge sources (Morrison & Vancouver, 2000; Tan & Zhao, 2003). This leads to the following hypothesis:

H5: Management support is negatively associated with perceived image cost of internal knowledge sourcing.

METHODOLOGY

We employed a web survey method to test the proposed model. In this section, we discuss the operationalization of constructs, data collection and data analysis.

Operationalization

This study adapted most of the measures from prior studies (see the Appendix). We changed the measures when necessary, mostly by adding expressions about internal and/or external knowledge to make them fit our research context. Where applicable, item responses were adjusted to

7-point Likert scales, ranging from 1 ("Strongly disagree") to 7 ("Strongly agree").

Three items measured accessibility of external knowledge. They gauged individuals' perceptions of the accessibility of knowledge from paid external sources (e.g., FACTIVA), free Internet services (e.g., Google) and external experts respectively. Also focusing on separate sources, we measured individuals' intentions to use external knowledge sources by asking them about their likelihood of using these sources respectively. In parallel, we measured individuals' intentions to use internal knowledge sources by asking them about their likelihood of using each of the available internal sources at the research site, namely an internal electronic knowledge repository, a designated research team, and colleagues. These items reflected the most-used knowledge sources at our research site as identified by the KM manager and three interviews with other individuals at the research site. The use of such measures, rather than overall perceptions or intentions towards groups of sources, enabled us to identify individuals' use patterns of specific knowledge sources, providing the opportunity to offer diagnostic assessments to managers at the research site, in addition to model testing.

These items were treated as formative indicators rather than reflective indicators due to the interrelationships among them. In our operationalization, each item pertained to an independent source of knowledge used by the employees at the research site. The accessibility of knowledge from one source and the accessibility of knowledge from another source were not necessarily correlated, nor was their use (Gray & Meister, 2004). For instance, a senior executive might have good connections in the profession, but limited skill in navigating through electronic resources; as a result, he or she might perceive external experts more accessible than the other two sources when seeking external knowledge. Reflective indicators are theoretically inappropriate in this case, because they imply a relationship between indicators as different ways to express the same overarching construct and they empirically assume a high correlation between each other (Cohen, Cohen, Teresi, Marchi, & Velez, 1990). Formative indicators, on the other hand, do not assume such correlations because they are different causes of the construct (Bollen, 1984). Therefore, formative indicators are more appropriate in our context.

We conducted a pre-test to verify the measures before the data collection. Questionnaires were sent out by email to 67 PhD students at a Canadian business school. This sample was to some extent representative of the target population – knowledge workers who have a wide array of knowledge resources at disposal – as most of the PhD students at the business school had MBA degrees and held knowledge-intensive (e.g., analytical or managerial) positions. A total of 23 questionnaires were returned. Analysis revealed that most items performed well while some needed re-wording. We rephrased the items that appeared to be repetitive, confusing or inaccurate, and consulted peer researchers to ensure the quality of the revisions.

Data Collection

We collected data at the Canadian offices of an international consulting firm using a convenience sample method. This firm specialized in strategic communications consultancy and provided public relations and public affairs support to governments and organizations in a variety of industries (e.g., healthcare, information technology, food, etc.). Faced with a fast-changing environment, the firm had invested a lot of effort in capturing internal knowledge as well as learning from external experiences, in order to support its employees' work. Internally, employees were able to seek knowledge from colleagues, a designated research team and an electronic knowledge repository. Externally, this firm had provided employees with access to a number of third-party services, such as FACTIVA and INFORMAT. Its employees constituted an appropriate sample for knowledge management related studies, because they were employed in a knowledge-intensive industry and were familiar with KM practices (Kankanhalli, Tan, & Wei, 2005a, 2005b).

We used a web survey rather than a traditional paper-based questionnaire because of its greater speed and lower cost (Stanton, 1998). The web survey was hosted on a secured server at the institution where two of the researchers were from; this minimized the possibility of leaking participants' confidential information. We followed the procedure suggested by Dillman (2000) to maximize response rate. First, a senior manager sent an introductory email to the survey recipients. Next, the manager distributed invitations to the web survey to the employees in multiple offices. Two days after the initial invitation, the senior manager sent a reminder upon our request to the participants who had not responded. To obtain more responses, a final reminder was sent four days after the first reminder to non-respondents.

After two reminders, 50 individuals who had consulting-type jobs completed the survey out of 167 survey recipients (30%) over a 2-month period. Among the 50 received responses, 70% were from female participants while 30% were from male respondents. Twenty-four percent of the respondents held master's degrees, 72% held bachelor's degrees or had some university education, and a few had only high school diplomas. The average age was 37 and the average industry experience was 11 years.

Data Analysis

We used Partial Least Squares (PLS) to analyze data in this study. As a second generation structural equation modeling technique, PLS is superior to the traditional regression and factor analysis because the items measuring a construct are assessed within the context of the theoretical model (Thompson, Higgins, & Howell, 1991). It has been widely used in behavioral research and

found to be effective with small samples (Chin & Newsted, 1999). Therefore, we adopted PLS and the computer program used was SmartPLS 2.0 M3 (Ringle, Wende, & Will, 2005).

In addition to the variables shown in Figure 1, we included several individual characteristics as well as task characteristics as control variables in the analysis. Gender and job tenure are thought to have effects on an individual's choice of knowledge sources (King & Lekse, 2006; Venkatesh & Morris, 2000). For example, senior employees who are very familiar with internal knowledge may perceive external knowledge as less flawed and more valuable, resulting in a preference for external knowledge sources (Menon & Pfeffer, 2003). In addition, men usually prefer external knowledge sources because they are more sensitive to image cost (Venkatesh & Morris, 2000). Besides individual characteristics, task nonroutineness may also play a role in knowledge source choice (Lin & Huang, 2008). When faced with a nonroutine task, such as new product development, individuals may give higher value to external knowledge where a larger proportion of novel ideas are thought to exist (Caloghirou, Kastelli, & Tsakanikas, 2004). Therefore, in this study we controlled for the effects of task nonroutineness and commonly used demographic characteristics such as age, gender, job tenure, education and industry experience.

The measurement model was assessed by item loadings, reliability, convergent validity and discriminant validity. Hypotheses were tested by examining the t-statistics of path coefficients. During the analysis, we dropped a few items that did not work well (see the Appendix). For instance, the reverse-coded items for perceived relative value of internal knowledge (PVI5) and management support (MS3) did not correlate with other items of respective constructs; they were thus taken out of the model. The results reported in the following section are the ones after dropping ineffective measures.

Table 1. Descriptive statistics, reliabilities and AVEs of constructs

Constructs	Mean	SD	AVE	Reliability
Satisfaction with internal knowledge (SIK)	4.95	1.22	0.74	0.90
Accessibility of external knowledge (AEK) *	4.94	0.70	N/A	N/A
Perceived relative value of internal knowledge (RVI)	4.01	1.27	0.79	0.94
Management support (MS)	5.74	0.77	0.71	0.88
Perceived image cost (PIC)	1.85	0.91	0.70	0.88
Intention to use internal knowledge sources (IUI) *	5.55	0.95	N/A	N/A
Intention to use external knowledge sources (IUE) *	5.75	0.88	N/A	N/A
Task nonroutineness (TNR)	4.68	1.02	0.63	0.78

* The noted constructs are formative.

RESULTS

Measurement Model

Table 1 presents the descriptive statistics, reliability and average variance extracted (AVE) of each construct. Table 2 reports the correlations between constructs with the square roots of AVEs on diagonal. In this study, we used formative items to measure accessibility of external knowledge, intention to use internal knowledge sources, and intention to use external knowledge sources, while other constructs were measured by reflective items. Table 3 summarizes the loading of each reflective item on its corresponding construct and cross-loadings on other constructs. Table 4 shows the weights of the formative items.

Our measures showed acceptable reliability. In PLS, composite reliability indicates measurement reliability (Barclay, Higgins, & Thompson, 1995). The threshold is that composite reliability is higher than 0.7 (Nunnally, 1978). As shown in Table 1, all the constructs measured by reflective measures reached this threshold, demonstrating acceptable reliability.

Table 2. Correlations of latent constructs with the square roots of AVEs on diagonal

	SIK	AEK	RVI	MS	PIC	IUI	IUE	TNR
Satisfaction with internal knowledge (SIK)	**0.86**							
Accessibility of external knowledge (AEK)	-0.37**	**N/A**						
Perceived relative value of internal knowledge (RVI)	0.50**	-0.46**	**0.88**					
Management support (MS)	0.68**	-0.36*	0.43**	**0.84**				
Perceived image cost (PIC)	-0.40**	0.12	-0.30*	-.54**	**0.84**			
Intention to use internal knowledge sources (IUI)	0.71**	-0.31*	0.42**	0.66**	-0.60**	**N/A**		
Intention to use external knowledge sources (IUE)	-0.38**	0.35*	-.052**	-0.31*	0.14	-0.12	**N/A**	
Task nonroutineness (TNR)	0.02	0.10	0.15	-0.03	-0.08	0.10	0.19	**0.79**

* P < 0.05, ** P < 0.01, *** P < 0.001

Table 3. Loadings and cross-loadings of reflective measures

	SIK	RVI	MS	PIC	TNR
SIK1	0.78	0.46	0.57	-0.44	0.16
SIK2	0.90	0.46	0.58	-0.29	-0.06
SIK3	0.90	0.33	0.60	-0.29	-0.06
RVI1	0.47	0.85	0.34	-0.28	0.09
RVI2	0.46	0.89	0.33	-0.19	0.15
RVI3	0.49	0.93	0.47	-0.34	0.10
RVI4	0.33	0.89	0.38	-0.24	0.18
MS1	0.53	0.45	0.84	-0.54	0.09
MS2	0.58	0.32	0.80	-0.40	-0.09
MS4	0.61	0.27	0.87	-0.38	-0.11
PIC2	-0.22	-0.17	-0.38	0.89	0.02
PIC4	-0.28	-0.18	-0.49	0.87	-0.10
PIC5	-0.49	-0.40	-0.45	0.75	-0.09
TNR3	0.02	0.13	0.03	-0.07	0.85
TNR4	0.01	0.10	-0.09	-0.05	0.74

Our measures demonstrated satisfactory convergent validity. Fornell and Larcker (1981) suggest that AVE indicates the convergent validity of a construct. The convergent validity of a construct is acceptable when its AVE is higher than 0.5 (Chin, 1998; Fornell & Larcker, 1981).

As shown in Table 1, the AVEs of all constructs with reflective measures were higher than 0.5, indicating satisfactory convergent validity.

The measurement model of this study also showed adequate discriminant validity. To achieve acceptable discriminant validity, the square root

Table 4. Weights of formative measures

Construct	Indicator	Weights
Accessibility of external knowledge (AEK)		
	AEK1	-0.38
	AEK2	0.29
	AEK3	0.88***
Intention to use internal knowledge sources (IUI)		
	IUI1	-0.12
	IUI2	1.00***
	IUI3	0.06
Intention to use external knowledge sources (IUE)		
	IUE1	-0.45
	IUE2	0.54*
	IUE3	0.67**

* $P < 0.05$, ** $P < 0.01$, *** $P < 0.001$

of the AVE of a construct should be greater than any other correlation related to this construct (Chin, 1998). In addition, no item should have a higher loading on another construct than it does on the construct it intends to measure (Barclay et al., 1995). As shown in Table 2 and Table 3, all measures satisfied these criteria, suggesting adequate discriminant validity.

Nevertheless, one observation from the data was that there were a few high (r>0.5) correlations among constructs. Besides the underlying theoretical relationships, a contributor to these correlations could be common method bias, because all the constructs were measured by a single method and at the same time (Podsakoff & Organ, 1986). While the results of our analysis on reliability and construct validity provided some confidence in not having this issue (Wixom & Todd, 2005), we conducted a Harmon one-factor test (Podsakoff & Organ, 1986; Podsakoff, MacKenzie, Lee, & Podsakoff, 2003), to further examine the existence of common method bias. Results from this test demonstrated there was not a single factor that could explain a large amount of variance in the measures, suggesting common method bias is unlikely to be a problem in this study.

For formative measures, it is inappropriate to apply conventional measurement indices, such as reliability and AVE, because they do not reflect the overarching constructs but form them (Chin, 1998; Cohen et al., 1990). They should be assessed by their weights, which indicate their contributions to the overarching constructs (Chin, 1998; Chin & Gopal, 1995). In Table 4, the significant weight of AEK3 reveals that external experts were perceived to be the major contributors of accessible external knowledge at the research site. As to the use of knowledge sources, the research team at the research site, free Internet services and outsiders appeared to be the most preferred ones.

One potential issue with formative measures is multicollinearity, which may make it impossible to interpret the influence of each individual indicator (Diamantopoulos & Winklhofer, 2001). Since

PLS does not provide tests for multicollinearity, we generated variance inflation factors (VIFs) in SPSS to detect whether our measures had this issue (Mathieson, Peacock, & Chin, 2001). The results showed that none of the VIFs was higher than 1.5. Compared to 10 as the threshold for serious multicollinearity (Myers, 1990), multicollinearity seems not to be an issue with these formative measures.

Structural Model

Figure 2 shows the path coefficients with significance levels. We used bootstrapping with 500 subsamples of size 50 to generate t-statistics.

As shown in Figure 2, our results supported Hypothesis 1a, 1b, 2a, 3, and 5. The path between accessibility of external knowledge and perceived relative value of internal knowledge was significant, but the coefficient was negative – opposite to the hypothesized positive relationship. Therefore, our results did not support Hypothesis 4. None of the control variables was significant.

A central premise of our theorizing was that internal and external knowledge sources compete with each other when both can help individuals cope with certain tasks (Hansen & Haas, 2001). The results supported this premise, with an indirect negative effect of accessibility of external knowledge on intention to use internal knowledge sources through perceived relative value of internal knowledge, after controlling for potential impacts of tasks. This suggests that external knowledge sources may indeed distract individuals from internal knowledge sources. In addition, we found that individuals who perceived internal knowledge as more valuable in improving performance would be more likely to use internal knowledge sources and less likely to use external knowledge sources, providing further evidence for the competition between internal and external knowledge sources.

In summary, our results supported the proposed model except for Hypothesis 2b and 4. The proposed model explained 46% of the variance in

Figure 2. Structural model results

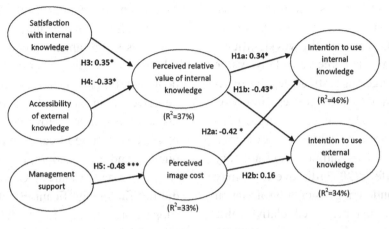

* P < 0.05, ** P < 0.01, *** P < 0.001

intention to use internal knowledge sources, 34% of the variance in intention to use external knowledge sources, 37% of the variance in perceived relative value of internal knowledge, and 33% of the variance in perceived image cost.

DISCUSSION

The objective of this study was to extend research on individuals' choice of knowledge sources, and to explore object-based beliefs and attitudes as well as extrinsic factors that organizations can draw upon to facilitate patterns of knowledge sourcing. We developed an integrated model by integrating research on consumer switching behavior into the existing research on knowledge source choice. As shown in the analysis, our model appears to be effective, explaining a considerable amount of variance in individuals' intentions to use internal and external knowledge sources and identifying a set of significant antecedents.

Our results underscore the role of satisfaction with internal knowledge in individual knowledge source choice. As expected, dissatisfied individuals perceived less relative value of internal knowledge, in turn lowering the intention to use internal knowledge sources and increasing the

intention to use external knowledge sources. This finding suggests that organizations need to pay attention to employees' satisfaction with internal KM to ensure the success of KM initiatives. Many organizations seek to improve performance by sharing best practices and re-using existing knowledge (Markus, 2001). However, easily accessible external knowledge sources may appear more competitive in knowledge markets (Hansen & Haas, 2001), and draw individuals away from using internal knowledge. As our analysis shows, dissatisfaction with internal knowledge is a major motive for individuals to reduce internal knowledge sourcing and potentially turn to alternative sources. Therefore, managers could focus on improving individual satisfaction with internal knowledge in order to foster knowledge reuse. This may require direct dialogue with front-line employees rather than turning to their superiors or job descriptions to understand their knowledge needs, as knowledge needs are often affected by personal experiences instead of generic job requirements. Such an approach would be able to help identify idiosyncratic needs, uncover potential reasons for dissatisfaction, and thus facilitate the design of proper tools and incentives that encourage knowledge sharing and reuse.

Though the results do not support our hypothesis about the effect of accessibility of external knowledge, they permit interesting observations. External knowledge accessibility was significant but negatively related to perceived relative value of internal knowledge, suggesting a negative indirect effect of external knowledge accessibility on intention to use internal knowledge sources and a positive indirect effect on intention to use external knowledge sources. This finding is contradictory to Menon and Pfeffer's (2003) discoveries. It may well be that a key underlying assumption of Menon and Pfeffer's work (that perceived relative value is driven by scarcity) might be less relevant in our context, or that this effect was swamped by the very real practical benefits of accessing external knowledge. Regardless these indirect effects are to some extent consistent with the findings in the research on consumer switching behavior. Studies on consumer switching behavior usually state a direct negative relationship between quality of alternatives and intention to stay with current products/services (Kim et al., 2006; Li et al., 2006). Our finding of the indirect effects of accessibility of external knowledge suggests that perceived relative value may actually be an intermediate step by which quality of alternatives may determine whether individuals will switch.

The effects of accessibility of external knowledge on individuals' use of internal and external knowledge sources provide an avenue to influence individuals' – and therefore eventually the organization's – knowledge acquisition patterns. By influencing the availability and ease of use of external sources (Davis, 1989), managers can create conditions where external sourcing is likely to increase and therefore see increased numbers of ideas imported to the organization. Similarly, not facilitating this access may encourage more internal sharing.

The analysis provides partial evidence for the effect of management support on knowledge source choice through perceived image cost. Menon and colleagues (2006) have argued that individuals may avoid internal knowledge sourcing so as to not appear incompetent in front of others. Our results support this argument by demonstrating the negative relationship between perceived image cost and intention to use internal knowledge sources. However, we can see that management support can reduce an individual's qualms, and thus encouraging him/her to use internal knowledge sources. On the other hand, we did not find a significant effect of management support on intention to use external knowledge sources because perceived image cost of internal knowledge sourcing was not significantly related to intention to use external knowledge sources, though there was a positive correlation between the two. This lack of support could be caused by the small sample size, which was only enough to detect strong effects (Cohen, 1988). While we could not make inferences based on the current results, we would expect to see such an effect in future research using larger samples, given the general competing relationship between internal and external knowledge sources supported in this study.

The effect of management support sheds light on another mechanism by which managers may influence knowledge integration. As the results show, with positive support, individuals tend to use internal knowledge sources more. On the other hand, a strong campaign for importing new ideas may discourage individuals to seek knowledge from internal sources and potentially turn to the outside. Therefore, managers can adjust managerial support to guide individuals to different knowledge sources.

Limitations and Future Research

This study has a few limitations that suggest opportunities for future research. First, we did not include the determinants of the object-based beliefs and attitudes in the model, partly due to the structure of the adopted theoretical framework and concern over the scope of this research. Future research including the effects of these determinants, such as the impact of service quality on satisfaction (Kettinger & Lee, 1994), could present a more complete picture and expand the explanatory power of the model. Second, while this research is based on an assumption common in the research on individual knowledge sourcing that internal and external knowledge sources compete with each other in helping accomplish a given task (Hansen & Haas, 2001; Menon & Pfeffer, 2003), we acknowledge that there may be situations where they complement each other (Cassiman & Veugelers, 2006). In some cases, individuals may need to consult both in order to work effectively. For instance, they may need to use multiple sources to solicit different perspectives, they may use external sources to benchmark internal knowledge/practices, or they may simply have to go to different sources for different knowledge. Research looking into these possibilities may contribute to the literature by presenting different use patterns of internal and external knowledge sources. Third, research has shown that there may be a discrepancy between intention and actual behavior (Straub, Limayem, & Karahanna-Evaristo, 1995). Future research could include actual usage of knowledge sources to see if this changes the results. Fourth, there is a need for more empirical evidence. Although 50 cases from a single research site are enough to test most of our hypotheses given the large effect sizes (Chin & Newsted, 1999; Cohen, 1988), we might not be able to detect potential small effects (e.g., the effect of perceived image cost of internal knowledge sourcing on intention to use external knowledge sources) and could not conclude that the findings apply to other settings. Research may benefit from further tests in various settings.

CONCLUSION

Individual knowledge source choice may influence knowledge integration in organizations and consequently long-term performance (Mitchell, 2006; Teigland & Wasko, 2003). This study explores the determinants of this important decision in the context where internal and external knowledge sources are comparable and compete for use. Our analysis reveals that an individual's choice between internal and external knowledge sources in volitional contexts is a function of his/her satisfaction with internal knowledge, perceived image cost associated with consulting internal knowledge sources, perceived relative value of internal knowledge, accessibility of external knowledge and management support for internal knowledge sourcing.

Our contributions to the literature are twofold. First, this study introduces a new perspective, namely the consumer switching behavior perspective, to research on knowledge sourcing. This perspective enables us to identify tangible factors that managers can influence to facilitate knowledge sourcing and the fusion of knowledge from various sources, resulting in an integrated model that complements existing research. Second, the results provide a clearer picture of the effects of a group of factors and thus further the understanding of individual knowledge source choice. Our empirical test of the determinants of individual knowledge source choice supports most of the hypotheses, but uncovers an effect of external knowledge accessibility that contrasts with existing research. External knowledge accessibility was negatively associated with perceived relative value of internal knowledge, and thus had a negative effect on intention to use internal knowledge sources and a positive effect on intention to use external knowledge sources. These findings help researchers and managers understand the phenomenon in question better, providing evidence for the effects of new factors as well as triangulation of the effects of the factors suggested by prior research.

REFERENCES

Ajzen, I., & Fishbein, M. (2005). The influence of attitudes on behavior. In Albarracin, D., Johnson, B. T., & Zanna, M. P. (Eds.), *The handbook of attitudes* (pp. 173–221). Mahwah, NJ: Lawrence Erlbaum Associates.

Alavi, M., & Leidner, D. E. (2001). Review: Knowledge management and knowledge management systems: Conceptual foundations and research issues. *Management Information Systems Quarterly, 25*(1), 107–136. doi:10.2307/3250961

Argote, L. (1999). *Organizational learning: Creating, retailing and transferring knowledge.* Norwell, MA: Kluwer Academic Publishers.

Barclay, D., Higgins, C., & Thompson, R. (1995). The partial least squares (PLS) approach to causal modeling: Personal computer adoption and use as an illustration. *Technology Studies, 2*(2), 285–309.

Becerra-Fernandez, I., & Sabherwal, R. (2001). Organizational knowledge management: A contingency perspective. *Journal of Management Information Systems, 18*(1), 23–55.

Bhattacherjee, A. (2001). Understanding information systems continuance: An expectation-confirmation model. *Management Information Systems Quarterly, 25*(3), 351–370. doi:10.2307/3250921

Bollen, K. A. (1984). Multiple indicators: Internal consistency or no necessary relationship? *Quality & Quantity, 18*(4), 377–385. doi:10.1007/BF00227593

Borgatti, S. P., & Cross, R. (2003). A relational view of information seeking and learning in social networks. *Management Science, 49*(4), 432–445. doi:10.1287/mnsc.49.4.432.14428

Caloghirou, Y., Kastelli, I., & Tsakanikas, A. (2004). Internal capabilities and external knowledge sources: Complements or substitutes for innovative performance? *Technovation, 24*(1), 29–39. doi:10.1016/S0166-4972(02)00051-2

Cassiman, B., & Veugelers, R. (2006). In search of complementarity in innovation strategy: Internal R&D and external knowledge acquisition. *Management Science, 52*(1), 68–82. doi:10.1287/mnsc.1050.0470

Chin, W. W. (1998). The partial least squares approach to structural equation modeling. In Marcoulides, G. A. (Ed.), *Modern methods for business research* (pp. 295–336). Mahwah, NJ: Lawrence Erlbaum Associates.

Chin, W. W., & Gopal, A. (1995). Adoption intention in GSS: Relative importance of beliefs. *The Data Base for Advances in Information Systems, 26*(2-3), 42–64.

Chin, W. W., & Newsted, P. R. (1999). Structural equation modeling analysis with small samples using partial least squares. In Hoyle, R. (Ed.), *Statistical strategies for small sample research* (pp. 307–341). Thousand Oaks, CA: Sage.

Cohen, J. (1988). *Statistical power analysis for the behavioral sciences* (2nd ed.). Philadelphia, PA: Lawrence Erlbaum Associates.

Cohen, P., Cohen, J., Teresi, J., Marchi, M., & Velez, C. N. (1990). Problems in the measurement of latent variables in structural equations causal models. *Applied Psychological Measurement, 14*(2), 183–196. doi:10.1177/014662169001400207

Constant, D., Sproull, L., & Kiesler, S. (1996). The kindness of strangers: The usefulness of electronic weak ties for technical advice. *Organization Science, 7*(2), 119–135. doi:10.1287/orsc.7.2.119

Cronin, J. J., Brady, M. K., & Hult, G. T. M. (2000). Assessing the effects of quality, value, and customer satisfaction on consumer behavioral intentions in service environments. *Journal of Retailing, 76*(2), 193–218. doi:10.1016/S0022-4359(00)00028-2

Darr, E. D., Argote, L., & Epple, D. (1995). The acquisition, transfer, and depreciation of knowledge in service organizations: Productivity in franchises. *Management Science, 41*(11), 1750–1762. doi:10.1287/mnsc.41.11.1750

Davis, F. D. (1989). Perceived usefulness, perceived ease of use, and user acceptance. *Management Information Systems Quarterly, 13*(3), 319–340. doi:10.2307/249008

Devaraj, S., Fan, M., & Kohli, R. (2002). Antecedents of B2C channel satisfaction and preference: Validation e-commerce metrics. *Information Systems Research, 13*(3), 316–333. doi:10.1287/isre.13.3.316.77

Diamantopoulos, A., & Winklhofer, H. (2001). Index construction with formative indicators: An alternative to scale development. *JMR, Journal of Marketing Research, 38*(2), 269–277. doi:10.1509/jmkr.38.2.269.18845

Dillman, D. A. (2000). *Mail and Internet surveys: The tailored design method*. New York, NY: John Wiley & Sons.

Einhorn, H., & Hogarth, R. (1981). Behavioral decision theory: Processes of judgment and choice. *Annual Review of Psychology, 32*(1), 53–88. doi:10.1146/annurev.ps.32.020181.000413

Fornell, C., & Larcker, D. F. (1981). Evaluating structural equation models with unobservable variables and measurement error. *JMR, Journal of Marketing Research, 18*(1), 39–50. doi:10.2307/3151312

Gilmour, D. (2003). How to fix knowledge management. *Harvard Business Review, 81*(10), 16–17.

Goodhue, D. L. (1998). Development and measurement validity of a task-technology fit instrument for user evaluations of information systems. *Decision Sciences, 29*(1), 105–138. doi:10.1111/j.1540-5915.1998.tb01346.x

Grant, R. M. (1996). Prospering in dynamically-competitive environments: Organizational capability as knowledge integration. *Organization Science, 7*(4), 375–387. doi:10.1287/orsc.7.4.375

Gray, P. H., & Meister, D. B. (2004). Knowledge sourcing effectiveness. *Management Science, 50*(6), 821–834. doi:10.1287/mnsc.1030.0192

Gray, P. H., & Meister, D. B. (2006). Knowledge sourcing methods. *Information & Management, 43*(2), 142–156. doi:10.1016/j.im.2005.03.002

Hansen, F. (1972). *Consumer choice behaviour*. New York, NY: Free Press.

Hansen, F. (1976). Psychological theories of consumer choice. *The Journal of Consumer Research, 3*(3), 117–142. doi:10.1086/208660

Hansen, M. T., & Haas, M. R. (2001). Competing for attention in knowledge markets: Electronic document dissemination in a management consulting company. *Administrative Science Quarterly, 46*(1), 1–28. doi:10.2307/2667123

Kanfer, R. (1990). Motivation theory in I/O psychology. In Dunnette, M. D., & Hough, L. M. (Eds.), *Handbook of industrial and organizational psychology* (pp. 75–170). Palo Alto, CA: Consulting Psychological Press.

Kankanhalli, A., Tan, B. C. Y., & Wei, K.-K. (2005a). Contributing knowledge to electronic knowledge repositories: An empirical investigation. *Management Information Systems Quarterly, 29*(1), 113–143.

Kankanhalli, A., Tan, B. C. Y., & Wei, K.-K. (2005b). Understanding seeking from electronic knowledge repositories: An empirical study. *Journal of the American Society for Information Science and Technology, 56*(11), 1156–1166. doi:10.1002/asi.20219

Kettinger, W. J., & Lee, C. C. (1994). Perceived service quality and user satisfaction with the information services functions. *Decision Sciences, 25*(5-6), 737–766. doi:10.1111/j.1540-5915.1994.tb01868.x

Kim, G., Shin, B., & Lee, H. G. (2006). A study of factors that affect user intentions toward email service switching. *Information & Management, 43*(7), 884–893. doi:10.1016/j.im.2006.08.004

King, W. R., & Lekse, W. J. (2006). Deriving managerial benefit from knowledge search: A paradigm shift? *Information & Management, 43*(7), 874–883. doi:10.1016/j.im.2006.08.005

Lewis, W., Agarwal, R., & Sambamurthy, V. (2003). Sources of influence on beliefs about information technology use: An empirical study of knowledge workers. *Management Information Systems Quarterly, 27*(4), 657–678.

Li, D., Browne, G. J., & Chau, P. Y. K. (2006). An empirical investigation of web site use using a commitment-based model. *Decision Sciences, 37*(3), 427–444. doi:10.1111/j.1540-5414.2006.00133.x

Li, D., Browne, G. J., & Wetherbe, J. C. (2007). Online consumers' switching behavior: A buyer-seller relationship perspective. *Journal of Electronic Commerce in Organizations, 5*(1), 30–42. doi:10.4018/jeco.2007010102

Lin, C. Y., Kuo, T. H., Kuo, Y. K., Ho, L. A., & Kuo, Y. L. (2007). The KM chain - Empirical study of the vital knowledge sourcing links. *Journal of Computer Information Systems, 48*(2), 91–99.

Lin, T.-C., & Huang, C.-C. (2008). Understanding knowledge management system usage antecedents: An integration of social cognitive theory and task technology fit. *Information & Management, 45*(6), 410–417. doi:10.1016/j.im.2008.06.004

Markus, M. L. (2001). Toward a theory of knowledge reuse: Types of knowledge reuse situations and factors in reuse success. *Journal of Management Information Systems, 18*(1), 57–93.

Mathieson, K., Peacock, E., & Chin, W. (2001). Extending the technology acceptance model: The influence of perceived user resources. *The Data Base for Advances in Information Systems, 32*(3), 86–112.

McQuitty, S., Finn, A., & Wiley, J. B. (2000). Systematically varying consumer satisfaction and its implications for product choice. *Academy of Marketing Science Review, 10*, 1–16.

Menon, T., & Pfeffer, J. (2003). Valuing internal vs. External knowledge: Explaining the preference for outsiders. *Management Science, 49*(4), 497–513. doi:10.1287/mnsc.49.4.497.14422

Menon, T., Thompson, L., & Choi, H.-S. (2006). Tainted knowledge vs. Tempting knowledge: People avoid knowledge from internal rivals and seek knowledge from external rivals. *Management Science, 52*(8), 1129–1144. doi:10.1287/mnsc.1060.0525

Miller, G. (1956). The magic number seven, plus or minus two: Some limits on our capacity for processing information. *Psychological Review, 63*(2), 81–97. doi:10.1037/h0043158

Mitchell, V. L. (2006). Knowledge integration and information technology project performance. *Management Information Systems Quarterly, 30*(4), 919–939.

Moore, G. C., & Benbasat, I. (1991). Development of an instrument to measure the perceptions of adopting an information technology innovation. *Information Systems Research, 2*(3), 192–223. doi:10.1287/isre.2.3.192

Morrison, E. W. (1993). Newcomer information seeking: Exploring types, modes, sources, and outcomes. *Academy of Management Journal, 36*(3), 557–589. doi:10.2307/256592

Morrison, E. W., & Vancouver, J. B. (2000). Within-person analysis of information seeking: The effects of perceived costs and benefits. *Journal of Management, 26*(1), 119–137. doi:10.1016/S0149-2063(99)00040-9

Myers, R. H. (1990). *Classical and modern regression with applications*. Boston, MA: PWS and Kent Publishing.

Nunnally, J. (1978). *Psychometric theory*. New York, NY: McGraw-Hill.

Podsakoff, P. M., MacKenzie, S. B., Lee, J. Y., & Podsakoff, N. P. (2003). Common method biases in behavioral research: A critical review of the literature and recommended remedies. *The Journal of Applied Psychology, 88*(5), 879–903. doi:10.1037/0021-9010.88.5.879

Podsakoff, P. M., & Organ, D. (1986). Self-reports in organizational research: Problems and prospects. *Journal of Management, 12*(4), 531–544. doi:10.1177/014920638601200408

Ringle, C. M., Wende, S., & Will, A. (2005). *SmartPLS 2.0 (beta)*. Hamburg, Germany: University of Hamburg. Retrieved from http://www.smartpls.de

Rogers, E. M. (2003). *Diffusion of innovation*. New York, NY: Free Press.

Rusbult, C. E., & Farrell, D. (1983). A longitudinal test of the investment model: The impact on job satisfaction, job commitment, and turnover of variations in rewards, costs, alternatives, and investments. *The Journal of Applied Psychology, 68*(3), 429–438. doi:10.1037/0021-9010.68.3.429

Sharma, R., & Yetton, P. (2003). The contingent effects of management support and task interdependence on successful information systems implementation. *Management Information Systems Quarterly, 27*(4), 533–555.

Stanton, J. M. (1998). An empirical assessment of data collection using the Internet. *Personnel Psychology, 51*(3), 709–725. doi:10.1111/j.1744-6570.1998.tb00259.x

Straub, D., Limayem, M., & Karahanna-Evaristo, E. (1995). Measuring system usage: Implications for IS theory testing. *Management Science, 41*(8), 1328–1342. doi:10.1287/mnsc.41.8.1328

Tan, H. H., & Zhao, B. (2003). Individual and contextual level antecedents of individual technical information inquiry in organizations. *The Journal of Psychology, 137*(6), 579–621. doi:10.1080/00223980309600637

Teigland, R., & Wasko, M. M. (2003). Integrating knowledge through information trading: Examining the relationship between boundary spanning communication and individual performance. *Decision Sciences, 34*(2), 261–286. doi:10.1111/1540-5915.02341

Tesser, A., Millar, M., & Moore, J. (1988). Some affective consequences of social comparison and reflection processes: The pain and pleasure of being close. *Journal of Personality and Social Psychology, 54*(1), 49–61. doi:10.1037/0022-3514.54.1.49

Thaler, R. (1985). Mental accounting and consumer choice. *Marketing Science, 4*(3), 199–214. doi:10.1287/mksc.4.3.199

Thompson, R. L., Higgins, C. A., & Howell, J. M. (1991). Personal computing: Toward a conceptual model of utilization. *Management Information Systems Quarterly, 15*(1), 125–143. doi:10.2307/249443

van Birgelen, M., de Jong, A., & de Ruyter, K. (2006). Multi-channel service retailing: The effects of channel performance satisfaction on behavioral intentions. *Journal of Retailing, 82*(4), 367–377. doi:10.1016/j.jretai.2006.08.010

Venkatesh, V., & Morris, M. G. (2000). Why don't men ever stop to ask for directions? Gender, social influence, and their role in technology acceptance and usage behavior. *Management Information Systems Quarterly, 24*(1), 115–139. doi:10.2307/3250981

Wixom, B. H., & Todd, P. A. (2005). A theoretical integration of user satisfaction and technology acceptance. *Information Systems Research, 16*(1), 85–102. doi:10.1287/isre.1050.0042

APPENDIX

Table A1. Measures of constructs

Constructs	Items
Satisfaction with internal knowledge (Becerra-Fernandez & Sabherwal, 2001)	SIK1. The available knowledge in my firm improves my effectiveness in performing my tasks.
	SIK2. I am satisfied with the management of knowledge I need in my firm.
	SIK3. I am satisfied with the availability of knowledge for my tasks in my firm.
Accessibility of external knowledge	AEK1. It is easy to get access to knowledge that I need by FACTIVA/INFOMART.
	AEK2. It is easy to get access to knowledge that I need by free Internet service, e.g., Google.
	AEK3. It is easy to get access to knowledge that I need by talking to experts and other people outside my firm.
Perceived relative value of internal knowledge (Moore & Benbasat, 1991)	RVI1. I can accomplish tasks more quickly if I use internal knowledge sources than if I use external knowledge sources.
	RVI2. The quality of my work is greater if I use internal knowledge sources than if I use external knowledge sources.
	RVI3. It is easier to do my job if I use internal knowledge sources than if I use external knowledge sources.
	RVI4. I am more effective in my job if I use internal knowledge sources than if I use external knowledge sources.
	Dropped: RVI5. I am more productive in my job if I use external knowledge sources than if I use internal knowledge sources (reverse-coded).
Intention to use internal knowledge sources	IUI1. How likely would you be to consult XXX.net (pseudo name for the knowledge repository at the research site) when you need knowledge in the future?
	IUI2. How likely would you be to consult the research team when you need knowledge in the future?
	IUI3. How likely would you be to consult colleagues when you need knowledge in the future?
Intention to use external knowledge sources	IUE1. How likely would you be to consult FACTIVA/INFOMART when you need knowledge in the future?
	IUE2. How likely would you be to consult free Internet services (e.g., Google) when you need knowledge in the future?
	IUE3. How likely would you be to consult people outside your firm when you need knowledge in the future?
Management support (Tan & Zhao, 2003)	MS1. Senior management in my organization encourages individuals to make use of internal resources to learn at work.
	MS2. Senior management in my organization is supportive of my efforts to acquire knowledge within my organization
	Dropped: MS3. It is unreasonable to try to get needed knowledge inside my organization at work because it will affect senior management's evaluation of my performance (reverse-coded)
	MS4. Senior management in my organization values internal knowledge inquiry and learning behavior of individuals

continued on following page

Table A1. Continued

Constructs	Items
Perceived image cost (Tan & Zhao, 2003)	Dropped: PIC1. I would not be nervous about seeking knowledge existing in my organization
	PIC2. It is not a good idea to seek knowledge within my organization; people might think I am incompetent
	Dropped: PIC3. It is embarrassing to ask for expertise in my organization
	PIC4. I think the source would think worse of me if I asked for knowledge from him/her
	PIC5. It is better to try and figure out the solution to problems on my own rather than ask for knowledge from internal sources
Task nonroutineness (Goodhue, 1998)	Dropped: TNR1. I frequently deal with ad hoc, nonroutine business problems
	Dropped: TNR2. I frequently deal with ill-defined business problems
	TNR3. Frequently, the business problems I work on involve answering questions that have never been asked in quite that form before
	TNR4. Frequently, in the mindset of using data to address some issues, I may decide to restate the problem and access slightly different data than I had at first planned

This work was previously published in the Journal of Organizational and End User Computing, Volume 23, Issue 2, edited by Mo Adam Mahmood, pp. 37-56, copyright 2011 by IGI Publishing (an imprint of IGI Global).

Chapter 4
Developers, Decision Makers, Strategists or Just End-Users? Redefining End-User Computing for the 21st Century:
A Case Study

Sandra Barker
University of South Australia, Australia

Brenton Fiedler
University of South Australia, Australia

ABSTRACT

The acceleration of technology in business since the 1980s suggests that traditional management techniques, systems, and strategies employed in a business environment should be challenged. As a consequence of this acceleration, end-user computing (EUC) and end-user development (EUD) have also grown. Definitions of EUC developed in the 1980s continue to be used by contemporary researchers without regard to the changing technological environment, user experience, and user needs. Therefore, the authors challenge traditional definitions of EUC developed and used by researchers to ascertain whether they meet the needs of management for the 21st century. There is a conflict among traditional definitions that has not been addressed since the early 1990s (Downey & Bartczak, 2005). In this regard, the authors proffer that the management strategies for end-user (EU) systems development in the 21st century should suggest a different and proactive role for users. This paper summarises key traditional definitions from the literature and evaluates their consonance with the technology and business system environment. The impetus for researchers to rethink the traditional definition of EUC is provided through a real world management project involving the development of a university staff workload database that investigated the role of end-users in system enhancement and development.

DOI: 10.4018/978-1-4666-2059-9.ch004

INTRODUCTION

End-users are defined as those who use computers for business requirements be they administrators, management, operators, information systems specialists, or a combination of all categories. "End-users are a diverse set. There is no single, stereotyped "end-user" with a single, defined set of characteristics" (Rockart & Flannery, 1983, p. 778) and over time it has been identified that academics and business have very different understandings of what exactly EUC is and who the end-users are (Amoroso, 1988). Research into EUC has been undertaken extensively over the past 25 years. This research has included investigation of various user satisfaction models (Doll & Torkzadeh, 1991; Etezadi-Amoli & Farhoomand, 1991) which have also impacted on the understanding of who end users are. The utilisation of these models has spanned more than two decades and therefore it is the opinion of the authors that these models were instrumental in developing an understanding of EUC when developed in the early 1990s. It has been recently identified however that there appears to have been little advancement in the understanding and definition of EUC over this period of time (Downey & Bartczak, 2005). During this time many researchers have identified key elements which defined EUC at a particular time in history. In the 10 years to 2000 almost 90% of the research in this area was empirical (Downey & Bartczak, 2005). The authors question whether the traditional definitions of EUC refer to all of the elements relevant to users in the 21st century as many of the academic papers covering this topic rely heavily on definitions constructed in the 1980s and 1990s. These dated definitions have not kept pace with the changes in technology and education which today see users that have a much higher level of computer literacy utilizing more advanced software (e.g. fourth generation language (4GL) software) to automate and complete their daily tasks. This creates the opportunity for research to be focussed in a manner which links IS/IT and management to create a more contemporary definition of EUC. Using a practical case study approach the authors identify and explore the relationship between elements of the traditional definitions of EUC and the actual activities being undertaken by today's users. The paper will investigate three hypotheses in order to explore these elements.

TRADITIONAL EUC DEFINITIONS

The evolution of EUC definitions has been relatively slow compared to the evolution in technology. There have been many definitions of EUC in the literature, most of which tend to involve the interaction of managers, professionals, and operational level users of application software within their own departments (Torkzadeh & Doll, 1993).

Alavi (1985, p. 171) stated that EUC could be defined as "...the user of the results of the computing also creates the software specifications necessary to effect the computing itself". This early definition of EUC is indirectly affected by the technology of the time. At this time, businesses had not committed to major investment in personal computers (PC) and very few users were directly involved in 'hands-on' computing. However, users were responsible for computing output and therefore prerequisite input into the specifications of the applications and the reports generated. On evaluating this early definition it can be seen that the impact of technology and technical competence of the user has been significant and as such these elements needs to be considered in any contemporary definition.

Davis and Olson (1985, p. 421) defined EUC as "... the capability of users to directly control their own applications and computing needs." This definition is passive in the sense that it provides for the capability of managing rather than actively enhancing information systems for improved personal and corporate objectives.

Brancheau and Brown (1993, p. 439) broadened previous definitions by concluding that EUC is the "... adoption and use of information technology by personnel outside the information systems department to DEVELOP software applications in support of organisational tasks." This statement implies initial development without considering the enhancement of existing applications. The definition has a focus on the technology rather than the task and thereby falls short of linking with corporate strategic directives.

Rainer and Harrison (1993, p. 1188) concluded that EUC was "... the direct, individual use of computers encompassing all the computer-related activities required or necessary to accomplish one's job". This is a generic definition outlining the basic uses of computers rather than looking at the strategic and tactical options for use, development, enhancement, and modification.

Garavan and McCracken (1993, p. 8) defined EUC as "... the managerial and professional use of computer power as compared with clerical tasks which use the computer hardware". This is clearly a dated generic definition that classifies end-user computing as a managerial task, in contrast to the data processing performed by clerical staff.

Downey and Bartczak (2005, p. 4) in their recent work defined EUC as "... the use and/or development of computing technology and software applications by end-users to solve organizational problems and assist in decision making." They also identified that "The themes of EUC, including the end-user, technology, and the organization, remain important and pervasive" (2005, p. 17). They recognised that collectively the definitions from the 1980s and 1990s came close to encapsulating the essence of EUC and as such have brought together key elements from these traditional definitions.

This gives the best definition of EUC since the mid-1990s. However, it fails to recognise the elements of enhancement and modification undertaken in the early 21st century such as the increased computer literacy of the users and the introduction of 4GL software. The definition is also slightly confusing as end-users do not generally develop computing technology, just applications.

From the Downey and Bartczak definition we can conclude that EUC involves:

- The use of computing technology in some form,
- The development of applications by end-users and IS departments to meet the needs of end-users,
- Support for organisational tasks and solving organizational problems,
- Support for managerial decision making.

While there is no disagreement with the tenet of these elements, the comments made above show a need to improve traditional definitions of EUC so that they may keep pace with the increasing functionality of technology and abilities of the users themselves. Updating the definition will allow the alignment of improved technology and functionality with contemporary organizational practice and provide opportunities for researchers and practitioners to reconsider the impact of EUC on operational environments.

HYPOTHESES

There are three key elements that arise from this discussion that warrant further investigation. Firstly, the authors hypothesise that the elements identified by Downey and Bartczak (2005) in their definition of EUC are still relevant components of a contemporary definition (H1).

Secondly, Barker (2007) identified, through a study of student perceptions of EUC definitions, that the elements of enhancement and modification are keys to contemporary EUC activities. To this end she defined EUC as follows:

End-user Computing (EUC) can now be defined as the use of computing technology and/or software applications, together with the enhancement and/ or development of information systems by end-users. On the other hand end-user development (EUD) is, more specifically, the development, modification and/or enhancement of information systems applications by end-users for individual, departmental or organisational use (Barker, 2007, p. 251).

From this the authors hypothesise that a contemporary definition of EUC should include enhancement and modification as key components (H2).

Although the Barker definition considers some of the missing elements discussed above it still does not include consideration of technology management, relationship to strategic management and the active role of users in the development process. It has been previously acknowledged that end-users can undertake a variety of tasks within an organizational setting, including administration, management and operations, and therefore it is considered necessary to include all elements outlined above into any definition of end-users and EUC.

The traditional EUC definitions above imply, or directly refer to, the use of computing technology or software applications, without referring to information technology management.

Kumar et al. (2007) provide a contemporary definition of information technology management as follows:

Ongoing advances in information technology (IT), along with increasing global competition, are adding complexity and uncertainty of several orders of magnitude to the organizational environment. Information technology management (or IT management) is a combination of two branches of study, information technology and management (p. 1).

The above infers that the impact of the advances in IT has had a significant impact on management and thus these areas need to be considered together instead of separately as they have been in the past. This means that there is a need to consider EUC as an integral part of management and not merely the utilisation of computing technology. It involves management strategies to develop and enhance information systems designed to achieve corporate objectives and improve corporate decision making. This also implies a connection between EUC and strategic management and planning. The principles of corporate strategic planning have long been established in the literature. Thibodeaux and Favilla (1995) state that established corporate strategic management principles (applied to the academic sector) allow an organization to:

initiate and influence (rather than just respond and react to) its environment and thus control its own destiny, providing an objective basis for allocating resources and for reducing internal conflicts that can arise when subjectivity or intuition alone is the basis for major decisions. In addition, strategic management allows an organization to take advantage of key environmental opportunities, to minimize the impact of external threats, to capitalize upon internal strengths, and to improve internal weaknesses (Thibodeaux & Favilla, 1995, p. 189).

Consistent with these principles, and to achieve success, much of the strategic planning effort must involve strategic input and planning from both senior management and end-users to achieve success. Traditional strategic management principles have approached the process from the *top-down*. This approach has historically limited the contribution of end-users in the planning process. We contend that the end-user should be an integral part of IS development, either as a developer or as an integral part of the project team, involved in the strategic planning process. Although this is not a new concept, it is the experience of the authors that end-users are often consulted in terms of new developments but are not usually included in project teams for small to medium scale developments.

The changing landscape of strategic management shows new recognition of the importance of end-users in the process. This *bottom-up* approach to strategic management involving all end-users is supported in contemporary literature. For example, Sminia and Van Nistelrooij (2006) state:

From an Organizational Development point of view ... a bottom-up approach with the full participation and active involvement of all employees is seen as essential for generating commitment and ensuring the strategic reorientation actually is realized (Sminia & Van Nistelrooij, p. 99).

Hence, one of the strategic factors critical to the process is the need to enhance IS. The impetus for such enhancement is generated by end-users working with the system on a daily basis. We consider that the traditional definition of EUC should be refined to include reference to IT and information management.

This led the authors to hypothesising that a contemporary definition of EUC should include the possibility of IT management and strategic management as key components (H3).

The testing of these hypotheses was undertaken through empirical research in the form of a case study.

THE CASE STUDY

The taxonomy of IS research developed by Galliers and Land (1987) acknowledges new interpretative approaches, including subjective, descriptive, and action research, as valid techniques for undertaking IS research.

Action research combines theory and practice (and researchers and practitioners) through change and reflection in an immediate problematic situation within a mutually acceptable ethical framework.

Action research is an iterative process involving researchers and practitioners acting together on a particular cycle of activities, including problem diagnosis, action intervention, and reflective learning (Avison, Lau, Myers, & Nielsen, 1999).

The changes proposed above to traditional definitions of EUC were tested using a project involving the development of a Workload Formula Management Information System for the management of academic staff within a university environment. The objective of the study was to test the validity of the Barker (2007) definition of EUC (as referred to earlier) within a community dominated by end-users of systems and the information they produce and use. The objectives of the case study were framed to test the practical elements of EUC involved in implementing a new Workplace Collective Agreement (WCA) to test bottom up involvement of end-users, the use of software applications, and the presence of enhancement and modification of systems. This case study is the first stage in the iterative action research process of redefining EUC.

Problem Statement for the Case Study

A hierarchy exists within the university's organizational structure which devolves much of the day-to-day management of staff to the division or faculty and subsequently to the school level. As a result, a number of universities have adopted workload models to undertake the strategic management of human resources and ensure equity for all academic staff (Sminia & Van Nistelrooij, 2006). This workload is an individual process where each staff member directly negotiates their workload, based on the standard workload template, with the head of department or school in consonance with school, division, and university strategic objectives which are embodied in the WCA. This information is kept at an individual level. The issue identified here is that there is a large volume of information which would be useful in the strategic management of human resources within the level but it is not accessible to those who make decisions.

The academic workload model focuses on information provided by individuals that is aggregated to improve the efficient allocation of academic staff, provide data for internal management information reports and allow a more accurate budgeting process. In this environment, the end-users are the academic staff, departmental managers, administrative staff, and senior management. Ordinarily, the hierarchy of universities would include senior management as end-users of information generated by the divisions and schools. However, one of the limitations of the project was that it was designed to accommodate division management rather than senior university management. The authors consider that the principles related to the definitions can be easily extrapolated to the dimensions of all levels of management. It has become imperative that all of the end-users have access to some, or all, information not currently stored in a central location. As universities involve many and varied users of information resources, they are useful environments for testing the impact of end-user application development on a broad range of end-users. By studying the impact on these end-users, conclusions can be drawn about the validity of the enhanced definition of EUC.

Implementation of the Project

Departmental (school) management made the decision that academic workload information needed to be refreshed whilst the reporting process needed to be modified and enhanced. Further to this, all aspects of human resource management needed to be challenged to accommodate new activity bases, data aggregation, and operational performance management data. The management problem was that a new WCA required the development of new workload formulae that complied with the WCA

but were tailored to the specific circumstances of individual divisions. The division developed such a formula to be implemented in all four schools throughout the division. The School of Commerce operationalized this formula by allocating a point value to specific academic tasks, aimed at meeting the objectives and output allocations for teaching research and administration activities referred to in the WCA.

The school identified an academic colleague from a neighbouring school who is also an experienced end-user developer with no formal IT/IS training and asked them to lead the development project. The project team consisted of the end-user developer and a senior member of the School of Commerce management team (also academic staff member) and they worked with management, academics, and administrative staff using the workload model developed by the school. Both members of the project team were technically proficient end users and approached the task from a managerial perspective. No professional IT staff members were involved in this project at any stage. A database application was constructed to meet strategic objectives including a focus on key performance indicators, financial budgeting, human resource scheduling and management, individual workload performance, compliance with the WCA, and school and division strategic objectives. Critical to the development of the database were the needs for access to (and update of) data by administrative staff, and data input and reporting for academic staff and management. In undertaking this case study, an analysis was made of the tasks involved in operationalizing the workload formula and developing the database. In essence, however, the authors view the development of the database as a test of the strategic and technology management impacts of the introduction of a new WCA.

Strategic management processes highlight a hierarchy where the corporate strategic plan lays the foundation for the division strategic plan which in turn informs the school in their strategic planning process (Thibodaux & Favilla, 1995). Similarly the development of workload principles at the school level needs to be consistent with both the division and university corporate strategic requirements. The new WCA introduced altered workload requirements for individual staff and engaged the development of new key performance indicators, based on an implementation plan. This changed the reporting and accountability requirements for control both at the end-user level (academic and administrative staff) as well as the different management levels within the university. Input was required to enhance and modify existing school reporting systems to integrate with new strategic reporting requirements that followed the chain of strategic command from academic disciplines within schools (e.g. Accounting) to school management to division management to university management. Through observation of academic and administrative staff in the School of Commerce, the authors identified a number of key tasks related to the implementation of the workload formula database. These tasks were documented in the table in the Appendix. Initially all the elements in the traditional definition were included in the table to confirm the continued existence of these elements when defining EUC (H1). Four additional elements relevant to hypotheses 2 and 3 were added to the table in order to test for their presence during the observation of staff tasks.

FINDINGS

The classification of each task undertaken by the end-users and the relationship to the elements of both the traditional and contemporary definitions of EUC are summarised in the Appendix.

Hypothesis 1

Our first hypothesis was that the elements identified by Downey and Bartczak (2005) in their definition of EUC are still relevant components of a contemporary definition. Their definition stated that "… the use and/or development of computing technology and software applications by end-users to solve organizational problems [task support] and assist in decision making [decision support]" (Downey & Bartczak, 2005, p. 4).

The elements that were included in this definition and confirmed by the case study are discussed below and summarised in the 'Traditional Definition Elements' columns of the Table A1 in the Appendix.

1. **Use and/or development of computing technology:** University records revealed that standard desktop and laptop computers are used extensively by end-users. From the observations of the authors none of the end-users involved in this study were developers of this technology.

2. **Use and/or development of software applications:** Software applications including the Microsoft Office suite, timetable software and university student data management systems were all used by participants in the case study. Applications were developed by participants using MS Excel and MS Access.

3. **Solve organizational problems (task support):** The applications developed by end-users: in the form of spreadsheets and databases have assisted the various levels of university users in understanding, reporting, planning and operationalization of the issues arising from the introduction of the WCA.

4. **Assist in decision making process (decision support):** The use of these applications developed by the end-users has provided mechanisms to support users at all levels with their decision making in terms of human resource requirements for the academic workload of the school.

The case study has shown that all elements in the existing traditional definitions, as best summarised by Downey and Bartczak, are still present in today's EUC environment.

Hypothesis 2

In this we hypothesised that a contemporary definition of EUC should include enhancement and modification as key components.

A major focus of the case study approach was to investigate the presence or otherwise of the elements of enhancement and modification as they relate to end-user developed software applications. The existence of these elements relative to each end-user in the study, their activity and the information they required or used is included in the Appendix as additional elements beyond those of the traditional definitions.

The case study revealed that in response to the WCA, at all levels of user, there were enhancements required to existing systems which also lead to modifications of those system to improve the internal reporting and compliance processes. Enhancements in the case study included:

- Enhancement of course web pages through use of Word or Frontpage templates,
- Combining existing spreadsheet and databases to form one fully incorporated database to handle all workload issues of the school,
- Addition of reports to the new database as required,
- Improved flexibility to manipulate data as required,
- Collection of data at course and discipline level,
- Support for administrative staff to determine actual academic staff members working on a particular course,
- Budgeting information for the identification of sessional staff usage across the school,
- Transparency to all academic members of staff in relation to the equity of workload,
- Eventual provision of staff workload information to divisional management.

In addition, modification of existing systems provided users with the following outcomes:

- Modified spreadsheets to comply with the new WCA,
- Internal reporting systems to take account of changes in Key Performance Indicators and new school/division/university strategies,
- Internal reports modified to show variances with the output requirements of each staff member in accordance with the new WCA,
- Teaching allocations were modified to take account of the new target activity outputs from staff members and schools.

The elements identified above solidify the argument for the inclusion of enhancement and modification in the Barker's (2007) definition of EUC. Previously, none of this information was easily accessible through existing information systems. The increased flexibility of the enhanced information system has given school management and academics considerably more access to the required information.

Hypothesis 3

This final hypothesis challenged whether a contemporary definition of EUC should include information technology management and strategic management as key components. The case study involved all staff categories from administrative staff to senior division staff. The strategic objectives involved in the case related to the objective of organizational compliance with the WCA, and to enhance the existing information system. An integral part of the process of seeking compliance was the strategic opportunity to improve management processes using the data capabilities of the workload database. The database provided detailed information relating to staff work activities by discipline area, specific functional teaching/research/administrative activities. The changes in the WCA responsibilities of division and school management provided the opportunity to enhance the internal reporting and management systems for decision making. As documented in the Appendix, the information allowed:

- Division management to:
  - Secure and review compliance with the new WCA (this data was not previously available); and
  - To enhance its activity reporting for planning and performance management tasks;
- School management to:
  - Undertake detailed activity reporting to ensure individual workloads were satisfactory;
  - To provide strategic data to Program Directors to ensure proper staffing of academic programs;
  - Revise monthly management reporting and enhance sessional staff and course financial budgeting; and
- Administrative staff to:
  - Collect relevant data to improve ad-hoc reporting for academics and School and division Management.

The case study provided a number of enhancements to the system that existed prior to the change in the WCA. Rich data was now available for financial and human resource planning. Evidence was now available to report compliance with the university's WCA. Information was now readily available for ad-hoc reporting and improved strategic planning processes. In all instances the new database improved the efficiency of data collection, added new dimensions to data combinations relevant to management decision making and strategic planning, and enhanced the planning capability of the school and division. The activities undertaken involved inherent improvements of existing systems, determined through consultation with all end-users of the system and based

on their strategic information requirements and job responsibilities. Current technological environments tend to indicate that an enhancement in data availability will encourage end-users to enhance their strategic management systems for planning and reporting, and the case study confirms this. Accordingly, in relation to hypothesis 3, we conclude that a contemporary definition of EUC should include information technology management and strategic management as key components.

LIMITATIONS OF THE STUDY

The case study was undertaken in a tertiary education environment and may not be generalisable to the circumstances of other corporate and non-corporate entities. Further, the impact of the changes has been measured over a relatively short period of time, and the full use of the enhanced system has not yet been assessed in relation to its impact on a complete strategic planning review. In addition, it must be recognized that the case study has been used for only one application of the university's information management system and may not be generalisable to all university systems. The presence of the range of activities identified in this case study provides a foundation for further research to investigate their presence in other EUC applications.

CONCLUSION AND FUTURE RESEARCH

It is clear from the case study that the traditional elements of the EUC definition continue to be present in 21st century users. However, the case study confirmed that the traditional definitions of EUC can now be challenged. Users are actively involved in the use, development, modification and enhancement of software applications as integral components of information management and corporate decision making. The role of users at all levels is to develop solutions to problems and be actively involved in strategic management activities rather than accept a *top-down* strategic development model. The case study used in this paper clearly demonstrates that EUC in the academic environment requires enhancement and modification of information systems to add to the traditional elements of EUC. Future research should challenge the traditional definitions of EUC. The case study referred to in this paper provided evidence that suggests the traditional definitions of EUC require refreshment to refer to the technology, management practices and processes in the new millennium.

We propose a new contemporary definition of EUC would be as follows:

End-User Computing is the use of computing technology and/or software applications, together with the enhancement, modification and/or development of information systems by end-users for individual, departmental or organisational use.

Anecdotally, we believe that studies beyond academia will provide similar results to the study in this paper. However, further empirical evidence is needed. Then, with evidence, we can confirm the above contemporary definition of EUC within the evolving framework of information technology, information systems and strategic management.

It is the reading of the authors that EUC has traditionally been the domain of the IS department in any organization and subsequently research in the area has been limited to the IS perspective. The case study suggests that EUC has an impact on the organization as a whole and therefore needs to be considered from an organizational management perspective. Although there is some seminal research on management of end user computing (Gerrity & Rockart, 1986; Brancheau & Brown, 1993) this has been developed from the IS/IT perspective. Redefining EUC as a construct which introduces the concept of the impact on management effectiveness provides ample opportunity for the research to be considered from a management

perspective. Accordingly opportunities for further research exist in the collaboration between IS/IT and management disciplines encouraging a more focussed approach to EUC across all facets of the organization. This will also be applicable to those researchers who are investigation the narrower concept of end user development which the authors consider to be a subset of EUC.

REFERENCES

Alavi, M. (1985). End-user computing: The MIS manager's perspective. *Information & Management, 8*(3), 171–178. doi:10.1016/0378-7206(85)90046-1

Amoroso, D. L. (1988). Organizational Issues of End-user Computing. *ACM SIGMIS Database, 19*(3-4), 49–58. doi:10.1145/65766.65773

Avison, D., Lau, F., Myers, M., & Nielsen, P. A. (1999). Action research. *Communications of the ACM, 42*(1), 94–97. doi:10.1145/291469.291479

Barker, S. K. (2007). End-user Computing and End-user Development: Exploring Definitions for the 21st century. In *Proceedings of the 2007 Information Resources Management Association International Conference*, Vancouver, BC, Canada (pp. 249-252). Hershey, PA: IGI Global.

Brancheau, J. C., & Brown, C. V. (1993). The Management of End-user Computing: Status and Directions. *ACM Computing Surveys, 26*(4), 437–482. doi:10.1145/162124.162138

Chang, P.-L., & Shen, P.-D. (1997). A conceptual framework for managing end-user computing by the total quality management strategy. *Total Quality Management, 8*(1), 91–102. doi:10.1080/09544129710477

Davis, G. B., & Olson, M. H. (1985). *Management Information Systems: Conceptual Foundations, Structure, and Development* (2nd ed.). New York, NY: McGraw-Hill.

Doll, W. J., & Torkzadeh, G. (1991). The measurement of end-user computing satisfaction: Theoretical and methodological issues. *Management Information Systems Quarterly, 15*(1), 5–10. doi:10.2307/249429

Downey, J. P., & Bartczak, C. A. (2005). End-user Computing Research Issues and Trends (1990-2000). In Mahmood, M. A. (Ed.), *Advanced Topics in End-user Computing* (*Vol. 4*, pp. 1–20). Hershey, PA: Idea Group.

Etezadi-Amoli, J., & Farhoomand, A. (1991). On End-User Computing Satisfaction. *Management Information Systems Quarterly, 15*(1), 1–4. doi:10.2307/249428

Galliers, R. D., & Land, F. F. (1987). Choosing Appropriate information systems research methodologies. *Communications of the ACM, 30*(11), 900–902. doi:10.1145/32206.315753

Garavan, T. N., & McCracken, C. (1993). Introducing End-user Computing: The Implications for Training and Development-Part 1. *Industrial & Commercial Training, 25*(7), 8–14. doi:10.1108/00197859310042443

Gerrity, T. P., & Rockart, J. F. (1986). End-user computing: Are you a leader or a laggard? *Sloan Management Review, 27*(4), 25–34.

Kumar, B. P., Selvam, J., Meenakshi, V. S., Kanthi, K., Suseels, A. L., & Kumar, L. K. (2007). Business Decision Making Management and Information Technology. *Ubiquity, 8*(8), 1. doi:10.1145/1226690.1232401

Rainer, R. K., & Harrison, A. W. (1993). Toward development of the end-user computing construct in a university setting. *Decision Sciences, 24*(6), 1187–1202. doi:10.1111/j.1540-5915.1993. tb00510.x

Rockart, J. F., & Flannery, L. S. (1983). The management of end-user computing. *Communications of the ACM, 26*(10), 776–784. doi:10.1145/358413.358429

Sminia, S., & Van Nistelrooij, A. (2006). Strategic Management and Organization Development: Planned Change in a Public Sector Organization. *Journal of Change Management, 6*(1), 99–113. doi:10.1080/14697010500523392

Thibodeaux, M. S., & Favilla, E. (1995). Strategic management and organizational effectiveness in colleges of business. *Journal of Education for Business, 70*(4), 189–196. doi:10.1080/0883232 3.1995.10117748

Torkzadeh, G., & Doll, W. J. (1993). The place and value of documentation in end-user computing. *Information & Management, 24*(3), 147–158. doi:10.1016/0378-7206(93)90063-Y

APPENDIX

Table A1. Elements of definition (H1, H2 & H3 – refer to the hypotheses identified in this paper)

End-User	Key Task	Information (Required/ Used)	Traditional Definition Elements			Contemporary Definition Elements				
			Use (H1)	Development (H1)	Task Support (H1)	Decision Support (H1)	Enhancement (H2)	Modification (H2)	IT Management (H3)	Strategic Management (H3)
Senior Management (division level)	• Implementation of Workplace Collective Agreement Principles	• Collective Agreement	✓	✓			✓	✓		
	• Operationalization of workload principles	• Division reports on staff workload that reflect implementation of Collective Agreement workplace principles	✓		✓	✓		✓		
	• Resource Management and Planning/ Information Technology Management/ Strategic Management	• Reports on staff utilization and productivity	✓		✓			✓	✓	
		• Reports of School academic staff activity at program level	✓		✓	✓				

continued on following page

Table A1 Continued

End-User	Key Task	Information (Required/Used)	Traditional Definition Elements		Contemporary Definition Elements					
			Use (H1)	Development (H1)	Task Support (H1)	Decision Support (H1)	Enhancement (H2)	Modification (H2)	IT Management (H3)	Strategic Management (H3)
		• Activity Reporting for decision making and future planning	✓		✓	✓	✓	✓		✓
Departmental Management (school level)	• Resource Management and Planning/ Information Technology Management/ Strategic Management	• Activity Reporting for decision making and future planning	✓		✓	✓	✓	✓		✓
		• Reports of individual academic staff activity	✓		✓	✓			✓	
		• Data for strategic planning activities for academic programs	✓		✓	✓	✓	✓		✓
		• Activity reports for school performance management	✓		✓	✓		✓		✓

continued on following page

Table A1 Continued

End-User	Key Task	Information (Required/Used)	Traditional Definition Elements			Contemporary Definition Elements					
			Use (H1)	Development (H1)	Task Support (H1)	Decision Support (H1)	Enhancement (H2)	Modification (H2)	IT Management (H3)	Strategic Management (H3)	
		• Reports and online data enquiry by program, course, discipline, and individual for day-to-day management and strategic planning	✓		✓	✓		✓	✓		
	• Financial Management/ Information Technology Management/ Strategic Management	• Monthly management reporting and program/course profitability analysis	✓		✓	✓	✓			✓	
		• Budgeting/ Financial data for strategic planning	✓		✓	✓	✓	✓		✓	
Academic Staff	• Implementation of School Workload Policy consistent with Division policy	• Operationalization of school and division strategic plans		✓		✓				✓	

continued on following page

Table A1 Continued

End-User	Key Task	Information (Required/Used)	Traditional Definition Elements		Contemporary Definition Elements					
			Use (H1)	Development (H1)	Task Support (H1)	Decision Support (H1)	Enhancement (H2)	Modification (H2)	IT Management (H3)	Strategic Management (H3)
	• Human Resource Management/ Information Technology Management/ Strategic Management	• Reports on planned activities for academic team resource planning and management	✓		✓	✓		✓		✓
		• Input data for sessional staff budgeting	✓		✓	✓				
		• Individual performance management data	✓		✓	✓			✓	
Admin-istrative Staff	• Financial Reporting	• Data for financial budgeting	✓		✓	✓			✓	
		• Informa-tion to support ca-sual contract preparation	✓		✓	✓			✓	
	• Manage-ment of student enquiries	• Informa-tion related to academic and casual staff teach-ing time-table	✓		✓	✓				

continued on following page

Table A1 Continued

End-User	Key Task	Information (Required/Used)	Traditional Definition Elements				Contemporary Definition Elements			
			Use (H1)	Development (H1)	Task Support (H1)	Decision Support (H1)	Enhancement (H2)	Modification (H2)	IT Management (H3)	Strategic Management (H3)
	• Preparation of statistical reports for school/division management	• Ad-hoc reports for school management	✓		✓	✓	✓	✓		

Note: Key tasks undertaken by each user type rely on the information produced by the Workload Information Management System to support decisions to fulfil their organizational roles.

Chapter 5
Expert and Novice End-User Spreadsheet Debugging:
A Comparative Study of Performance and Behaviour

Brian Bishop
Dundalk Institute of Technology, Ireland

Kevin McDaid
Dundalk Institute of Technology, Ireland

ABSTRACT

The reliability of end-user developed spreadsheets is poor. Research studies find that 94% of 'real-world' spreadsheets contain errors. Although some research has been conducted in the area of spreadsheet testing, little is known about the behaviour or processes of individuals during the debugging task. In this paper, the authors investigate the performance and behaviour of expert and novice end-users in the debugging of an experimental spreadsheet. To achieve this aim, a spreadsheet debugging experiment was conducted, with professional and student participants requested to debug a spreadsheet seeded with errors. The work utilises a novel approach for acquiring experimental data through the unobtrusive recording of participants' actions using a custom built VBA tool. Based on findings from the experiment, a debugging tool is developed, and its effects on debugging performance are investigated.

INTRODUCTION

Spreadsheet Prevalence

Commercial off-the-shelf spreadsheet programs are one of the most popular end-user programming environments in use today (Burnett, Cook, & Rothermel, 2004), and have become ubiquitous, and in many cases indispensable software tools throughout industry and within all levels of the business world. Some of the reasons spreadsheets are so popular in end-user computing is that they are easy to develop and modify, they are highly scalable, and they support numeric data analysis without the specific need to use a 'behind the scenes' programming environment. Spreadsheet

DOI: 10.4018/978-1-4666-2059-9.ch005

programs are also extremely flexible in terms of size and complexity: the 2007 version of Microsoft Excel can support over 1 million rows and 16 thousand columns per worksheet, with a maximum formula length of 8 thousand characters per cell.

It has been estimated that by 2012 there will be 90 million end-users in the U.S. alone and of these 13 million will be end-user programmers; which is significantly higher than the estimated 3 million professional programmers in the U.S. by the same year (Scaffidi et al., 2005). Purser and Chadwick (2006) stated that *"commonly known – though unpublished and intentionally unconfirmed by Microsoft – Excel has an install base of 90% on end-user desktops"* (p. 185).

The reported usage of spreadsheet programs spans a wide variety of job functions, purposes and industries. In a survey of nearly 1600 respondents, Baker et al. (2006) found that spreadsheets were used by end-users in various job functions including finance, engineering, manufacturing, marketing, sales, HR etc., and for many different purposes, such as maintaining lists, analysing data, tracking data, determining trends etc. Maybe more so than of any other industry, spreadsheets are of critical importance to the finance sector (Croll, 2005). In a study on the use of spreadsheets in organisations in the City of London, Croll (2005) found that with regard to the financial markets *"Excel is utterly pervasive. Nothing large (good or bad) happens without it passing at some time though Excel"* (p. 3).

Spreadsheet Reliability

Worryingly, the reliability of end-user developed spreadsheets has been shown to be poor following empirical and anecdotal evidence collected from studies investigating operational (real world) spreadsheet error rates, most of which are detailed in previous studies (Panko, 1998; Panko, 2000; Powell, Baker, & Lawson, 2008). The most recent study that investigated spreadsheet error rates was conducted by Powell, Baker and Lawson (2007),

who reported that of the 50 real-world operational spreadsheets they audited, 94% contained errors. Due to the nature of spreadsheet use, it is the case that when failures do occur the results can be quite significant. Many real world examples of the impact of errors in spreadsheets are published on the European Spreadsheet Risks Interest Group website (www.eusprig.org). The following is just a small selection of these reported errors:

- A government housing authority had to pay $216,352 to cover expenses incurred as a result of a spreadsheet data-entry error that overpaid landlords.
- A spreadsheet error caused the Nevada City 2006 budget to show a deficit of $5 million for a water fund.
- In 2005, a Kodak employee was mistakenly accrued an $11 million severance. A Kodak spokesman said the severance error was traced to a faulty spreadsheet, and the money was not paid out.
- Stock prices for an online retailer tumbled by 25% in one day, and the Chief Financial Officer resigned due to a single number mis-recorded in one cell of a spreadsheet.
- Numbers entered as text lost a school £30,000 from its budget.
- A finance company was undercharged $25,652 for rent in 1999 due to an error in a spreadsheet formula.
- A Legislative Auditor found a government department's key performance indicator value overstated by $58 million (45%). The miscalculation was due to a spreadsheet formula error.

The reported and published spreadsheet errors are just the tip of the iceberg. Very few organisations report embarrassing spreadsheet errors and their impacts, and details of these errors only enter the public domain when it becomes unavoidable. From correspondence with professionals that cannot be published, there appears to be many instances of such errors throughout industry that remain behind closed doors.

Spreadsheet Debugging

Relatively little research has been conducted on the debugging and error detection process for spreadsheets. The emphasis of spreadsheet research available has been on the prevention of spreadsheet errors through spreadsheet design and testing methodologies. The notable exceptions to this are (Galletta et al., 1993, 1996; Panko, 1999; Howe & Simkin, 2006) in which studies on error-finding performance, the effect of spreadsheet presentation in error detection, applying code inspection to spreadsheet testing and the factors affecting the ability to detect spreadsheet errors were undertaken respectively. Importantly, none of these papers, unlike our work, deal with the behaviour or cell-by-cell processes involved in finding and correcting spreadsheet errors. Galletta et al. (1996) concludes that an increased understanding of the error-finding process could help avert some of the well publicised spreadsheet errors.

Of the studies mentioned above, only (Galletta et al., 1993) utilised professional participants; as opposed to the student population. Panko (1999) notes that the use of student subjects is always a concern, but says that the findings from (Galletta et al., 1993) suggest that "*students should be fairly good surrogates of experienced spreadsheet developers*" (p. 1). A limiting factor that may have contributed to the (Galletta et al., 1993) findings relating to debugging performance was letting the subjects know in advance the maximum number of errors in each spreadsheet (zero to two), as this would undoubtedly have affected their debugging behaviour. Also, the subjects from (Galletta et al., 1993) were asked to highlight (on paper) any errors found, but were not asked to correct the errors. Teo and Lee-Partridge (2001) noted that the use of undergraduate students was a limitation in their spreadsheet debugging study and that "*it is practically impossible to persuade a large group*

of practitioners to participate in spreadsheet experiments" (p. 453). What can be taken from this is that there is a need for more spreadsheet debugging experiments involving professional spreadsheet users, and that the difference between professionals and students needs to be explored in greater depth. Based on these findings, a key objective of this study was to recruit professional subjects to partake in a spreadsheet debugging experiment, and to analyse and compare the debugging performance and behaviour of professional and student spreadsheet users.

With regard to software tools that aid the spreadsheet debugging process, quite a few are available as both commercial and freeware applications, but in order to develop tools that complement end-users' natural debugging behaviour, some understanding of that behaviour would be required. The concept of human-centered development in software engineering could be applied to the development of spreadsheet debugging tools. Norman (1999) stated that "*At its core, human-centered product development requires developers who understand people and the tasks they wish to achieve. It means starting by observing and working with users*" (p. 185). Only a few studies have observed and empirically recorded the behaviour of end-users while debugging spreadsheets (Chen & Chan, 2000; Prabhakararao et al., 2003; Beckwith et al., 2005), with the latter two having done so within the framework of a testing tool and methodology (WYSIWYT), using a non-commercial research-based spreadsheet application (Forms/3). Also, in these studies participants behaviour was recorded using videotaping and think-aloud protocols; data acquisition methods that are very noticeable to the subjects under observation. Such data acquisition methods considerably alter a subjects' behaviour (Spannagel, Gläser-Zikuda, & Schroeder, 2005). In order to unobtrusively record Excel end-user behaviour during the debugging process, a single less visible data acquisition method would be required such as mouse-and-keystroke recording, but to date, no spreadsheet debugging studies have done this. To this end, the authors developed a novel data acquisition tool for spreadsheet debugging experiments, which was an integral tool in answering the research questions detailed in the next section.

RESEARCH QUESTIONS AND OBJECTIVES

The primary objectives of the research are to investigate expert and novice spreadsheet users' debugging performance and behaviour. The spreadsheet debugging *performance* aspect of the research is addressed by Research Question 1 (RQ1) and the subsidiary Research Question 2 (RQ2). The spreadsheet debugging *behaviour* aspect of the research is addressed by Research Question 3 (RQ3) and 4 (RQ4).

RQ1: Will expert spreadsheet users outperform novice spreadsheet users in detecting and correcting spreadsheet errors?

RQ2: Will expert spreadsheet users outperform novice spreadsheet users in detecting and correcting certain types of spreadsheet errors?

RQ3: What differences can be identified in the spreadsheet debugging behaviour of expert and novice spreadsheet users?

RQ4: Does a relationship exist between the number of spreadsheet cells inspected and debugging performance?

Based on the findings from Research Question 4, the following supplementary research question was derived:

RQ5: What effect would the availability and use of a debugging tool, that gives feedback on cell coverage during the debugging process, have on debugging performance?

An experiment was conducted in order to provide the data required for answering research questions 1, 2, 3 and 4. The participants consisted of a number of expert and novice spreadsheet users. The experiment participants were instructed to find and correct errors seeded within an experimental spreadsheet model. A quantitative method was employed in answering RQ1, RQ2 and RQ4. Due to its exploratory nature, RQ3 was answered using a more qualitative method. In order to answer RQ5, a second experiment was conducted consisting of a control group (no tool) and a test group (debugging tool available). The effects of the debugging tool on debugging performance and the number of spreadsheet cells inspected were investigated, along with characteristics of how the tool was used by each participant.

RESEARCH METHODOLOGY

The method employed in answering RQ1, RQ2 and RQ4 relied primarily on quantitative techniques for data acquisition, data analysis and experiment reporting. Due to the exploratory nature of RQ3, an interpretive method was employed which contained some elements of grounded theory. A comparative experiment was conducted to answer these research questions, which is herein referred to as the *Spreadsheet Debugging Experiment* - Experiment 1.

In order to answer RQ5, a simple spreadsheet debugging tool was developed, the main function of which was to give feedback to users on the cells that had been inspected (or not inspected) during the debugging process. An experiment was conducted consisting of a control group (no tool) and a test group (debugging tool available), and is herein referred to as the *Debugging Tool Experiment* - Experiment 2.

Spreadsheet Debugging Experiment (1) Method

Experimental Spreadsheet Model. A spreadsheet model was developed consisting of three worksheets seeded with errors. The names and functions of each of the three worksheets were as follows: *Payroll*, compute typical payroll expenses; *Office Expenses*, compute office expenses; *Projections*, perform a 5-year projection of future expenses. Each worksheet had different error characteristics. Payroll had data entry, rule violation and formula errors; Office Expenses had clerical, data entry and formula errors; Projections had mostly formula errors. The error categories and number of seeded errors within each category were as follows. *Clerical/Non Material errors* (4): errors that do not affect the bottom-line value of the spreadsheet, such as spelling errors. *Rule Violation errors* (4): are cell entries that violate company policy, for example paying an employee overtime when that employee is not eligible for overtime. The rule violation errors could be identified by reading the 'rules' for each worksheet as detailed in the instructions sheet supplied to participants. *Data Entry errors* (8): incorrect numeric data input values, for example the entering of a number as a negative by mistake. *Formula errors* (26): errors in formulas cells, such as inaccurate range references, illogical formulas etc (see Appendix 1 for colour coded diagrams of the errors within each worksheet).

Participants were instructed to debug the spreadsheet, and each error found was to be corrected directly in the spreadsheet itself. The spreadsheet model was adapted from a model used in a previous study (Howe & Simkin, 2006), and is presented and described in detail in a more recent paper (Bishop & McDaid, 2007). It should be mentioned that the distribution and number of seeded errors is not truly reflective of real-world spreadsheets. Indeed, this is the case for most spreadsheet debugging experiments. For future experiments, the author would suggest the use

of slightly larger, more poorly structured spreadsheets with less seeded errors that would be more reflective of real-world spreadsheets.

Time-stamped Cell Activity Tracking Tool (T-CAT). The data acquisition method for Experiment 1 was facilitated with the use of a custom built 'time-stamped cell activity tracking tool' (T-CAT). The T-CAT tool was developed in VBA, and makes use of MS Excel's macro programming environment (see flowchart in Figure 1). The time-stamped data recorded (in seconds and milliseconds) by T-CAT during the debugging process is as follows: overall debugging time taken by participants, cell values edited/re-edited, order in which sheets and cells were inspected, number and addresses of cells inspected, the exact time taken to inspect each cell, and use of drag-and-fill for copying logical areas. The data recorded by T-CAT is printed to a hidden worksheet within the experimental spreadsheet when the spreadsheet is closed.

An issue highlighted during initial pilot-studies related to MS Excel's error checking feature. The types of potential errors flagged include inconsistent formulas, numbers stored as text, formulas referring to empty cells, etc. This would not have been a problem if the numerous versions of MS Excel had the same error checking options, and if all of the experiment participants utilised this feature in the same manner, but that was not the case. MS Excel 1997 and 2000 do not have an error checking feature, whereas MS Excel 2002 and 2003 have 7 and 8 error checking options respectively. To 'level the playing field' so to speak, T-CAT temporarily disabled the error checking feature. Without error checking disabled, certain participants might have had a slight advantage e.g. some participants would have been alerted to spelling errors, while others would not.

Sample. The Experiment 1 sample consisted of 13 professional participants and 34 student participants. The sampling process used can best be described as *purposive sampling*. Purposive sampling is conducted with a purpose in mind,

Figure 1. T-CAT flowchart

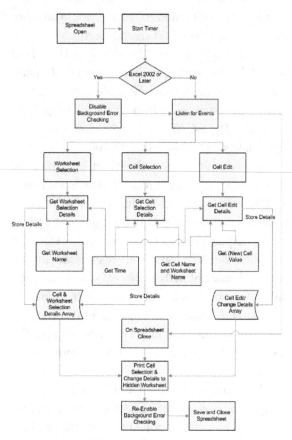

usually where one or more specific predefined groups are being sought (Trochim & Donnelly, 2006).

The professional participants chosen were considered expert spreadsheet users. All the professional participants had an industry-based working knowledge of spreadsheet use/development. The sampling of the expert spreadsheet user group also involved some *snowball sampling*. A snowball sampling process involves the identification of an initial number of subjects who meet the criteria for inclusion in a sample. These subjects are then asked to recommend other potential subjects who they know also meet the criteria. Prior to their partaking in the spreadsheet debugging experiment, it was known by the author, through personal contact and recommendation, that the each of the 13 professional subjects were *industry-based spreadsheet*

developers, who regularly worked with complex and important spreadsheets.

The novice spreadsheet users selected for Experiment 1 were students in their 2nd year of a four year Accounting & Finance degree. All of the student subjects had taken MS Excel skills classes in their first year, in which the basics of spreadsheet development and use had been taught. The students were asked about their spreadsheet experience by the researchers, and it was found that the student's spreadsheet experience was limited to the spreadsheet skills classes that were taken in 1st year of college, and any exercises in spreadsheet use and development that those classes entailed. This low level of (classroom only based) experience met the criteria required by the author for a subject's inclusion in the *novice* sample. In some of the spreadsheet debugging literature reviewed by the author, student subjects appeared to have been selected primarily because they were an easily accessible sample. This is known as *convenience sampling*, and the main disadvantage of this method of sampling is that it is not at all possible to determine how representative the sample is of the population as a whole. Panko (1999) justifies using students as fairly good surrogates of experienced spreadsheet developers by referring to the findings of (Galletta et al., 1993). This was the one of the reasons that the *purposive sampling* method, used for Experiment 1, targeted the student population. Not only could the question of spreadsheet experience relating to debugging performance be answered (given that student subjects were *novice* spreadsheet users), but the notion that 'student spreadsheet users' are a good replacement for experienced 'in-the-field' spreadsheet users in debugging experiments could be investigated.

Process. For Experiment 1, the student subjects partook in the experiment in a computer laboratory environment. The subjects were given a copy of the experimental spreadsheet along with an instructions page. A short introduction on the instructions page explained the purpose of the task, namely to investigate how effectively spreadsheet users discover and correct errors. Subjects were asked to correct any errors found directly on the spreadsheet itself. The instructions page also contained some rules with regards to the data in the worksheets e.g. only employees with codes B or C are eligible to receive overtime pay.

Both the spreadsheet and the instructions were e-mailed to each of the industry-based professional subjects after they had been contacted and had agreed to take part.

The student subjects were given the opportunity during a single 60 minute class period to participate in the study. No time limit was given to the professional subjects. As pre-tests suggested, the 60 minute time limit was sufficient. The task was completed by student subjects in an average 36 minutes and by professional subjects in an average 28 minutes.

Debugging Tool Experiment (2) Method

In traditional software development, verification and validation (V&V) is a process used to determine if the software is being built correctly and if it the correct software is being built respectively (Jenkins et al., 1998). The average lines of code inspected and average inspection rates are key metrics used in the code inspection phase of software V&V (Barnard & Price, 1994). The number of cells inspected during debugging in the spreadsheet paradigm is somewhat equivalent to the lines of code inspected metric used in traditional software V&V. RQ5 was derived from both a finding of the Spreadsheet Debugging Experiment that a correlation exists between cell coverage and debugging performance, and from the use of lines of code inspected metrics in traditional software V&V. The main aim of Experiment 2 was to investigate if feedback on cell coverage (via a highlighting tool) would positively affect debugging performance.

The control group in Experiment 2 was given the same spreadsheet and instructions as the participants from Experiment 1. The data acquisition tool, T-CAT, was also incorporated in the spreadsheet, just as in Experiment 1. The test group were given the same spreadsheet as the control group (with T-CAT also incorporated), but each worksheet had a 'Highlight' button to activate the debugging tool. The Highlight button as it appeared on the test group's worksheets can be seen in Figure 2. The debugging tool is described in the next section.

Debugging Tool. Each worksheet of the experimental spreadsheet model had a button attached titled 'Highlight' (see Figure 2). When a user clicked on this button, any cells that had not yet been edited, or selected for a minimum specified time of 0.3 seconds, became highlighted. The subjects could then inspect the highlighted cells if they so wished. The minimum specified time of 0.3 seconds was based on findings from analysis of the minimum time it took the participants to inspect a cell. If the Highlight button was clicked again, the highlighting would be updated to reflect any extra cells that had been inspected, and any cells that had still not been inspected. This process could be repeated as many times as a subject wanted. Only cells containing formula or data values were highlighted. A flowchart for the design of the debugging tool can be seen in Figure 3.

Sample. The sample for Experiment 2 consisted of 16 subjects. The subjects were evenly and randomly assigned to a control group (no debugging tool), and a test group (debugging tool available). All 16 subjects were in their final year of a four year Software Development degree course. All of the subjects had some experience of spreadsheet use, but none of the subjects described themselves as being above a novice to intermediate level.

Process. Experiment 2 was conducted in a computer laboratory environment. The sample was divided randomly between two groups, a control group, and a test group. The control group was given the spreadsheet model (with no debugging tool) and instructions. The test group was given the same spreadsheet model but with the debugging tool incorporated into to, and an instructions sheet with an extra paragraph detailing how the debugging tool was to be used. As in Experiment 1, subjects were asked to correct any errors found directly on the spreadsheet itself.

EVALUATION

Debugging Performance: Expert vs. Novice

The measure used in answering RQ1 was the average debugging performance for both the expert and novice participant groups, and standard significance tests were applied to the resulting data. The debugging performance for each of the error-categories was calculated for the novice and expert users and the results were used in answering RQ2. One sided hypothesis tests examining whether the mean performance of professionals exceeded the mean performance of students in discovering the four types of errors were examined through the usual Student T-test using a significance level of 5%. Given the relatively low number of professional subjects and the uncertainty as to the nature of the distribution of the number of errors found by individual participants, one sided tests of a difference in the measure of centrality were also conducted using the non parametric Mann-Whitney U-test.

The overall average debugging performance of the industry-based professionals (experts) was 72%, which was 14% higher than the student subjects (novices), who corrected 58% of all seeded errors. The p-values indicate that professionals outperform students in spreadsheet debugging and that the difference is statistically significant.

The results from Table 1 show a clear distinction between performances of industry based professionals and students for Rule Violation and Formula errors, with professionals correcting 16% more Formula errors than students and 20% more Rule Violation errors (both results being statistically significant). Although the professionals corrected more Clerical and Data Entry errors, the results are not statistically significant.

To further investigate the difference in performance between expert and novice subjects in Formula error debugging, the formula error category was further sub-categorised as follows:

Figure 2. Debugging tool: Highlighted cells were not inspected

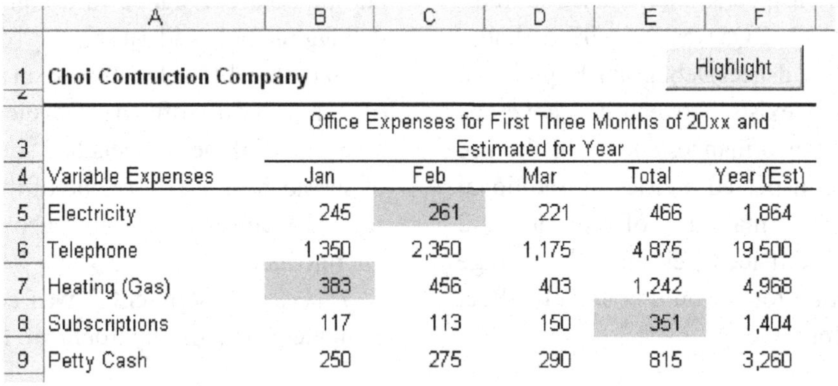

	A	B	C	D	E	F
1	**Choi Contruction Company**					Highlight
2						
3		Office Expenses for First Three Months of 20xx and Estimated for Year				
4	Variable Expenses	Jan	Feb	Mar	Total	Year (Est)
5	Electricity	245	261	221	466	1,864
6	Telephone	1,350	2,350	1,175	4,875	19,500
7	Heating (Gas)	383	456	403	1,242	4,968
8	Subscriptions	117	113	150	351	1,404
9	Petty Cash	250	275	290	815	3,260

Figure 3. Flowchart for debugging tool

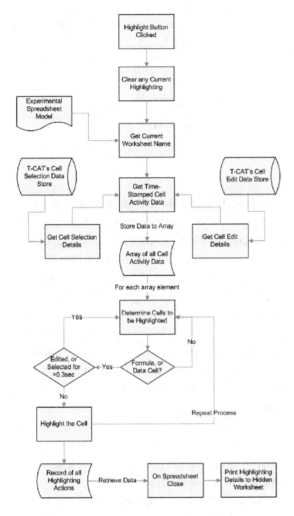

- **Logic Errors:** a formula is used incorrectly, leading to an incorrect result.
- **Cell Reference Error:** a formula contains one or more incorrect references to other individual cells.
- **Range Reference Error:** a formula contains one or more incorrect references to a range of multiple cells.
- **Remote Reference Error:** a formula contains an incorrect reference to a cell on a different sheet.

It was found that the expert subjects outperformed the novice subjects in each of the formula error sub categories (see Table 2), but that combined T-test and Mann-Whitney results were statistically significant for Logic and Remote Reference errors only, with both p-values for these two categories being lower than the alpha level of 0.05.

Debugging Behaviour

Overall Cell Coverage vs. Debugging Performance. An important facet of end-users' natural spreadsheet debugging behaviour, which had not been previously investigated, is that of cell coverage. The only studies that refer to cell coverage, found during an extensive literature review of available spreadsheet research, deal with cell coverage within the context of testing tools and associated testing methodologies, not necessarily within the context of end-users' natural debugging behaviour. One such study (Rothermel et al., 2001) addressed the area of cell coverage within the framework of the 'What You See Is What You Test' (WYSIWYT) testing methodology and testing tool.

With the data recorded during the spreadsheet debugging process of the expert and novice subjects, it was possible to identify any cells that were inspected or edited. It was also possible to determine the number of times each cell was inspected and to determine the time spent inspecting each cell. An important research question, RQ4, was to determine if there was a correlation between the number of cells inspected and debugging performance. To answer this question, analysis was conducted to identify, for each subject, the number of individual cells inspected or edited during the debugging process. A cell was considered inspected/checked if that cell was selected for a specified minimum time or if the cell value or formula was edited or changed directly, and if

Table 1. Debugging performance: 13 professionals & 34 students

Error Type	Number of Seeded Errors	% Corrected Professionals	% Corrected: Students	T-test P-value	Mann Whitney P-value
Clerical/Non-Material	4	17%	11%	0.15106	0.1532
Rules Violation	4	85%	65%	0.02421*	0.0047**
Data Entry	8	68%	63%	0.10121	017376
Formula	26	79%	63%	0.00031***	0.0086**
Total	42	72%	58%	0.000063***	0.00131**

the cell was within a specified range of cells. The specified ranges of cells for this analysis were cells that contained formulas or values. Blank cells and column/row headings were not included.

Figure 4 shows a scatterplot for errors corrected versus coverage including a linear regression model for professionals and students, where the minimum time specified for a cell to be considered inspected/checked was >0.3 seconds. The statistical outliers are circled; these are participants that had standard residuals of <-2 or >2. The R^2 value of 0.6199 indicates a moderate-strong correlation. To determine what effect the *statistical outliers* were having on the correlation, the same scatterplot was recreated with the same specified minimum time of 0.3, but with the three *statistical outliers* circled in Figure 4 removed. The R^2 value increased to 0.7107.

Coverage per Cell Analysis. In order to determine the main areas within the spreadsheet that the participants focused on during the debugging experiment, a *coverage per cell* data analysis tool was developed which calculates, for each cell in the spreadsheet, the percentage of participants that inspected that cell. A minimum time is specified for which a cell has to be selected for it to be deemed inspected. If an individual cell was edited or if a range of cells were edited, then these cells are also considered inspected.

The data generated during the *coverage per cell* analysis was collated and is represented in the form of colour coding within each of the three worksheets that make up the experimental

spreadsheet. The colour coding makes it easier to recognise the groups of cells that participants inspected and the difference between the student and professional participant's behaviour.

An example of the colour coding (grayscale here for printing purposes) can be seen for the Projections worksheet in Figure 6 and Figure 7 (professional and student coverage respectively) with the colour coding key in Figure 5. Formula cells are denoted with an 'F', and data value cells with a 'D'. Seeded error cells also have 'Error' written in them. For example 'F error' denotes a formula error, 'C Error' a clerical error and 'D Error' a data error. Cells that contained column or row headings have been left unmarked, so the structure of the spreadsheet is clear.

These following findings contribute to the answer for RQ3. For all worksheets and both participant groups, formula cells took precedence over data cells, but a greater percentage of professional participants looked at each cell. Summation formula cells and bottom-line value cells received more attention than other formula cells. Formula cells that outputted text values were not inspected as much as formula cells that outputted numeric values. This was the case for both the professionals and students.

A key finding concerned the inspection rates for some of the logically equivalent groups of formula cells. Participants checked the first cells in these groups, either the topmost cells in the case of vertically oriented groups or the leftmost cells for horizontally oriented groups, and a distinct drop-

Table 2. Comparing 13 professional & 34 student formula error correction results

Formula Error Sub Categories	Number of Seeded Errors	%Corrected by Professionals	% Corrected by Students
Logic	9	77%	63%
Cell Reference	7	85%	68%
Range Reference	7	86%	71%
Remote Reference	3	56%	28%
Total	**26**	**79%**	**63%**

off in inspection rates could be seen for the rest of the cells in these logically equivalent groups. This was the case for both the professional and student participants, but this behaviour was more evident for the student (novice) sample.

Debugging Tool Evaluation

The sample for this experiment was made up of 16 fourth year Software Development students. The students were randomly selected and assigned to one of two groups. The 8 participants in the control group, Group-A, debugged the spreadsheet without the tool. The cell coverage tool was made available to the 8 participants in the test group, Group-B, and instructions were given to them on how to use the tool. The participant worked on similar computers using Microsoft Excel 2000.

Debugging Tool Overall Performance. The overall debugging performance categorised by error type for the eight participants not using the debugging tool, Group-A, and the eight Group-B participants to whom the debugging tool was available can be seen in Table 3. The test group participants who were using the *cell coverage tool* corrected slightly more errors overall, 62%, than the control group, Group-A, correcting 59% of the errors.

The test group corrected 9% more *clerical* errors, 6% more *rule violation* errors and 9% more *data entry* errors. The same number of *formula* errors was corrected by both groups. The table also shows the p-value for the appropriate one-sided statistical T-test with unequal variances which examines whether those students with the tool outperformed those students not given access to the tool to an extent that could be considered as statistically significant. Based on a 5% significance level the values show that there is no statistically significant evidence that the tool aids students to find Clerical/Non-Material, Rules Violation, Data Entry or Formula errors.

It is also interesting to compare the eight 4th year Software Development participants' results from the control group, Group-A in Table 3, with the professional and student results from Experiment 1, the Spreadsheet Debugging Experiment as seen in Table 1. Both the software and finance student groups had very similar overall correction rates, just 1% in the difference, whereas the experts significantly outperformed both.

Effect of the Debugging Tool on Debugging Performance. The debugging tool was designed to record any cells that become highlighted when the 'Highlight' button is clicked and the times associated with each highlighting action. This data is then available for analysis. Along with this data, the VBA recording tool makes a record of the time and detail of all cell selection and cell change actions of individuals while debugging a spreadsheet. This allows for detailed analysis of any effect that the coverage tool had on the debugger's performance. The results of this analysis can be seen in Table 4.

Overall, 25 more errors were detected and corrected in cells identified by the tool after par-

Figure 4. All Participants: Errors corrected over cell coverage (>0.3 sec)

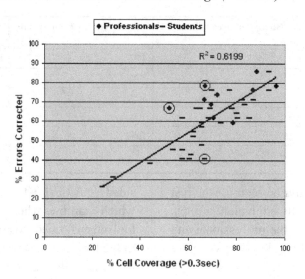

ticipants had chosen to use the tool on a sheet. This amounted to a 7.4% increase in error correction overall. Of course, these errors may have been found without the use of the tool. Of the 25 extra errors corrected, 22 were formula errors, 2 were data entry errors and 1 was a clerical error.

The extra errors found are detailed in Table 5. Where more than one participant found a certain error due to using the debugging tool, these cells are denoted with an asterisk. Of particular interest is 'Office Expenses F20', where 3 of the test group inspected and corrected this formula after it became highlighted. The formula in 'Office Expenses F20' is unusual in that it outputs a text value rather than a numeric value.

It is important to note that this analysis must be considered in light of any change in debugging

Figure 5. Coverage per cell colour coding key

90-100%	
80-90%	
70-80%	
50-70%	
<50%	

behaviour of the user due to the tool. Some users may consult the tool very regularly and thus most cells will be highlighted and any errors found may be considered as ones identified by the tool. This work next investigates the impact of the tool on the behaviour of the debugger.

Debugging Tool & Cell Coverage – Group-A and Group-B. In the previous two sections findings were presented on the overall error correction performance of both groups and the extra errors corrected by the test group. In this section the rates of cell coverage for the control and test groups, Group-A and Group-B, are discussed and compared.

Cell coverage analysis was carried out for all 16 of the 4th year Software Development students. As was mentioned earlier, a cell is considered inspected/checked if that cell was selected for a specified minimum time or if the cell value or formula was edited or changed directly, and if the cell was within a specified range of cells. The results are presented in terms of the control group and the test group. The cell coverage analysis was carried out with minimum specified times for a cell to be considered inspected of >1 second and >0.3 seconds. Again, individual and multiple

Figure 6. Professionals: Projections sheet coverage

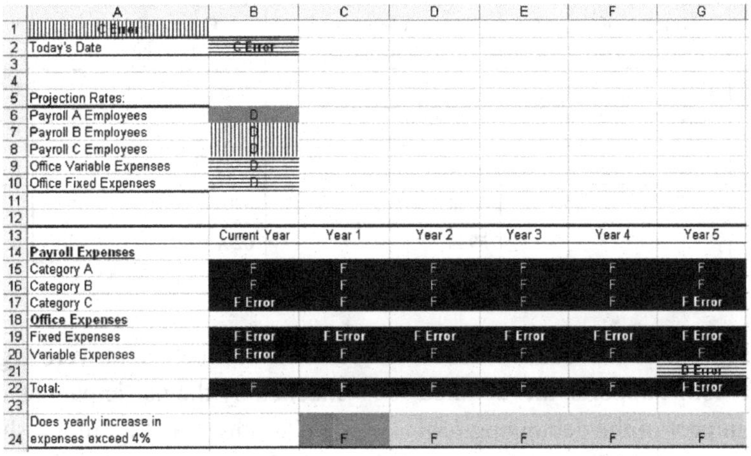

cells that have been edited or changed are also considered as inspected.

The cell coverage for participants in Group A and B can be seen in the boxplot in Figure 8. The minimum time specified for cell selections was >1 second. It is clear from this boxplot that the students using the cell coverage debugging tool inspected more cells than the control group. The overall average cell coverage for Group A was 52% compared to the overall average cell coverage for Group B of 71%.

As can be seen in Figure 9, the test group using the cell coverage tool achieved a signifi-

cantly higher cell coverage rate than the control group. The minimum time specified for a cell to be considered was >0.3 seconds. The overall average cell coverage for Group B was 90%, 26% higher than Group A, who achieved an overall average cell coverage of 64%.

Characteristics of Behaviour in Debugging Tool Use. Although the 8 fourth year software development students who formed the experimental group in Experiment 2 were given the same instructions for using the cell coverage debugging tool during the debugging process, there was some differences in the way each student used the tool.

Figure 7. Students: Projections sheet coverage

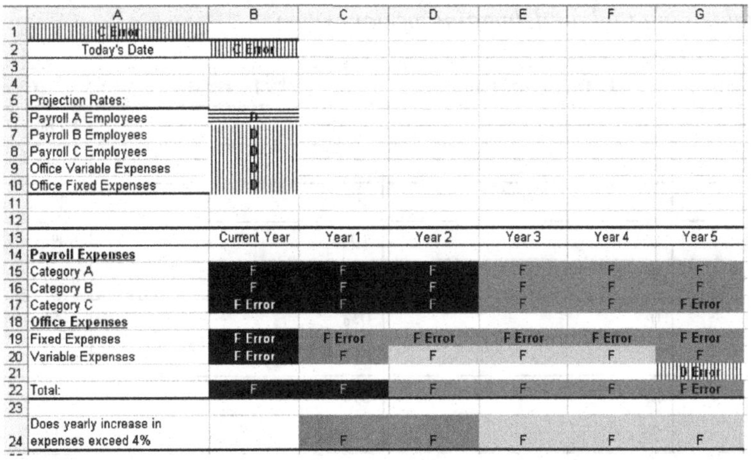

Table 3. Debugging performance: Group-A and group-B (group-B using coverage tool)

Error Category	Number of Seeded Errors	Group-A (no coverage tool)	Group-B (using coverage tool)	P-Value
Clerical/Non Material	4	13%	22%	0.062
Rules Violation	4	66%	72%	0.356
Data Entry	8	66%	75%	0.145
Formula	26	63%	63%	0.483
Overall Average		**59%**	**62%**	**0.356**

The scope and exactness of the data recorded through the T-CAT tool supports the detailed examination of the impact of the debugging tool on behaviour. The following figures represent three examples of different participant behaviour in using the tool and serve to investigate the impact of the tool on the structure of the debugging activity.

Figure 10 shows the behaviour of one of the participants using the debugging tool. What can be seen is that the participant clicked the 'Highlight' button on the 'Payroll' sheet after 54.7 minutes and 8 cells became highlighted. This process was repeated for the 'Office Expenses' and 'Projections' sheet. The participant in this case had inspected almost all of the cells in the spreadsheet, and when the highlight buttons were clicked, only a few cells became highlighted. In this case, the participants debugging behaviour does not seem to have been affected by the tool,

and the few cells that were highlighted could be checked by the participant as desired.

The chart in Figure 11 shows a debugging behaviour that appears to have been influenced by the use of the debugging tool. When the highlight button on the 'Payroll' sheet was clicked, 18 cells became highlighted. This process was repeated for the second and third sheets of the spreadsheet, with 5 and 0 cells being highlighted respectively. As the participant debugged each sheet and used the cell coverage tool, a conscious effort appears to have been made to inspect all the cells. The cell inspection rate increased from the first to the third sheet. Note that no check was made that all cells had been covered for the first two sheets. As with the previous participant the tool was only used once on each sheet.

The behaviour of the third participant, as represented in Figure 12, appears to involve the use of the cell coverage debugging tool to a

Table 4. Effect of debugging tool on debugging performance

Test Group Participants	Extra Errors Found	% Increase	Overall Performance
1	7	17%	69%
2	2	5%	71%
3	3	7%	38%
4	8	19%	69%
5	3	7%	40%
6	0	0%	69%
7	0	0%	69%
8	2	5%	71%
Total	25	7.4%	62%

Table 5. Extra error cells corrected

Extra Errors Corrected		
Sheet/Cell	Number of Participants	Error Type
Payroll		
F9*	2*	Formula
F10	1	Formula
F11*	2*	Formula
Office Expenses		
F5*	2*	Formula
F10	1	Formula
F18*	2*	Formula
F20*	3*	Formula
C14	1	Data
D16	1	Data
D12	1	Clerical
Projections		
D19: G19 (4 cells)	1	Formula
G17*	2*	Formula
G22*	2*	Formula

greater extent than the participants in the previous two examples to aid in the debugging process. The figure shows that the highlight button on the 'Payroll' sheet was clicked 4 times. It also shows that the participant re-inspected the 'Payroll' and 'Office Expenses' sheets, using the coverage tool each time.

Overall, analysis of the data collected by T-CAT showed that subjects interacted with the debugging tool in many different ways and to different extents.

CONCLUSION

The motivation for this research arises from the fact that although spreadsheets are prevalent throughout industry and their reliability is known to be poor, there is still little information available regarding end-user spreadsheet debugging behaviour and performance.

Research Questions Revisited

RQ1: Will expert spreadsheet users outperform novice spreadsheet users in detecting and correcting spreadsheet errors?

The overall average debugging performance of the industry-based professionals (experts) was 72%, which was 14% higher than the student subjects (novices), who corrected 58% of all seeded errors. In order to determine if the overall difference between the expert and novice subjects was statistically significant, t-tests were performed on the percentage of errors corrected by all subjects from both samples. This resulted in a P-value of 0.000063, strongly indicating that expert spreadsheet users outperform novice spreadsheet users in spreadsheet debugging and the difference in performance is statistically significant.

Figure 8. Cell coverage boxplot: Group A and B. min. time >1 second

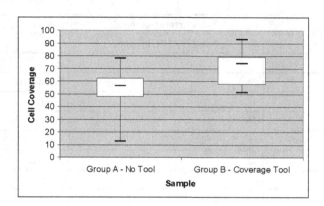

RQ2: Will expert spreadsheet users outperform novice spreadsheet users in detecting and correcting certain types of spreadsheet errors?

Clerical/Non Material Errors. There were four clerical errors within the experimental spreadsheet model. Expert participants corrected 17% of the clerical errors, and novice participants corrected 11% of the clerical errors.

Rule Violation Errors. The experts corrected 85% of the rule violation errors, 20% more than the novices. To correct the rule violation errors, participants needed to read the instructions and then check that the rules detailed in the instructions had been applied to the spreadsheet.

Data Entry Errors. The expert and novice samples achieved a similar result overall for data entry errors, with performances of 68% and 65% respectively.

Formula Errors. Formula errors are the most important error category with regards to spreadsheet reliability. The expert subjects performed considerably better than the novice subjects in Formula error debugging, correcting 79% of the errors; 16% higher than the novice subject's performance of 63%. With a t-test p-value of 0.00031, the lowest from all four error categories, and a Mann-Whitney p-value of 0.00869, this result is very significant.

Figure 9. Cell coverage boxplot: Group A and B. min. time >0.3 seconds

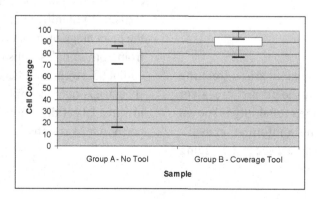

Figure 10. Debugging tool user behaviour: Subject 1

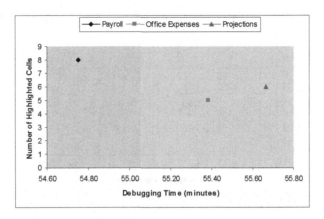

RQ3: What differences can be identified in the spreadsheet debugging behaviour of expert and novice spreadsheet users?

There were potentially many approaches that could be taken in identifying and comparing debugging behaviour, and it was decided that one aspect of participant behaviour would be analysed in terms of coverage per cell, where charts were generated from to show patterns and intensities of participants' cell focus during the experiment.

For all worksheets and both participant groups, formula cells took precedence over data cells, but a greater percentage of professional participants looked at each cell. Summation formula cells and bottom-line value cells received more atten-

tion than other formula cells. Formula cells that outputted text values were not inspected as much as formula cells that outputted numeric values. This was the case for both the professionals and students. A key finding concerned the inspection rates for some of the logically equivalent groups of formula cells. Participants checked the first cells in these groups, either the topmost cells in the case of vertically oriented groups or the leftmost cells for horizontally oriented groups, and a distinct drop-off in inspection rates could be seen for the rest of the cells in these logically equivalent groups. This was the case for both the professional and student participants, but this behaviourism was more evident for the student (novice) sample.

Figure 11. Debugging tool user behaviour: Subject 2

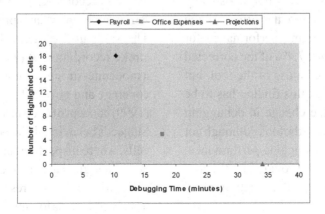

Figure 12. Debugging tool user behaviour: Subject 3

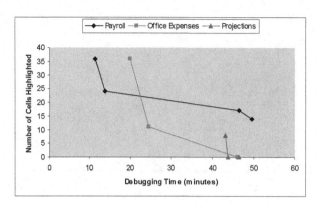

RQ4: Does a relationship exist between the number of spreadsheet cells inspected and debugging performance?

Overall, it was found that there is a moderate to strong relationship between the number of cells inspected and debugging performance. All participants combined (experts and novices) resulted in a correlation with an R2 value of 0.6199, moderate to strong.

RQ5: What effect would the availability and use of a debugging tool, that gives feedback on cell coverage during the debugging process, have on debugging performance?

The test group participants who were using the debugging tool corrected slightly more errors overall than the control group; 62% and 59% respectively, but the result was not statistically significant. When examining the effect of the debugging tool on debugging performance for the test group participants, 7.4% of the corrected errors were attributed to the use of the tool, but it is important to note that this finding has to be considered in light of the change in debugging behaviour of the users due to the tool. Although not significantly increasing debugging performance, the cell coverage feedback provided by the tool did dramatically increased cell inspection rates.

Contributions and Implications

The study confirms that experienced professional spreadsheet users are significantly more effective spreadsheet debuggers than students, and that this is especially the case with formula errors. This finding has implications for the use of student subjects as reasonable surrogates for professional practitioners in debugging experiments. Future research could build on this study by recruiting more professional subjects for spreadsheet debugging experiments.

To the authors' knowledge, this is the first study that unobtrusively records and analyses the behaviour of professional and student subjects debugging a MS Excel spreadsheet. Previous studies (Chen & Chan, 2000; Prabhakararao et al., 2003; Beckwith et al., 2005) used obtrusive recording methods with student subjects, such as video recording, with the latter two also using non-commercial spreadsheet applications. This study utilised a novel VBA mouse and keystroke recording tool (T-CAT). The finding that a moderate-strong correlation exists between cell coverage and error detection supports the Panko (1999) concept of code inspection in spreadsheets. Subject's behaviour in inspecting logical groups of cells, where inspection rates drop off towards the end of a logical group, and the lack of inspection of formula cells that result in a text value, may have implications for future debugging tool design.

There is also potential for the development of the cell-coverage based Debugging Tool as a commercial auditing utility, where feedback on spreadsheet cells inspected could be presented in real-time or retrospectively. Future research studies, especially those involving the development of spreadsheet debugging tools, could benefit from a greater understanding of end-user debugging behaviour, which could ultimately lead to a reduction in spreadsheet errors.

REFERENCES

Baker, K. R., Powell, S. G., Lawson, B., & Foster-Johnson, L. (2006). Comparison of Characteristics and Practices among Spreadsheet Users with Different Levels of Experience. In *Proceedings of the European Spreadsheet Risks Interest Group Conference*.

Barnard, J., & Price, A. (1994). Managing Code Inspection Information. *IEEE Software, 11*(2), 59–69. doi:10.1109/52.268958

Beckwith, L., Sorte, S., Burnett, M., Wiedenbeck, S., Chintakovid, T., & Cook, C. (2005, September). Designing Features for Both Genders in End-User Software Engineering Environments. In *Proceedings of the IEEE Symposium on Visual Languages and Human-Centric Computing* (pp. 153-160).

Bishop, B., & McDaid, K. (2007). An Empirical Study of End-User Behaviour in Spreadsheet Error Detection & Correction. In *Proceedings of the European Spreadsheet Risk Interest Group Conference*, Greenwich, UK.

Burnett, M., Cook, C., & Rothermel, G. (2004). End User Software Engineering. *Communications of the ACM, 47*(9), 53–58. doi:10.1145/1015864.1015889

Chen, Y., & Chan, H. (2000). An Exploratory Study of Spreadsheet Debugging Processes. In *Proceedings of the 4th Pacific Asia Conference on Information Systems* (pp. 143-155).

Croll, G. (2005). The importance and criticality of spreadsheets in the city of London. In *Proceedings of the European Spreadsheet Risks Interest Group Conference*.

Galletta, D. F., Abraham, D., El Louadi, M., Lekse, W., Pollailis, Y. A., & Sampler, J. L. (1993). An Empirical Study of Spreadsheet Error-Finding Performance. *Journal of Accounting, Management, and Information Technology, 3*(2), 79–95. doi:10.1016/0959-8022(93)90001-M

Galletta, D. F., Hartzel, K. S., Johnson, S., Joseph, J., & Rustagi, S. (1996). An Experimental Study of Spreadsheet Presentation and Error Detection. In *Proceedings of the 29th Annual Hawaii International Conference on System Sciences* (Vol. 2, p. 336).

Howe, H., & Simkin, M. G. (2006). Factors Affecting the Ability to Detect Spreadsheet Errors. *Decision Sciences Journal of Innovative Education, 4*(1), 101–122. doi:10.1111/j.1540-4609.2006.00104.x

Jenkins, R., Deshpande, Y., & Davison, G. (1998). Verification and validation and complex environments: a study in service sector. In *Proceedings of the 1998 Winter Simulation Conference* (Vol. 2, pp. 1433-1440).

Norman, D. A. (1999). *The Invisible Computer*. Cambridge, MA: MIT Press.

Panko, R. (1998). What We Know About Spreadsheet Errors. *Journal of End User Computing, 10*(2), 15–21.

Panko, R. (1999). Applying Code Inspection to Spreadsheet Testing. *Journal of Management Information Systems, 16*(2), 159–176.

Panko, R. (2000, July 17-18). What We Know About Spreadsheet Errors. In *Proceedings of the Spreadsheet Risk Symposium,* Greenwich, UK.

Powell, S. G., Baker, K. R., & Lawson, B. (2007). *Errors in Operational Spreadsheets*. Retrieved November 1, 2008, from http://mba.tuck.dartmouth.edu/spreadsheet/product_pubs.html

Powell, S. G., Baker, K. R., & Lawson, B. (2008). A critical review of the literature on spreadsheet errors. *Decision Support Systems, 46*(1), 128–138. doi:10.1016/j.dss.2008.06.001

Prabhakararao, S., Cook, C., Ruthruff, J., Creswick, E., Main, M., Durham, M., & Burnett, M. (2003, October). Strategies and Behaviors of End-User Programmers with Interactive Fault Localization. In *Proceedings of the IEEE Symposium on Human-Centric Computing Languages and Environments*, Auckland, New Zealand (pp. 15-22).

Purser, M., & Chadwick, D. (2006). Does an awareness of different types of spreadsheet errors aid end-users in identifying spreadsheet errors? In *Proceedings of the European Spreadsheet Risks Interest Group Conference*, London, UK.

Rothermel, G., Burnett, M., Li, L., Dupuis, C., & Sheretov, A. (2001). A Methodology for Testing Spreadsheets. *ACM Transactions on Software Engineering and Methodology, 10*(1), 110–147. doi:10.1145/366378.366385

Scaffidi, C., Shaw, M., & Myers, B. (2005). *The 55M End-User Programmers Estimate Revisited* (Tech. Rep. No. CMU-ISRI-05-100). Pittsburgh, PA: Institute for Software Research, Carnegie Mellon University.

Spannagel, C., Gläser-Zikuda, M., & Schroeder, U. (2005). Application of Qualitative Content Analysis in User-Program Interaction Research. *Forum Qualitative Sozial Forschung, 6*(2).

Teo, T., & Lee-Partridge, J. (2001). Effects on error factors and prior incremental practice on spreadsheet error detection: An experimental study. *Omega, 29,* 445–456. doi:10.1016/S0305-0483(01)00037-8

Trochim, W., & Donnelly, J. (2007). *Research Methods Knowledge Base* (3rd ed.). Mason, OH: Atomic Dog Publishing.

APPENDIX

Figure 13. Error type colour key

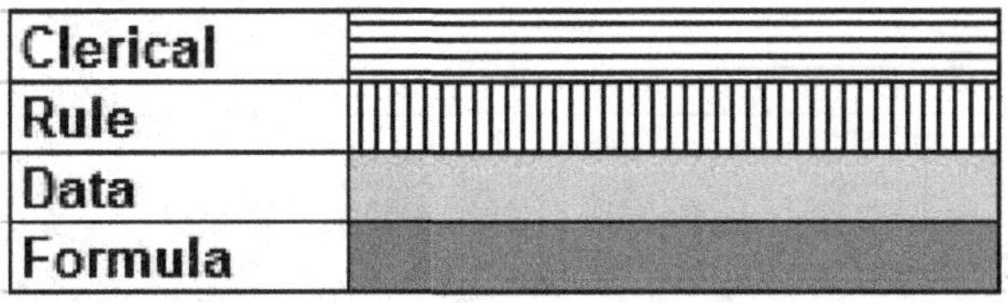

Figure 14. Payroll worksheet

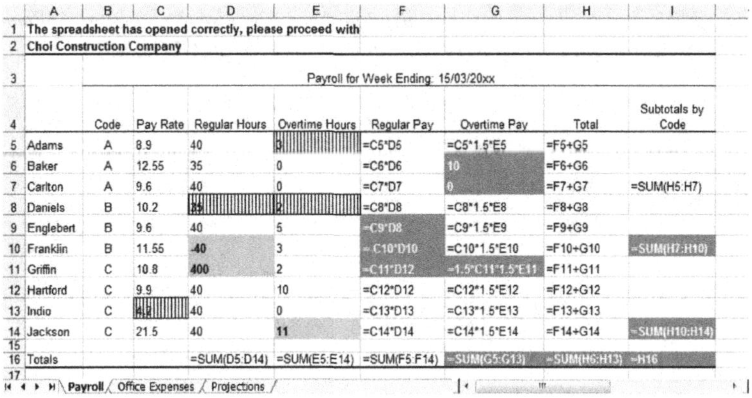

Figure 15. Office expenses worksheet

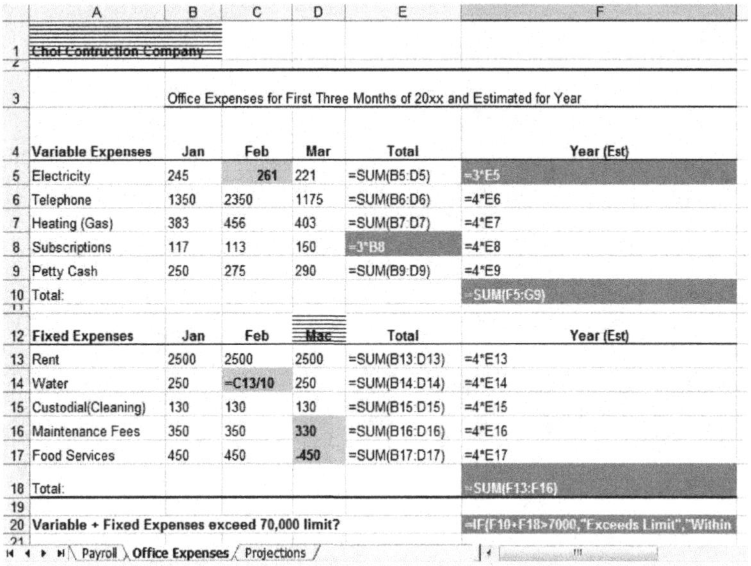

Figure 16. Projections worksheet

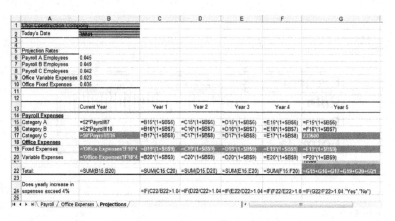

This work was previously published in the Journal of Organizational and End User Computing, Volume 23, Issue 2, edited by Mo Adam Mahmood, pp. 57-80, copyright 2011 by IGI Publishing (an imprint of IGI Global).

Chapter 6
Exploring the Dimensions and Effects of Computer Software Similarities in Computer Skills Transfer

Yuan Li
Columbia College, USA

Kuo-Chung Chang
Yuan Ze University, Taiwan

ABSTRACT

Computer software similarities play important roles in users' skills transfer from one application to another. Despite common software attributes recognized in extant literature, a systematic understanding of the components and structure of software similarities has not been fully developed. To address the issue, a Delphi study was conducted to explore the underlying dimensions of software similarities. Inputs gathered from 20 experienced Information Systems instructors show that Computer Software Similarity is a multi-dimensional construct made up of interface similarity, function similarity, and syntax similarity. Each dimension consists of software attributes that users perceive to be transferable in learning new applications. A field study was carried out to test the impact of the construct. Results from a survey on students' learning two software applications confirm the expectation that Computer Software Similarity facilitates the students' skills transfer between the applications. These studies provide a basis to better design training programs for improved training performance.

INTRODUCTION

Computer skills transfer is an important area of research in end user computer training (Powell & Moore, 2002). Also known as the carryover effect (Agarwal, Sambamurthy & Stairs, 2000) and transfer of training (Salas & Cannon-Bowers, 2001), it refers to the reuse of a person's computer skills acquired from previously learned software applications in new applications (Singley & Anderson, 1985). For scholars doing research in this area, an important mission is to understand why

DOI: 10.4018/978-1-4666-2059-9.ch006

it happens and how to facilitate skills transfer. A popular explanation is that computer skills transfer occurs when a to-be-learned software application shares similar attributes with previously learned applications, and the extent of similarity determines the amount of transfer. A number of common software attributes were recognized, such as menu items (Smelcer & Walker, 1993), functional keys (Polson, Bovair, & Kieras, 1987), user commands (Singley & Anderson, 1988), and dialog structures (Foltz, Davies, Polson, & Kieras, 1988); more recent studies helped to uncover additional items, including the database knowledge in Enterprise Resource Planning (ERP) system training (Coulson, Zhu, Stewart, & Rohm, 2004) and the knowledge structure in computer gaming (Schuelke, Day, McEntire, Boatman, Boatman, Kowollik, & Wang, 2009).

Progress has been made in this area, but notably a satisfying solution has yet to be developed to adequately specify what the common software attributes are in computer skills transfer. Firstly, most research has been focused on only one type of software attribute, but a cross-validation of the impact of multiple attributes was seldom made, questioning the validity of previous findings in the existence of multiple attributes. Secondly, most studies were based upon a single area of software, such as databases (Shayo & Olfman, 1998) or text editors (Singley & Anderson, 1988), but limited attention was paid to the software attributes that are shared across application areas. Due to these limitations, the extant literature suffers from the lack of a more comprehensive and generalizable list of attributes. The consequence is that when new generations of computer applications are introduced to a networked, ubiquitous computing environment (Olson & Olson, 2003), new features of the applications may not be properly integrated in existing knowledge frameworks to effectively predict the transferability of skills (Urbaczewski & Wheeler, 2001). Considering the importance of computer skills transfer for both research and practice (Agarwal et al., 2000), the quest for common software attributes should be continued.

This research attempts to achieve two objectives: to search for a more comprehensive and generalizable list of common software attributes to measure computer software similarity, and to find empirical evidence of the impact on computer skills transfer. To this end, two interrelated studies were conducted (Poston & Royne, 2008): the first study applied an inductive approach in search of a set of common software attributes due to the lack of sufficient theoretical basis. Specifically, the Delphi method (Schmidt, 1997) was applied to solicit opinions from a group of experienced computer instructors to aid in the recognition of common software attributes and the discovery of the underlying dimensions of the items. Based on the outcomes, the second study was conducted to test the impact of the common software attributes on computer skills transfer. The results show a three-dimensional model of computer software similarity (including function similarity, interface similarity, and syntax similarity), with each dimension containing a number of software attributes that are predictive of skills transfer. The list of items covers various aspects of software features and is not restricted to a single area of applications, making it more comprehensive and generalizable than those recognized in previous studies. This has significant meanings for further research and practice.

The article is structured as follows. First, past research in this area is briefly summarized, with achievements and limitations recognized. Next, the approach to the research questions is introduced, followed by the description of the Delphi study. Then, the empirical test is presented to verify the impact of the common software attributes on computer skill transfer. Findings are finally summarized and the limitations and implications of the research are discussed.

LITERATURE REVIEW

Multiple factors are thought to have an impact on computer skills transfer, including trainees' characteristics and prior knowledge, training methods, task characteristics, software attributes, and the roles of trainers (Bostrom, Olfman, & Sein, 1990; Compeau, 2002; Mykytyn, 2007; Salas & Cannon-Bowers, 2001; Spitler, 2005; Yamnill & McLean, 2001). Among the factors, the attributes of software applications play important roles. According to the theory of identical elements (also known as the theory of common elements) (Singley & Anderson, 1985), a person's skills learned from a previous software application can be transferred to a new application only if the two share common attributes. For instance, Singley and Anderson (1988) studied students learning two text editors and showed that the common text editing commands between the editors helped to reduce the time required to learn the second editor. Similarly, Polson, Muncher, and Engelbeck (1986) applied the theory to the study of computer utility tasks; the results also confirmed the impact of common software attributes on skills transfer.

Although alternative theories exist to interpret computer skills transfer, such as the principles theory and the near/far transfer theory (Yamnill & McLean, 2001), the common element theory receives most popularity. Studies based on the theory have been extended to various areas of software training, including databases (Shayo & Olfman, 1998), Windows operating systems and applications (Agarwal et al., 2000), programming (Scholtz & Wiedenbeck, 1990; Urbaczewski & Wheeler, 2001), and the ERP systems (Coulson et al., 2004). A large number of software attributes have been recognized from the studies; Table 1 provides a summary of the attributes.

It can be inferred from Table 1 that the common elements theory does not provide a clear prescription of the common software attributes in computer training, as the empirical studies to date have recognized distinctive features. Other theories, such as the ACT (standing for Adaptive Character of Thought) theory (Singley & Anderson, 1985), were developed to provide further guidance. The ACT theory suggests that the production-rule set of a task, referring to the basic actions or skills to perform the task, is the basis of transfer: the higher the degree of overlapping between the production-rule sets of two tasks, the higher the transferability of the skills. Key to the ACT theory is the accurate counting of the common production rules between two tasks, which may work best in relatively simple or well structured tasks; nevertheless, applications of the theory in complex tasks face challenges, as those tasks usually contain a large number of activities that may not be accurately enumerated, such as ERP training. Other theories, such as Ye's (1991) five-level knowledge hierarchy and Sein, Bostrom, and Olfman's (1999) six-level knowledge framework, also did not provide a satisfying solution. Sein et al.'s (1999) framework, for instance, describes a hierarchical model of Information Systems (IS) knowledge from lower levels of commands to higher levels of business motivation and meta-cognition. Although this model is helpful in general IS education, its application in computer skills transfer is limited; interestingly, Coulson et al. (2004) found that transfer of higher-level business knowledge did not happen in ERP training, and only the transfer of lower-level commands took place. Ye's (1991) framework suffers from the same limitation.

Empirical studies so far have their drawbacks as well. As mentioned above and also illustrated in Table 1, the studies were limited to a small number of software attributes within the same software domains. This is probably because most of the studies were based on experimental research, which could only manipulate a small number of software attributes in experimental design. After all, the fundamental question still remains: What are the common software attributes? The answer to this question is of critical importance to end user computer training, as the lack of a clear list

Table 1. Common elements analyzed in computer training research

Literature	Common Elements	Primary Findings
Coulson et al., 2004	Database application knowledge	Subjects' knowledge of the database systems has no impact on ERP training as a whole, but is related to specific ERP skills such as data retrieval and graphic user interface.
Foltz et al., 1988	Dialog structure and lexical attributes of menus	Changes to the dialog structure of the menu system are not detrimental, while changes to menu lexical attributes hinder user performance.
Polson et al., 1986	Production rules (or skills) used to perform utility tasks, such as loading a diskette, printing a document, and changing file names.	Learning time is determined by the new rules required to master a utility task, such as diskette duplication or document printing. A new task requires less time to learn if it shares more rules with other tasks.
Polson et al., 1987	Production rules (or skills) in text editing, such as inserting, copying, and moving texts.	Training time is a function of the number of unique rules needed to perform a text-editing task; similarities in the task structure and common methods and functions facilitate skills transfer and save time.
Scholtz & Wiedenbeck, 1990	Syntactic, semantic, and planning level of programming knowledge	Syntactic and semantic knowledge is easier to transfer to a new programming language than planning knowledge.
Singley & Anderson, 1985	Commands in text-editing software (editors)	Positive transfer between line editors that share most commands is significant; so is the skills transfer from the line editors to the screen editor, although not as strong.
Smelcer & Walker, 1993	Menu organization and command naming	Knowledge of the functional organization transfers from one application menu to another, thereby reducing required search times for any second application. When users do not know the exact name of the commands, the functional organization leads to shorter search time.

of common attributes has severely undermined the ability to predict the transferability of a person's computer skills. People used to believe that sequence matters when teaching a variety of software applications due to the various extents of similarities between the applications (Agarwal et al., 2000). However, empirical studies in the areas of programming (Urbaczewski & Wheeler, 2001) and ERP training (Coulson et al., 2004) did not find full support of the claim. To effectively address the issue, we need a list of common software attributes that are both comprehensive and generalizable so as to make further progress in this area.

While earlier studies tended to focus on rudimentary levels of software attributes such as menu items and keystrokes, more recent studies began to shift attention to higher levels of features, including procedures, interfaces (Besnard & Cacitti, 2005), and macro knowledge (Harvey & Rousseau, 1995; Scholtz & Wiedenbeck, 1990). Recent theories in computer training – such as assimilation theory (Bostrom et al., 1990), mental model theory (Compeau, 2002; Vandenbosch & Higgins, 1996), domain-specific strategic knowledge theory (Kontogiannis & Shepherd, 1999) and the analogical conceptual model theory (Mao & Brown, 2005) – suggest that perceptions and knowledge of higher-level software features are developed through the daily use of software and then applied to assimilate new or similar features in other software. Such a higher-level conceptualization of software, in forms of aggregated features rather than individual elements, portrays a more comprehensive view of software similarities and is helpful in explaining the transfer of skills across applications. In this research we focus on the aggregate level, or dimensions, of software attributes. To prevent any confusion, we hereafter use Computer Software Similarity

(CSS) to indicate the extent to which two software applications share common features.

Due to the limitations in theories and empirical studies, we adopted an inductive approach, the Delphi method, in this research. This method is well known for its utility in gathering opinions to address issues in areas that are not well explored, and has been widely used in the IS field as a means to identify issues and assist in concept development (Okoli & Pawlowski, 1995; Schmidt, 1997). We used the method to solicit opinions from experienced computer instructors to explore the domain of CSS and to recognize a generalizable list of common attributes. Once this work was done, the second study was conducted to test the impact of CSS on computer skills transfer via a survey. These two studies are closely related, as the first study is fundamental to addressing the research questions, while the second is necessary to substantiate the outcome. Following the approach by Poston and Royne (2008), we describe the two studies in the following sections.

STUDY 1: EXPLORING THE DOMAIN OF COMPUTER SOFTWARE SIMILARITY

In this study, opinions were collected from a group of experienced computer instructors whose knowledge was an important source for the recognition of the components as well as the structure of software similarities. Normally a Delphi study consists of three major steps: brainstorming, narrowing down factors, and ranking factors (Okoli & Pawlowski, 1995; Schmidt, 1997). In this study, however, ranking the elicited common elements was not the primary concern; instead, we were interested in understanding the underlying structure of software similarities. Therefore, following Tojib and Sugianto's (2007) approach, we replaced the third step of the Delphi study with the analysis of empirical data in order to meet this objective. The three steps of the study are described in the next sections.

Step 1: Brainstorming

A group of 28 computer instructors from the universities the authors affiliate with were invited to this study. A questionnaire was sent out electronically to the panelists, requesting them to enumerate the common software features that may influence the transfer of computer skills and to provide a detailed description of each item. Demographic information and records of teaching experience were also requested. A cover letter and a consent form were enclosed in the questionnaire, explaining the research purpose and privacy policies. Two rounds of follow-up notes were sent out within three weeks of the initial invitation. For those who did not respond, a phone call was made a week after the last email reminder. Of the 28 instructors invited, eight declined to participate, yielding a total of 20 responses. Table 2 shows the profiles of the respondents.

The invited panelists contributed 82 items, and the authors generated 15, producing a total of 97 items in the first round. Each item consists of a label and a brief description, for example, "*Keyboard shortcuts: the extent to which the application follows conventional keyboard shortcuts helps students master the new application faster*". Each item was verified carefully by the authors for the face validity based on a key criterion: Is the item about the similarities between software applications? In total 38 items that did not comply with this criterion were recognized; although these items have the potential to facilitate learning of a new software application, they do not fit the current research purpose of recognizing the similarities between applications and were therefore eliminated. These items include:

- 20 items about the unique characteristics of a software application, such as response time, graphics used, auto hints, help for novices, and availability of examples. Although these features make an application easy to learn, they are related to the features of a single software package and

Table 2. Profile of respondents

Demographic Information	Count	%
Gender:		
Male	16	80%
Female	4	20%
Age:		
Less than 30	2	10%
30-40	9	45%
41-50	5	25%
51-60	2	10%
More than 60	2	10%
Computer-related courses teaching experience:		
Less than 1 year	1	5%
Between 1-3 years	3	15%
Between 4-6 years	5	25%
Between 7-9 years	3	15%
More than 9 years	8	40%
Specific teaching experience with respect to particular software applications:		Avg. years
Office productivity (e.g., spreadsheet, word processing, & presentation)	16	5.7
Programming (e.g., Java, VB, & ASP)	17	5.6
Database (e.g., Access, Oracle, & mySQL)	13	5.7
Others	15	8.1

do not necessarily measure the similarities between applications.

- 6 items referring to the training methods, such as the same method of teaching and learning, and the use of flowcharts to teach. These items are not pertinent to software attributes.
- 4 items referring to personal learning styles, for examples, "some students are attracted to colored pictures while other students like animation". These factors are about learners and learning styles (Bostrom et al., 1990), which are distinctive from software attributes.
- 4 items referring to personal knowledge or prior experience, such as user background, knowledge of related fields, and knowledge about the information processing cy-

cle. These are about individuals' computer experiences rather than software attributes.

- 4 duplicate items identical to those on the list, including the appearance of dialog boxes, message boxes, the task performance process, and the task concept.

After the removal of the disqualified items, the remaining 59 items were examined for content uniqueness, as the authors noticed that several items possessing different labels or descriptions actually referred to the same attributes. For instance, a panelist mentioned "similar response to mouse events", while another suggested "mouse clicks". According to the descriptions, both refer to a mouse-click response, such as opening a shortcut menu by right-clicking, so that these two items should be counted as one. An item-to-item comparison was conducted on all the 59 items, and 16 of them were found to be overlapping with others. The overlapping items were merged accordingly, with their labels and descriptions adjusted. After the two rounds of purification, 43 items survived for the next step.

Step 2: Narrowing Down the Elements

With the 43 items recognized, a second survey was conducted to narrow down the list of attributes and to discover the preliminary structure of CSS. Due to the fact that different levels of categorization may exist, a hypothetical categorization scheme was developed by the authors to provide guidance to the panelists. The scheme was used to show how the items could be categorized into groups according to their closeness to each other. To do so, the sorting process described by Moore and Benbasat (1991) was used, in which the index cards of the items were prepared and sorted into categories, and then a label was given to each category to best describe its items. The authors worked independently to collate the items and developed their own schemes outlining the structure of software similarity. Next, the two independently constructed schemes were likened and reconciled by the authors working in tandem.

Five categories surfaced out of the 43 items within the resultant scheme, including functional attributes (i.e., features of the software constructed to perform specific tasks or action), interface attributes (i.e., the appearance and the structure of how the functionalities of an application are organized), process attributes (i.e., procedures involved in performing a series of functions or steps to complete a task), syntax attributes (i.e., the set of rules that constitutes the statements or structure of programming in an application), and user document attributes (i.e., user files that are created in an application with certain features, structures, appearance, and components). The decision regarding which category each item was assigned to was based on the authors' consensus judgment. The hypothetical scheme was then sent to the 18 responding panelists from Step 1 for further validation.

A new survey was developed consisting of the 43 items and the five hypothetical categories, and the panelists were asked to 1) validate the categorization scheme and modify it if necessary, 2) validate the placement of each item, and 3) modify the item pool by adding or removing items if necessary. Based on the feedback provided by several panelists during the first round of study, the label and the description of each item were merged into a single statement within the survey. For instance, the "user document attributes" item was labeled as *File components* and defined as: *If the components, for example, tables, charts, graphs, and clip arts, etc., can be used in different user documents, they are easier to learn.*

Out of the 18 panelists, 16 responded in this round of survey. The responses show that most panelists chose not to make significant changes. The removal of several items from the list was suggested, but there was no unanimity for any of the items. In addition, a few new categories were proposed, but most panelists chose the categorization scheme provided by the authors. However, some of the feedback obtained during this round of surveying aroused our attention: several panelists pointed out that two suggested categories, function and interface, were not very distinctive. As one panelist stated, software functions are performed through interfaces; therefore, items that were placed in the function category or the interface category seem to carry similar characteristics, making it difficult to distinguish between them. To address this issue, we employed an Exploratory Factor Analysis in the third step to empirically investigate the underlying structure of the items.

Further analysis focused on the verification of the placement scheme, for which the hit rate was used. Hit rate refers to the extent to which an item falls into its designated group (Moore &

Table 3. Revised pool of common attributes

1. Window layout, such as the arrangement of the components in the window.
2. Programming language statements, such as variable declarations.
3. Integrated development environment that combines various features of the software, such as the execution of commands and provision of feedback.
4. The tasks that standardized icons can perform, such as the printer icon executing the printing feature.
5. Keyboard shortcuts, such as Ctrl+C for Copy and Ctrl+V for Paste.
6. Toolbar options, such as the options on a standard toolbar.
7. Toolbars, such as the standard toolbar and a formatting toolbar.
8. Formats of user files that are supported by different applications.
9. Class library, such as a math library.
10. Information feedback, such as the information shown on the status bar or confirmation dialog.
11. Menu key words that indicate the grouping of menu items, such as the Tools menu.
12. Icons constructed to perform the same functions, such as the Save icon and the Open icon.
13. User file components, such as tables, charts, and clip art.
14. Object structures of the programming language, such as user defined forms or dialogs.
15. Open standards of programming.
16. Programming language terminology.
17. Events related to the programming language, such as responses to a confirmation dialog.
18. Problem-solving processes, such as sending a document by email.
19. User file structures, such as the body area and header/footer areas.
20. File manipulation features, such as saving, opening, editing, or formatting a file.
21. Message boxes to provide feedback to users, such as error messages or confirmation messages.
22. Menu structure, such as the Format menu that contains all the formatting features.
23. The processes of turning on the software, such as creating a new file via wizard.
24. Programming algorithms, such as logic comparisons and mathematical calculations.
25. Error detection, such as spell checkers and auto-correction.

Benbasat, 1991). It is calculated by dividing the number of panelists who assigned an item to its target group by the total number of the panelists. A high hit rate suggests that the panelists generally believe an item to rightly characterize the construct it represents. A threshold of 0.5 is recommended, and items with a hit rate lower than 0.5 (Moore & Benbasat, 1991) are discarded from research. After the analysis, 25 items had hit rates above the recommended level, and were kept for further analysis.

The remaining items were evaluated by a group of 10 experienced computer users, who were IS graduate students with a minimum of 5 years of training in computer applications, to gauge the understandability and appropriateness of the items from users' perspective. Descriptions of several items were adjusted based on the feedback, after which another 5 experienced users, who were IS graduate students as well, were invited to further validate the appropriateness of the items. The final purification process was completed with no further changes recommended. Table 3 shows the resulting list of items.

Step 3: Recognizing the Structure of the CSS Construct

The retained items were empirically tested to explore the underlying structure of CSS. Data were gathered at a private university in northern Taiwan. The students chosen were enrolled in a computer course covering two media production and web development applications consecutively: first *Macromedia Flash*, then *Director*. To correctly capture the students' perceptions of any similarities between the two applications, the survey was administered after the students received training in both applications. The students were asked to compare the two applications and to assess their similarities with regard to the common software attribute items in Table 3. A total of 102 students participated in the survey, from which 78 complete observations were made.

In the survey, a 7-point Likert scale was used to measure each item to provide consistency for responses, and the scale was anchored on 1 for great dissimilarity and on 7 for great similarity. An Exploratory Factor Analysis with SPSS was

Table 4. Factor structure of the computer software similarity construct

Item Description	Factor Structure
Function Similarity (FS):	
4. The tasks that standardized icons can perform, such as the printer icon executing the printing feature.	FS1
18. Problem-solving processes, such as sending a document by email.	FS2
20. File manipulation features, such as saving, opening, editing, and formatting a file.	FS3
Interface Similarity (IS):	
7. Toolbars, such as the standard toolbar and a formatting toolbar.	IS1
11. Menu key words that indicate the grouping of menu items, such as the Tools menu.	IS2
12. Icons, such as the Save icon and the Open icon.	IS3
22. Menu structure, such as the Format menu that contains all the formatting features.	IS4
Syntax Similarity (SS)	
2. Programming language statements, such as variable declaration.	SS1
9. Class library, such as a math library.	SS2
15. Open standards of programming.	SS3
24. Programming algorithms, such as logic comparison and mathematical calculations.	SS4

performed to identify the underlying structure of the construct. The results showed a three-factor solution. Items with loadings below 0.4 were dropped (Hair, Anderson, Tatham, & Grablowsky, 1998), and the iterative process continued until all items loaded on certain factors above the suggested guideline. Table 4 shows the resultant factor structure.

The first dimension, labeled as Function Similarity (FS), consists of three items that describe application features constructed to perform a specific task or a series of actions. The second dimension, labeled as Interface Similarity (IS), contains four items associated with the appearance and structure of how the functionalities in an application are organized. The third dimension was labeled as Syntax Similarity (SS), given that its four items depict the statements or structures of the supporting languages used in an application. All 11 remaining items were loaded on their corresponding factors, and no high cross-loadings (over 0.5) were found.

DISCUSSION

Contributing to the literature, the results from the Delphi study showed a three-dimensional model of CSS, with each dimension comprising a number of software attributes that reflect the extent to which two software applications resemble each other. Although further validations are necessary to confirm the structure, the current results depicted a generalizable and comprehensive structure of CSS, as the three dimensions cover a broad spectrum of software attributes independent of software domains.

As we assume that the new list of items should incorporate findings from previous studies, it is not surprising to notice that some of the items unveiled in this study are to some extent similar to those recognized in previous studies. For instance, Item No. 20, *file manipulation features,* is a particular example of file operation skills that contains such activities as saving, opening, editing, and formatting a file. There features are similar to the text editing rules analyzed by Polson et al. (1987) and also the utility task rules studied by Polson et al. (1986). Item No. 11, *menu key words*, resembles the naming styles analyzed by Teasley (1994) and also the lexical attributes analyzed by Foltz et al. (1988), as both describe how the menu items or dialog elements were named and comprehended. Finally, the items within the SS dimension were part of the computer knowledge studied by Scholtz and Wiedenbeck (1990), although the latter research was conducted from the programmers' perspective. Our results renovate many previous studies, indicating strong content validity of the recognized items.

Adding to the literature, the study uncovered several new attributes that were not discussed in previous studies, such as *problem-solving process* (Item No. 18), *menu structure* (Item No. 22), and *class library* (Item No. 9). These items reflect the attributes of contemporary generations of software applications, especially the Object Oriented (OO) programs. The addition of these components to those listed in the previous literature helps to broaden our knowledge regarding common software attributes across application areas. In other words, our goal of searching for a comprehensive list of software attributes was accomplished.

Moreover, the structure of the software similarities highlights the fact that the common software features are closely related to each other, allowing computer users to develop a higher level cognition of those features. Function Similarity, for instance, consists of items that are related to the functionalities of a software application including task-performing and problem-solving. This dimension determines the basic features of a software application that a user can learn to use. Also, Interface Similarity, including the elements that a user sees on the screen, influences users' ability to perform tasks on the software. Finally, Syntax Similarity, from an application development perspective, indicates how the features of an application are actually performed through the components. All three dimensions are fundamen-

tal to software applications, and a person who is learning a new software application can compare it to those previously learned applications based on these three dimensions.

The results from the Delphi study lead to our second question: How effective are the common attributes and dimensions of CSS in interpreting the transferability of a person's computer skills? Related to this question is the need to clarify the relationship between the three dimensions of CSS. To answer this question, an empirical test was conducted, as described next.

Study 2: An Empirical Test on the Effect of Computer Software Similarity

As implied in the previous sections, the focal construct in this study, Computer Software Similarity/ CSS, was conceptualized as a reflective latent construct consisting of three dimensions. The purpose of the empirical study was to validate the structure of the CSS construct and to test its impact on computer skills transfer. Based on the common elements theory and the extant literature, we hypothesized that *the extent of CSS has a positive relationship with the extent of Computer Skills Transfer (CST)*.

In literature, a number of instruments are available to measure the extent of computer skills transfer, including time used to perform an operation (Singley & Anderson, 1985), time used to search for a command (Smelcer & Walker, 1993), self-reported course grades (Urbaczewski & Wheeler, 2001), and direct comparison by judges (Scholtz & Wiedenbeck, 1990). These instruments were developed from particular research contexts but were not extendable to a general context. Therefore, a new scale to fit the purpose of this research was required. To fulfill this demand, a 10-item scale was developed to measure the transferability of computer skills, including various aspects such as ease of learning, readiness of application, and reusability, as shown in Table 5. This subjective

Table 5. Measurement items of the computer skills transfer construct

CST1. My skills learned from _a_ are helpful in learning how to use _b_.
CST2. My knowledge of _a_ saves effort in learning _b_.
CST3. I can apply the skills I learned from _a_ to the use of _b_.
CST4. The skills I learned from _a_ can be reused in _b_.
CST5. My knowledge of _a_ can be applied to my understanding of the features of _b_.
CST6. The knowledge I possess from learning _a_ enables me to spend less time to learn _b_.
CST7. My skills in _a_ are helpful in recognizing the features of _b_.
CST8. My skills in _a_ enable me to find the needed features of _b_ to solve a problem.
CST9. Without further training, I can apply the skills learned from _a_ to _b_ to perform simple tasks. (dropped)
CST10. I have a better understanding of _b_ if I learn more about _a_. (dropped)

Note: "a" refers to the first software application learned, and "b" refers to the second application.

measure of users' perceptions was believed to yield comparable information to an objective measure.

A new survey was developed from the CSS items (Table 4) and the new items generated for CST (Table 5). Similar to the CSS items, each of the CST items was measured on a 7-point Likert scale, ranging from 1 indicating strong disagreement to 7 denoting strong agreement. Data were collected from another 98 college students who completed the same software courses (*Flash* and *Director*), from which 62 complete observations were made.

DATA ANALYSIS AND RESULTS

The statistical properties of the measurement items are shown in Table 6. The mean values and standard deviations of the items show that even though we collected data from a single research setting, the subjects differ in their perceptions of the similarities between the applications and skills transfer. To further test the potential common method bias, we conducted Harman's single-factor test (Podsakoff, MacKenzie, Lee, & Podsakoff, 1986) to identify the factor that explains most of the variances in the data set. An Exploratory Factor Analysis with all the items from the survey showed that more than one factor was extracted, and the first factor accounted for 37% of the total variances, far less than the suggested cutoff value of 50% (Podsakoff et al., 1986), suggesting that

common method bias is not a major concern in this study.

We then examined the psychometric properties of the items and the latent constructs, as shown in Table 7. Since the CSS construct contained three dimensions, each dimension was treated as an individual construct during the test – a common approach applied in contemporary studies (Agarwal & Karahanna, 2000). Cronbach's alpha was used to examine the reliability. For the CST construct, the reliability was relatively low, as two items (#9 and #10) showed low item-to-the-total correlations (0.44 and 0.46, respectively); these two items were dropped from further analysis. The remaining items showed a reliability of 0.92. For the Function Similarity, the reliability was not very high, either, but greater than the minimum level of 0.60 for a newly developed scale. The other two dimensions, Interface Similarity and Syntax Similarity, both displayed acceptable reliabilities. Composite Reliabilities (CR) were also tested, ranging from 0.81 to 0.94, and were above the recommended guideline of 0.70 (Agarwal & Karahanna, 2000). All of the reliability measures are listed in Table 7.

The square root of the Average Variance Extracted (AVE) was calculated to test the convergent validity. Shown on the diagonal of Table 7, the square root of the AVE of each construct was higher than the threshold value of .707 (Chin, 1998), indicating good convergent validity. Meanwhile, the square root of the AVE of each construct

Table 6. Statistical properties and the factor structure matrix

Item	Mean	Stdev	Loadings and Cross-loadings			
			FS	IS	SS	CST
FS1	4.00	1.46	**0.887**	0.407	0.023	0.170
FS2	3.56	1.40	**0.652**	0.245	0.027	0.170
FS3	4.23	1.69	**0.755**	0.437	0.109	0.316
IS1	3.76	1.47	0.380	**0.845**	0.228	0.087
IS2	3.89	1.43	0.436	**0.858**	0.202	0.122
IS3	4.07	1.50	0.343	**0.583**	0.181	0.234
IS4	3.60	1.46	0.278	**0.722**	0.195	0.305
SS1	3.02	1.30	-0.186	0.125	**0.839**	0.235
SS2	3.34	1.32	0.101	0.331	**0.893**	0.302
SS3	3.11	1.33	0.038	0.192	**0.863**	0.204
SS4	3.66	1.33	0.310	0.224	**0.701**	0.399
CST1	4.13	1.42	0.240	0.258	0.377	**0.859**
CST2	4.19	1.69	0.317	0.232	0.346	**0.787**
CST3	4.10	1.48	0.226	0.208	0.478	**0.843**
CST4	4.02	1.55	0.340	0.228	0.169	**0.814**
CST5	4.19	1.62	0.164	0.116	0.231	**0.871**
CST6	3.98	1.62	0.192	0.133	0.152	**0.831**
CST7	3.87	1.35	0.197	0.313	0.275	**0.799**
CST8	3.60	1.30	0.170	0.124	0.174	**0.753**

Note: FS – Function Similarity, IS – Interface Similarity, SS – Syntax Similarity, CST – Computer Skills Transfer.

was greater than the correlations between that construct and other constructs (the off-diagonals in Table 7), indicating strong discriminant validity. Additionally, the factor structure matrix in Table 6 indicates that all items exhibited acceptable loadings of 0.5 or higher (Falk & Miller, 1992; Hair et al., 1998) on their respective constructs, and no strong cross-loading (over 0.5)

was observed. Overall, the self-reported measurement instruments exhibited sufficiently strong psychometric properties (Chin, 1998; Fornell & Larcker, 1981).

The partial least square (PLS) method was chosen for the hypotheses testing, for which the PLS Graph version 3.0 was used. Unlike the covariance-based structural equation modeling

Table 7. Results of psychometric property tests

Latent Variables	Cronbach's α	CR	1	2	3	4
1. FS	0.64	0.81	**0.77[a]**			
2. IS	0.80	0.88	0.48[b]	**0.80[a]**		
3. SS	0.83	0.90	0.07[b]	0.26[b]	**0.82[a]**	
4. CST	0.92	0.94	0.28[b]	0.25[b]	0.34[b]	**0.82[a]**

Note: a – the square root of the Average Variance Extracted (AVE); b – correlation between latent factors; CR – Composite Reliability.

Table 8. Results of PLS analysis: loadings

Constructs and Indicators	Loadings
Systems Similarity	0.660[a]
Functional Similarity (FS)	0.798[b]
Interface Similarity (IS)	0.718[b]
Syntax Similarity (SS)	0.880[b]
Computer Skills Transfer	0.823[b]
CST1	0.869[b]
CST2	0.799[b]
CST3	0.840[b]
CST4	0.793[b]
CST5	0.798[b]
CST6	0.729[b]
CST7	
CST8	

Note: a – p < 0.01; b – p < 0.001

technique (e.g., LISREL) whose objective is to maximize the fit between the observed covariance structure and the hypothesized structure, PLS assumes a component-based method and is more prediction-oriented, seeking to maximize the variance explained in constructs (Chin, 1998). It does not depend on having multivariate normal distributions or a large sample size (Yi & Im, 2007). Although PLS can directly test the second-order construct using repeated indicators (Wetzels, Odekerken-Schroder, & van Oppen, 2009), according to Wilson and Henseler (2007), a two-step approach outperforms the repeated indicator approach in small samples with fewer indicators. Therefore, the latter approach was adopted. The factor scores of the three dimensions of CSS were first calculated based on the raw data and then used as input in the PLS model to test the hypothesis (Agarwal & Karahanna, 2000). The bootstrapping process with 200 resamples was performed to examine the significance of factor loadings and the path coefficient. The results are presented in Table 8 and Figure 1.

The results showed that the path coefficient between CSS and CST was positive and significant at the 0.001 level ($\beta = 0.439$, t = 6.33), indicating a strong relationship between these two constructs. Variance in CST explained by CSS reached approximately 19%, suggesting that a substantial amount of skills transferred were due to the similarities between the two applications. All three dimensions of CSS showed high loadings, indicating that they are important components of the higher-order CSS construct. In general, the empirical test provided strong evidence of the impact of CSS on CST.

DISCUSSION

This study provided answers to the questions raised in the Delphi study regarding how effectively CSS explains the transferability of users' computer skills and what structure this construct comprises. In terms of the first question, the results showed a strong positive relationship between CSS and CST. Even though the variance explained was only 19%, it was significant for a single predictor; the addition of other antecedent factors (Bostrom et al., 1990; Shayo & Olfman, 1998) may increase the explanatory power and yield a better understanding of the dependent variable.

In terms of the structure of CSS, the results confirmed that the three dimensions of CSS were reflective indicators of a higher-order construct with each dimension containing a homogenous set of common attributes. The covariances between the dimensions evidence a general pattern or

Figure 1. PLS test of the research model

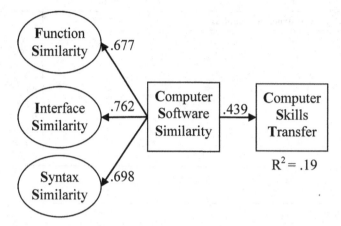

frame of knowledge on software use (Besnard & Cacitti, 2005), as users tend to develop an "action pattern" for a certain interface and apply the same pattern of action in a new interface if the two share same features. In other words, users develop an overall assessment of the similarity between software applications based on the association between interface, function, and syntax, rather than relying on the individual elements. In doing so, they match an interface item (i.e., what it looks like) with the underlying function (i.e., what it does) and also the semantic and syntactic knowledge of executing the function (i.e., how it does) in the new software. For example, *Record New Macro* is a tool available in most Microsoft Office applications, which contains a universal interface (i.e., the record button), a common function (i.e., recording a new macro), and standard syntax (i.e., the rules or procedure of recording the macro). Comprehension of this feature and reuse in other Office applications demand knowledge in all three aspects.

Moreover, the results also confirm that the three similarity dimensions are manifestations of the CSS construct such that changes in the CSS construct will cause changes in the three dimensions, while changes in a single dimension is not sufficient to alter the whole construct if other dimensions do not alter accordingly. For instance,

similar interfaces between two applications are not adequate to establish the similarities between the applications if the underlying functions differ; and if two applications share the similar functions but different interfaces or syntax, the users who have learned the first application would have trouble identifying those functions in the second. On the other hand, if two applications are similar in certain aspect, the three components should all be similar in that aspect for the skills to be successfully transferred. The causality between the CSS construct and the three dimensions and also the covariances among the dimensions suggest that the three dimensions of CSS are reflective indicators of the higher-order construct rather than its formative indicators (Jarvis, Mackenzie, & Podsakoff, 2003).

CONCLUSION

This research advances our knowledge on computer skills transfer in several ways. First, a comprehensive list of common software attributes has been recognized, ranging from functional keys to interface elements and then to syntactical components. This list of attributes comprises a wide range of software features that are domain-independent, making it more generalizable to various software applications as compared to other items recognized in previous studies. It enables scholars to gain deeper insight in software applications and conduct further research on skills transfer. It also promotes the common element theory to a higher level in the computer training area. In addition, the conceptualization of the CST construct and the development of a measurement instrument enable researchers to conduct further research in this area. For instance, researchers may tailor the context-free measurement of CST (Table 5) in particular research contexts without developing unique measures, which allows researchers to compare the results horizontally.

LIMITATIONS

Several limitations exist in this research. First, this research was primarily based upon a common elements theory perspective, although other theories are available to interpret computer skills transfer. In addition, the research focused solely on the impact of software attributes on computer skills transfer, but other factors that may have an influence (Bostrom et al., 1990; Compeau, 2002; Salas & Cannon-Bowers, 2001) were not included. Some recent studies identified additional sources of skills transfer, such as the impact of master users (Spitler, 2005) and the problem-based learning style (Mykytyn, 2007). Incorporating these theoretical lenses and factors will surely enhance our knowledge in this area.

The second limitation relates to the list of software attributes recognized. Although a large group of experienced computer instructors participated in the study, and two data sets were collected to empirically test the structure of the CSS construct, there were potentials that other software attributes could have been overlooked. We do not expect that the exclusion of those items poses serious concerns to the results, since we shifted the focus from individual items to a higher order CSS construct so that the exclusion of some individual

items would not impair the overall structure of the construct. For the purpose of refining the research results, other studies could be conducted to extend the list of items to capture the features of newer generations of software.

The third limitation resides in our selection of the two software applications in this research: both *Flash* and *Director* are within the same software category and are distinctive from other popular applications such as office productivity software and ERP systems. Readers may question what an ERP user would perceive the three dimensions of CSS. Since the research findings were not verified in multiple application areas, its generalizability needs to be further tested before conclusions can be made.

Limitation also exists in the sampling of both surveys in that the respondents came from a single university, restricting the applicability of the findings to other training contexts. In recent years, computer skills transfer research has moved beyond students and begun to focus on industry users, represented by a large group of research on ERP skills transfer (Coulson et al., 2004; Ko, Kirsch, & King, 2005). Although the principles remain the same between students and industry users, further validations would be necessary to generalize the results across training targets.

Additionally, the relatively small sample size in this research may limit its power. To explore the potential impact of the sample size on the power of the study, we conducted an ad hoc analysis in which we combined the data sets collected from both studies and re-tested the model in Figure 1 using repeated indicators of the CSS construct. The results are comparable: the path coefficient between CSS and CST is .461 (t=7.22), and the factor loadings are between .70 and .90. Even though this ad hoc analysis is not conclusive, it does suggest that the sample size may not be a major concern in the current study. Of course, further research is needed to test the model with larger sample sizes and across multiple application areas.

IMPLICATIONS FOR RESEARCH AND PRACTICE

This research has a number of implications. For academia, the structure and the components of the CSS construct establish a universal standard to measure software similarities in a variety of end-user computing fields. Researchers could use this standard as the basis to analyze skills transfer in different software domains, without struggling to develop context-specific measures. They may also use this standardized list to compare research results across software domains, for instance, from text editing to graphic editing, and from user-developed software to off-the-shelf software.

The CSS construct also provides a more comprehensive view on software attributes, allowing researchers to develop a holistic frame of reference (Shaw, Lee-Partidge, & Ang, 2003) of software applications to improve training. The frame of reference refers to a group of assumptions, expectations, and knowledge about applications that end-users use to evaluate new applications, which has a strong impact on user satisfaction. Such a frame of reference was not developed in the computer training area; scholars therefore could start with the three dimensions of the CSS construct and develop a reference system based on the functions, interface, and syntax features in order to evaluate or learn new applications.

Another implication comes from the development of the CST measurement scale. As mentioned earlier, previous studies suffered from the lack of common methods to measure skills transfer. With the help of the new, context-free instrument, it is possible to measure CST across multiple application contexts and compare the results directly. For instance, researchers may measure the extent to which a person's skills in electronic spreadsheet software (i.e., application *a* in Table 5) is transferred to other software such as accounting software or accounting modules in ERP systems (i.e., application *b* in Table 5). This increases the comparability of studies across contexts.

This research has practical meanings for computer trainers and IS educators. It provides a framework to rethink the traditional ways of training design and helps to develop better training programs to improve the process and performance. Prior to this study, scholars tended to apply some simplified criteria to judge the similarity between applications, such as OO programming versus non-OO programming (Urbaczewski & Wheeler, 2001). This study suggests that training design can be more effective if the functions, interface, and syntax used in the applications are considered simultaneously. For instance, compared to word processing software (such as *Microsoft Word* and *OpenOffice Writer*), electronic spreadsheet software (such as *Microsoft Excel* and *OpenOffice Calc*) is more similar to database software (such as *Microsoft Access* and *OpenOffice Base*) in terms of data manipulation functions, syntax and the corresponding interface, so that the skills learned in electronic spreadsheet software would be more helpful in learning database software (Kruck, Maher, & Barkhi, 2003). Using the three dimensions of CSS as the guidance, computer trainers could better manage the specific content to be covered and the sequence within the training programs (Agarwal et al., 2000).

For end users, a better understanding of software similarities in terms of the three dimensions provides guidelines to learn new applications by rationalizing their personal training programs. When a new software application is to be learned, the user should ask these questions: what functions are available in this application, compared to other applications that I have learned? How is the interface compared to other software interfaces? What underlying syntax is used in this application, or is it compatible with other applications? Answers to these questions help the learner to gain better knowledge of the features in a new software application, to prepare to re-use the features learned, and to focus on the new features that need to be explored. This would make it easier to master new applications so as to improve quality of use (Boudreau & Seligman, 2005).

Lastly, understanding computer skills transfer also helps system developers manage the system development process to design a system that is easy to learn and use. To do so, they need to put usability engineering (Carroll, 1997; Komischke & Burmester, 2000) into action and follow the user-centered development principles in order to design a system with needed features. The current research adds to the principles of user-centered design by directing attention to the three dimensions of CSS. It suggests that system developers should carefully choose the kinds of functions, interface, and syntax in their systems that are most compatible with the target systems that they replace, or the target users who may have had certain computer experience. To do so, the developers should prepare a usability specification (Carroll, 1997) by listing what functions are included, what interfaces are available, and what syntax are used for those functions and interfaces, and conduct a usability test (Komischke & Burmester, 2000) based on the features. A balanced view of these three aspects would help eliminate negative transfer in learning and enhance positive transfer. Meanwhile, system developers may also think about using these dimensions as a basis to prepare the business software specifications (Wang, 2005) for clients' references. Such specifications should clearly indicate functions, interface, and syntax in their systems in order to reduce ambiguity and facilitate skills transfer.

ACKNOWLEDGMENT

This research is partly funded by National Science Council at Taiwan (ROC) under the project NSC94-2420-H-155-030.

REFERENCES

Agarwal, R., & Karahanna, E. (2000). Time flies when you're having fun: cognitive absorption and beliefs about information technology usage. *Management Information Systems Quarterly*, *24*(4), 665–694. doi:10.2307/3250951

Agarwal, R., Sambamurthy, V., & Stair, R. M. (2000). The Evolving relationship between general and specific computer self-efficacy – An empirical assessment. *Information Systems Research*, *11*(4), 418–430. doi:10.1287/isre.11.4.418.11876

Besnard, D., & Cacitti, L. (2005). Interface changes causing accidents - An empirical study of negative transfer. *International Journal of Human-Computer Studies*, *62*(1), 105–125. doi:10.1016/j.ijhcs.2004.08.002

Bostrom, R. P., Olfman, L., & Sein, M. K. (1990). The importance of learning style in end-user training. *Management Information Systems Quarterly*, *14*(1), 101–119. doi:10.2307/249313

Boudreau, M. C., & Seligman, L. (2005). Quality of use of a complex technology: a learning-based model. *Journal of Organizational and End User Computing*, *17*(4), 1–22. doi:10.4018/joeuc.2005100101

Carroll, J. M. (1997). Human-computer interaction: psychology as a science of design. *Annual Review of Psychology*, *48*(1), 61–83. doi:10.1146/annurev.psych.48.1.61

Chin, W. W. (1998). Issues and opinion on structural equation modeling. *Management Information Systems Quarterly*, *22*(1), 7–16.

Compeau, D. R. (2002). The role of trainer behavior in end user software training. *Journal of End User Computing*, *14*(1), 23–32. doi:10.4018/joeuc.2002010102

Coulson, T., Zhu, J., Stewart, W., & Rohm, T. (2004). The importance of database application knowledge in successful ERP system training. *Communications of the IIMA, 4*(3), 95–122.

Falk, R. F., & Miller, N. B. (1992). *A Primer for Soft Modeling*. Akron, OH: The University of Akron.

Foltz, P., Davies, S. E., Polson, P. G., & Kieras, D. E. (1988). Transfer between menu systems. In E. Soloway, D. Frye, & S. B. Sheppard (Eds.), *Proceedings of the ACM SIGCHI Conference on Human Factors in Computing Systems*, Washington, DC (pp. 107-112).

Fornell, C., & Larcker, D. F. (1981). Evaluating structural equation models with unobservable variables and measurement error. *JMR, Journal of Marketing Research, 18*(1), 39–50. doi:10.2307/3151312

Hair, J. F., Anderson, R. E., Tatham, R. L., & Grablowsky, B. J. (1998). *Multivariate Data Analysis* (5th ed.). New York, NY: Prentice Hall.

Harvey, L., & Rousseau, R. (1995). Development of text-editing skills: from semantic and syntactic mappings to procedures. *Human-Computer Interaction, 10*(4), 345–400. doi:10.1207/s15327051hci1004_1

Jarvis, C. B., Mackenzie, S. B., & Podsakoff, P. M. (2003). A critical review of construct indicators and measurement model misspecification in marketing and consumer research. *The Journal of Consumer Research, 30*(2), 199–218. doi:10.1086/376806

Ko, D., Kirsch, L. J., & King, W. R. (2005). Antecedents of knowledge transfer from consultants to clients in Enterprise Systems implementations. *Management Information Systems Quarterly, 29*(1), 59–85.

Komischke, T., & Burmester, M. (2000). User-centered standardization of industrial process control user interface. *International Journal of Human-Computer Interaction, 12*(3-4), 375–386. doi:10.1207/S15327590IJHC1203&4_8

Kontogiannis, T., & Shepherd, A. (1999). Training conditions and strategic aspects of skill transfer in a simulated process control task. *Human-Computer Interaction, 14*(4), 355–393. doi:10.1207/S15327051HCI1404_1

Kruck, S. E., Maher, J. J., & Barkhi, R. (2003). Framework for cognitive skill acquisition and spreadsheet training. *Journal of End User Computing, 15*(1), 20–37. doi:10.4018/joeuc.2003010102

Mao, J., & Brown, B. R. (2005). The effectiveness of online task support vs. instructor-led training. *Journal of Organizational and End User Computing, 17*(3), 27–46. doi:10.4018/joeuc.2005070102

Moore, G. C., & Benbasat, I. (1991). Development of an instrument to measure the perception of adopting an information technology innovation. *Information Systems Research, 2*(3), 192–222. doi:10.1287/isre.2.3.192

Mykytyn, P. (2007). Educating our students in computer application concepts: a case for problem-based learning. *Journal of Organizational and End User Computing, 19*(1), 51–61. doi:10.4018/joeuc.2007010103

Okoli, C., & Pawlowski, S. D. (1995). The Delphi method as a research tool: An example, design considerations and application. *Information & Management, 42*(1), 15–29.

Olson, G. M., & Olson, J. S. (2003). Human-computer interaction: psychological aspects of the human use of computing. *Annual Review of Psychology, 54*(1), 491–516. doi:10.1146/annurev.psych.54.101601.145044

Podsakoff, P. M., MacKenzie, S. B., Lee, J., & Podsakoff, N. (1986). Common method biases in behavioral research: a critical review of the literature and recommended remedies. *The Journal of Applied Psychology, 99*(5), 879–903.

Polson, P. G., Bovair, S., & Kieras, D. (1987). Transfer between text editors. *ACM SIGCHI Bulletin, 18*, 27–32. doi:10.1145/1165387.30856

Polson, P. G., Muncher, E., & Engelbeck, G. (1986). A test of a common elements theory of transfer. In M. Mantei & P. Orbeton (Eds.), *Proceedings of the ACM SIGCHI Conference on Human Factors in Computing Systems,* Boston, MA (pp. 78-83).

Poston, R. S., & Royne, M. B. (2008). Rating scheme bias in e-commerce: preliminary insights. *Journal of Organizational and End User Computing, 20*(4), 45–73. doi:10.4018/joeuc.2008100103

Powell, A., & Moore, J. E. (2002). The focus of research in end user computing: where have we come since the 1980s? *Journal of End User Computing, 14*(1), 3–22. doi:10.4018/joeuc.2002010101

Salas, E., & Cannon-Bowers, J. A. (2001). The science of training: A decade of progress. *Annual Review of Psychology, 52*(1), 471–499. doi:10.1146/annurev.psych.52.1.471

Schmidt, R. C. (1997). Managing Delphi surveys using nonparametric statistical techniques. *Decision Sciences, 28*(3), 763–774. doi:10.1111/j.1540-5915.1997.tb01330.x

Scholtz, J., & Wiedenbeck, S. (1990). Learning second and subsequent programming languages: A problem of transfer. *International Journal of Man-Computer Interaction, 2*(1), 51–72. doi:10.1080/10447319009525970

Schuelke, M. J., Day, E. A., McEntire, L. E., Boatman, P. R., Boatman, J. E., Kowollik, V., & Wang, X. (2009). Relating indices of knowledge structure coherence and accuracy to skill-based performance: Is there utility in using a combination of indices? *The Journal of Applied Psychology, 94*(4), 1076–1085. doi:10.1037/a0015113

Sein, M. K., Bostrom, R. P., & Olfman, L. (1999). Rethinking end-user training strategy: applying a hierarchical knowledge-level model. *Journal of End User Computing, 11*(1), 32–39.

Shaw, N., Lee-Partidge, J., & Ang, J. S. K. (2003). Understanding the hidden dissatisfaction of users toward end-user computing. *Journal of End User Computing, 15*(2), 1–22. doi:10.4018/joeuc.2003040101

Shayo, C., & Olfman, L. (1998). The role of conceptual models in formal software training. In S. Poltrock & J. Grudin (Eds.), *Proceedings of the 1998 ACM Conference on Computer Personal Research (SIGCPR),* Seattle, WA (pp. 242-253).

Singley, M. K., & Anderson, J. R. (1985). The transfer of text-editing skill. *International Journal of Man-Machine Studies, 22*(4), 403–423. doi:10.1016/S0020-7373(85)80047-X

Singley, M. K., & Anderson, J. R. (1988). A keystroke analysis of learning and transfer in text editing. *Human-Computer Interaction, 3*(3), 223–274. doi:10.1207/s15327051hci0303_2

Smelcer, J. B., & Walker, N. (1993). Transfer of knowledge across computer command menus. *International Journal of Human-Computer Interaction, 5*(2), 147–165. doi:10.1080/10447319309526062

Spitler, V. (2005). Learning to use IT in the workplace: mechanisms and masters. *Journal of Organizational and End User Computing, 17*(2), 1–25. doi:10.4018/joeuc.2005040101

Teasley, B. E. (1994). The effects of naming style and expertise on program comprehension. *International Journal of Human-Computer Studies, 40*(5), 757–770. doi:10.1006/ijhc.1994.1036

Tojib, D. R., & Sugianto, L. F. (2007). The development and empirical validation of the B2E portal user satisfaction (B2EPUS) scale. *Journal of Organizational and End User Computing, 19*(3), 43–63. doi:10.4018/joeuc.2007070103

Urbaczewski, A., & Wheeler, B. C. (2001). Do sequence and concurrency matter? An investigation of order and timing effects on student learning of programming languages. *Communications of the AIS, 5*(1).

Vandenbosch, B., & Higgins, C. (1996). Information acquisition and mental modes: An investigation into the relationship between behavior and learning. *Information Systems Research, 7*(2), 198–214. doi:10.1287/isre.7.2.198

Wang, S. (2005). Business software specifications for consumers: towards a standard format. *Journal of Organizational and End User Computing, 17*(1), 23–37. doi:10.4018/joeuc.2005010102

Wetzels, M., Odekerken-Schroder, G., & van Oppen, C. (2009). Using PLS path modeling for assessing hierarchical construct models: guidelines and empirical illustration. *Management Information Systems Quarterly, 33*(1), 177–195.

Wilson, B., & Henseler, J. (2007). Modeling reflective higher-order constructs using three approaches with PLS path modeling: a Monte Carlo comparison. In M. Thyne, K. R. Deans, & J. Gnoth (Eds.), *Proceedings of the Australian and New Zealand Marketing Academy Conference* (pp. 791-800).

Yamnill, S., & McLean, G. N. (2001). Theories supporting transfer of training. *Human Resource Development Quarterly, 12*(2), 195–208. doi:10.1002/hrdq.7

Ye, N. (1991). *Development and Validation of a Cognitive Model of Human Knowledge System: Toward an Effective Adaptation to Differences in Cognitive Skills*. Unpublished doctoral dissertation, Purdue University, Lafayette, IN.

Yi, M. Y., & Im, K. S. (2004). Predicting computer task performance: personal goal and self-efficacy. *Journal of Organizational and End User Computing, 16*(2), 20–37. doi:10.4018/joeuc.2004040102

This work was previously published in the Journal of Organizational and End User Computing, Volume 23, Issue 3, edited by Mo Adam Mahmood, pp. 48-66, copyright 2011 by IGI Publishing (an imprint of IGI Global).

Chapter 7

Preparing IS Students for Real-World Interaction with End Users Through Service Learning:
A Proposed Organizational Model

Laura L. Hall
University of Texas at El Paso, USA

Roy D. Johnson
University of Pretoria, South Africa

ABSTRACT

Although teaching the technical skills required of Information Systems (IS) graduates is a straightforward process, it is far more difficult to prepare students in the classroom environment for the challenges they will face interacting with end users in the real world. The ability to establish a successful relationship with end users is a critical success factor for any IS project. One way to prepare students for interaction with end users is through the implementation of service learning projects. Service learning projects provide a rich environment for students to experience real world interactions with users. This paper presents an organizational model to guide the implementation of service learning projects in IS curriculums. Service learning projects better prepare students to assume important management positions by giving them experience in applying the system development life cycle to an IS project and working with people. This organizational model uses the system development life cycle approach to integrate typical curriculum and service learning models. The organizational model is grounded in anecdotal evidence from prior experiences with IS students in service learning environments.

INTRODUCTION

It is of tremendous importance for Information Technology (IT) professionals to establish successful relationships with end users. Difficulties in working with end users are well documented (Perrin, 2007; Summerfield, 2006). End users often view IT professionals as enemies and are distrustful of them. IT professionals have a reputation for being, "aloof, geeky, and non-communicative" (Dubie, 2007, p. 1). Leonard (2000, p. 492) described the relationship between

DOI: 10.4018/978-1-4666-2059-9.ch007

IT professionals and end users as, "intriguing, complex, and should be seen and managed as multi-dimensional entities." It is therefore very difficult in a controlled classroom environment to prepare IT students to interact with end users in a real-world environment. The multidimensional aspects of the real world cannot be duplicated in a traditional classroom setting. One way to help prepare students for these challenges is through the implementation of service learning programs in Information Systems (IS) curriculums. By implementing service learning projects, students can be exposed to the mechanisms of working directly with end users and to be better prepared for the challenges they will encounter as IT managers.

Service learning is also known as collaborative learning, cooperative learning, or community-based learning. Although closely related, there are subtle differences in the meaning of these terms. The National Service Learning Clearinghouse (2008, p. 1) defines Service learning as, "a teaching and learning strategy that integrates meaningful community service with instruction and reflection to enrich the learning experience, teach civic responsibility, and strengthen communities." Collaborative learning refers to a variety of approaches in education that involve joint ventures by students or students and faculty (Smith & MacGregor, 1992). Cooperative learning refers to teaching strategies where teams increase understanding of subject material by implementing learning activities. Students typically have different levels of ability and team members cooperate by helping other team members learn. The difference between cooperative learning and collaborative learning is that in a cooperative learning environment each person is responsible for a portion of a team's work. In collaborative learning environments, participants work together to solve a problem. Community-based learning may be collaborative or cooperative but specifically refers to the type of project. In community-based learning, students, faculty and community focus on solving a pressing community problem or effecting social change

(Strand, Marullo, Cutforth, Stoecker & Donohue, 2003). Service learning has a strong focus on the reflection aspects of the process.

All of these approaches focus on action learning and were popularized by John Dewey (1859-1952), a well-known educator and philosopher. He was considered an unconventional educator and avoided traditional ideas of instructional methods of the day. Dewey believed that focusing on community based projects and applying a strong emphasis on citizenry would result in social and educational reforms (Ryan & Cooper, 1998). In the 1980s, a national strategy was initiated to promote civic responsibility and volunteerism at which time Universities began funding and staffing service learning programs (Elsner, 2000). Often these programs included training and allocation of resources to support service learning projects. Some campuses made service learning projects mandatory for specific courses (Elsner, 2000).

By the 1990s, the National and Community Service Act established funding guidelines for school based service learning and the Corporation for National Services (CNS) was created (Scales & Roehlkepartain, 2004). The CNS linked national service with education. In 1999, the W. K. Kellogg Foundation established a four-year, $13,000,000 investment into service learning for K-12. By 2004, over 10% of all K-12 public school students and 28% of all K-12 public institutions were involved in some type of service learning, reaching approximately 4.7 million K-12 students in 23,000 public schools (Scales & Roehlkepartain, 2004).

In the last several years, service learning has captured attention at all levels of education. A 2000 study by Billig showed that 64% of all public schools and 83% of all public high schools had implemented service learning projects. In addition, statistics from the report on *Community Service in K-12 Public Schools* by the National Center of Education Statistics (1999) show that approximately one third of all schools and half of public high schools had invested in programs that link service learning projects with school curriculums.

At the University level, in a 2008 survey by Campus Compact, over 1100 campuses showed that 31% of faculty offered service learning courses and over 24,471 service learning courses were offered. Eighty-five percent of college freshmen reported having experienced service learning projects in high school (Kielsmeier, 2000). The most visible advancements of service learning policy have come in the form of state or district level service mandates. Some states have adopted an hours-based or project-based service requirement as a way to interest students in service learning.

Given the need of local community groups for assistance with technology and students who are in need of real-world experiences that relate to their course material, service learning projects provide a better understanding of user issues by providing an opportunity to work with real users (Lazar & Norcio, 2000). By getting real-world experience, students can get a sampling of the ethical and political issues that can occur in a workplace (Lazar & Norcio, 2000). Through their real-world projects, students may also make contacts, and develop professional networks that can help them in their careers (Shneiderman, 1998).

In addition, students frequently do not grasp the complexity of implementing an information system from start to finish. By going through the complete process of the system development lifecycle, working with users to gather requirements, developing a system, and then implementing that system, students can experience firsthand the level of complexity involved in implementing a system from start to finish.

LITERATURE REVIEW

Technology trends (e.g., use of the Internet, wireless capabilities, availability of specialized software, accessibility of websites designed for user interaction, increased computer literacy, and mobile applications) have given IT professionals a much broader environment to manage than just a few years ago. Clark (2006, p. 66) notes important computer trends including expanding the variety of use, "related to mobile computing and alternative office arrangements".

End user computing covers a wide range of environments, which may run from unofficial developments by an autonomous end user to integrating several levels of an organization. Regardless of the depth and role of end using computing in the organization, IT professionals must be capable of efficient communication across all levels. The Information System Specialist serves as an interpreter between the user and the computer system. Figure 1 illustrates the important role that Information Specialists play in the relationship between the users and computers.

The end user relationship with IT professionals is an important determinant of success in planning, designing, and implementing IS projects (Leonard, 2000). Brennan (2009, p. 69) states that while, "potential benefits of information systems that provide greater access and immediacy, businesses tend to focus on the technical challenges of the implementation and on the first-level effects of the improvements." The relationship is integral for the successful communication and collaboration efforts that must take place between the stakeholders. The success or failure of these relationships depends on IT professionals to manage

Figure 1. Role of the information system specialist

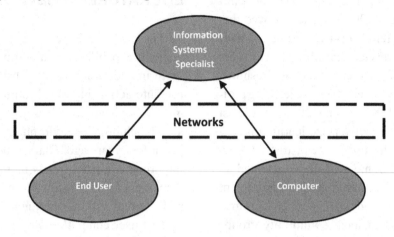

this relationship. End users have described IT professionals as rude arrogant, patronizing (Bohlmann, 2007), secretive, the enemy, aloof, geeky and non-communicative (Dubie, 2007). End user relationships have been defined in terms of physical and abstract dimensions (Leonard, 2000). The physical dimension focuses on elements, which enable contact between the developer and end users. The abstract dimension includes the "softer side" such as sensitivity to change, supportive culture, co-operation, commitment, holistic nature, dynamicism, knowledge base, and sustainability (Leonard, 2000).

The majority of research articles concerning service learning appear in the education literature although application papers appear in a myriad of disciplines. Shelley H. Billig (2000) of RMC

Research Corporation developed an important brief as part of the W.K. Kellogg Foundation's Learning In Deed Initiative. The brief categorizes service learning (Figure 2) into six major areas of impact on 1) student personal and social development, 2) civic responsibility, 3) student academic learning, 4) career exploration and aspirations, 5) schools, and 6) communities. Many studies exist that support this framework.

Research by Astin, Vogelgesang, Ikeda and Yee (2000) show service learning programs have significant positive effects on student GPA, writing and critical thinking skills, commitment to activism, diversity understanding, interpersonal and leadership skills, and choice of a more service oriented career.

Figure 2. Categories of service learning impact

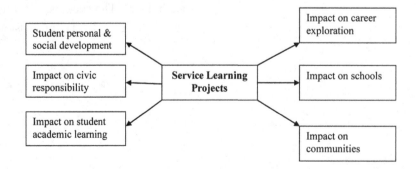

Research in marketing has also shown positive impacts for students, clients, instructors and Universities involved in service learning projects (Clark & Kaminski, 1986; de los Santos & Jensen, 1985).

In the field of information systems, there is a plethora of applied research in the implementation of service learning programs. Information systems are particularly well suited for service learning projects because there is a widespread need of technological resources in community-based organizations, non-profits, and school systems. There is a serious lack of resources to support these technology needs. These groups have little or no budget, are seriously understaffed and personnel are seriously over-extended. Often school systems cannot afford training programs or the qualified people to manage their technology (Lazar & Norcio, 2000).

Kesling (1989) describes similar projects where students work with community organizations to recommend computing equipment. He reported very favorable feedback from both students and the organizations. Harris (1994) presents an outline of the mechanics for a capstone systems development course, which incorporates service learning projects. His findings consider the course the most important course in the IS major's program and that the course strengthened oral and written communication skills. Cougar (1995) also emphasized the importance of this type of course in improving creativity in systems analysis and design. He stated that the highest level of knowledge or understanding attainable in an academic setting is that of application and encourages applying creativity techniques in a comprehensive systems analysis and design class project, such as one for a local company. Schuldt (1994) also discusses the importance of real world projects in a systems development course. He comments on the students' perceived value of the project and the potential for creating a positive university-community relationship. Other studies have addressed issues such as team effectiveness,

performance, and satisfaction (Werner & Lester, 2001) and the importance of peer review (Dyrud, 2001). Tanniru and Agarwul (2002) examined student team project success factors such as project complexity, team experience, and coordination and stress the importance of providing opportunities for students to work on real projects with real companies.

Additional topics such as group dynamics, interpersonal and team skills as they apply to technology have been addressed in the literature (Jessup, 1991, 1992; Rooney, 2000). Studies specifically focusing on team projects have examined issues such as managing interpersonal conflict (Barki & Hardwick, 2001) and cross-functional teams (McDonough, 2000; Sethi, Smith & Whan, 2000), but these studies focused on professionals rather than students. One study of graduate students involved in a software project examined perceptions of their contribution to the project (Rajlich, Syed, & Martinez, 2000). Watson, Johnson, and Merritt (1998) examined various issues pertaining to diversity in student teams, but not with a systems development project (Astin et al., 2000). Although these studies have provided applied applications of service learning in the classroom, they have not proposed a framework to organize the process of planning, designing, and implementing service learning projects. The following proposed organizational model provides a framework to organize these phases of development for service learning projects in IS classrooms.

CURRICULUM MODELS

Curriculum models are generally of two types, process models and content models. A process model describes the important phases within a process. A process model provides a framework of how things should or could proceed throughout a project. The process model should estimate what the process will actually entail (Rolland & Pernici,

Figure 3. Process model

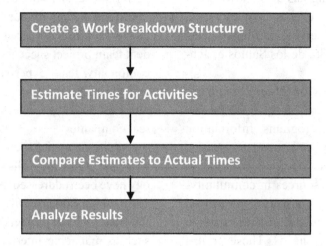

1998). An example of a process model of a project management assignment is seen in Figure 3.

Content models are specific and describe the actions that should be included in each phase of the model. A service learning content model would include the elements of the service activity as well as a reflection phase. A typical service learning model is shown in Figure 4.

ORGANIZATIONAL MODEL

The following proposed organizational model (Figure 5) uses the system development life cycle approach to integrate curriculum development models with service learning development approaches. The purpose of this integration is to organize a "how-to" framework for implementing service learning projects into IS curriculums. The model is important because it provides a structured mechanism for faculty to follow in order

Figure 4. Typical content model

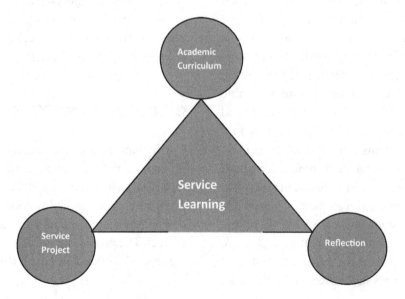

Figure 5. Organizational model

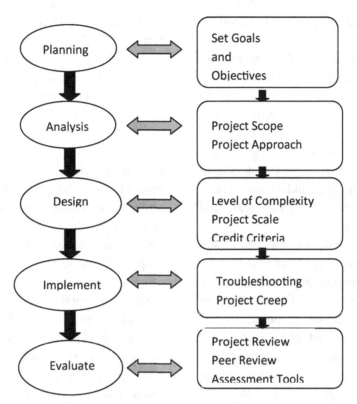

to implement service learning programs in their courses. Using the system development life cycle structure reinforces the processes needed for any IS project. This model combines the structure of the process models in curriculum development literature and the content models of the service learning literature. The model documents five major phases of implementing service learning projects. It is important to consider the iterative characteristics of the model. Projects that are organized around these five phases have been successful in many instances.

PLANNING

Set Goals and Objectives

The first action in the planning phase of the proposed model is to set the goals and learning objectives for the course. Goals and learning objectives (often called performance objectives or competencies) are brief, clear, specific statements of what learners will be able to perform at the conclusion of instructional activities. It is important to articulate goals and objectives for many different reasons. From a student's perspective, goals and objectives help them understand what is expected from them and the focus of the course. Goals and objectives outline standards and expectations for the course and communicate the thinking of the professor to the students. Well thought out goals and objectives give the students direction and certainty about what will be important concepts and developments throughout the course.

From a faculty perspective, the goals and objectives of the course clarify what is expected from the students and what skills, competencies, and abilities they are expected to develop throughout the course. Goals and objectives are also useful for the development of assessments by identifying the types of tasks that students need to produce to demonstrate understanding. Goals and objectives should drive curriculum planning and can guide the formation of instructional activities. They are also useful as a basis for a framework to evaluate understanding and progress of students. The goals and objectives should serve as a guide for the development of service learning projects. Once the goals and objectives for the class have been established, projects should be planned and developed to encompass the stated goals and objectives of the course.

From an organizational perspective, goals and objectives are important to create a framework for evaluating overall effectiveness of an educational program and provide evidence of student learning to be utilized by accreditation agencies.

Important sources of goals and objectives for information system courses can be found in "What every Business Student Needs to Know About Information Systems" (Ives et al., 2002) and "Implementing the Seven Principles: Technology as Lever" (Chickering & Ehrmann, 1996). These key articles outline basic knowledge that students in information systems courses should know.

These goals and objectives are generalized for most IS courses and but can serve as a guide for developing more specific goals and objectives. Figure 6 shows examples of goals from a service learning based Expert Systems course at the University of Texas at El Paso. The students in this course build robots and demonstrate their creations in local elementary schools.

ANALYSIS

Project Scope

The first decision to make concerning implementation of a service learning project is the scope or breadth of the project. Service learning projects run the gamut from optional, low impact projects to full credit high impact projects. A good heuristic for novices is to begin with small projects and work into the larger ones. Scope must be written down in the project documentation and agreed to with all parties affected by the project. It is critical to set the scope of the projects during planning. The scope defines the beginning and ending of the project and outlines what will be included in the project. Project scope should not be generalized and should clearly specify the task and the timeline for the project. All participants in the project should know their job responsibility and be aware of what is included in the project and what is not included. Confusion over the terms or scope of the project can lead to problems for the project team and dissatisfaction with the outcome of the project by the end users. The project scope serves as the boundary that defines the limits of the tasks involved in the project. Many times there is confusion about what falls inside the boundary of a specific project and what does not. Developing a solid project scope and socializing it with the project team, sponsors and key stakeholders is critical. Research with 327 project improvement teams showed that the definition of scope and objectives was in the top three most important start-up activities (BPR Online Learning Center, 2010).

A common mistake made by project designers is to define their project scope in general terms. The lack of definition causes key stakeholders to make assumptions related to their own processes or systems falling inside or outside of the scope of the project. Then later, after significant work has been completed by the project team, some managers are surprised to learn that their assumptions were not correct, resulting in problems for the project team.

In the Expert Systems course at the University of Texas at El Paso, the curriculum is project based, meaning that the curriculum is designed around the project. The project consists of student teams who build robots to demonstrate in local elementary classrooms. Students, in groups of five, assume

Figure 6. CIS 4330 expert systems goals and objectives

COGNITIVE SKILLS
Objectives:
● To develop a systems view of the role of AI in technology
● To understand the metaphor of brains and computers
● To conceptualize directions in technology for the future
● To synthesize material from a variety of sources
ANALYTICAL SKILLS
Objectives:
● To judge reliability, integrity, and accuracy of sources of information
● To perform statistical analysis on data sets
● To design, analyze, and implement complete projects
● To evaluate peer performance and presentation of information
TECHNICAL SKILLS
Objectives:
● To learn basics of robotic construction
● To communicate electronically in both synchronous and asynchronous environments
● To develop a high level of technical and creative writing skills
● To design and develop a full multimedia presentation
CREATIVE SKILLS
Objectives:
● To strengthen creative writing skills
● To create effective, informative presentations to the appropriate audience
● To communicate in cyberspace
● To network electronically
PROFESSIONAL SKILLS
Objectives
● To develop impression management skills
● To demonstrate professional presentation skills
● To create and manage contacts

roles for which they must submit resumes and apply for positions. The roles include a Team Leader, Project Manager, Robotics Engineer, Presentation Manager, and Multimedia Developer. Each group uses the system development life cycle and uses Pert networking techniques to plan and schedule their projects. Classroom topics are organized in modules and modules are available just-in-time as students need the knowledge. For example, the beginning module is a Project Management module and focuses on Pert networking applied to information systems. The module prepares the group for planning and estimating time and costs of their classroom projects.

Within student groups, scope of the classroom project is set between the teacher in the classroom and the student groups. Groups must communicate the scope of the project with teachers through meetings and discussions. Establishing this contact and communication channels are often the first time that students have engaged in professional interactions.

Project Approach

There are three primary approaches to planning a service learning based curriculum. By far the simplest is to recreate documented projects. In this approach, it is not necessary to reinvent the wheel. Use existing examples and follow the methods and practices already developed. Many applied research projects provide sufficient documentation to recreate projects.

Another approach is to modify existing projects. This is a good way to customize a course for the students, curriculum, or environment. The advantage of this approach is that a faculty member is able to target student weaknesses or focus on remedial areas in planning the service learning activities. Existing projects are an excellent source for ideas to apply to course goals and objectives. Assessment tools can then be adapted to the needs of the course.

The most difficult approach is to create a project from scratch. This is necessary in circumstances in which no suitable similar projects can be found. Given the room for creativity and innovation in planning service learning curriculums it may be necessary to create individualized course requirements to accomplish the goals and objectives of the course.

DESIGN

Level of Complexity

Time increases complexity in any task. An important consideration in choosing a service learning project is the time frame in which the project must occur. A common mistake in projects is to schedule too much work. The project must be completed over the time frame available for the project. This can be achieved through the planning process for the project. Another important consideration is the level of expertise of the faculty member. Inexperienced faculty members should begin with very controlled environments.

Project Scale

Projects can range from very simple to very complex projects. Very large projects can be very difficult to organize, manage, and implement. The work outputs are less clear the longer the time line and it is harder to make accurate estimates for effort, duration and cost. Large projects are usually long-term as well and that makes them very difficult to plan successfully. Conditions can change over time and it is hard to maintain organizational enthusiasm and support over long periods of time. It is also very difficult to predict resource requirements and availability far into the future. Long-term projects can be broken into separate smaller projects based on the life cycle. One semesters' project may consist of projects, which address the Planning Phase of the system development life cycle. The next semesters' students can pick up at the Analysis Phase, and the next at the Design Phase, until the long-term project is completed.

Set Credit Criteria

The service learning projects can range from a very simple project to the main focus of the semester. The service learning projects can be required (as a stated part of the course grading), or they can be optional. For an optional service learning project, the students can substitute a service learning project for other course requirements, or the students can earn an extra credit ("the for-credit option") by taking part in a service learning experience (Sanderson & Vollmar, 2000). Although service-learning is traditionally used in conjunction with classroom-based courses, service learning could also be implemented as a co-op or internship program. Students can apply their knowledge from a number of different classes to assist a non-profit organization with their technology needs. Of particular value are internships, which allow students to be involved in the complete process of analyzing, designing, implementing, delivering, and maintaining a system. Alternatively, students can learn a great deal from in-depth immersion in one or more of these development phases. In some universities where service learning is a frequent occurrence, there is a designated service learning coordinator who can help find appropriate community partners (Sanderson & Vollmar, 2000). At other universities there might be community service coordinators, an entire office of community service, or an office of community partnerships, which can be a valuable resource for the faculty member looking for community partners (Lamb et al., 1998). These offices not only assist those who are leading service learning classes, but also promote the concept of service learning within the organization (Lamb, Swinth, Vinton & Lee, 1998). In addition, some colleges of business have the infrastructure in place for faculty to work with local for-profit and non-profit organizations.

IMPLEMENT

Troubleshooting

Troubleshooting is an ongoing process, which extends throughout the life of the project. It is the faculty member's responsibility to monitor service learning projects through the development process to head off problems that may occur. Troubleshooting begins with checking and rechecking the plans that have been put in place. Plans developed by students should be carefully scrutinized and examined for barriers and pitfalls. Deliverables should be carefully planned in order to monitor the progress of the project. It is also critical to establish communication between team leaders and faculty members. Students must be encouraged to share information before problems arise. Common problems that may occur during a service learning project include finding appropriate community partners and motivating students. Experience will also aid in anticipating areas where problems commonly occur.

Project Creep

Project creep occurs when a projects scope is overrun by additional unplanned requirements from the end users. This is a common occurrence and can be difficult to cope with for students. In a service learning project at the University of Texas at El Paso, a group of students planned a robotic demonstration for a fourth grade class of 20 students. Other teachers quickly asked if their students could attend the presentation and before it was over the students presented to over 200 fourth graders. Planning a project for 20 and extending it to over 200 is an example of how project creep can easily occur.

Common causes of project creep include poor planning for the scope of the project, lack of understanding and communication with end users, and underestimating a projects' level of complexity (Haughey, 2009). Service learning projects require adherence to the agreement of the scope of the project and any changes should be properly documented, considered, reviewed, approved, and resourced.

EVALUATE

The evaluation phase of the model is a critical part of the process. Evaluation is the mechanism for measuring the impact and effectiveness of the service project to all stakeholders in the project. The evaluation for service learning projects should be multidimensional. Evaluations are for service learning projects are often multi-layered in order to measure the results from many different perspectives. An increasingly important dimension for evaluation of service learning programs is for accreditation purposes. The focus of current accreditation bodies is to determine whether or not programs are meeting their stated goals.

Project Review

Service learning projects can be helpful in program assessment by showing how well students are able to perform in real-life situations. Real-world projects by students can demonstrate that students, upon completing a program of study, have mastered the subject material and are able to apply their knowledge in a real-world setting to solve real-world problems. Successfully completed service learning projects can be an additional outcome measure to evaluate the effectiveness of the curriculum in an IS program. For instance, many universities perform internal evaluations of their academic programs on a regular basis. These service learning projects provide assessment of students' individual and group work, which provides important feedback on the effectiveness of an academic program.

Peer Review

Peer review is important within the evaluation schema. Peer review allows students to set evaluation standards and to develop effective measurement of those standards. Brown, Rust and Gibbs (1994) found that owning the assessment process motivated students. At the University of Texas at El Paso in CIS 4330, 70% of the overall project grade is based on peer review by team members. Team members must evaluate each team member as well as the project and the course. The evaluations are based on measurements determined in the course objectives.

Assessment Tools

The evaluation process should flow naturally from the learning objectives developed in the beginning stage of the model. The evaluations may be either quantitative, such as service learning logs, hours of service, number of people served, gains in test scores, or surveys, or qualitative, such as observation, reflection logs, or interviews with key participants. The evaluation process may also

be a combination of quantitative and qualitative methods (University of Southern California, n.d.). If objectives are not met, then it is important to carefully review the structure of the course, the quality of the placements and the interplay between course material and community experience to see what needs modification.

CONCLUSION

Working with end users is one of the greatest challenges that IT professionals face. Developing information systems based service learning projects for IS curriculums are a mechanism for preparing future IT professionals for the challenges they will most certainly encounter in their professional careers. Experience in applying the system development life cycle to a real world project and experience in collaborating and communicating with end users will prepare students for those very same activities in their careers. Organizations will benefit by having IT professionals with experience dealing with end users. Many students stated that when asked by interviewers to describe a team project that they were involved in reported that they learned the most from their service learning experiences. Service learning is a useful technique for incorporating real-world experiences into the curriculum. The proposed organizational model offers a framework to organize, plan, and implement service learning projects in information-systems based curriculums. The importance of this model is to provide a how-to approach for educators with a methodology to begin creating service learning components to their curriculums. The model is important to IS educators who want to provide students with the competitive advantage of real-world experience.

This paper contributes to the literature by providing an integrated model of current process and control models structured around service learning. The advantages of this model is that it provides a structured approach with the flexibility and agility needed for implementation of IS projects.

REFERENCES

Astin, A. W., Vogelgesang, L. J., Ikeda, E. K., & Yee, J. A. (2000). *How service learning affects students*. Los Angeles, CA: Higher Education Research Institute.

Barki, H., & Hartwick, J. (2001). Interpersonal Conflict and its Management in Information System Development. *Management Information Systems Quarterly*, *25*(2), 195–228. doi:10.2307/3250929

Billig, S. H. (2000). Research on K-12 school-based service-learning: The evidence builds. *Phi Delta Kappan*, *81*, 658–664.

Bohlmann, M. (2007). *Managing relationships with end-users*. Retrieved March 22, 2010, from http://bmighty.informationweek.com/network/showArticle.jhtml?articleID=202200206

BPR Online Learning Center. (2010). *Module 2: The importance of scope and objectives*. Retrieved March 11, 2010, from http://www.prosci.com/tutorial-project-plan-mod2.htm

Brennan, L. L. (2009). A strategic approach to IT-enabled access and immediacy. *Journal of Organizational and End User Computing*, *21*(4), 63–72. doi:10.4018/joeuc.2009062604

Brown, J. D. (1995). *The elements of language curriculum: A systematic approach to program development*. Boston, MA: Heinle & Heinle.

Brown, S., Rust, C., & Gibbs, G. (1994). *Strategies for Diversifying Assessments in Higher Education*. Oxford, UK: Oxford Centre for Staff Development.

Campus Compact. (2008). *Service Statistics 2008*. Retrieved March 22, 2010, from http://www.compact.org/wp-content/uploads/2009/10/2008-statistics1.pdf

Carol, C. (2006). End user computing ergonomics: Facts or Fads? *Journal of Organizational and End User Computing*, *18*(3), 66–76. doi:10.4018/joeuc.2006070104

Chickering, A. W., & Ehrmann, S. C. (1996). Implementing the Seven Principles: Technology as Lever. *AAHE Bulletin*, 3-6.

Clark, G. L., & Kaminski, P. (1986). Bringing a reality into the classroom: The team approach to a client-financed marketing research project. *Journal of Education for Business*, *62*(2), 61–65.

Cougar, J. D. (1995). Implied creativity no longer appropriate for IS curriculum. *Journal of IS Education*, *7*(1), 12–13.

de los Santos, G., & Jensen, T. D. (1985). Client-sponsored projects: Bridging the gap between theory and practice. *Journal of Marketing Education*, *7*(2), 45–50. doi:10.1177/027347538500700207

Dubie, D. (2007). Can IT and end users get along? *Network World*. Retrieved March 11, 2010, from http://www.networkworld.com/news/2007/121207-it-end users-relationships.html

Dyrud, M. A. (2001). Group projects and peer review. *Business Communication Quarterly*, *64*(4), 106–112. doi:10.1177/108056990106400412

Elsner, P. (2000). *Civic responsibility and higher education*. Phoenix, AZ: Oryx Press.

Gartner Research. (2001). *U.S. Syposium/ITxpo 2001*. Retrieved May 15, 2010, from http://www.courtesycomputers.com/Networking/enduser%20computing%20best%20practices.pdf

Harris, A. L. (1994). Developing the systems project course. *Journal of Information Systems Education*, *6*(4), 192–197.

Haughey, D. (2009). *Stop scope creep running away with your project*. Retrieved October 15, 2008, from http://www.projectsmart.co.uk

Ives, B., Valacich, J., Watson, R. T., Zmud, R., Alavi, M., & Baskerville, R. (2002). What every business student needs to know about information systems. *Communications of the AIS, 9*, 467–477.

Jessup, H. (1991). A model for workteam success. *Journal for Quality and Participation, 14*(3), 70–74.

Jessup, H. (1992). The road to results for teams. *Training & Development, 46*(9), 65–68.

Kesling, G. (1989). A community project approach to teaching management information systems. *Journal of Education for Business*, 341–344. doi:10.1080/08832323.1989.10117386

Kielsmeier, J. C. (2000). A time to serve, a time to learn: Service learning and the promise of democracy. *Phi Delta Kappan, 81*, 652–657.

Lamb, C. H., Swinth, R. L., Vinton, K. L., & Lee, J. B. (1998). Integrating service learning into a business school curriculum. *Journal of Management Education, 22*(5), 637–655. doi:10.1177/105256299802200506

Lazar, J., & Norcio, A. (2000). Service-research: Community partnerships for research and training. *Journal of Informatics Education and Research, 2*(3), 21–25.

Lazar, J., & Preece, J. (1999). Implementing service learning in an online communities course. In *Proceedings of the International Academy for Information Management 1999 Conference* (pp. 22-27).

Leonard, A. C. (2000). The importance of having a multidimensional view of IT-end user relationships for the successful restructuring of IT departments. In Khosrow-Pour, M. (Ed.), *Challenges of Information Technology Managers in the 21st Century* (pp. 492–495). Hershey, PA: Idea Group.

Margevicius, M. (2001). *End user computing best practices.* Retrieved January 5, 2010, from http://www.courtesycomputers.com/Networking/enduser%20computing%20best%20practices.pdf

McDonough, E. F. III. (2000). Investigation of factors contributing to the success of cross-functional teams. *Journal of Product Innovation Management, 17*(3), 221–235. doi:10.1016/S0737-6782(00)00041-2

McLeod, R. Jr, & Schell, G. P. (2000). *Management information systems* (8th ed.). Upper Saddle River, NJ: Pearson-Prentice Hall.

National Center for Education Statistics. (1999). *Community Service in K-12 Public Schools.* Washington, DC: U.S. Department of Education.

National Service Learning Clearinghouse. (2008). *What is service-learning?* Retrieved February 10, 2010, from http://www.servicelearning.org/what-service-learning

Perrin, C. (2007). Work with end users -- Not against them -- To improve security. *TechRepublic.* Retrieved March 11, 2010, from http://blogs.techrepublic.com.com/security/?p=290

Rajlich, V., Syed, W. A., & Martinez, J. (2000). Perceptions of contribution in software teams. *Journal of Systems and Software, 54*(1), 61–63. doi:10.1016/S0164-1212(00)00026-1

Rolland, C., & Pernici, C. T. (1998). A comprehensive view of process engineering. In *Proceedings of the 10th International Conference CAiSE'98* (LNCS 1413, pp. 1-24).

Rooney, P. (2000). Constructive Controversy: A new approach to designing team projects. *Business Communication Quarterly, 63*(1), 53–61. doi:10.1177/108056990006300106

Ryan, K., & Cooper, J. (1998). *Those who can, Teach.* Boston, MA: Houghton Mifflin.

Sanderson, P., & Vollmar, K. (2000). A primer for applying service learning to computer science. In *Proceedings of the ACM Conference on Computer Science Education (SIGCSE)* (pp. 222-226).

Scales, P. C., & Roehlkepartain, E. C. (2004). *Community service and service-learning in U.S. public schools, 2004: Findings from a National Survey*. St. Paul, MN: National Youth Leadership Council. Retrieved March 11, 2010, from http://www.nylc.org/objects/inaction/initiatives/2004G2G/2004G2GCompleteSurvey.pdf

Schuldt, B. A. (1991). 'Real-world' versus 'simulated' projects in database instruction. *Journal of Education for Business, 67*(1), 35–39. doi:10.1080/08832323.1991.10117514

Sethi, R., Smith, D. C., & Whan, C. (2000). Cross-functional product development teams, creativity, and the innovativeness of new consumer products. *JMR, Journal of Marketing Research, 38*(1), 73–85. doi:10.1509/jmkr.38.1.73.18833

Shneiderman, B. (1998). Relate-Create-Donate: a teaching/learning philosophy for the cyber-generation. *Computers & Education, 31*, 25–39. doi:10.1016/S0360-1315(98)00014-1

Smith, B. L., & MacGregor, J. T. (1992). *What is Collaborative Learning?* University Park, PA: National Center on Postsecondary Teaching, Learning, and Assessment, Pennsylvania State University.

Strand, K., Marullo, S., Cutforth, N., Stoecker, R., & Donohue, P. (2003). *Community-based research and higher education: Principles and practices*. San Francisco, CA: Jossey-Bass.

Summerfield, B. (2006). Working with end users toward an effective solution. *Certification Magazine*. Retrieved March 11, 2010, from http://www.certmag.com/read.php?in=1869# Tanniru, M. R., & Agarwal, R. (2002). Applied technology in business program. *e-Service Journal*, 5-23.

University of North Carolina. (2009). *Service-learning overview*. Retrieved March 22, 2010, from http://olsl.uncg.edu/svl/about/

University of Southern California. (n.d.). *Service learning theory and practice*. Retrieved March 11, 2010, from http://college.usc.edu/service-learning-theory-practice/eval.htm

Watson, W. E., Johnson, L., & Merritt, D. (1998). Team orientation, self orientation, and diversity in task groups. *Group & Organization Management, 23*(2), 161–188. doi:10.1177/1059601198232005

Werner, J. M., & Lester, S. W. (2001). Applying a team effectiveness framework to the performance of student case teams. *Human Resource Development Quarterly, 12*(4), 385–402. doi:10.1002/hrdq.1004

This work was previously published in the Journal of Organizational and End User Computing, Volume 23, Issue 3, edited by Mo Adam Mahmood, pp. 67-80, copyright 2011 by IGI Publishing (an imprint of IGI Global).

Chapter 8
Construct Validity Assessment in IS Research:
Methods and Case Example of User Satisfaction Scale

Dewi Rooslani Tojib
Monash University, Australia

Ly-Fie Sugianto
Monash University, Australia

ABSTRACT

Valid and reliable measures are critical to theory development as they facilitate theory testing in empirical research. Efforts in scale development have been put on ensuring aspects of validity. In this paper, the authors address a specific topic of construct validity assessment in scale development. Using data from the five leading IS journals between 1989-2008, in this paper, the authors determine if and how the field has advanced in construct validity assessment. Findings suggest that the proportion of studies reporting construct validity had increased and Confirmatory Factor Analysis (CFA), Exploratory Factor Analysis (EFA), and Multi-Trait Multi-Method (MTMM) were the three most common methods of construct validity assessment. The authors also apply a popular method from psychology and exemplify how the correlation analysis technique can be used to measure construct validity.

INTRODUCTION

The development of new measures is important in research studies that push the boundaries in order to enrich the domain in theory building. Consequently, it is essential that new measures be reliable and valid to ensure advancement in the body of knowledge. From the positivist standpoint, discussions on aspects of validity draw on the empirical position of quantitative science for justification. Academic enquiry in the Information Systems (IS) discipline is greatly influenced by this empiricist

DOI: 10.4018/978-1-4666-2059-9.ch008

paradigm. In this paper, we examine the literature landscape regarding the assessment of construct validity when developing new measures.

In the IS field, many research efforts have gone into developing new instruments to help us gain a better understanding of IS constructs and enable us to explore new paths of IS research. For instance, Doll and Torkzadeh (1988) with their End User Computing Satisfaction (EUCS) instrument have successfully guided other IS researchers to understand user satisfaction with many different types of Information Technology (IT) applications. Practitioners, on the other hand, regard instruments as practical measures that can be directly adopted to evaluate their particular area of interest. For instance, Barnes and Vidgen's (2000) E-Qual (previously called WebQual) has been increasingly gaining attention from industry, particularly from those who would like to assess the quality of the usability, information, and service interaction of their Internet websites. Having acknowledged the important role of instruments within the academic and industry communities, it is essential to ensure that any developed instruments accurately and reliably assess what they purport to measure in order to ensure the legitimacy of the results. Consequently, the notion of construct validity assessment is crucial for any instrument developers.

Many researchers in other disciplines, particularly in the field of psychology, have long acknowledged this type of validity when validating new measures (Nunnally, 1978). In the IS field, such validity concerns began to attract more attention when Straub (1989) initially raised the issue of the lack of validation within IS research. In 2001, Boudreau, Gefen and Straub replicated the latter's first study and found a considerable improvement in construct validity assessment compared with the initial study. However, these two studies did not provide a detailed explanation of the construct validity, which is essential in order to highlight the importance of, and to encourage

researchers to better appreciate, construct validity assessment. This paper presents the theoretical background of construct validity and explores the changes (if any) that have been made to construct validity assessment since 1989. Focusing only on instrument development studies published in the five leading IS journals between 1989-2008, this paper highlights various methods that have been utilized to assess construct validity of the newly developed instruments. Our findings reveal that IS researchers commonly assess construct validity of their scales internally through the employment of Confirmatory Factor Analysis (CFA), Exploratory Factor Analysis (EFA), and Multi-Trait Multi-Method (MTMM). This paper demonstrates that construct validity can also be assessed externally via correlation analysis with other measures. A case example showing how this method works is then presented. The theoretical and practical implications of the study, as well as directions for future research, are discussed in the concluding sections of this paper.

CONSTRUCT VALIDITY IN GENERAL

Construct validity might seem complicated for those who are not familiar with instrument validity. Cronbach and Meehl (1955) provide an often-cited, easy-to-understand definition. They describe construct validity as a condition whereby items measuring one particular construct are considered together and provide a reasonable operationalization for that particular construct (compared with other latent constructs). Thus, an instrument demonstrates construct validity if, in measuring an intended construct, it measures the concept it purports to measure regardless of any other established instruments of other constructs (Nunnally, 1978; Zaltman, Duncan, & Holbek, 1973). Such assessment is important since, knowing the constructs are properly measured and measure what they are intended to measure, (1) will increase our confidence in the overall empirical results, and (2) will show how well researchers translate their theories into actual measures.

The complexity of construct validity increases since different researchers categorize this type of validity differently. Such complexity arises because researchers attempt to associate the meaning of their empirical observations with the research context. Bagozzi (1980) considers three types of construct validity, namely: convergent, discriminant, and nomological validity. Cronbach (1990) and Rogers (1995) include predictive and concurrent validity as a subtype of construct validity. Factorial validity is also introduced as a component of construct validity by Straub, Boudreau and Gefen (2004). In this study, we focus our investigation on the most commonly accepted conventional classification of construct validity, namely convergent and discriminant validity (Anastasi, 1968).

Convergent validity and discriminant validity are opposites. The former refers to the extent to which the items measuring one construct appear to be indicators of that single underlying construct (Salisbury et al., 2002). Convergent validity is about discovering whether items measuring one factor are reflective of that respective factor only. We recently developed an instrument to measure user satisfaction with employee portals. After a series of refinements, the final instrument now consists of five dimensions, namely: *Confidentiality*, *Ease of Use*, *Convenience of Access*, *Usefulness* and *Portal Design*. We found that items measuring each of these dimensions strongly and significantly loaded on to their respective dimensions, suggesting convergent validity of the scale (Sugianto, Tojib & Burstein, 2007). On the other hand, the concept of discriminant validity lies in the principle that items measuring one factor can be differentiated from items measuring the other factors (Bagozzi & Phillips, 1982). Deriving from our examples when developing the user satisfaction measure, an examination of factor correlations revealed that all five dimensions were moderately

correlated (Sugianto, Tojib & Burstein, 2007). This finding suggests that these dimensions were distinct from one another, thereby indicating the discriminant validity of the instrument.

Methods for assessing construct validity vary. Straub, Boudreau and Gefen (2004) point out that the Multi-Trait Multi-Method (MTMM) matrix, Principle Component Analysis (PCA), Confirmatory Factor Analysis (CFA), Q-sorting, Average Variance Extracted (AVE) can be employed to examine construct validity. However, to date, it is not clear to what extent IS researchers have utilized these methods and whether some methods are preferred to the others. Moreover, there still remains the issue of whether or not IS researchers have paid more attention to construct validity since 2001. We explore these matters in the following literature analysis.

CONSTRUCT VALIDITY IN IS RESEARCH

To investigate construct validity assessment rigorously, we reviewed articles from *MIS Quarterly (MISQ)*, *Information and Management (IM)*, *Journal of Management Information Systems (JMIS)*, *Management Science (MS)*, and *Information Systems Research (ISR)*. We selected these journals because previous researchers (Boudreau, Gefen & Straub, 2001; Straub, 1989) used these journals in their own validity assessment studies. Furthermore, they are still considered as top tier journals in the IS field and thus we believe that articles from these leading journals should reflect common research practices in our field.

In our assessment, we included only those articles pertaining to instrument development published between January 1989 and December 2008. We chose 1989 as the starting point because the issue of whether IS researchers were sufficiently validating their instruments was initially raised in that year (Straub, 1989). By selecting December 2008 as the end of the assessment period, we

Table 1. Percentage of Studies Reporting Construct Validity Assessment

	1989-1998 (n=21)	1999-2002 (n=18)	2003-2008 (n=23)
Construct Validity (Both Convergent & Discriminant Validity)	66.7%	72.2%	74%
Only Convergent Validity	9.5%	0%	4.3%
Only Discriminant Validity	14.3%	5.6%	13%
Not Mention At All	9.5%	22.2%	8.7%

hope to present the most up-to-date results. We also limited our samples to instrument development studies, which employed survey as the data collection method, for two reasons. Firstly, it is imperative that researchers assess construct validity when developing new instruments. If they initially find the instrument to be construct valid, other researchers may have a higher level of confidence in adopting the instrument and they may not revalidate it for practical reasons. Thus, we believe that we can obtain the required information by limiting the sample articles to instrument development papers. Secondly, the administration of the instrument is generally taken in the form of survey (Boudreau, Gefen, & Straub, 2001).

A total of 5452 articles were initially collected and they were then filtered out by using keywords, namely: scale development, instrument development, measure, assessment, survey, validity, and construct validity. Only 62 articles met our criteria and were subsequently used for the literature analysis. Among the articles reviewed, 11 articles originated in ISR, 3 in MS, 12 in JMIS, 12 in MISQ, and 24 in IM. To be able to see the trend of construct validity assessment in the IS field, we categorized the 62 articles according to specific assessment periods. We simulated different potential combinations, keeping in mind that the articles should be evenly distributed among the assessment periods to avoid skewness of findings. The best possible one is to assess these articles in three time periods: 1989-1998, 1999-2002, and 2003-2008. For each article, whether or not the authors conducted construct validity assessment

is noted. Furthermore, the method used to assess construct validity (i.e., convergent or discriminant validity) was identified. The content analysis of past literature is presented in Appendix 1 and the key summary of our findings is presented in the following paragraphs.

As shown in Table 1, there is an overall increase in the number of IS researchers who investigated the construct validity of their new scales within the three periods of analysis. There is a slight decrease in the proportion of studies reporting only convergent or discriminant validity from the first to the second period of analysis, but the percentage increases again in the third period. Nevertheless, the majority of the studies already reported both convergent and discriminant validity assessment in 2003-2008. While some studies made no mention of construct validity at all, the percentage of these decreased during the last period of analysis. Such a positive improvement shows that IS researchers have taken the construct validity issue into greater consideration. It also shows the robustness of the newly developed instruments and hence, replication studies can be performed with a higher level of evidence, and practitioners can also confidently use these as practical measures.

Table 2 and Table 3 show the methods that have been employed to assess construct validity (Please note that we did not calculate the percentage as one study sometimes can employ more than one method of construct validity assessment). A detailed explanation of these methods is presented in Appendix 1. To summarize, the EFA method includes the assessment of eigenvalues of 1, item loading of at least 0.40, and no cross-loading of items above 0.40. CFA is commonly performed through Structural Equation Modeling (SEM) by investigating model fit measures, chi-square difference tests, and significant t-values of item loadings. The MTMM approach, on the other hand, is a classic method proposed by Campbell and Fiske (1959), which emphasizes the analysis of the MTMM matrix. Another method that has been utilized is the analysis of the correlation among the dimensions of the instrument, although this is not as popular as the first three. Furthermore, some researchers also looked at sorting procedure, conceptual discriminant process, and Average Variance Extracted (AVE) as other possible alternatives to assess construct validity. Surprisingly, reliability analysis through the calculation of coefficient alpha and/or composite reliability was mentioned several times as a method of construct validity assessment. This finding is unexpected, as reliability assessment must not be treated as validity assessment (Leedy & Ormrod, 2005). Reliability and validity are two totally different concepts. The former concerns the extent to which an instrument measures the same construct with consistent and error free results, while the latter focuses on whether an instrument measures what it is supposed to measure (Bailey & Pearson, 1983). While ideally an instrument has to be reliable and valid, we could not claim that a reliable instrument should involuntarily be valid and vice versa.

Hence, no claim can be made for construct validity if only the reliability of the scale is assessed. A closer look at the articles in question revealed that, in addition to reliability analysis, they also employed other common methods, namely CFA (Bhattacherjee, 2002; Kulkarni, Ravindran & Freeze, 2006) to confirm the construct validity of the scales. This research practice should not be further encouraged as it may create confusion among IS researchers. Researchers should make a distinction between reliability and validity assessment.

CFA currently seems to be the most popular method for construct validity assessment. Table 4 shows item loading assessment has been more frequently employed when examining convergent validity. In this method, convergent validity is achieved when items load more strongly on their corresponding factors than on other factors in the model (Kankanhalli, Tan & Wei, 2005; Shi, Kunnathur & Ragu-Nathan 2005; Van der Heijden & Verhagen, 2004). On the other hand, chi-square difference test turns out to be the most popular method when investigating discriminant validity (Balasubramanian, Konana & Menon, 2003; Shi, Kunnathur & Ragu-Nathan, 2005; Xia & Lee, 2005). In this method, the χ^2 statistic of the unconstrained CFA model (where all factors are freely correlated) is compared with that of constrained models where the correlation between pairs of factors is set to one (Bhattacherjee, 2002). A significant difference in the χ^2 between the

Table 2. Methods of Convergent Validity Assessment

	1989-1998	1999-2002	2003-2008
EFA	4	3	2
CFA	5	10	8
MTMM	2	3	4
Correlation Analysis	5	0	1
Reliability	0	1	2
AVE	0	1	3
Sorting Procedure	1	0	0

Table 3. Methods of Discriminant Validity Assessment

	1989-1998	1999-2002	2003-2008
EFA	5	2	2
CFA	6	10	15
MTMM	3	3	4
Correlation Analysis	4	0	2
Reliability	1	0	0
AVE	1	1	2
Sorting Procedure	1	0	0
Conceptual Discriminant Process	0	1	0

unconstrained and constrained models provides proof of discriminant validity between the constrained pair of constructs.

This finding does not suggest that CFA is the best construct validity assessment method. Depending on the purpose of the study, other methods such as PCA, MTMM, and correlation analysis are also powerful and appropriate. For instance, researchers conducting exploratory studies tend to employ PCA and MTMM, while those who perform confirmatory studies often utilize CFA. Hence, researchers should carefully select the best available method in order to achieve the most interpretable and reasonable results.

Our findings also revealed that correlation analysis is not a popular method of construct validity assessment in IS research. If such a method is utilized, IS researchers commonly measure correlations among dimensions constituting an instrument to be investigated. This is different from the psychology literature where this technique can be found in many of published

instrument development studies and their focus is on correlating the investigated instrument with another established measure (e.g., Bagby, Taylor, & Parker, 1994; Chung, Kim, & Abreu, 2004; Mullins & Kopelman, 1988; Raskin & Hall, 1981; Thomas & Freeman, 1990). Specifically, to assess convergent validity, the investigated scale should correlate positively and substantially with another established and related measure (Chatterji, 2003). For instance, Dukes, Discenza, and Couger (1989) conducted a study to assess convergent validity of instruments measuring computer anxiety. They employed four different anxiety measures, namely *the Attitude Toward Computers (ATC), the Computer Anxiety Index (CAIN), the Computer Anxiety Scale (CAS), and the Blomberg-Erickson-Lowery Computer Attitude Tasks (BELCAT)*. The intercorrelations among these instruments were found to be positive and statistically significant beyond the 0.001 level using two-tailed critical values with values between 0.75 and 0.89. These high degrees of correlations among the four instruments sug-

Table 4. CFA Methods of Construct Validity Assessment

CFA Methods	Convergent Validity	Discriminant Validity
Assessment of model fit	5	3
Item loadings	16	6
Chi-square difference tests	1	20
CFA was mentioned as a method to assess construct validity but particular methods were not identified.	4	2

gest that all utilized instruments are measuring the computer anxiety construct, thereby demonstrating the robustness of the construct itself. Hence, convergent validity of these instruments is highly supported. On the other hand, in order to claim discriminant validity, the investigated instrument should be weakly correlated with measures of constructs known to be distinctly different (Chatterji, 2003; Sekaran, 2003). For instance, when Chung, Kim, and Abreu (2004) developed *the Asian American Multidimensional Acculturation Scale (AAMAS-AA)*, they tested its discriminant validity by correlating it with *the Rosenberg Self Esteem Scale (RSES)* (Rosenberg, 1968). A low correlation between these two instruments is expected since each measures different constructs. Specifically, the AAMAS-AA measures acculturation and the RSES is a measure of self-esteem. The result yielded a non-significant coefficient of 0.03 for the correlations between these two instruments, confirming the discriminant validity of the AAMAS-AA.

As previously described, the method of correlation analysis using external measures has been widely utilized in the psychology discipline and probably in other areas as well. While EFA, CFA, and MTMM focus on establishing construct validity within the measure and its sub-measures only, this alternative method of construct validity assessment provides a different perspective to our understanding of construct validity. It highlights the importance of a new instrument to be related with existing instruments measuring similar constructs, and emphasizes the importance of having a new instrument that is unrelated to existing instruments measuring different constructs. This broad view is certainly essential for instrument developers who want reassurance that their newly developed instruments indeed measure what they are supposed to measure and are not replicating currently available instruments. To assist us in better understanding this technique, we demonstrate how this method would work when applied in the IS field. Using a case example of the recently developed Business-to-Employee Portal User Satisfaction (B2EPUS) instrument (see Sugianto, Tojib, & Burstein, 2007), a detailed explanation of this approach is given in the following section.

RESEARCH METHODOLOGY

Procedure and Participants

We invited ten Australian universities, which have implemented staff portals for at least one year, to participate in this research study. Four universities agreed to participate in the study. Since we were not allowed to obtain email addresses of the university employees, the non-probability convenience sampling method was used to recruit the respondents. The contact person for each participating university, on our behalf, sent an email invitation to all staff. Staff participation was entirely voluntary; confidentiality and anonymity were assured. Two hundred and six responses were collected. The sample characteristics are outlined in Table 5.

Measures

For the purpose of construct validity assessment, three different measures were used in the study, namely the B2EPUS instrument (Sugianto, Tojib, & Burstein, 2007), the Staff Portal Anxiety (SPA) instrument (Tojib, 2007), and the End User Computing Satisfaction (EUCS) instrument (Doll & Torkzadeh, 1988). Please refer to Appendix 2 for a complete list of items used in this study.

The B2EPUS instrument is an eighteen-item, self-report questionnaire designed to assess user attitude towards the staff portal. These items measure five dimensions of the B2EPUS construct, namely *Convenience of Access, Usefulness, Ease of Use, Portal Design*, and *Confidentiality*. A previous study reported that this instrument had a very good internal consistency and it also demonstrated the convergent, discriminant, criterion-related, and

Table 5. Sample Characteristics for the Field Study

Gender	Frequency	Percentage (%)
Male	77	37.4
Female	129	62.6
Age		
18-25	13	6.3
26-35	57	27.7
36-45	66	32.0
46 and above	70	34.0
Job Categories		
Academic staff	46	22.4
Researcher	8	4.4
General staff	151	73.2

nomological validity of the scale (Tojib, Sugianto, & Sendjaya, 2008).

The EUCS instrument is a measure of five dimensions of user satisfaction construct, namely *Format, Timeliness, Accuracy, Ease of Use,* and *Content.* This twelve-item instrument was selected to assess the convergent validity of the B2EPUS instrument because it is an established measure that has been widely used and validated in a large number of studies (e.g., Glorfeld & Cronan, 1992; McHaney & Cronan, 1998; Palvia, 1996; Zviran, Pliskin, & Levin, 2005). Furthermore, it was initially developed to measure user satisfaction, a similar construct measured by the B2EPUS instrument. Therefore, positive moderate to high correlations among the sub-dimensions of the EUCS and the B2EPUS instruments are expected to show the convergent validity of the B2EPUS instrument.

The SPA instrument is a five-item measure of anxiety towards the B2E portal, which were adapted from the Computer Anxiety Rating Scale (Heinssen, Glass, & Knight, 1987), Computer Attitude Scale (Loyd & Gressard, 1984), and Computer Anxiety Index (Maurer, 1983). The items were generated from different anxiety measures because: (1) it was not possible to obtain the complete items from the authors as their contact details were not provided in the published articles;

and (2) some of the items were not applicable to the B2E portal environment.

The SPA instrument was considered a suitable measure for assessing the discriminant validity of the B2EPUS instrument for several reasons. Firstly, a number of researchers believe that the 'attitude toward computer' construct and 'computer anxiety' construct are two different, unrelated constructs (Heissen, Glass, & Knight, 1987; Popovitch, Hyde, & Zakrajsek, 1987). The former construct is very much similar to the B2EPUS construct as they both measure attitudes towards particular systems. The latter is also very much related to the SPA instrument since both measure fear and apprehension felt towards the certain systems. Hence, the B2EPUS construct can be considered as unrelated to the SPA construct. Secondly, Igbaria and Nachman (1990) found a weak correlation between 'user satisfaction' and 'computer anxiety' constructs, thereby suggesting that these two constructs are actually distinct. Furthermore, the fact that the B2EPUS instrument does not take into account the anxiety felt towards the portal makes the selection of the SPA measure as the competing instrument for discriminant validity assessment most appropriate. Consequently, discriminant validity of the B2EPUS instrument would be demonstrated if a negative low relationship existed among the sub-dimensions of the B2EPUS instrument and the SPA measure.

Prior to conducting the correlation analysis, it should be noted that it is utterly important to pre-process the data and pay attention to the characteristics of the sample data, such as checking the skewness, kurtosis and standardized scores, to ensure that the correlation coefficient is not misleading. Discrepancy in data, including outliers, can have significant effect on the correlation coefficient, leading to incorrect conclusion on the strength of the relationship between the variables. Once we are satisfied with the data quality, we can then perform the analysis with high confidence. Our findings are presented in the next section.

ANALYSIS AND RESULTS

Reliability

Reliability analysis was performed through the assessment of Cronbach Alpha and composite reliability (Hair et al., 1998). Table 6 shows that the coefficient alphas and composite reliability values were all well above 0.70, suggesting that

they are indeed highly reliable measures. Thus, the validity assessment can then be performed.

Convergent Validity

To assess convergent validity, the correlation analysis between the B2EPUS instrument and the EUCS measure was performed. While there is no consensus on the cut-off point to determine convergent validity, we follow Pett, Lackey, and Sullivan (2003) which categorized correlation coefficient of 0.30 – 0.49 as low, 0.50 – 0.69 as moderate, and 0.70- 0.89 as strong. Table 7 presents the correlations between all sub-dimensions measured by these two scales. There are five high correlations found among the results. The *Confidentiality* dimension is found to be highly correlated with three sub-dimensions of the EUCS, namely *Content*, *Timeliness*, and *Accuracy*. This finding is reasonable as the *Confidentiality* dimension, to a certain extent, shares similar themes, namely quality of information. In particular, the three sub-dimensions measure whether a system presents precise, relevant, reliable, sufficient, up-to-date, and accurate information, while the *Confidentiality* dimension captures the notions

Table 6. Reliability assessment of the B2EPUS, EUCS, and SPA instruments

	Cronbach Alpha	Composite Reliability
B2EPUS - Usefulness	0.93	0.93
B2EPUS – Ease of Use	0.79	0.81
B2EPUS – Confidentiality	0.88	0.88
B2EPUS – Convenience of Access	0.75	0.75
B2EPUS – Portal Design	0.91	0.92
EUCS – Content	0.92	0.92
EUCS – Timeliness	0.80	0.80
EUCS – Accuracy	0.90	0.90
EUCS – Ease of Use	0.88	0.88
EUCS – Format	0.85	0.84
SPA	0.93	0.93

of *Security, Confidentiality, Information Content* and *Timeliness* (see Figure 4 in Sugianto and Tojib (2007) where we documented the evolution of the dimensions of the B2EPUS scale from conceptualization to empirical validation). Clearly, all dimensions highlight the quality of the information presented by the evaluated system from different point of views. The *Portal Design* dimension is found to be highly correlated with the *Ease of Use* sub-scale of the EUCS. This is also understandable as the *Portal Design* dimension is concerned with the characteristics and attractiveness of the portal site to facilitate the *Ease of Use* of the portal. The effectiveness of the portal design may allow users to explore the usability of the portal easily. In fact, the high correspondence between these two dimensions was already expected in the original study when the authors combined these two dimensions together when initially developing the B2EPUS conceptual model but these dimensions turned out to be separated after conducting factor analysis (Tojib & Sugianto, 2007). The two scales have the one same-labeled dimension, namely *Ease of Use*, and they are also found to be highly correlated. This finding is certainly expected considering these two dimensions tap into a similar theme, which is the user friendliness of the system. Although these five correlation relationships tend to be high, they are still far below 0.80, and hence, still within the acceptable range of claiming convergent validity (Bagozzi, 1994).

The other correlation relationships show low to moderate correlations among the dimensions of these two scales. Overall, all findings are desirable as they show that the B2EPUS and the EUCS scales are related, which further demonstrates that the B2EPUS is a measure of user satisfaction.

Discriminant Validity

In order to assess discriminant validity, correlation analysis between the B2EPUS instrument and the SPA measure was performed. Similar to convergent validity assessment, there is no consensus on the cut-off point to determine discriminant validity; hence, we also follow the same rules as those introduced by Pett, Lackey, and Sullivan (2003). As shown in Table 8, the data indicates that the sub-dimensions of the B2EPUS instruments are weakly correlated with the SPA measure. As theoretically predicted, the sub-dimensions of the B2EPUS instrument are negatively related to the SAP measure. The findings of weak negative correlations provide strong support for the discriminant validity of the B2EPUS instrument. It is also found that the *Usefulness* dimension of the B2EPUS scale, although negatively correlated with the SAP scale, did not quite reach a level of significance. This finding is somewhat unexpected; nevertheless, the full B2EPUS scale and its other four dimensions all correlated negatively and significantly with the SAP measure. To con-

Table 7. Correlations among the sub-dimensions of the B2EPUS and EUCS

		B2EPUS				
		Usefulness	*Confidentiality*	*Ease of Use*	*Convenience of Access*	*Portal Design*
EUCS	*Content*	0.41*	0.65*	0.48*	0.44*	0.54*
	Format	0.41*	0.55*	0.57*	0.44*	0.59*
	Timeliness	0.29*	0.63*	0.49*	0.42*	0.45*
	Ease of Use	0.43*	0.53*	0.70*	0.51*	0.67*
	Accuracy	0.26*	0.73*	0.39*	0.39*	0.38*

*) p< 0.001

clude, the findings suggest that the B2EPUS and SAP scales are not related, and hence, indicate generally satisfactory discriminant validity.

DISCUSSIONS AND LIMITATIONS

This paper addresses two major topics related to construct validity. Firstly, through a review of 62 instrument development articles, we have explored how rigorously IS researchers have construct-validated their instruments and which method they utilize to perform construct validity assessment. Secondly, we present a case example describing the use of the correlation analysis technique adopted from the psychology literature to assess construct validity of the B2EPUS instrument. Further detailed discussion on each topic is given separately in the following paragraphs.

Construct Validity in IS Research

While there is a welcome improvement in the proportion of studies reporting construct validity assessment, 8.7% of the sample articles within the last assessment period, did not report either the convergent or discriminant validity assessment. When reviewing these articles, it was discovered that one study did not mention anything about construct validity, but explained the assessment of item loadings (Muylle, Moenaert, & Despontin, 2004). This may be seen as a sign of convergent validity assessment. We also found that several papers (e.g., Chen, Soliman, Mao, & Frolick, 2000; Joshi, 1989; Lewis, Snyder, & Rainer, 1995; Saarinen, 1996; Templeton, Lewis, & Snyder, 2002;

Torkzadeh & Lee, 2003) indeed mentioned the construct validity assessment but did not separate it into convergent or discriminant validity. In response to this, researchers are encouraged to write the construct validity assessment section in more detail. Classifying construct validity into convergent and discriminant validity is crucial as this helps readers to better understand the difference between these two sub-types of construct validity. It will also demonstrate the rigor of one's studies, which in the end will enhance the confidence level of future researchers to adapt or adopt any findings.

As previously shown in Table 4, there are also some studies which reported the use of CFA to examine convergent and discriminant validity of the instrument, but did not specifically mention the methods of assessment. While this seems to be much better than no mention at all, this is again undesirable. Researchers should provide a description of the methods, as this will help other researchers to learn different ways of assessing construct validity. This will also encourage other researchers, who have no previous experience in validity assessment, to start validating their instruments.

Researchers should not be considered as the main culprits in poor current practice. Publishers/editors who increasingly limit the length of final submissions may play a major role as well. It is quite common for researchers to be asked to reduce the length of their submissions in order to accommodate the requirements of the publishers/editors. Researchers might end up omitting parts that they consider to be not as important (i.e., describing assessment procedures) but which indeed may be

Table 8. Correlations between the sub-dimensions of B2EPUS and SAP

	B2EPUS				
	Usefulness	*Confidentiality*	*Ease of Use*	*Convenience of Access*	*Portal Design*
SAP	- 0.13	- 0.25*	-0.39*	-0.45*	- 0.35*

*) p<0.001

highly relevant for others. To address and remedy this trend, two suggestions are proposed. Firstly, the development of a valid and reliable instrument requires a number of sequential stages, so the full paper can be quite long. Hence, publishers/editors should be flexible about the number of submitted and published pages for papers in this area. This will allow researchers to provide a better, more detailed description of the process. Secondly, publishers/editors should allow researchers to divide a complete version of a scale development paper into two or more shorter versions. For instance, researchers are given the opportunity to write and publish an article focusing on the construct validity assessment only. It is indeed not common for such a paper to be published in the IS field, although it is widely published in other research streams, for instance Mullins and Kopelman (1988) as well as Bagby, Taylor, and Parker (1994).

Construct Validity of the B2EPUS Scale

The results of the present study provide strong evidence of convergent and discriminant validity of the B2EPUS instrument as a measure of user satisfaction with B2E portals. As predicted, the sub-dimensions of the B2EPUS instrument are positively and strongly related to the EUCS instrument, supporting the convergent validity of the B2EPUS instrument. The fact that sub-dimensions of the B2EPUS instrument and the EUCS instrument are not entirely highly correlated demonstrates that these two instruments are related, yet distinct. In other words, these two instruments are indeed a measure of user satisfaction; however, they measure the user satisfaction construct from a different perspective. In particular, the EUCS is intended to measure user satisfaction with specific IT applications in the End User Computing environment, while the B2EPUS instrument captures user attitudes towards staff portals, which are operated in the Web-based environment. It is also revealed that the sub-dimensions of the B2EPUS instrument are negatively and weakly related to the SPA measure. Such a finding is desirable as it

demonstrates that these two scales are unrelated, each measuring a separate construct. Thus, the result confirms the discriminant validity of the B2EPUS instrument. To conclude, the B2EPUS instrument has been proven to be construct-valid.

The main aim of describing the adopted correlation analysis method in this paper is to enrich our body of knowledge, particularly in the area of construct validity assessment. The proposed technique also provides an additional exposure to a more typical construct validity assessment, such as CFA, EFA, and MTMM. IS researchers should now appreciate that construct validity of an instrument can be determined not only by its relationship with its sub-dimensions, but also with other related and unrelated external measures. While this alternative method seems to be straightforward, it is a real challenge for IS researchers to find the most appropriate competing instruments (i.e., the related and unrelated measures) to correlate with the scale under investigation as we do not have as many instruments as there are in the psychology discipline. Consequently, there is the possibility that this method may be applicable only to a particular area within the IS field, such as user satisfaction for which many similar measures have been developed.

Nevertheless, this study contains limitations that could be addressed in future studies. Firstly, we included instrument development articles from only five leading IS journals and investigated only construct validity assessments. Future research may replicate and expand upon this study by: (1) increasing the size of the sample journals; (2) including articles, which validated previously developed scales; and (3) exploring other validity types. Secondly, this study focuses mainly on exploring the construct validity assessment methods that have been employed by IS researchers. Future research could examine the relationship between these methods and the research findings. Specifically, it may explore whether there are differences in research findings between studies employing PCA, CFA, MTMM, or other methods. Thirdly, although the present study provides the evidence of convergent and discriminant validity of the B2EPUS instrument, this must be regarded as preliminary, as the data was obtained from employees within the higher education industry only. Future research should test the B2EPUS instrument with new data sets in other settings. For instance, researchers could perform cross-cultural studies using larger samples from different industries.

CONCLUSION

In this paper, we have shown how rigorously IS researchers have construct-validated their instruments and the approach that they most often utilize for the construct validity assessment. A review of 62 articles between 1989 and 2008 revealed that the proportion of studies reporting both convergent and discriminant validity has increased steadily but is still only at 74%. It was also found that CFA, in particular the item loading assessment and chi-square difference tests, are the most commonly used methods for assessing convergent and discriminant validity respectively. The employment of the correlation analysis approach using external measures for assessing content validity of the B2EPUS instrument has been described in this paper. It is hoped that this study will motivate IS researchers to conduct and report more comprehensive construct validity procedures and to consider the correlation analysis technique for construct validity assessment.

REFERENCES

Abdul-Gader, A. H., & Kozar, K. A. (1995). The Impact of Computer Alienation of Information Technology Investment Decisions: An Exploratory Cross-National Analysis. *Management Information Systems Quarterly, 19*(4), 535–559. doi:10.2307/249632

Agarwal, R., & Karahanna, E. (2000). Time Flies When You're Having Fun: Cognitive Absorption and Beliefs about Information Technology Usage. *Management Information Systems Quarterly, 24*(4), 665–694. doi:10.2307/3250951

Agarwal, R., & Prasad, J. (1998). Conceptual and Operational Definition of Personal Innovativeness in the Domain of Information Technology. *Information Systems Research, 9*(2), 204–215. doi:10.1287/isre.9.2.204

Agarwal, R., & Venkatesh, V. (2002). Assessing a Firm's Web Presence: A Heuristic Evaluation Procedure for the Measurement of Usability. *Information Systems Research, 13*(2), 168–186. doi:10.1287/isre.13.2.168.84

Aladwani, A. M., & Palvia, P. C. (2002). Developing and validating an instrument for measuring user-perceived Web quality. *Information & Management, 39*, 467–476. doi:10.1016/S0378-7206(01)00113-6

Anastasi, A. (1968). *Psychological Testing* (3rd ed.). New York: Macmillan.

Bagby, R. M., Taylor, G. J., & Parker, J. D. A. (1994). The twenty-item Toronto alexithymia scale-II: convergent, discriminant, and concurrent validity. *Journal of Psychosomatic Research, 38*(1), 33–40. doi:10.1016/0022-3999(94)90006-X

Bagozzi, R. P. (1980). Evaluating structural equation models with unobservable variables and measurement error: a comment. *Journal of Marketing Research, 13*, 375–381.

Bagozzi, R. P. (1994). Measurement in marketing research: basic principles of questionnaire design. In Bagozzi, R. P. (Ed.), *Principles of Marketing Research*. Oxford, UK: Blackwell Publishers.

Bagozzi, R. P., & Phillips, L. W. (1982). Representing and testing organizational theories: a holistic construal. *Administrative Science Quarterly, 27*(3), 459–489. doi:10.2307/2392322

Bailey, J. E., & Pearson, S. W. (1983). Development of a tool for measurement and analysing computer user satisfaction. *Management Science, 29*(5), 530–545. doi:10.1287/mnsc.29.5.530

Balasubramanian, S., Konana, P., & Menon, N. M. (2003). Customer Satisfaction in Virtual Environments: A Study of Online Investing. *Management Science, 49*(7), 871–889. doi:10.1287/mnsc.49.7.871.16385

Barki, H., & Hartwick, J. (1994). Measuring User Participation, User Involvement, and User Attitude. *Management Information Systems Quarterly, 18*(1), 59–82. doi:10.2307/249610

Barki, H., Rivard, S., & Talbot, J. (1993). Toward an Assessment of Software Development Risk. *Journal of Management Information Systems, 10*(2), 203–225.

Barnes, S., & Vidgen, R. (2000). WebQual: an exploration of website quality. In *Proceedings of the 8th European Conference on Information Systems*, Vienna, Austria.

Bassellier, G., & Benbasat, I. (2004). Business Competence of Information Technology Professionals: Conceptual Development and Influence on IT-Business Partnership. *Management Information Systems Quarterly, 28*(4), 673–694.

Bhattacherjee, A. (2002). Individual Trust in Online Firms: Scale Development and Initial Test. *Journal of Management Information Systems, 19*(1), 211–241.

Boudreau, M. C., Gefen, D., & Straub, D. W. (2001). Validation in information systems research: a state-of-the-art assessment. *Management Information Systems Quarterly, 25*(1), 1–16. doi:10.2307/3250956

Byrd, T. A., & Turner, D. E. (2000). Measuring the Flexibility of Information Technology Infrastructure: Exploratory Analysis of a Construct. *Journal of Management Information Systems, 17*(1), 167–208.

Campbell, D. T., & Fiske, D. W. (1959). Convergent and discriminant validity by the multitrait-multimethod matrix. *Psychological Bulletin, 56*(2), 81–105. doi:10.1037/h0046016

Chang, J. C.-J., & King, W. R. (2005). Measuring the Performance of Information Systems: A Functional Scorecard. *Journal of Management Information Systems, 22*(1), 85–115.

Chatterji, M. (2003). *Designing and using tools for educational assessment.* Upper Saddle River, NJ: Pearson Education.

Chen, L., Soliman, K. S., Mao, E., & Frolick, M. N. (2000). Measuring user satisfaction with data warehouses: an exploratory study. *Information & Management, 37*(3), 103–110. doi:10.1016/S0378-7206(99)00042-7

Chin, W. W., Gopal, A., & Salisbury, W. D. (1997). Advancing the Theory of Adaptive Structuration: The Development of a Scale to Measure Faithfulness of Appropriation. *Information Systems Research, 8*(4), 342–366. doi:10.1287/isre.8.4.342

Chiou, J.-S. (2004). The antecedents of consumers' loyalty toward Internet Service Providers. *Information & Management, 41*, 685–695. doi:10.1016/j.im.2003.08.006

Chung, R. H. G., Kim, B. S., & Abreu, J. M. (2004). Asian American Multidimensional Acculturation Scale: Development, Factor Analysis, Reliability, and Validity. *Cultural Diversity & Ethnic Minority Psychology, 10*(1), 66–80. doi:10.1037/1099-9809.10.1.66

Cronbach, L. J. (1990). *Essentials of Psychological Testing* (5th ed.). New York: Harper-Row.

Cronbach, L. J., & Meehl, P. E. (1955). Construct validity in psychological tests. *Psychological Bulletin, 55*(4), 281–302. doi:10.1037/h0040957

D'Ambra, J., & Rice, R. E. (2001). Emerging factors in user evaluation of the World Wide Web. *Information & Management, 38*, 373–384. doi:10.1016/S0378-7206(00)00077-X

Davison, R. (1997). An instrument for measuring meeting success. *Information & Management, 32*, 163–176. doi:10.1016/S0378-7206(97)00020-7

Doll, W. J., & Torkzadeh, G. (1988). The measurement of end user computing satisfaction. *Management Information Systems Quarterly, 12*(2), 259–275. doi:10.2307/248851

Doll, W. J., & Torkzadeh, G. (1998). Developing a multidimensional measure of system use in an organizational context. *Information & Management, 33*(4), 171–185. doi:10.1016/S0378-7206(98)00028-7

Dukes, R. L., Discenza, R., & Couger, J. D. (1989). Convergent Validity of Four Computer Anxiety Scales. *Educational and Psychological Measurement, 49*, 195–203. doi:10.1177/0013164489491021

Ferratt, T. W., Short, L. E., & Agarwal, R. (1993). Measuring the Information Systems Supervisor's Work-Unit Environment and Demonstrated Skill at Supervising. *Journal of Management Information Systems, 9*(4), 121–144.

Gatignon, H., Tushman, M. L., Smith, W., & Anderson, P. (2002). A Structural Approach to Assessing Innovation: Construct Development of Innovation Locus, Type, and Characteristics. *Management Science, 48*(9), 1103–1122. doi:10.1287/mnsc.48.9.1103.174

Glorfeld, K. D., & Cronan, T. P. (1992). Computer information satisfaction: a longitudinal study of computing systems and EUC in a public organisation. *Journal of End User Computing, 5*(1), 27–36.

Govindarajulu, C., & Reithel, B. J. (1998). Beyond the Information Center: An Instrument to Measure End User Computing Support for Multiple Sources. *Information & Management, 33*, 241–250. doi:10.1016/S0378-7206(98)00030-5

Hair, J. F., Tatham, R. L., Anderson, R. E., & Black, W. (1998). *Multivariate Data Analysis* (5th ed.). Upper Saddle River, NJ: Prentice Hall.

Heinssen, R. K., Glass, C. R., & Knight, L. A. Jr. (1987). Assessing Computer Anxiety: Development and Validation of the Computer Anxiety Rating Scale. *Computers in Human Behavior, 3*(1), 49–59. doi:10.1016/0747-5632(87)90010-0

Huang, M.-H. (2005). Web performance scale. *Information & Management, 42*, 841–852. doi:10.1016/j.im.2004.06.003

Igbaria, M., & Baroudi, J. J. (1993). A Short-Form Measure of Career Orientations: A Psychometric Evaluation. *Journal of Management Information Systems, 10*(2), 131–154.

Igbaria, M., & Nachman, S. A. (1990). Correlates of user satisfaction with end user computing: an exploratory study. *Information & Management, 19*(2), 73–82. doi:10.1016/0378-7206(90)90017-C

Joshi, K. (1989). The Measurement of Fairness or Equity Perceptions of Management Information Systems Users. *Management Information Systems Quarterly, 13*(3), 343–358. doi:10.2307/249010

Kankanhalli, A., Tan, B. C. Y., & Wei, K. K. (2005). Contributing Knowledge to Electronic Knowledge Repositories: An Empirical Investigation. *Management Information Systems Quarterly, 29*(1), 113–143.

Karahanna, E., Agarwal, R., & Angst, C. M. (2006). Reconceptualizing compatibility beliefs in technology acceptance research. *Management Information Systems Quarterly, 30*(4), 781–804.

Kim, K. K., & Umanath, N. S. (2005). Information transfer in B2B procurement: an empirical analysis and measurement. *Information & Management, 42,* 813–828. doi:10.1016/j.im.2004.08.004

Ko, D.-G., Kirsch, L. J., & King, W. R. (2005). Antecedents of Knowledge Transfer from Consultants to Clients in Enterprise System Implementations. *Management Information Systems Quarterly, 29*(1), 59–85.

Kulkarni, U. R., Ravindran, S., & Freeze, R. (2006). A Knowledge Management Success Model: Theoretical Development and Empirical Validation. *Journal of Management Information Systems, 23*(3), 309–347. doi:10.2753/MIS0742-1222230311

Lee, K. C., Lee, S., & Kang, I. W. (2005). KMPI: measuring knowledge management performance. *Information & Management, 42,* 469–482. doi:10.1016/j.im.2005.10.003

Lee, Y. W., Strong, D. M., Kahn, B. K., & Wang, R. Y. (2002). AIMQ: a methodology for information quality assessment. *Information & Management, 40,* 133–146. doi:10.1016/S0378-7206(02)00043-5

Leedy, P. D., & Ormrod, J. E. (2005). *Practical research: Planning and Design* (8th ed.). Upper Saddle River, NJ: Prentice Hall.

Lewis, B. R., Snyder, C. A., & Rainer, R. K. (1995). An Empirical Assessment of the Information Resource Management Construct. *Journal of Management Information Systems, 12*(1), 199–223.

Loyd, B. H., & Gressard, C. (1984). Reliability and factoral validity of computer attitude scales. *Educational and Psychological Measurement, 44*(2), 501–555. doi:10.1177/0013164484442033

Mak, B. L., & Sockel, H. (2001). A confirmatory factor analysis of IS employee motivation and retention. *Information & Management, 38,* 265–276. doi:10.1016/S0378-7206(00)00055-0

Maurer, M. (1983). *Development and validation of a measure of computer anxiety.* Unpublished master's thesis, Iowa State University.

McHaney, R., & Cronan, T. P. (1998). Computer Simulation Success: On the Use of the End-User Computing Satisfaction Instrument. *Decision Sciences, 29*(2), 525–536. doi:10.1111/j.1540-5915.1998.tb01589.x

McKinney, V., Yoon, K., & Zahedi, F. (2002). The Measurement of Web-Customer Satisfaction: An Expectation and Disconfirmation Approach. *Information Systems Research, 13*(3), 296–315. doi:10.1287/isre.13.3.296.76

McKnight, D. H., Choudhury, V., & Kacmar, C. (2002). Developing and Validating Trust Measures for e-Commerce: An Integrative Typology. *Information Systems Research, 13*(3), 334–359. doi:10.1287/isre.13.3.334.81

Molla, A., & Licker, P. S. (2005). eCommerce adoption in developing countries: a model and instrument. *Information & Management, 42,* 877–899. doi:10.1016/j.im.2004.09.002

Moore, G. C., & Benbasat, I. (1991). Development of an Instrument to Measure the Perceptions of Adopting an Information Technology Innovation. *Information Systems Research, 2*(3), 192–222. doi:10.1287/isre.2.3.192

Mullins, L. S., & Kopelman, R. E. (1988). Toward an Assessment of the Construct Validity of Four Measures of Narcissism. *Journal of Personality Assessment, 52*(4), 610–625. doi:10.1207/s15327752jpa5204_2

Muylle, S., Moenaert, R., & Despontin, M. (2004). The conceptualization and empirical validation of Web site user satisfaction. *Information & Management, 41,* 543–560. doi:10.1016/S0378-7206(03)00089-2

Nunnally, J. C. (1978). *Psychometric Theory* (2nd ed.). New York: McGraw-Hill.

Osmundson, J. S., Michael, J. B., Machniak, M. J., & Grossman, M. A. (2003). Quality management metrics for software development. *Information & Management, 40,* 799–812. doi:10.1016/S0378-7206(02)00114-3

Palmer, J. W. (2002). Web Site Usability, Design, and Performance Metrics. *Information Systems Research, 13*(2), 151–167. doi:10.1287/isre.13.2.151.88

Palvia, P. C. (1996). A model and instrument for measuring small business user satisfaction with information technology. *Information & Management, 31,* 151–163. doi:10.1016/S0378-7206(96)01069-5

Palvia, P. C. (1997). Developing a model of the global and strategic impact of information technology. *Information & Management, 32,* 229–244. doi:10.1016/S0378-7206(97)00023-2

Peace, A. G., Galletta, D. F., & Thong, J. Y. L. (2003). Software Piracy in the Workplace: A Model and Empirical Test. *Journal of Management Information Systems, 20*(1), 153–177.

Pett, M. A., Lackey, N. R., & Sullivan, J. J. (2003). *Making sense of factor analysis: the use of factor analysis for instrument development in health care research.* Beverly Hills, CA: Sage.

Popovich, P. M., Hyde, K. R., & Zakrajsek, T. (1987). The Development of the Attitudes Toward Computer Usage Scale. *Educational and Psychological Measurement, 47*(1), 261–269. doi:10.1177/0013164487471035

Raghunathan, B., Raghunatan, T. S., & Tu, Q. (1999). Dimensionality of the Strategic Grid Framework: The Construct and its Measurement. *Information Systems Research, 10*(4), 343–355. doi:10.1287/isre.10.4.343

Raskin, R., & Hall, C. S. (1981). The Narcissistic Personality Inventory: Alternate Form Reliability and Further Evidence of Construct Validity. *Journal of Personality Assessment, 45*(2), 159–162. doi:10.1207/s15327752jpa4502_10

Rogers, T. B. (1995). *The Psychological Testing Enterprise.* Belmont, CA: Brooks/Cole.

Rosenberg, M. (1968). *Society and the adolescent self-image.* Princeton, NJ: Princeton University Press.

Saarinen, T. (1996). An expanded instrument for evaluation information system success. *Information & Management, 31,* 103–118. doi:10.1016/S0378-7206(96)01075-0

Salisbury, W. D., Chin, W. W., Gopal, A., & Newsted, P. R. (2002). Research Report: Better Theory Through Measurement-Developing a Scale to Capture Consensus on Appropriation. *Information Systems Research, 13*(1), 91–103. doi:10.1287/isre.13.1.91.93

Saunders, C. S., & Jones, W. J. (1992). Measuring Performance of the Information Systems Function. *Journal of Management Information Systems, 8*(4), 63–82.

Segars, A. H., & Grover, V. (1998). Strategic Information Systems Planning Success: An Investigation of the Construct and Its Measurement. *Management Information Systems Quarterly, 22*(2), 139–163. doi:10.2307/249393

Sekaran, U. (2003). *Research Methods for Business: A Skill Building Approach.* New York: John Wiley & Sons.

Sethi, V., & King, W. R. (1994). Development of Measures to Assess the Extent to Which an Information Technology Application Provides Competitive Advantage. *Management Science, 40*(12), 1601–1627. doi:10.1287/mnsc.40.12.1601

Seyal, A. H., Rahim, M. M., & Rahman, M. N. A. (2000). Computer attitudes of non-computing academics: a study of technical colleges in Brunei Darussalam. *Information & Management, 37*, 169–180. doi:10.1016/S0378-7206(99)00045-2

Shi, Z., Kunnathur, A. S., & Ragu-Nathan, T. S. (2005). IS outsourcing management competence dimensions: instrument development and relationship exploration. *Information & Management, 42*, 901–919. doi:10.1016/j.im.2004.10.001

Smith, H. J., Milberg, S. J., & Burke, S. J. (1996). Information Privacy: Measuring Individual's Concerns about Organisational Practices. *Management Information Systems Quarterly, 20*(2), 167–196. doi:10.2307/249477

Straub, D. W. (1989). Validating instruments in MIS research. *Management Information Systems Quarterly, 13*(2), 147–169. doi:10.2307/248922

Straub, D. W., Boudreau, M. C., & Gefen, D. (2004). Validation guidelines in IS positivist research. *Communications of the Association for Information Systems, 14*, 380–426.

Sugianto, L. F., & Tojib, D. R. (2007). Portal Power. *Monash Business Review, 3*(1), 25. doi:10.2104/mbr07016

Sugianto, L. F., Tojib, D. R., & Burstein, F. (2007). A practical measure of employee satisfaction with B2E portals. In *Proceedings of the 28th International Conference on Information Systems,* Montreal, Quebec, Canada.

Templeton, G. F., Lewis, B. R., & Snyder, C. A. (2002). Development of a Measure for the Organisational Learning Construct. *Journal of Management Information Systems, 19*(2), 175–218.

Teo, H. H., Wei, K. K., & Benbasat, I. (2003). Predicting Intention to Adopt Interorganisational Linkages: An Institutional Perspective. *Management Information Systems Quarterly, 27*(1), 19–49.

Thomas, C. D., & Freeman, R. J. (1990). The Body Esteem Scale: Construct Validity of the Female Subscales. *Journal of Personality Assessment, 54*(1-2), 204–212. doi:10.1207/s15327752jpa5401&2_20

Tojib, D. R. (2007). *Development and Validation of the Business-to-Employee Portal User Satisfaction (B2EPUS) Scale.* Unpublished doctoral dissertation, Monash University, Australia.

Tojib, D. R., & Sugianto, L. F. (2007). The Development and Empirical Validation of B2E Portal User Satisfaction (B2EPUS) Scale. *Journal of End User Computing, 19*(3), 1–18.

Tojib, D. R., Sugianto, L. F., & Sendjaya, S. (2008). User Satisfaction with Business-to Employee (B2E) Portals: Conceptualization and Scale Development. *European Journal of Information Systems, 17*(6), 649–667. doi:10.1057/ejis.2008.55

Torkzadeh, G., & Dhillon, G. (2002). Measuring factors that influence the success of internet commerce. *Information Systems Research, 13*(2), 187–204. doi:10.1287/isre.13.2.187.87

Torkzadeh, G., & Lee, J. (2003). Measures of perceived end-user computing skills. *Information & Management, 40*, 607–615. doi:10.1016/S0378-7206(02)00090-3

Van der Heijden, H., & Verhagen, T. (2004). Online store image: conceptual foundations and empirical measurement. *Information & Management, 41*, 609–617. doi:10.1016/S0378-7206(03)00100-9

Wang, Y.-S. (2003). Assessment of learner satisfaction with asynchronous electronic learning systems. *Information & Management, 41*, 75–86. doi:10.1016/S0378-7206(03)00028-4

Webster, J., & Martocchio, J. J. (1992). Microcomputer Playfulness: Development of a Measure with Workplace Implications. *Management Information Systems Quarterly, 16*(2), 201–226. doi:10.2307/249576

Xia, W., & Lee, G. (2005). Complexity of Information Systems Development Projects: Conceptualization and Measurement Development. *Journal of Management Information Systems*, *22*(1), 45–83.

Yang, Z., Cai, S., Zhou, Z., & Zhou, N. (2005). Development and validation of an instrument to measure user perceived service quality of information presenting Web portals. *Information & Management*, *42*, 575–589. doi:10.1016/S0378-7206(04)00073-4

Zaltman, G., Duncan, R., & Holbek, J. (1973). *Innovations and Organizations*. New York: Wiley.

Zhu, K., & Kraemer, K. L. (2002). E-commerce Metrics for Net-Enhanced Organisations: Assessing the Value of e-Commerce to Firm Performance in the Manufacturing Sector. *Information Systems Research*, *13*(3), 275–295. doi:10.1287/isre.13.3.275.82

Zviran, M., Pliskin, N., & Levin, R. (2005). Measuring user satisfaction and perceived usefulness in the ERP context. *Journal of Computer Information Systems*, *45*(3), 43–52.

APPENDIX 1. A SUMMARY OF CONSTRUCT VALIDITY ASSESSMENT IN IS RESEARCH

Table 1A.

Year	Author	Scale	Convergent Validity	Discriminant Validity	Remark
Management Science					
1994	Sethi & King	A measure of competitive advantage provided by an IT application	CFA (method not specified)	CFA (Chi-square difference test)	CFA was utilized to assess convergent validity of the scale. However, there was no clear explanation on how this validity was achieved
2002	Gatignon et al.	A measure of innovation's locus, type, and characteristics	CFA (Chi-square difference test)	CFA (Chi-square difference test)	For convergent validity, the correlations between pairs of factors was set to zero; while for discriminant validity, they were set to one
2003	Balasubramanian et al.	A measure of customer satisfaction in virtual environments	Not mentioned	CFA (Chi-square difference test)	The article did mention the assessment of convergent validity through the use of CFA but it was only for one construct not for the scale
Journal of Management Information Systems					
1992	Saunders & Jones	A measure of IS function performance.	Not mentioned	Not mentioned	No other validity assessments were explained in the paper
1993	Igbaria & Baroudi	A measure of career orientations	Not mentioned	Correlation analysis; Modified MTMM	For correlation analysis, discriminant validity is achieved when items of one constructs correlate more highly with their own construct than with other constructs in the model
1993	Barki et al.	A measure of software development risk	Not mentioned	Not mentioned	Only face, content, and criterion validity were reported
1993	Ferratt et al.	A measure of IS supervisors' supervising skills	PCA (method not specified)	PCA (method not specified)	No clear explanation on how these validities were achieved through the use of EFA
1995	Lewis et al.	Assessment of the information resource management construct	PCA (method not specified)	PCA (method not specified)	The construct validity assessment was not separated into convergent and discriminant validity. PCA was performed to assess construct validity, however, to what extent the construct validity has been achieved was not clearly explained.
2000	Byrd & Turner	A measure of IS infrastructure flexibility	CFA (Item loading)	CFA (Chi-square difference tests)	Item loading should be > 0.30 to achieve convergent validity
2002	Bhattacherjee	A measure of individual trust in online firms	CFA (Item loading), Reliability, the AVE	CFA (Chi-square difference tests)	Item loading should be > 0.70 to achieve convergent validity. Reliability assessment such as cronbach alpha and AVE was assumed as a method of convergent validity assessment
2002	Templeton et al.	A measure for the organizational learning construct	PCA (method not specified)	PCA (method not specified)	The construct validity assessment was not separated into convergent and discriminant validity. PCA was performed to assess construct validity, however, to what extent the construct validity has been achieved was not clearly explained.

continued on following page

Table 1A. Continued

Year	Author	Scale	Convergent Validity	Discriminant Validity	Remark
2003	Peace et al.	Software piracy in the workplace scale	Composite reliability; the AVE	CFA (item loading)	Reliability assessment such as composite reliability and AVE was assumed as a method of convergent validity assessment
2005	Xia & Lee	A measure of IS development project complexity	CFA (Model fit measures)	CFA (Chi-square difference tests)	The convergent validity assessment was measured by identifying the model fit of each construct (with its respecting items). If the fit measures are adequate, the convergent validity of the construct is supported.
2005	Chang & King	A measure of IS functional scorecard	CFA (Item loading)	CFA (Chi-square difference tests)	
2006	Kulkarni et al.	A measure of knowledge management success model	CFA (item loading); the Cronbach alpha	CFA (Squared correlation between a pair of constructs)	The CFA procedure followed Anderson & Gerbing (1988). Reliability assessment such as the Cronbach alpha was assumed as a method of convergent validity assessment
MIS Quarterly					
1989	Joshi	A measure of fairness or equity perceptions of MIS users	Correlation analysis; EFA (method not specified)	Correlation analysis; EFA (method not specified)	The author did not separate construct validity into convergent and discriminant validity. Construct validity was assessed through correlating two different measures of the same construct. Factor analysis was employed as well but to what extent the construct validity has been achieved was not clearly explained.
1992	Webster & Martocchio	A measure of computer playfulness	Not mentioned	Comparing the operational measures in terms of form and content; combined reliability analysis; item correlation analysis; comparing scale reliabilities with interscale correlations.	Discriminant validity was assessed with respect to other two related measures. Four step procedure proposed by Schmitt (1991)
1994	Barki & Hartwick	A measure of user participation, user involvement, and user attitude.	Correlation among factors constituting the scale	Correlation among different constructs assessed in the study.	Strong and significant correlations among factors constituting the scale represent convergent validity. To claim discriminant validity, correlations among different constructs must be significantly less than 1.00.
1995	Abdul-Gader & Kozar	A computer alienation measurement scale	Correlation analysis	Not mentioned	Convergent validity was assessed by correlating Computer Alienation scale and overall satisfaction item. Reason why overall satisfaction item was used is not convincing.
1996	Smith et al.	A measure of the individual's concerns about organizational information privacy practices.	Correlation analysis; CFA (item loading)	Correlation analysis; AVE	Reliability assessment such as AVE was assumed as a method of discriminant validity assessment

continued on following page

Table 1A. Continued

Year	Author	Scale	Convergent Validity	Discriminant Validity	Remark
1998	Segars & Grover	A measure of strategic information systems planning success	CFA (Model fit measure and item loading)	CFA (Chi-square difference test)	
2000	Agarwal & Karahanna	A measure of cognitive absorption	CFA (Item loading)	CFA (item loading); AVE	Reliability assessment such as AVE was assumed as a method of discriminant validity assessment
2003	Teo et al.	A measure of intention to adopt financial electronic data interchange	CFA (method not specified)	CFA (Chi-square Difference Test)	Method used to assess convergent validity is not clearly explained.
2004	Bassellier & Benbasat	A scale to measure business competence of IT professionals.	Not mentioned	CFA (item loading); AVE	Convergent validity assessment was not mentioned but item loadings are analyzed and all items have significant loadings above 0.7.
2005	Ko et al.	A measure of knowledge transfer from consultants to clients in enterprise system implementations	AVE	CFA (item loading)	
2005	Kankanhalli et al.	A measure of contributing knowledge to electronic knowledge repositories	CFA (Item loading)	CFA (Item loading)	
2006	Karahanna et al.	A measure of compatibility beliefs in technology acceptance	Not mentioned	CFA (item loading); AVE	
Information Systems Research					
1991	Moore & Benbasat	A measure of perceptions of adopting IT Innovation.	Sorting procedure	Sorting procedure	If an item was consistently placed within a particular category, then it was considered to demonstrate convergent validity with the related construct and discriminant validity with the others.
1997	Chin et al.	The faithfulness of appropriation scale	CFA (item loading and model fit measures)	CFA (chi-square difference tests); correlation analysis	To assess, discriminant validity, two external scales were correlated with the developed scale.
1998	Agarwal & Prasad	Personal innovativeness in the domain of IT	PCA (item loading) & CFA (model fit measures)	PCA (item loading) & CFA (model fit measures)	A classic approach (i.e., PCA) and a contemporary approach (i.e., CFA) were utilized in assessing both convergent and discriminant validity. However, it was not clearly explained how convergent and discriminant validity of the scale were obtained.

continued on following page

Table 1A. Continued

Year	Author	Scale	Convergent Validity	Discriminant Validity	Remark
1999	Raghuna-than et al.	A measure for the dimensionality of the strategic grid framework.	Not mentioned	CFA (Chi-square difference test)	The method utilized to assess convergent validity was not clearly explained and the result of this assessment was not mentioned either.
2002	McKnight et al.	Trust measures for e-Commerce	CFA (item loading)	CFA (chi-square difference test)	Discriminant validity was also tested using constraint analysis method.
2002	Zhu & Kraemer	e-Commerce metrics for net-enhanced organizations	CFA (item loading)	Conceptual discrimination process	The conceptual discrimination process compares the extracted construct variance with correlations among constructs (Campbell & Fiske,1959)
2002	Salisbury et al.	A scale to capture consensus on appropriateness	CFA (item loading)	CFA (chi-square difference test)	
2002	McKinney et al.	A measure of Web-customer satisfaction	MTMM; CFA (item loading)	MTMM; CFA (item loading)	
2002	Agarwal & Venkatesh	The measurement of website usability.	Not mentioned	Not Mentioned	
2002	Palmer	Website usability, design, and performance metrics	MTMM; CFA (method not specified)	MTMM; CFA (method not specified)	A modification of the MTMM analysis (Campbell & Fiske, 1959) was utilized (simpler and more straight-forward analysis).
2003	Torkzadeh & Dhillon	A measure of internet commerce success	MTMM	MTMM	
Information & Management					
1996	Saarinen	A measure of IS success	PCA (item loading)	PCA (item loading)	Construct validity was not separated into convergent and discriminant validity.
1996	Palvia	A measure of small business user satisfaction with IT	MTMM	MTMM	
1997	Davison	A measure of meeting success	CFA (method not specified)	CFA (method not specified)	The construct validity assessment was not separated into convergent and discriminant validity. CFA was performed to assess construct validity, however, to what extent the construct validity has been achieved was not clearly explained.
1997	Palvia	A measure of the global and strategic impact of IT	MTMM	MTMM	
1998	Govinda-rajulu & Reithel	An instrument to measure end user computing support from multiple sources	Correlation analysis	Not Mentioned	Correlation analysis was conducted to find out how closely the items in a single construct are correlated with each other. High correlations indicate strong convergent validity.

continued on following page

Table 1A. Continued

Year	Author	Scale	Convergent Validity	Discriminant Validity	Remark
1998	Doll & Torkzadeh	A measure of system-use in an organizational context	Not mentioned	CFA (chi-square difference test)	
2000	Seyal et al.	An instrument to measure computer attitudes of non-computing academics	Not mentioned	Not Mentioned	No validity assessment was conducted
2000	Chen et al.	A measure of user satisfaction with data warehouses	PCA	PCA	The construct validity assessment was not separated into convergent and discriminant validity. PCA was performed to assess construct validity, however, to what extent the construct validity has been achieved was not clearly explained.
2001	D'Ambra & Rice	User evaluation of the WWW	Not mentioned	Not Mentioned	No validity assessment was conducted
2001	Mak & Sockel	A measure of IS employee motivation and retention	Normed-Fit Index > 0.90	CFA (Chi-square difference test)	
2002	Lee et al.	Information quality assessment	Not Mentioned	Not Mentioned	No validity assessment was conducted. Only scale reliability is reported.
2002	Aladwani & Palvia	An instrument for measuring user-perceived Web quality	PCA; MTMM	MTMM	PCA results demonstrated that all factors had adequate convergent validity since each factor's items converged cleanly on the same factor representing these items.
2003	Osmundson et al.	Quality management measure for software development	Not mentioned	Not Mentioned	
2003	Wang	A measure of learner satisfaction with asynchronous electronic learning systems	MTMM	MTMM	
2003	Torkzadeh & Lee	A measure of perceived end-user computing skills	Correlation analysis; PCA (method not specified)	Correlation analysis; PCA (method not specified)	The construct validity assessment was not separated into convergent and discriminant validity. PCA was performed to assess construct validity, however, to what extent the construct validity has been achieved was not clearly explained.
2004	Muylle et al.	Web site user satisfaction scale	Not mentioned	Not Mentioned	Assessment of convergent and discriminant validity was not mentioned. However, the study examined item loading and found high and significant loading for all items related to each respective factor. This can be seen as a sign of convergent validity.
2004	Van der Heijden & Verhagen	A measure of the components of an online store image	PCA (Item loading)	PCA (Item loading)	When all items load higher on their own factors than on the other factors, this represents convergent and discriminant validity.

continued on following page

Table 1A. Continued

Year	Author	Scale	Convergent Validity	Discriminant Validity	Remark
2004	Chiou	A model examining the antecedents of consumer loyalty toward ISPs.	CFA (item loading)	CFA (model fit measures, chi-square difference test)	
2005	Kim & Umanath	A measure for electronic information transfer	CFA (item loading)	CFA (model fit measures, correlation analysis)	In assessing discriminant validity, differential correlations between two constructs
2005	Huang	Web performance scale	MTMM	MTMM; CFA(chi square difference test)	
2005	Molla & Licker	A measure of eCommerce adoption in developing countries	MTMM	MTMM	
2005	Lee et al.	A measure of knowledge management performance	Method not specified	Not Mentioned	Claimed that the convergent validity of the scale was achieved but not clear which method was used.
2005	Yang et al.	A measure of user perceived service quality of information presenting Web portals	AVE	CFA (Chi-square difference test)	
2005	Shi et al.	IS Outsourcing Management Competence scale	CFA (model fit measure, item loading)	CFA (Chi-square difference test)	

APPENDIX 2. ITEMS FOR THE B2EPUS AND EUCS INSTRUMENTS

Table 2A.

B2EPUS*	
Usefulness	The staff portal enables me to share or exchange project/task information with my team member colleagues
	The staff portal facilitates my collaboration work with all colleagues
	The staff portal enables me to discuss work or project issues with my immediate work colleagues
	The staff portal enables me to share general information via email or on website with other colleagues within the whole organization
Confidentiality	I feel confident in submitting personal information through the staff portal because it will be properly used by authorized people
	The information presented on the portal can be trusted
	Information presented on the portal is dependable
	I feel the staff portal is secure

Tale 2A. Continued

B2EPUS*	
	I can rely on the information presented on the staff portal to carry out my tasks
Ease of Use	No training on the use of the staff portal is necessary as the portal use is self- explanatory
	The staff portal is easy to navigate, both forward and backward
	When I am navigating the staff portal, I feel that I am in control of what I am doing
Convenience of Access	The staff portal is accessible from my home through internet connection
	Gaining access to the staff portal is easy
	The staff portal is accessible 24 hours a day, 7 days a week
Portal Design	The staff portal is aesthetically designed
	The design of the staff portal is attractive
	The staff portal is user friendly with many help functions and useful buttons and links
EUCS**	
Content	The system provides the precise information you need
	The information content meets your needs
	The system provides reports that seem to be just about exactly what you need
	The system provides sufficient information
Accuracy	The system is accurate
	I am satisfied with the accuracy of the system
Format	I think the output is presented in a useful format
	The information is clear
Ease of Use	The system is user friendly
	The system is easy to use
Timeliness	I get the information I need in time
	The system provides up-to-date information
SPA	
I feel apprehensive about using the portal	
I have avoided the staff portal because it is unfamiliar and somewhat intimidating to me	
I hesitate to use the staff portal for fear of making mistakes I cannot correct	
The staff portal makes me feel uncomfortable	
I sometimes get nervous just thinking about the staff portal	

*) Adopted from Tojib, Sugianto, & Sendjaya (2008)
**) Adapted from Doll & Torkzadeh (1988)

This work was previously published in the Journal of Organizational and End User Computing, Volume 23, Issue 1, edited by Mo Adam Mahmood, pp. 38-63, copyright 2011 by IGI Publishing (an imprint of IGI Global).

Chapter 9
A Social Capital Perspective on IT Professionals' Work Behavior and Attitude

Lixuan Zhang
Augusta State University, USA

Mary C. Jones
University of North Texas, USA

ABSTRACT

Attracting and retaining information technology (IT) professionals is a current concern for companies. Although research has been conducted about the job behavior and attitudes of IT professionals over the past three decades, little research has explored the effect of IT professionals' social capital. The primary research question that this study addresses is how social capital affects IT professionals' work attitude and behavior, including job satisfaction and job performance. Data were collected from 128 IT professionals from a range of jobs, organizations and industries. Results indicate that the strength of the ties an IT professional has in his or her organization is positively related to job satisfaction. The number of ties that an IT professional has outside the organization is also positively related to job performance. Several implications for research and practice are offered based on these findings.

INTRODUCTION

Despite the prevalence of outsourcing and globalization, information technology (IT) executives rank attracting, developing and retaining IT professionals as their top concern. Successful retention of IT professionals is critical to organizational computing in order to provide continuity in IT development, support, and vendor relationships. Much research has been conducted about how to manage, recruit, and retain IT professionals with regard to a variety of topics including motivation,

DOI: 10.4018/978-1-4666-2059-9.ch009

job satisfaction, knowledge and skill sets, career anchors, and turnover (Agarwal & Ferratt, 2001, 2002; Rutner, Hardgrave & McKnight, 2008).

Research in this area has been drawn largely from theories based in psychology and has examined the constructs of interest from an individual, or cognitive, perspective. Extensive research has been conducted about what motivates IT professionals, but very little examines the effect of social relationships on IT professionals' behavior and attitude. Some researchers suggest that IT professionals have lower social needs than other professionals (Couger, Oppermann & Amoroso, 1994), yet others find that there is no difference between IT and non-IT professionals regarding social needs (Wynekoop & Walz, 1998). Unfortunately, much of the research in this vein has stopped with examining the extent to which social needs are important to IT professionals. Little research has explored the extent to which the social component may impact IT professionals' work behavior and attitudes.

Because IT is becoming increasingly integrated with the business departments and functions of organizations, IT professionals are expected to work effectively with many different groups of people. It is critical that they develop and build relationships with their business partners/clients. Furthermore, IT professionals work not only in face-to-face teams, but are increasingly working in self-managed or virtual teams. Thus, the role of social factors may be critical in shedding light on IT professionals' work behavior and attitudes. Theories grounded in social capital may provide an important and useful lens through which to examine this phenomenon. Complementing the traditional focus on individual attributes, the social capital perspective focuses on the relationships among individuals. Social capital theory provides a fine-grained analysis of how social ties affect individuals' work attitude and behavior through a variety of mechanisms. However, this theoretical lens has not been widely used to examine the attitudes and behavior of IT professionals.

The purpose of this study is to examine how the social capital IT professionals build within their organizations and within the IT profession impact their work behavior and attitudes. The primary research question addressed is how social capital affects IT professionals' work attitude and behavior including job satisfaction and job performance. The study examines the influence of two aspects of social capital on IT professionals' job attitude and work behavior: tie strength and the number of ties.

The rest of the paper is organized as follows. First, a rationale is provided for why it is important to extend prior work on social capital influences on the IT profession. Next, the theoretical foundation for the study is provided, and hypotheses are formulated. The methodology is then discussed, the data analysis is provided, and the results are discussed. Finally, implications of the research are presented along with implications for practice and for future research.

IT PROFESSIONALS

This study provides insight specifically into the role of social capital in the work attitude and behavior of IT professionals. It is important to examine IT professionals apart from other professionals for several reasons. A strong IT staff with technical skills, business understanding and problem solving orientation is a valuable asset for an organization and can contribute significantly to developing long-term competitiveness. Despite the importance of this critical resource, attracting and retaining IT staff remains a key and difficult issue with which industry and researchers alike still grapple. The IT profession is faced with higher turnover rates than many other professions, and it is increasingly difficult for organizations to retain IT employees (Agarwal, Brown, Ferratt & Moore, 2006).

Another reason to focus on IT professionals is that they are identified as a distinct occupational group, and this distinction is thought to impact their work related attitudes and behavior (Guzman, Stam & Stanton, 2008). IT professionals must also deal with a somewhat unique facets of their job; that of continually eroding competencies and skills. In most other professional fields, competencies and skill increase with experience over time. However, the dynamic evolution of new technologies, languages, and platforms means that an IT professional's competencies and skills can erode in a relatively short period of time (Ang & Slaughter, 2000). Therefore, they must continually develop new skills and competencies to remain valuable in their organizations and their professions. Social capital can be an invaluable resource in this process.

Researchers have studied the social needs of IT professionals and the importance of social support. Social needs of the IT professionals include supportive working relationships with colleagues and understanding from supervisors. In addition, social support from supervisors and colleagues, defined as the availability and quality of helping relationship, is directly related to IT professionals' job satisfaction, especially for those with high social affiliation needs, such as system analysts and IT managers (Lee, 2004).

Social capital not only has psychological effects on IT professionals, but it also impacts job performance. For example, the success of the highest performing computer professionals in Bell Telephone Laboratories is attributed to their ability to maintain and leverage their networks (Kelly & Caplan, 1993). A study investigating co-op engineers finds that high performers have significantly more work-related contacts and more informal social ties with other professional staff than low performers (Lee, 1994). IT professionals with a higher level of business competence also have a large social network and a directory of the location of knowledge in that network (Bassellier & Benbasat, 2004).

IT professionals are members of organizations in which they are employed. They are more likely to form relationships with each other in their work groups and in their departments. In addition, their jobs may require them to be involved in boundary-spanning activities. The interaction with people from different departments may yield benefits such as access to information not available within their own department. An IT professional is also a member of IT profession. IT professionals can interact with other colleagues through "professional associations, conferences, training classes, and graduate programs at universities, as well as social, community, school and church functions"(Moore & Burke, 2002, p.75). The informal communication may convey the attitudes and behaviors of fellow IT professionals.

Although there is research that indicates that supportive social relationships in the workplace are important to IT professionals, there is conflicting evidence about whether it is as important to IT professionals as it is to their non-IT counterparts. Furthermore, there is little guidance about whether IT people define supportive relationships in the same way as others do. Thus, we have limited knowledge about the role of the social component in IT professionals' work environments. Prior research suggests, however, that social ties of IT professionals may play an important role in these environments. Examining IT professionals apart from other knowledge workers in this context may provide insight into their work attitudes and behavior that other research has not yet tapped. Although the web of social ties of IT professionals may influence their attitude and work behavior, these ties may exert their influence in different ways than they do for non-IT professionals.

THEORETICAL BACKGROUND

Social capital theory has received extensive attention from researchers in a number of disciplines. In this paper, the definition of social capital given by is adopted: "the sum of actual and potential resources embedded with, available through, and derived from the network of relationship possessed by an individual or social unit" (Nahapiet & Ghoshal, 1998, p.243). Therefore, social capital consists of relationships among these individuals and assets or resources that may be brought by these relationships. This definition is chosen because it is comprehensive in the sense that it focuses on both the substance and sources of social capital (Adler & Kwon, 2002).

Many demographic factors play a major role in social capital theory including social status, gender and race. According to Bourdieu (1986), social capital is inherited from the past and is accumulated continuously, providing the members in a network the backing of the collectivity-owned capital. Thus, individuals with high social status have a greater capacity to take advantage of this capital (Bourdieu, 1986). Gender and race also may affect an individual's social capital (Helliwell & Putnam, 2007). For example, racial inequalities with regard to social capital may be as great as inequalities in financial and human capital (Putnam, 1993). Evidence shows that both network size and density are larger for Caucasians than either Hispanics or African Americans (Marsden, 1988). Research also indicates that men have larger social networks than women, and men are more likely to be located in core units of the organization (Lin, 2000).

There has been an increased focus on social capital in organizational research. Employees of organizations are viewed as objects embedded in organizational networks and the properties of these networks help to explain organizational outcomes including performance, innovation, and promotion (Totterdell, Wall, Holman & Epitropaki, 2004).

In this study we investigate how social capital influences job satisfaction and work performance.

HYPOTHESES

In this study we examine the impacts of two aspects of social capital IT professionals' job satisfaction and job performance: tie strength and the number of ties.

Tie strength: Granovetter pioneered the concept of weak ties. He argues that the strength of a tie is a function of "the amount of time, the emotional intensity, the intimacy (mutual confiding), and the reciprocal services which characterize the tie" (Granovetter, 1973, p. 1361). Weak ties are less emotionally intense, less frequent and less intimate than strong ties. Weak ties can provide people with information and resource access that extend beyond what is available to them in their immediate social circles. A greater number of weak ties has also been shown to provide opportunities for greater job mobility as well as more innovative ideas (Granovetter, 1982; Rodan & Galunic, 2004).

On the other hand, those with whom we have strong ties are more motivated to help us and are typically easier to access. Strong ties refer to family, friends and other contacts with whom an individual has strong bonds whereas weak ties refer to distant relationships such as acquaintances. Strong ties are likely to lead to redundant information since strong ties exist among similar people, whereas weak ties serve as a bridge across disconnected groups and provide individuals access to novel information. Krackhardt (1992) stressed the importance of strong relationships where people have known each other for a long time and have developed affection with each other. Strong ties can be developed not only in a work group, but they may also be developed among colleagues in the same department or organization. Co-worker statuses and overlapping organizational membership are predictors of tie strength (Marsden & Campbell, 1984).

Strong ties may increase individuals' job satisfaction. Cranny, Smith and Stone (1992) define job satisfaction as an affective reaction to a person's job, resulting from his/her comparison of actual outcomes with those that are desired. The quality of relationships with coworkers and supervisors is one of the most important predictors of job satisfaction (Lee, 2004). Strong ties can help build relationships by providing the exchange of career and psychosocial support and by facilitating the sharing of general organizational information such as organizational norms and culture (Shah, 1998). Strong ties can lead to close-knit networks, where individuals get the most social and career support (Ibarra, 1995). For example, the strong ties between mentor and protégé yield consistent guidance and increase the protégé's work satisfaction and organizational commitment (Aryee & Chay, 1994). Newcomers with small networks composed of strong ties have a greater sense of organizational commitment, which is closely related to job satisfaction (Morrison, 2002). Strong ties are particularly valuable in uncertain or crisis situations, providing timely assistance and trust. Therefore, IT professionals who have strong ties are likely to be satisfied with their job. Based on the discussion above, the following is hypothesized:

H1: An IT professional's tie strength in his/her organization is positively related to that IT professional's job satisfaction.

Number of ties: Number of ties refers to an individual's direct ties across departmental and organizational boundaries. Empirical evidence suggests that number of ties is related to job performance. Job performance can be defined as a product of obtaining the right information to solve novel and challenging problems in a knowledge-intensive work context (Cross and Cummings, 2004). According to social capital theory, individuals can improve their job performance by leveraging their relationships. Individuals with many ties have more access to information and resources (Seibert, Kraimer & Liden, 2001; Rosenthal, 1996). The number of ties of an individual outside the department, spanning physical barriers and to higher hierarchical levels is positively related to individual's performance rating (Cross & Cummings, 2004). For example, managers with many non-redundant contacts and those operating in closed groups tend to achieve high performance (Moran, 2005). IT professionals now may find ties especially valuable. Many IT professionals are heavily engaged in boundary-spanning activities, especially with the implementation of enterprise systems. The enterprise-level projects create opportunities for IT professionals to develop ties with other employees from different departments (Jones, Cline & Ryan, 2006). Maintaining good relationships with them may facilitate IT professionals' work processes. Therefore the following are hypothesized:

H2a: The number of ties an IT professional has in his/her organization is positively related to his/her job performance.

One way to keep abreast of new technical developments is to have formal or informal contacts with peers in other organizations. These informal social channels are effective for individuals to acquire timely, current and soft information. IT professionals can gain new knowledge and stay informed of technological trends in the industry from their peers in other organizations. Furthermore, many external business partners are involved in current IT projects such as hardware, software and service vendors and application service providers, and maintaining good relationships with them also helps IT professionals' job performance. Therefore, the following is hypothesized:

H2b: The number of ties an IT professional has outside his/her organization is positively related to his/her job performance.

METHODOLOGY

A survey was designed to gather data for the study. The survey was derived from prior research and existing measures, and it was refined through expert assessment and a pilot test. Six academic experts in the IT field, the psychology field and the social network field reviewed the survey in order to assess face validity. They were asked to evaluate items for ambiguity, sequencing and flow, as well as for appropriateness of the questions. Modifications were made based on their comments. Next, a pilot test was conducted in which 20 IT professionals were asked to complete the revised survey. The 20 IT professionals were from an IT department of a large railroad company in southern United States. They were asked to provide comments about the clarity and content of the survey in addition to completing it. The instrument was finalized based on changes suggested by the outcome of the pilot test.

Measures

Dependent Variables

- **Job Satisfaction:** Items measuring job satisfaction were selected from the 16 aspects of job satisfaction in Flap and Völker's study (Flap & Völker, 2001). Eight items that refer to the instrumental and the social aspects of the job are selected. Items include satisfaction with social climate, cooperation with supervisors, cooperation with colleagues, income, career opportunities, certainty of the job in the future, utilization of human capital and clarity of expectation.
- **Job Performance:** Items for job performance were taken from the task component of performance rating scale because this study intends to only measure task-related job performance (Touliatos, Bedeian, Mossholder & Barkman, 1984). Items in-

clude ratings of ability, judgment, accuracy, job knowledge, creativity and promotability. In previous studies, the performance rating was provided by supervisors. In this study self-rating of performance was used to increase the response rate. IT research indicates that managerial performance rating is highly correlated with self-rating and self-evaluation of performance has been widely adopted in IT research (Rasch & Tosi, 1992; Jones & Harrison, 1996).

Independent Variables

The name generator methodology is a type of network measure used to elicit the names of people in one's network of relationships and then characterize respondents' relations. Respondents were asked to identify people in their work group, in the department and in the organization and outside their current organization with whom they interact most frequently for job or task-related purposes or other purposes. To protect the anonymity of network members, respondents were asked to provide their contacts' initials.

Tie Strength: Tie strength was measured by closeness between the respondent and his or her contacts. This measure has been demonstrated to be a more reliable indicator of tie strength than others such as frequency and duration of contact which tend to overestimate the strength of the ties in a workplace environment (Mardsen & Campbell, 1984). Tie strength was measured for each person identified in a respondent's network using a 4-point social distance scale ranging from 1 (distant) to 4 (very close). Following Brüderl and Preisendörfer (1998), tie strength in measured by a summative score by adding the closeness score of each tie of a respondent.

Number of ties: Respondents were asked to identify up to fifteen people who have acted to help his/her career by speaking on his/her behalf, providing him/her with information, career opportunities, advice or psychological support with whom he/she has regularly spoken regarding difficulties at work, alternative job opportunities or long-term career goals in his/her current organization. Respondents were given 15 blank lines to list people in their organization. To protect the anonymity of network members, respondents were asked to provide their contacts' initials. Then for each respondent, the total number of ties in the organization was calculated. Then the respondents were asked to list up to 15 contacts with the same description, but outside their current organization. The total number of ties outside organization was then calculated.

In order to help eliminate possible alternative explanations for findings, several individual characteristics were controlled. These were organizational tenure, educational level, industry experience and position. One organizational characteristic, organization size, was also used as a control variable. Although there are a number of variables that could be use as controls, these are the most relevant to the relationships tested in our hypotheses. For example, it is reasonable to posit that organizational size and tenure might influence the amount of social capital because the larger the organization or the longer he/she has been in the organization, the more opportunities the person has to build social capital. Similarly, people with more industry experience may have had greater opportunities to build social capital outside the organization. People with higher positions in the organization or with higher levels of education may also have greater opportunities to leverage their social capital, or may have done so to arrive at their positions. Therefore, we include these items as control variables.

Data Collection

To contact IT professionals for the study, a list of 1500 IT professionals with names, addresses and organizational affiliations was obtained from a market research company. The company maintains a database of over a million of professionals in the United States. The 1500 names and addresses used from the database were randomly selected from individuals classified in this database as an "IT professional." The list represented a range of jobs, organizations and industries. Several steps were taken to ensure a good response rate (Dillman, 2000). First, a personalized cover letter was sent along with the instrument, explaining the research purpose, instructions to the respondents and an appeal to the respondents to complete the survey. Second, a self-addressed, stamped envelope was included with the instrument. Third, a promise of confidentiality was included in the cover letter.

A follow-up card was sent to 750 of the original list three weeks after the first mailout. Of the 1,500 surveys, 1,314 were delivered and the remaining 186 were returned with incorrect address information. The first mailing resulted in 109 responses. The follow-up cards brought 25 more responses. Therefore, there were a total of 134 responses for a 10.20% response rate. Six responses were discarded because of the incomplete information. Although there has been a decline in the response rate of surveys in general over the years (Hambrick, Geletkanycz, & Frederickson, 1993), this response rate is consistent with those in other exploratory studies with IT respondents (e.g., Mabert, Soni & Venkatramanan, 2000; Jones & Young, 2006).

Nonresponse bias was assessed to determine whether there were significant differences between respondents and nonrespondents. This was done by comparing responses from the first mailing and the responses from the follow-up cards after three weeks (Karahana, Straub & Chervany, 1999; Ryan, Harrison & Schkade, 2002). T-tests and Chi-Square tests revealed no significant differences between the two groups on the basis of demographic variables, tie variables or the dependent variables. Therefore, this sample of respondents is consistent with that of the sampling frame and we can draw conclusions from this sample that are generalizable to the broader population of IT professionals.

Profile of Respondents

The average age of respondents was approximately 46 years. Their average organizational tenure was 10.28 years and their average IT experience was 15.50 years. The majority of the respondents (73.6%) were males. In addition, the majority of the respondents were Caucasian (72.9%), followed by foreign-born IT professionals living in the United States (12.4%), Asian American (4.5%), African American (3.8%), Hispanic American (2.3%) and Native American (1.5%). The respondents were well educated and nearly 74% of the respondents had at least a four-year college degree. Furthermore, the respondents represented a broad sample in terms of organization size and organization type. They came from organizations ranging from those with less than 100 employees to those with more than 10,000 employees. A significant number of respondents had managerial positions such as chief information officers, IS directors and IS/IT managers. Other respondents were developers, administrators, analysts, programmers and technicians. All respondents belong to the IT industry.

DATA ANALYSIS

Dependent Variables

The dependent variables for the study are job satisfaction and job performance. Although measures of these variables have been demonstrated to have acceptable measurement properties in previous studies, this study adapted and synthesized measures from a variety of studies. Thus, it is appropriate to examine the measurement properties of the items as measured in this study.

Principle component factor analysis using a Varimax rotation was used to assess the dimensionality of the items in this study. The criterion of an eigenvalue of at least one was used to assess the number of factors to extract, and the dimensionality of each of the factors extracted was assessed by examining factor loadings. For this set of data, the factor analysis yielded two distinct factors based on the Eigenvalue >= 1.0 criteria, and all items loaded on their respective factors with a loading of at least 0.50 (Table 1). The total variance explained by the model was 51.25%. The second aspect of construct validity assessed was reliability of the measures. One of the most commonly used indicators of reliability is internal consistency, which assesses how consistently individuals respond to items within a scale. Cronbach's alphas for all two factors extracted were above 0.80, which indicates the measures were internally consistent. Table 1 shows the factor analysis results along with the Cronbach's alpha and factor means and standard deviations.

Independent Variables

The independent variables for the study are tie strength in the organization, number of ties in the organization and number of ties outside the organization. Table 2 lists the descriptive statistics about the tie variables.

Control Variables

The control variables for the study are education, organization size, organizational tenure, industry experience and position. In this study, education is coded "1" if a respondent has a graduate degree or higher or "0" if otherwise. Position is a control variable because managers may have more ties and more diverse ties than other employees. For this study, position is coded "1" if a respondent is a manager or "0" if otherwise.

Table 1. Factor analysis of dependent variables

	Job Satisfaction	Job Performance
Variance Explained	29.64%	21.61%
Eigenvalue	4.43	2.75
Cronbach's alpha	0.854	0.8
sa3*	**0.790**	0.061
sa8	**0.778**	0.068
sa5	**0.775**	0.084
sa7	**0.767**	0.055
sa1	**0.684**	0.038
sa2	**0.674**	-0.036
sa6	**0.575**	-0.012
sa8	**0.6597**	0.137
perf3	-0.021	**0.763**
perf1	0.025	**0.739**
perf2	0.212	**0.720**
perf54	-0.085	**0.712**
perf5	0.064	**0.689**
perf6	0.321	**0.598**

* Sa = Job Satisfaction; Perf = Job Performance; The numbers after each variable correspond to the question used to measure the items.

RESULTS

Ordinary Least Squares Regression is used to test all hypotheses. To test each hypothesis, a dependent variable is regressed on an independent variable as well as all control variables including education, industry experience, organizational experience, position and organization size. Hypothesis 1 was supported ($\beta = 0.170$, p=0.045). An IT professional's tie strength in the organization is positively related to his/her job. However, the adjusted R square is only 5%. Hypothesis 2a was not supported. We did not find the number of ties an IT professional has in his/her organizations is positively related to job performance. Hypothesis 2b posited that number of ties an IT professional has outside his/her organization is positively related to job performance. This hypothesis was supported ($\beta = 0.337$, p=0.035). The adjusted R square is 4%. The results are presented in Tables 3, 4, and 5.

DISCUSSION

The study examines the relationship between social capital variables and work attitude and behavior of IT professionals. Findings indicate that IT professionals are more satisfied with their job when their ties with other people in the organization are strong. This finding is consistent

Table 2. Tie strength and number of ties

Variables	Mean	Standard Deviation	Min	Max
Tie strength in the organization	14.38	7.17	1	42
Number of ties in the organization	5.66	3.35	1	14
Number of ties outside the organization	4.01	3.21	1	13

Table 3. Tie strength in organization and job satisfaction

Variables	B	t-value	p-value
Education	-0003	-0.22	0.826
Organization size	-0.018	-0.431	0.668
Organizational tenure	-0.002	-0.248	0.801
Industry experience	-0.142	-1.499	0.137
Position	0.014	0.150	0.881
Tie strength in the organization	0.170	2.031	0.045**

Table 4. Number of ties in organization and job performance

Variables	B	t-value	p-value
Education	-0.034	-0.351	0.726
Organization size	0.011	0.107	0.915
Organizational tenure	-0.068	-0.700	0.485
Industry experience	0.005	0.048	0.962
Position	0.029	0.292	0.771
Number of ties in the organization	0.125	1.213	0.228

with what theory predicts about the role of these aspects of social capital in job satisfaction. In addition to being theoretically indicated, this finding is intuitively appealing because people that have strong relationships within their organizations are expected to have formed bonds that make the workplace more appealing or satisfactory for them. The positive relationship between number of ties an IT professional has outside his/her organization and his/her job performance (H2b) is empirically supported, yet it is somewhat disturbing, particularly in light of the lack of support for

H2a. An IT professional's number of ties within the organization is not significantly related to his/her job performance. These findings indicate that what an IT professional gains from his/her ties with people outside the organization is more important to job performance than what they gain from ties within the organization. One disturbing interpretation of this is that there is something lacking in internal support or facilitation of IT job performance. Another explanation, however, may be found in the fact that the ties outside the organization are measured as ties with other IT

Table 5. Number of ties outside the organization and job performance

Variables	B	t-value	p-value
Education	0.102	-0.48	0.632
Organization size	-0.012	-0.43	0.665
Organizational tenure	-0.001	-0.21	0.832
Industry experience	0.005	0.83	0.411
Position	-0.027	-0.26	0.798
Number of ties outside the organization	0.037	2.14	0.035**

professionals. IT professionals may turn to others in their field for ideas and help in solving IT problems they encounter in their own organizations for which IT professionals in the organization have no experience. Professional IT organizations such as the Association for Computing Machinery (ACM) or the Association of Information Technology Professionals (AITP) provide many avenues for these ties to be built and utilized though professional meetings, special interest groups, trade publications, and certification support.

LIMITATIONS

This study is subject to several possible limitations in terms of internal and external validity. First, the cross-sectional design does not allow for causal inferences to be drawn about the effects of measured variables. For example, rather than concluding that the number of ties leads to job satisfaction, it is more appropriate to conclude that number of ties is positively related to IT professionals with higher job satisfaction. Thus, only correlational inferences can be drawn.

Common-method variance is another possible limitation of the study. Common method variance refers to the fact that potential respondent biases might constitute a systematic error. This is common when using survey responses from the same source because a single respondent for each survey can only yield one perspective. Thus, there might be spurious correlation (Bagozzi, 1980). Several precautions were taken to minimize the effects of common-method variance. The dependent variables and independent variables were separated into different sections of the survey instrument. Different question formats were used for each set of variables. Common method variance may also be minimized because social capital variables are behavioral and work attitude and behavior are attitudinal. Thus, two types of perspectives from each respondent are tapped.

Another possible limitation is social desirability bias. Social desirability bias happens when respondents engage in over-reporting positive behaviors and underreporting negative behaviors. The construct in the instrument that may be subject to social desirability bias is job performance. The two constructs could have different values if peer-rated or supervisor-rated measures were used. However, self-reported measures are shown to be useful (Wade & Parent, 2001), and other ratings may introduce other types of bias.

Another possible limitation is one that could be addressed by future research. This study does not investigate the role of virtual social capital. Social capital can occur in the new forms emerging from social networking websites, blogs, virtual teams and web communities. With easy access to Internet, IT professionals may be more likely to develop social ties with people that they never meet face to face. Future research should investigate how virtual social capital affects IT professionals' work attitude and behavior.

THEORETICAL A ND PRACTICAL CONTRIBUTIONS

Consistent with social capital theory, the results of the study indicate that tie strength in an organization is significantly correlated with IT professionals' job satisfaction. The stronger the ties in an organization that IT professionals have, the more satisfied they are with their jobs. In addition, the number of ties that IT professionals have outside of their organizations is positively related to their job performance. However, the low R squares indicate that neither of these social variables explain much variance in IT professionals' work attitude and behavior. The study also does not provide support for a significant relationship between an IT professional's number of ties in an organization and his/her job performance. Research about the importance of social needs among IT professionals is mixed. Some studies indicate that IT professionals have lower social needs whereas others indicate they have social needs equal to their non-IT counterparts (Wynekoop & Walz, 1998). Our findings are more consistent with the former, and as such have implications future research and practice. For example, one area that future research could explore is how IT professionals perceive, build, and use social capital. Our findings are based on *a priori* assumptions grounded in social capital research, yet not much is known in this area specifically about IT professionals. Given that they may be different than their non-IT counterparts, a more in-depth view of how social capital is constructed in the IT community might provide greater insight into the work behaviors and attitudes of IT professionals and into the application of social capital theory.

Another contribution of the study is the identification of social capital for IT professionals both within an organization and outside an organization. The study empirically assesses the number of these ties and the strength of these ties. Prior research has largely overlooked the importance of relationships that IT professionals have within the larger IT community, yet those relationships are prevalent through memberships in professional organizations, vendor relationships, and electronic linkages. The study indicates that the ties outside an organization are significantly related to an IT professional's job performance. This demonstrates that relationships in the large IT community may be valuable assets for IT professionals.

The study also provides guidance for practitioners. For organizational executives, the study provides a possible avenue for increasing job satisfaction of IT professionals. Although the variance explained suggests that social capital is not an overriding force in job satisfaction and performance, findings do indicate that it is significant. For example, the significant link between ties outside the organization and job performance indicates that these ties may be of benefit to the organization and should be supported. Similarly, because job satisfaction is positively related to ties in the organization, managers might undertake

activities such as integration-oriented training so IT professionals have connections to other professionals within the organization. IT managers and business managers can work closely to help IT professionals develop personal connections not only in the work group and department, but also in the organization. Human resource managers can implement community building strategies as suggested by Ferratt et al.(2005). This includes practices such as using newsletters and intranets for employees and providing social activities as well as activities to help employees build social responsibility. These practices support information sharing and building a sense of community among IT professionals, which may lead to a greater number of strong ties among IT professionals.

However, managers also should be cautious about investing too heavily in community-building practices. This study shows that the number of ties in the department and in the organization is not significantly related to job performance. The non-significant impact of social capital on IT professionals may make people wonder how useful these practices are. The reality may be that it is the market, not the organization, that ultimately determines the movement of employees. Executives and managers can make their own organizations as appealing and rewarding as possible. However, if the market draw on employees is strong enough, then talented IT professionals may be lured away by attractive opportunities and aggressive offers. Therefore, Cappelli (2000) suggests a market-driven approach to retain talented employees. Instead of stopping the outflow of talent entirely, human resource managers should have a new goal: to influence who leaves and when by using such mechanisms as compensation, job design, hiring and location. Organizations should also adapt to attrition by outsourcing the required skills or using organizational memory to retain knowledge.

The study presents a theory-driven examination of how social capital is related to work attitude and behavior of IT professionals. This is one of the first studies that is undertaken to apply the social capital theory to these constructs in the context of IT professionals. While previous research on IT professionals examines the IT professionals' social needs, relationship with supervisors and social support, little research has investigated their social ties, the strength of these ties and structure of the ties. The study contributes to filling this research gap and investigates the influence of tie strength and number of ties on IT professionals' work attitude and behavior, and contributing to both research and practice.

REFERENCES

Adler, P. S., & Kwon, S.-W. (2002). Social capital: prospects for a new concept. *Academy of Management Review, 27*(1), 17–40. doi:10.2307/4134367

Agarwal, R., Brown, C., Ferratt, T., & Moore, J. E. (2006). Five mindsets for retaining IT staff. *MIS Quarterly Executive, 5*(3), 137–150.

Agarwal, R., & Ferratt, T. W. (2001). Crafting an HR strategy to meet the need for IT workers. *Communications of the ACM, 44*(7), 58–64. doi:10.1145/379300.379314

Agarwal, R., & Ferratt, T. W. (2002). Enduring practices for managing IT professionals. *Communications of the ACM, 45*(9), 73–79. doi:10.1145/567498.567502

Ang, S., & Slaughter, S. A. (2000). The missing context of information technology personnel: a review and future directions for research. In Zmud, R. W. (Ed.), *Framing the Domains of IT Management* (pp. 305–327). Cincinnati, OH: Pinnaflex Educational Resources.

Aryee, S., & Chay, Y. W. (1994). An examination of the impact of career-oriented mentoring on work commitment attitudes and career satisfaction among professional and managerial employees. *British Journal of Management, 5*(4), 241–249. doi:10.1111/j.1467-8551.1994.tb00076.x

Bagozzi, P. (1980). *Causal methods in marketing*. New York: John Wiley & Sons.

Bassellier, G., & Benbasat, I. (2004). Business competence of information technology professionals: conceptual development and influence on IT-business partnerships. *Management Information Systems Quarterly, 28*(4), 673–694.

Bourdieu, P. (1986). The forms of capital. In Richardson, J. G. (Ed.), *Handbook of Theory and Research for the Sociology of Education* (pp. 241–258). New York: Greenwood.

Brüderl, J., & Preisendörfer, P. (1998). Network support and the success of newly founded business. *Small Business Economics, 10*(3), 213–225. doi:10.1023/A:1007997102930

Cappelli, M. (2000). A market-driven approach to retaining talent. *Harvard Business Review*, 103–111.

Couger, J. D., Oppermann, E. B., & Amoroso, D. L. (1994). Changes in motivation of IS managers – Comparison over a Decade. *Information Resources Management Journal, 7*(2), 5–13.

Cranny, C. J., Smith, P. C., & Stone, E. F. (1992). *Job satisfaction: how people feel about their jobs and how it affects their performance*. New York: Lexington Press.

Cross, R., & Cummings, J. N. (2004). Tie and network correlates of individual performance in knowledge-intensive work. *Academy of Management Journal, 47*(6), 928–937. doi:10.2307/20159632

Dillman, D. A. (2000). *Mail and Internet Surveys: the Tailored Design*. New York: John Wiley & Sons.

Ferratt, T. W., Agarwal, R., Brown, C., & Moore, J. E. (2005). IT Human resource management configurations and IT turnover: theoretical synthesis and empirical analysis. *Information Systems Research, 16*(3), 237–255. doi:10.1287/isre.1050.0057

Flap, H., & Völker, B. (2001). Goal specific social capital and job satisfaction effects of different types of networks on instrumental and social aspects of work. *Social Networks, 23,* 297–320. doi:10.1016/S0378-8733(01)00044-2

Granovetter, M. S. (1973). The strength of weak ties. *American Journal of Sociology, 78*(6), 1360–1380. doi:10.1086/225469

Granovetter, M. S. (1982). The strength of weak ties: a network theory revisited. In Mardsen, P. V., & Lin, N. (Eds.), *Social Structure and Network Analysis* (pp. 105–130). Beverly Hills, CA: Sage.

Guzman, I. R., Stam, K. R., & Stanton, J. M. (2008). The occupational culture of IS/IT personnel within organizations. *The Data Base for Advances in Information Systems, 39,* 33–50.

Hambrick, D. C., Geletkanycz, M. A., & Frederickson, J. W. (1993). Top executive commitment to the status quo: Some tests of its determinants. *Strategic Management Journal, 14,* 401–418. doi:10.1002/smj.4250140602

Helliwell, J., & Putnam, R. (2007). Education and social capital. *Eastern Economic Journal, 33,* 1–19. doi:10.1057/eej.2007.1

Ibarra, H. (1995). Race, Opportunities, and diversity of social circles in managerial networks. *Academy of Management Journal, 38,* 673–703. doi:10.2307/256742

Jones, M. C., Cline, M., & Ryan, S. (2006). Exploring knowledge sharing in ERP implementation: an organizational culture framework. *Decision Support Systems, 41*(2), 411–434. doi:10.1016/j.dss.2004.06.017

Jones, M. C., & Harrison, A. W. (1996). Is project team performance: an empirical assessment. *Information & Management, 31*(2), 57–65. doi:10.1016/S0378-7206(96)01068-3

Jones, M. C., & Young, R. (2006). ERP usage in practice: An empirical investigation. *Information Resources Management Journal, 19*(1), 23–42.

Karahanna, E., Straub, D. W., & Chervany, N. L. (1999). Information technology adoption across time: a cross-sectional comparison of pre-adoption and post-adoption beliefs. *Management Information Systems Quarterly, 23*(2), 183–213. doi:10.2307/249751

Kelly, R., & Caplan, J. (1993). How Bell Labs creates star performers. *Harvard Business Review, 71*(4), 128–139.

Krackhardt, D. (1992). The strength of strong ties: the importance of philos in Organizations. In Nohria, N., & Eccles, R. (Eds.), *Networks and Organizations: Structures, Form and Action* (pp. 216–239). Boston: Harvard Business School Press.

Lee, D. M. S. (1994). Social ties, task-related communication and first job performance of young engineers. *Journal of Engineering and Technology Management, 11*(3-4), 203–229. doi:10.1016/0923-4748(94)90010-8

Lee, P. C. B. (2004). Social support and leaving intention among computer professionals. *Information & Management, 41*(3), 323–334. doi:10.1016/S0378-7206(03)00077-6

Lin, N. (2000). Inequality in social Capital. *Contemporary Sociology, 29*(6), 785–795. doi:10.2307/2654086

Mabert, V. A., Soni, A., & Venkatramanan, M. A. (2000). Enterprise resource planning survey of US manufacturing firms. *Production and Inventory Management, 41*(2), 52–58.

Marsden, P. V. (1988). Homogeneity in confiding networks. *Social Networks, 10,* 57–76. doi:10.1016/0378-8733(88)90010-X

Marsden, P. V., & Campbell, K. E. (1984). Measuring tie strength. *Social Forces, 63*(2), 482–501. doi:10.2307/2579058

Moore, J. E., & Burke, L. S. (2002). How to turn around 'turnover culture' in IT. *Communications of the ACM, 45*(2), 73–78. doi:10.1145/503124.503126

Moran, P. (2005). Structural vs. relational embeddedness: social capital and managerial performance. *Strategic Management Journal, 26*, 1129–1151. doi:10.1002/smj.486

Morrison, E. W. (2002). Newcomers' relationships: the role of social network ties during socialization. *Academy of Management Journal, 45*(6), 1149–1160. doi:10.2307/3069430

Nahapiet, J., & Ghoshal, S. (1998). Social capital, intellectual capital and the organizational advantage. *Academy of Management Review, 23*(2), 242–266. doi:10.2307/259373

Putnam, R. D. (1993). The Prosperous community: social capital and public life. *The American Prospect, 13*, 35–42.

Rasch, R. H., & Tosi, H. L. (1992). Factors affecting software developers' performance: an integrated approach. *Management Information Systems Quarterly, 16*(3), 395–413. doi:10.2307/249535

Rodan, S., & Galunic, C. (2004). More than network structure: how knowledge heterogeneity influences managerial performance and innovation. *Strategic Management Information, 25*, 541–562.

Rosenthal, E. A. (1996). *Social Network and Team Performance*. Unpublished doctoral dissertation, University of Chicago.

Rutner, P. S., Hardgrave, B. C., & McKnight, D. H. (2008). Emotional dissonance and the information technology professionals. *Management Information Systems Quarterly, 32*(3), 635–652.

Ryan, S. D., Harrison, D. W., & Schkade, L. L. (2002). Information-technology investment decisions: when do costs and benefits in the social subsystem matter? *Journal of Management Information Systems, 19*(2), 85–127.

Seibert, S., Kraimer, M. L., & Liden, R. (2001). A social capital theory of career success. *Academy of Management Journal, 44*(2), 219–237. doi:10.2307/3069452

Shah, P. P. (1998). Who are employees' social referents: using a network perspective to determine referent others. *Academy of Management Journal, 41*(3), 249–268. doi:10.2307/256906

Totterdell, P., Wall, T., Holman, D., & Epitropaki, O. (2004). Affect network: A structural analysis of the relationship between work ties and job–related affect. *The Journal of Applied Psychology, 89*(5), 854–867. doi:10.1037/0021-9010.89.5.854

Touliatos, J., Bedeian, A. G., Mossholder, K. W., & Barkman, A. I. (1984). Job-related perceptions of male and female government, industrial and public accountants. *Social Behavior and Personality, 12*, 61–68. doi:10.2224/sbp.1984.12.1.61

Wade, M. R., & Parent, M. (2001). Relationships between job skills and performance: a study of webmasters. *Journal of Management Information Systems, 18*(3), 71–76.

Wynekoop, J. L., & Walz, D. B. (1998). Revisiting the perennial question: are IT people different? *The Data Base for Advances in Information Systems, 29*(2), 62–72.

This work was previously published in the Journal of Organizational and End User Computing, Volume 23, Issue 1, edited by Mo Adam Mahmood, pp. 64-78, copyright 2011 by IGI Publishing (an imprint of IGI Global).

Chapter 10
Gender Differences in E-Learning:
Communication, Social Presence, and Learning Outcomes

Richard D. Johnson
University at Albany – SUNY, USA

ABSTRACT

Although previous research has suggested that women may be at a learning disadvantage in e-learning environments, this study examines communication differences between women and men, arguing that women's communication patterns may provide them with a learning advantage. Using data from 303 males and 252 females, this paper discusses gender differences in course communication processes and course outcomes in a web-based introductory information systems course. Results indicate that women communicated more, perceived the environment to have greater social presence, were more satisfied with the course, found the course to be of greater value, and had marginally better performance than men. Despite the challenges facing women in e-learning environments, the results of this study suggest that e-learning environments that allow for peer to peer communication and connectedness can help females overcome some of these disadvantages. Implications for research and practice are also discussed.

INTRODUCTION

Estimates suggest that organizations are annually spending over $40 billion on technology based training (American Society for Training and Development, 2007). In addition, nearly 90% of universities are providing e-learning opportunities in which millions of students are enrolling (Wirt & Livingston, 2004). Some researchers have even argued that these training initiatives are "part of the biggest change in the way our species conducts training since the invention of the chalkboard"

DOI: 10.4018/978-1-4666-2059-9.ch010

(Horton, 2000, p. 6). Although e-learning has been argued to provide on-demand training to employees around the globe (Salas, DeRouin, & Littrell, 2005), learners with increased flexibility and control (Piccoli, Ahmad, & Ives, 2001), and organizational cost savings (Salas et al., 2005), e-learning still has several shortcomings. For example, it is difficult for courses to keep trainees engaged and learners tend to feel isolated when participating in these courses (Salas et al., 2005).

For organizations and universities to realize the benefits and overcome the limitations, it is important for researchers to not only investigate technology and processes involved in e-learning, but also to understand the characteristics of those who are participating in e-learning initiatives and how that may affect e-learning outcomes. Research has suggested that women make up the majority of those participating in e-learning initiatives, exceeding 60 percent in some countries (Kramarae, 2001; Sehoole & Moja, 2003). Thus, researchers are becoming more interested in e-learning gender issues. For example, researchers have found that the online environment can be openly hostile toward females, with men often dominating discussions and making negative comments about females (cf. Blum, 1999; Kramarae, 2001). In addition, research has suggested women's participation in e-learning is often related to how easy or difficult they find the software to use (Ong & Lai, 2006). Thus, it has been argued that women may be at a disadvantage in e-learning because they have lower experience or confidence in the use of computers (cf. Thompson & Lynch, 2003).

Despite these acknowledged disadvantages, evidence suggests that the nature of women's communication patterns in e-learning courses may provide them with a potential advantage. For example, female student's communication in e-learning environments tends to be more interactive and socially oriented than men (Barrett & Lally, 1999; Blum, 1998). Because peer interaction and the facilitation of a shared learning space is central to effective e-learning (cf. Gunawardena,

Lowe, Constance, & Anderson, 1997; Hiltz, 1994; Hiltz, Zhang, & Turoff, 2002; Tu & McIsaac, 2002), this higher social orientation may allow women to be more connected to other learners, have a higher sense of peer and instructor presence, and have better outcomes. Although there have been studies focusing on gender differences in e-learning (cf. Arbaugh, 2000; Barrett & Lally, 1999; Blum, 1999), we are not aware of research that has specifically focused on gender differences in communication, perceptions of peer presence, and learning outcomes. Therefore, the goal of this study was to investigate gender differences in communication, perceptions of social presence, and e-learning outcomes.

The remainder of the paper is organized as follows. In the next section, the relationship between gender, communication, and social presence are discussed. Following this, gender and e-learning outcomes are discussed. In the third section, the research setting, participants, and methods are discussed. In the fourth section, results are discussed. Finally the manuscript closes with a discussion of the findings and implications for e-learning researchers and practitioners.

LITERATURE REVIEW

E-Learning Communication Processes

Learning has long been conjectured to have a social component (cf. Vygotsky, 1978), with researchers arguing that the most effective learning occurs when learners are able to actively interact with content, peers, and the instructor (cf. Hiltz, 1994; Hiltz et al., 2002; Schmidt & Ford, 2003). Pedagogically, interaction with content encompasses the constructivist notion that learning takes place only by being actively engaged with the material to be learned, by incorporating it into existing mental models, and individually creating knowledge. Interaction with peers embodies the concepts found in social constructivism and collaborative learning where it is believed that knowledge is created within a social context. Active discussions between students are thought to engage students and create deeper learning. Interaction with the instructor reflects the approaches of objectivist learning where the learner is thought to receive knowledge imparted by the instructor. These types of interaction processes are argued to form, the most "'natural' way for people to learn" (Hiltz, 1994, p. 22).

The challenge is that the e-learning environment inserts a layer of mediation between those involved in the course, creating a potential barrier to the interactive and shared social context in which learning takes place. A shared learning environment is only created as learners interact with each other mediated through technology. Despite these challenges, a shared learning space can be created in these environments as trainees structure the environment through their interactions meditated via the technology (Hillman, Willis, & Gunawardena, 1994). Specifically, there are two aspects of communication which are critical to the creation of this shared learning environment: interaction and social presence (cf. Hiltz et al., 2002; Tu & McIsaac, 2002).

Interaction is defined by researchers as "the exchange of information between the various stakeholders in the course (e.g. peers, instructors, and other support staff)" (Johnson, Hornik, & Salas, 2008, p. 360). Typical forms of interaction studied by e-learning researchers include asynchronous tools such as discussion threads and email, and synchronous tools such as the electronic classroom, chat, and instant messaging (cf. Alavi, Marakas, & Yoo, 2002; Gunawardena, 1995; Hiltz, 1994; Johnson et al., 2008).

But interaction only captures a portion of the communication processes. Communication and learning is further enhanced when learners are part of a shared learning context. Learning in a social context can change the way in which individuals interact and the way they learn and

think (Vygotsky, 1978) and can lead to greater learning than when only information exchange occurs (Kuhl, Tsao, & Liu, 2003; Rosenzweig & Bennett, 1978). Therefore, within an e-learning environment, the technology must be able to support the creation of a shared learning environment.

In e-learning, a critical factor to the development of these shared learning environments is social presence. Social presence has been defined as "the degree of salience of the other person in the interaction and the consequent salience of the interpersonal relationships" (Short, Williams, & Christie, 1976, p. 65). Essentially, for the e-learning environment, social presence reflects the extent to which learners perceive the technology as enabling them to create an environment in which they feel is warm, personal, sociable, and active and allows them to be connected in a shared learning space (Biocca, Harms, & Burgoon, 2003). As with interaction, social presence has been theorized to be critical to the success of e-learning (cf. Gunawardena, 1995; Johnson et al., 2008). This is because individuals who perceive themselves in a shared context exchange greater and more complex information (McGrath, Arrow, Gruenfeld, Hollingshead, & O'Connor, 1993). Individuals may feel that they can more effectively judge the quality of peer inputs, leading them to place higher value on peer messages and process these messages to a greater extent (Sahay, 2004), than when they are isolated.

Gender and E-Learning Communication Processes

Evidence suggests that women and men may differ in how they interact, how much they interact, and how connected they are in the e-learning environment. For instance, Gender Role Theory (Eagly, 1987) suggests that women are more attuned to the socially oriented aspects of communication. In addition, they are more concerned with creating and maintaining relationships and intimacy in those relationships (Tannen, 1990; Wood &

Rhodes, 1992). Finally, women are more likely to use communication to develop and maintain connections and community (Edelsky, 1981; Tannen, 1993).

We believe that this tendency for women to be network focused in their communication can lead to increased interaction and stronger perceptions of peer presence. For example, the tendency for women to emphasize social interaction in communication may lead to greater course interaction as women seek to develop relationships with instructors and peers. They may take fuller advantage of the collaborative nature of e-learning environments through asynchronous and synchronous communication tools. Second, women have been found to use technology for the development and maintenance of social relationships (cf. Boneva, Kraut, & Frohlich, 2001; Gorriz & Medusa, 2000). Thus, we believe women's communication patterns should also be more socially oriented, with the goal of developing a shared and more intimate learning context. This in turn should lead to women having greater perceptions of social presence than men. Therefore, the following hypotheses were investigated.

H1: In an e-learning environment, women will have higher levels of interaction than men.
H2: In an e–learning environment, women will have a higher level of perceived social presence than men.

Gender and E-Learning Outcomes

Although there are many types of course outcomes, e-learning research has traditionally focused only on learning performance and satisfaction. Satisfaction is important in e-learning design because individuals who are less satisfied with their experiences or view the system negatively are less likely to continue to participate in e-learning initiatives (Carswell & Venkatesh, 2002; Lim, 2001). Using the levels of learning outlined by Kirkpatrick (1976), this study also focuses on instrumentality.

Instrumentality is a first level learning outcome (e.g. reactions to training) reflecting judgments of the utility of training such as the whether the knowledge gained in the course can be used after training and the motivation of the trainee to do so. Instrumentality is an important outcome for e-learning researchers to study because it is often a better predictor of actual learning transfer than course performance (Alliger, Tannenbaum, Bennett, Traver, & Shotland, 1997).

We argue that because women should communicate more and feel higher social presence they should also have better e-learning outcomes than men. For example, as discussed earlier, socially connected learning with interaction between peers, content, and instructors has been argued to be the most effective and natural way to learn. This type of learning allows individuals to share information, to receive feedback and to more readily evaluate progress (Piccoli et al., 2001) and can change the way in which individuals interact and the way they learn and think (Vygotsky, 1978), lead to greater and more complex information sharing (McGrath et al., 1993), and lead to greater learning than when only information exchange occurs (Kuhl et al., 2003; Rosenzweig & Bennett, 1978).

For example, through the use of asynchronous communication tools such as discussion threads, ongoing trainee responses allow individual learners to post comments, review comments made since they previously posted, and respond to these comments. Through the use of synchronous communication tools such as chat, learners can gain immediate feedback and evaluation of their comments and questions.

Ultimately the shared learning space and shared peer connections can lead to a broader and deeper sharing and processing of information and active engagement in the learning process than when individuals are isolated. This can occur because learners will be more likely to attend to messages more deeply and see the value of the messages and to value the ideas and contributions of others (Mackie, Worth, & Asuncion, 1990). This in turn should lead to improved learning and improved perceptions of instrumentality. In fact, previous research has already found that in e-learning environments, increased interaction has led to improved course learning and performance (cf. Alavi et al., 2002; Gunawardena et al., 1997; Hiltz et al., 2002; Schmidt & Ford, 2003). Overall, these results suggest that women should have greater opportunities to more broadly and deeply process course information than men, improving learning outcomes.

H3: Women will have higher perceptions of e-learning instrumentality than will men.
H4: Women will have higher e-learning performance than men.

In addition, these higher levels of interaction and presence suggest that women should have higher affective reactions to the learning (e.g. satisfaction). For example, previous research has found that isolation in e-learning environments has been one of the largest complaints about these initiatives (Moore, 2002), with research finding that only 3% of individuals prefer learning alone (Linne & Plers, 2002).

As individuals interact and socially connect, they should begin to feel less isolated and more connected with others. Timely feedback and interaction with the instructor can help learners feel that they are valued as well as providing them information in a more timely fashion that they need to learn. In addition, social presence can overcome these potential negative reactions by enabling stronger peer connections, reducing feelings of isolation and strengthening feelings of psychological connection and community. Finally with greater connectedness improving course performance, learners should be more positively inclined toward the course (cf. Arbaugh, 2001; Gunawardena et al., 1997). Together these results suggest that the greater interaction and connectedness of women should lead increased satisfaction.

H5: Women will have higher e-learning satisfaction than men.

METHOD

Research Setting & Procedures

The study was conducted in a required management information systems (IS) fundamentals course at a large university in the United States which was taught exclusively online using WebCT. The course was taught over 15-weeks and was divided into 6 modules. Each module lasted approximately two weeks and focused on different topical areas, such as the strategic use of information and technology, e-commerce, decision support, etc. The structure required individuals to read material from an introductory IS textbook (and supplemental web-links) and then to post answers to questions from an end of chapter case. To leverage the social context of communication between learners, the students were then required to discuss each other's answers using the asynchronous capabilities of WebCT (i.e. a threaded discussion). To facilitate communication and to better manage the course, students were separated into groups of approximately 30. Although chat and email capabilities were available, their use was not required. The course was managed by one instructor and two graduate assistants (GAs), who communicated exclusively online, and one GA held office hours both online and in person[1].

Data were collected from multiple sources. Perceptual data were collected using an online survey made available to participants for one week during the last course module. Interaction data were obtained from WebCT system logs, and course grades (e.g. performance) were obtained from the instructor. Students were able to opt-in to the study by completing the online survey and agreeing to have their course interaction and performance data used as part of the study. Participating students received nominal course credit for participating.

Research Participants

A total of 719 students participated in the study, of which usable data were obtained from 555. The sample consisted of 303 males and 252 females. The average age was 24.0 (SD=5.5), with a range of 18-56. Of those participating in the study nearly 75% were in their first e-learning course, 19% in their second, and 6% in their third online course. Participants were computer and Internet literate, with over 50% indicating that they had high levels of experience in both.

Measures

This study included one independent variable, gender, and several dependent variables which are discussed below.

- **Interaction**: Interaction was assessed in three ways using WebCT logs: the number of discussion postings read, the number of original discussion postings, and the number of follow-up discussion posts. Note that the measurement of interaction in this fashion is consistent with previous research (Hornik, Johnson, & Wu, 2007; Johnson et al., 2008).
- **Social Presence:** Social presence was measured with a 5-item scale developed by Short et al. (1976). For each question, respondents evaluated the characteristics of the environment using a 5-point, Likert-type scale with anchors such as "unsocia-ble-sociable" and "impersonal-personal." The coefficient alpha reliability estimate for this scale was .80. Note that this scale has been used in many different studies on technology mediated communication and e-learning (cf. Burke & Chidambaram, 1999; Hornik et al., 2007; Yoo & Alavi, 2001).
- **e-Learning Effectiveness:** As discussed earlier, Kirkpatrick (1976) identifies four

levels at which the effectiveness of training can be evaluated. Of these, the first two levels (reactions and learning) are of interest in this study. Reactions to the training can further be divided into affective reactions (i.e. satisfaction) and judgments of the utility or instrumentality of the training.

- **Satisfaction:** Measured with a 7-item Likert-type scale developed by Biner (1993). The scale used a 7-point strongly disagree to strongly agree response format. The coefficient alpha reliability estimate for this scale was .85. A sample item included, "I am satisfied with the clarity with which the class assignments were communicated." Note that this scale has been used in previous research on e-learning (Biner, Welsh, Barone, Summers, & Dean, 1997; Hornik, Saunders, Li, Moskal, & Dzuiban, 2008).
- **Instrumentality:** Measured using a 6-item scale from Alavi et al. (1994), which was designed to assess the extent to which train-ees believed that the course enhanced their skills. This scale used a 7-point, Likert type scale anchored by strongly disagree and strongly agree, and the wording was updated to reflect the current context. The coefficient alpha reliability estimate for this scale was .89. A sample item included, "I improved my ability to critically think about Information Technology." Note that this scale has been used in previous research (Alavi et al., 2002; Alavi, Yoo, & Vogel, 1997).
- **Performance (learning):** The final grade obtained in the course, which could vary from an F to an A (+/- scales were used), converted to a 0.0 - 4.0 scale.
- **Control Variables:** We also collected data on several other factors which could affect course processes and outcomes, including age, self-reported GPA, work experience, experience in e-learning courses, computer

experience, Internet experience, and general computer self-efficacy (GCSE). GPA was measured in the same way as performance (e.g. 0.0 – 4.0 scale), work experience was measured with a single question that asked how long (in years) that the individual has been working, and experience in e-learning courses was measured with a single item asking the respondent to list the number of previous e-learning courses they had taken. Computer and Internet experience were each measured using a single item scale that asked individuals to assess their previous database skills along a 5-point scale anchored by "no experience" and "a lot of experience". GCSE was measured with a 7-item scale developed Marakas, Johnson, and Clay (2007) and used previously by Johnson, Marakas, and Palmer (2006). The scale used a response format of 0 (Cannot Do) - 100 (Totally Confident) in increments of 10. A sample item included "I believe I have the ability to identify and correct common operational problems with a computer." The coefficient alpha reliability for this construct was .91. A complete listing of scale items can be found in the Appendix.

Preliminary Analysis

An initial set of analysis of variances (ANOVA) were run to determine whether there were gender differences in any of the control variables. Results indicated that the group means did not differ for men and women on several control variables: age ($F = 1.39$, $p = .24$), computer experience ($F = 1.45$, $p = .23$), experience in distance education courses ($F = 0.44$, $p = .51$), and work experience ($F = .17$, $p = .68$). Conversely there were differences between males and females on GPA ($F = 7.90$, $p < .01$), Internet experience ($F = 4.49$, $p < .05$), and general computer self-efficacy ($F = 52.18$, $p < .001$). Females ($M = 3.14$) had higher GPAs

than males ($M = 3.03$), but males reported more Internet experience ($M = 4.31: 4.01$) and higher GCSE ($M = 78.70: 67.49$). Thus, these variables were used as covariates in the analysis.

Table 1 shows the means and standard deviations of the learning process and outcomes based upon gender of the learner and Table 2 shows the correlations among the dependent variables. As can be seen, the dependent variables were correlated ($p < .05$ to $p < .001$), so a multivariate analysis of covariance (MANCOVA) was run to assess the effects of gender on the variables of interest in the study. The overall test was significant (Wilks' lambda $F(8,483) = 4.17$, $p < .001$), which allowed for an individual analysis of covariance (ANCOVA) to be performed for each process and outcome variable.

Analysis

To test each hypothesis, a separate ANCOVA was run, with GPA, GCSE, and Internet experience as covariates.

RESULTS

Support was found for H1, which theorized that women would interact more than men. Females were more likely than men to read others' postings ($M = 429.65: 262.48$, $F(1,549) = 20.01$, $p \leq .001$), make original postings ($M = 6.97: 6.44$, $F(1,549) = 7.12$, $p \leq .01$), and make follow-up postings ($M = 7.18: 4.96$, $F(1,550) = 17.07$, $p \leq .001$). GPA was a significant covariate ($p \leq .001$) for both the number of original posts and follow-up postings made.

Females also had higher perceptions of social presence than did males ($M = 2.97: 2.71$, $F(1,525) = 9.61$, $p \leq .01$), providing support for H2. There were no significant covariates with social presence. Partial support was found for H3, with females ($M = 5.13$) having higher perceptions of instrumentality than males ($M = 4.91: F(1,550) =$

Table 1. Means and standard deviations of learning process and outcomes

Variable	Gender	M	SD
Postings Read	Male	245.53	418.68
	Female	406.51	483.58
	Total	318.76	456.06
Postings Made	Male	6.29	2.74
	Female	6.95	2.75
	Total	6.60	2.76
Follow-Up Posts	Male	4.58	7.26
	Female	6.89	6.30
	Total	5.63	6.93
Social Presence	Male	2.73	0.79
	Female	2.93	0.81
	Total	2.82	0.81
Instrumentality	Male	4.80	1.19
	Female	4.91	1.12
	Total	4.85	1.16
Performance	Male	3.34	0.66
	Female	3.46	0.58
	Total	3.40	0.62
Satisfaction	Male	4.39	1.15
	Female	4.71	1.26
	Total	4.54	1.21

Table 2. Correlations among the dependent variables

	1.	2.	3.	4.	5.	6.	7.
1. Postings Read	---						
2. Postings Made	.26***	--					
3. Follow Up Postings	.53***	.42***	--				
4. Social Presence	.15***	.10**	.16***	--			
5. Instrumentality	.10**	.16***	.14***	.43***	--		
6. Performance	.16***	.33***	.27***	.17***	.13***	--	
7. Satisfaction	.18***	.11**	.17***	.55***	.47***	.14***	--

** $p \leq .01$ *** $p \leq .001$

Table 3. Summary results

Dependent Variable	F	p	Covariate	Hypothesis	Supported
Postings Read	20.01	≤.001	--	1	Y
Postings Made	7.12	≤.01	GPA (p <.001)	1	Y
Follow Up Postings	17.07	≤.001	GPA (p <.001)	1	Y
Social Presence	9.96	≤.01	--	2	Y
Instrumentality	5.60	≤.05	GCSE (p <.01)	3	Y
Performance	2.88	=.09	GPA (p <.001)	4	N
Satisfaction	11.65	≤.001	--	5	Y

$5.60, p ≤.05$), but female's performance (M=3.46) was only marginally better than males (M = 3.34; $F (1,550) = 2.88, p =.09$) suggesting lack of support for H4. GCSE was a significant covariate for instrumentality ($p ≤.01$) and GPA for performance ($p ≤.001$). Finally, in support of H5, women (M = 4.76) were more satisfied with the learning environment than men (M=4.37: $F (1,522) = 11.65, p ≤.001$). The results are summarized in Table 3.

DISCUSSION

The results of this study indicate that in e-learning environments gender differences exist in interaction, perceptions of social presence, instrumentality and satisfaction. Women were more likely to make postings, read the postings of others, to respond to others' postings, and to have higher perceptions of social presence. In addition, women were more satisfied with the course and found it to be of greater value (e.g. instrumentality). Contrary to the theorized relationships, women only performed marginally better than men (p =.09). Overall the results of this study suggest that communication differences between women and men in e-learning environments exist, and that these differences may provide females a richer, more connected, and more valuable learning experience than males.

Implications

This study has several implications for research and practice. First, the results suggest that e-learning environments which encourage peer to peer interaction may provide women with a learning advantage over men through increased communication and social presence. Because women tend to be more social, network, and intimacy oriented in their communication, they may find that they are able to more easily overcome the inherent isolation in e-learning environments. They may be able to leverage the environment to create the social connections necessary for a shared learning space. Conversely, the results suggest that males may actually be at a disadvantage in the online environment because they do not communicate as much and may be less likely to create the social connections necessary for effective learning.

Consistent with previous research (Hornik et al., 2007), there were strong relationships amongst interaction, social presence, and course outcomes. This reinforces the importance of creating connections amongst learners and a shared learning space. Therefore, as more e-learning initiatives begin to incorporate additional interaction opportunities into their programs, and enhanced communication tools provide additional interaction opportunities, the potential exists for even larger differences to emerge between women and men.

Men may not naturally take advantage of these communication opportunities, reducing learning potential, satisfaction and perceptions of course utility. Thus, in an organizational training setting, men may not be as likely to transfer their learning into the workplace, because of lower perceived learning value.

Ultimately, these differences in communication and outcomes suggest that there may be a need to explicitly design e-learning initiatives to not only encourage communication, but which also make communication a formal part of the learning process. This can create many challenges for organizations who are currently reducing training budgets and the majority of whose technology based training opportunities are often individually focused (Anonymous, 2008). The thought of having to invest additional resources into expanding interactive learning environments can seem daunting to firms.

Despite this potential challenge, organizations can and design e-learning to support and encourage communication and a shared learning environment without having to rely on expensive or sophisticated interfaces. For example, many universities are starting to use online immersive virtual environments to support learning (Johnson, 2008). Many of these are available at a fraction of the cost of self-developing a fully functional immersive environment within the organization. These environments can provide users with the sensation of "being there" and the creation of a shared learning space (Ducheneaut, Yee, Nickell, & Moore, 2006), enhancing opportunities for both women and men to feel more connected within the e-classroom.

In addition, designers of e-learning initiatives might consider developing small projects where learners must team together to complete a simple project. The advantage of such an approach is that it would both a) encourage closer ties among learners, and b) it should improve learning engagement as they work together. Another approach could

be to create a game type scenario where random credit is given for those who post valuable ideas and suggestions to a discussion group or listserv. This can be used to not only facilitate the learning process, but also to improve the connections between learners (Salas et al., 2005).

Although not the main focus of this study, the research findings reinforce the importance of self-efficacy in e-learning. GCSE was related to perceptions of instrumentality, suggesting that those with higher GCSE found more value in the course than those with lower GCSE. This may be because they are better able to leverage the communication tools to their advantage, focusing their energies in the content of their communication instead of expending cognitive resources on the navigation of the environment. Given the finding that women had lower GCSE and the broader findings that women often have less computer experience (cf. Whitley, 1997), it is important to better understand how this lower efficacy may affect gender differences in e-learning communication and outcomes. In addition, e-learning designers may benefit by providing training or games so that individuals can have early positive experiences with the technology (Salas et al., 2005).

Future Research

The findings in this study also suggest opportunities for future research. For example, future research could consider investigating how to better create environments which support not only the natural social communications of women, but also to develop environments which increase the social connections of men. In addition, with previous research on communication suggesting that multiple types of communication (task focused, team maintenance, and member support) are necessary for the development and effectiveness of teams (McGrath, 1991), research should also address whether the same types of communication patterns improve social connections and performance in e-learning. Finally, future research could systematically focus on how men and women's communication patterns differ in terms of the focus of communication (e.g. task focused, team maintenance, social maintenance, etc.) and how this affects learning outcomes.

Another interesting question is how a shared learning environment can be developed which supports behavioral modeling. The basic premise of behavioral modeling is that through seeing others model the behavior or completing the task, trainees can learn better how to complete the behavior or task. Given the importance of behavioral modeling techniques in training (cf. Compeau & Higgins, 1995; Johnson & Marakas, 2000; Latham & Saari, 1979), an interesting future research question is how to translate behavioral modeling techniques into an e-learning context where individuals are not in physical proximity. A related question is whether both women and men would equally benefit from the utilization of such modeling techniques mediated via technology.

Limitations

As with any study, there are several potential factors that limit the generalizability of these findings. First, the results from this study represent a specific technology implementation in a single course. Future research should replicate and extend the findings of this study with different technologies and different contexts. A second potential limitation is that the results are based upon student perceptions of the environment as opposed to a formal manipulation of the training environment. Although this limitation must be acknowledged, we believe that this limitation is somewhat mitigated because, as previous research has shown, it is not the actual learning environment which is critical, but instead it is the student interpretation of the environment that defines the environment (Stryker, 1980).

CONCLUSION

Previous research on gender differences in e-learning environments has often focused on the barriers women face when participating in e-learning initiatives and has described the e-learning environment as often being hostile toward women. Despite these previous findings, communication research and Gender Role Theory suggests that differences in communication patterns between women and men may actually provide women an advantage in e-learning initiatives. Thus, the goal of this study was to investigate gender differences in communication, perceptions of social presence, and learning outcomes in an e-learning environment. Results of the study indicated that females communicated more, perceived the environment to have greater social presence, believed the course to be of greater value, were more satisfied, and performed marginally better than males. Overall, this research suggests the importance of a delving more deeply into how men and women communicate, use technology, and perform in e-learning environments. Through this, e-learning initiatives can be developed that better support all types of learners irrespective of gender.

REFERENCES

Alavi, M. (1994). Computer-mediated collaborative learning: an empirical evaluation. *Management Information Systems Quarterly*, *18*(2), 159–174. doi:10.2307/249763

Alavi, M., Marakas, G. M., & Yoo, Y. (2002). A comparative study of distributed learning environments on learning outcomes. *Information Systems Research*, *13*(4), 404–415. doi:10.1287/isre.13.4.404.72

Alavi, M., Yoo, Y., & Vogel, D. R. (1997). Using information technology to add value to management education. *Academy of Management Journal, 40*(6), 1310–1333. doi:10.2307/257035

Alliger, G. M., Tannenbaum, S. I., Bennett, W. Jr, Traver, H., & Shotland, A. (1997). A meta analysis of the relations among training criteria. *Personnel Psychology, 50*, 341–358. doi:10.1111/j.1744-6570.1997.tb00911.x

American Society for Training and Development. (2007). *2007 ASTD State of the Industry Report.*

Anonymous,. (2008). 2008 Industry Report. *Training (New York, N.Y.), 45*(9), 16–34.

Arbaugh, J. B. (2000). An exploratory study of the effects of gender on student learning and class participation in an Internet-based MBA course. *Management Learning, 31*(4), 503–519. doi:10.1177/1350507600314006

Arbaugh, J. B. (2001). How instructor immediacy behaviors affect student satisfaction and learning in web-based courses. *Business Communication Quarterly, 64*(4), 42–54. doi:10.1177/108056990106400405

Barrett, E., & Lally, V. (1999). Gender differences in an on-line learning environment. *Journal of Computer Assisted Learning, 15*, 48–60. doi:10.1046/j.1365-2729.1999.151075.x

Biner, P. M. (1993). The development of an instrument to measure student attitudes toward televised courses. *American Journal of Distance Education, 7*(1), 63–73. doi:10.1080/08923649309526811

Biner, P. M., Welsh, K. D., Barone, N. M., Summers, M., & Dean, R. S. (1997). The impact of remote-site group size on student satisfaction and relative performance in interactive telecourses. *American Journal of Distance Education, 11*(1), 23–33. doi:10.1080/08923649709526949

Biocca, F., Harms, C., & Burgoon, J. K. (2003). Toward a more robust theory and measure of social presence: Review and suggested criteria. *Presence (Cambridge, Mass.), 12*(5). doi:10.1162/105474603322761270

Blum, K. (1998). Gender differences in CMC-based distance education. *Feminista, 2*(5).

Blum, K. (1999). Gender differences in asynchronous learning in higher education. *Journal of Asynchronous Learning Networks, 3*(1), 46–47.

Boneva, B., Kraut, R., & Frohlich, D. (2001). Using e-mail for personal relationships. *The American Behavioral Scientist, 45*(3), 530–549. doi:10.1177/00027640121957204

Burke, K., & Chidambaram, L. (1999). How much bandwidth is enough? A longitudinal examination of media characteristics and group outcomes. *Management Information Systems Quarterly, 23*(4), 557–579. doi:10.2307/249489

Carswell, A. D., & Venkatesh, V. (2002). Learner outcomes in a distance education environment. *International Journal of Human-Computer Studies, 56*(5), 475–494. doi:10.1006/ijhc.2002.1004

Compeau, D., & Higgins, C. A. (1995). Application of social cognitive theory to training for computer skills. *Information Systems Research, 6*(2), 118–143. doi:10.1287/isre.6.2.118

Ducheneaut, N., Yee, N., Nickell, E., & Moore, R. J. (2006). *"Alone together?" Exploring the social dynamics of massively multiplayer online games.* Paper presented at the ACM CHI 2006 Conference on Human Factors in Computing Systems, Montreal, Quebec, Canada.

Eagly, A. H. (1987). *Sex differences in social behavior: A social role interpretation.* Hillsdale, NJ: Earlbaum.

Edelsky, C. (1981). Who's got the floor? *Language in Society, 10.*

Gorriz, C. M., & Medusa, C. (2000). Engaging girls with computers through software games. *Communications of the ACM, 43*(1), 42–49. doi:10.1145/323830.323843

Gunawardena, C. N. (1995). Social presence theory and implications for interaction and collaborative learning in computer conferences. *International Journal of Educational Telecommunications, 1*(2-3), 147–166.

Gunawardena, C. N., Lowe, X., Constance, A., & Anderson, T. (1997). Analysis of a global debate and the development of an interaction analysis model for examining social construction of knowledge in computer conferences. *Journal of Educational Computing Research, 17*(4), 397–431. doi:10.2190/7MQV-X9UJ-C7Q3-NRAG

Hillman, D. C. A., Willis, D. J., & Gunawardena, C. N. (1994). Learner-interface interaction in distance education: An extension of contemporary models and strategies for practitioners. *American Journal of Distance Education, 8*(2), 30–42. doi:10.1080/08923649409526853

Hiltz, S. R. (1994). *The Virtual Classroom: Learning Without Limits via Computer Networks*. Norwood, NJ: Ablex Publishing Company.

Hiltz, S. R., Zhang, Y., & Turoff, M. (2002). Studies of effectiveness of learning networks. In *Elements of Quality Online Education*. Needham, MA: SCOLE.

Hornik, S. R., Johnson, R. D., & Wu, Y. (2007). When technology does not support learning: The negative consequences of dissonance of individual epistemic beliefs in technology mediated learning. *Journal of Organizational and End User Computing, 19*(2), 23–46.

Hornik, S. R., Saunders, C. S., Li, Y., Moskal, P. D., & Dzuban, C. D. (2008). The impact of paradigm development and course level on performance in technology-mediated learning environments. *Informing Science, 11*, 35–58.

Horton, W. K. (2000). *Designing web-based training: How to teach anyone anything anywhere anytime*. Hoboken, NJ: John Wiley & Sons.

Johnson, L. (2008). *NMC Virtual Worlds Announces Plans for 2008*. Retrieved June 8, 2009, from http://virtualworlds.nmc.org/wp-content/uploads/2008/01/press-release-nmc-virtual-worlds-2008-plans.pdf

Johnson, R. D., Hornik, S. R., & Salas, E. (2008). An empirical examination of factors contributing to the creation of successful e-learning environments. *International Journal of Human-Computer Studies, 66*, 356–369. doi:10.1016/j.ijhcs.2007.11.003

Johnson, R. D., & Marakas, G. M. (2000). The Role of Behavioral Modeling in Computer Skills Acquisition: Toward Refinement of the Model. *Information Systems Research, 11*(4), 402–417. doi:10.1287/isre.11.4.402.11869

Johnson, R. D., Marakas, G. M., & Palmer, J. W. (2006). Differential social attributions toward computing technology: An empirical examination. *International Journal of Human-Computer Studies, 64*(5), 446–460. doi:10.1016/j.ijhcs.2005.09.002

Kirkpatrick, D. L. (1976). Evaluation of training. In Craig, R. L. (Ed.), *Training and Development Handbook: A Guide to Human Resource Development*. New York: McGraw-Hill.

Kramarae, C. (2001). *The third shift: Women learning online*. Washington, DC: Association of American University Women.

Kuhl, P. K., Tsao, F.-M., & Liu, H.-M. (2003). Foreign-language experience in infancy: Effects of short-term exposure and social interaction on phonetic learning. *Proceedings of the National Academy of Sciences of the United States of America, 100*(15), 9096–9101. doi:10.1073/pnas.1532872100

Latham, G. P., & Saari, L. M. (1979). Application of social-learning theory to training supervisors through behavioral modeling. *The Journal of Applied Psychology, 64*(3), 239–246. doi:10.1037/0021-9010.64.3.239

Lim, C. K. (2001). Computer self-efficacy, academic self-concept, and other predictors of satisfaction and future participation of adult distance learners. *American Journal of Distance Education, 15*(2), 41–51. doi:10.1080/08923640109527083

Linne, J., & Plers, L. (2002). The DNA of an e-tutor. *ITTraining,* 26-28.

Mackie, D. M., Worth, L. T., & Asuncion, A. G. (1990). Processing of persuasive in-group messages. *Journal of Personality and Social Psychology, 58,* 812–822. doi:10.1037/0022-3514.58.5.812

Marakas, G. M., Johnson, R. D., & Clay, P. F. (2007). The evolving nature of the computer self-efficacy construct: An empirical investigation of measurement construction, validity, reliability, and stability over time. *Journal of the Association for Information Systems, 8*(1), 16–46.

McGrath, J. E. (1991). Time, interaction, and performance (TIP): A theory of groups. *Small Group Research, 22*(2), 147–174. doi:10.1177/1046496491222001

McGrath, J. E., Arrow, H., Gruenfeld, D. H., Hollingshead, A. B., & O'Connor, K. M. (1993). Groups, tasks, and technology: The effects of experience and change. *Small Group Research, 24,* 406–420. doi:10.1177/1046496493243007

Moore, M. G. (2002). Editorial: What does research say about the learners using computer-mediated communication in distance learning? *American Journal of Distance Education, 16*(2), 61–64. doi:10.1207/S15389286AJDE1602_1

Ong, C. S., & Lai, J. Y. (2006). Gender differences in perceptions and relationships among dominants of e-learning acceptance. *Computers in Human Behavior, 22,* 816–829. doi:10.1016/j.chb.2004.03.006

Piccoli, G., Ahmad, R., & Ives, B. (2001). Web-based virtual learning environments: a research framework and a preliminary assessment of effectiveness in basic IT skills training. *Management Information Systems Quarterly, 25*(4), 401–426. doi:10.2307/3250989

Rosenzweig, M. R., & Bennett, E. L. (1978). Experiential influences on brain anatomy and brain chemistry in rodents. In Gottlieb, G. (Ed.), *Studies on the Development of Behavior and the Nervous System* (pp. 289–387). New York: Academic Press.

Sahay, S. (2004). Beyond utopian and nostalgic views of information technology and education: Implications for research and practice. *Journal of the Association for Information Systems, 5*(7), 282–313.

Salas, E., DeRouin, R., & Littrell, L. (2005). Research-based guidelines for designing distance learning: What we know so far. In Gueutal, H. G., & Stone, D. L. (Eds.), *The Brave New World of e-HR: Human Resources Management in the Digital Age.* San Francisco: Jossey-Bass.

Schmidt, A. M., & Ford, J. K. (2003). Learning within a learner control training environment: The interactive effects of goal orientation and metacognitive instruction on learning outcomes. *Personnel Psychology, 56*(2), 405–429. doi:10.1111/j.1744-6570.2003.tb00156.x

Sehoole, C. T., & Moja, T. (2003). Pedagogical issues and gender in cyberspace education: Distance education in South Africa. *African and Asian Studies, 2*(4), 475–496. doi:10.1163/156920903773004022

Short, J., Williams, E., & Christie, B. (1976). *The Social Psychology of Telecommunications*. New York: John Wiley & Sons.

Stryker, S. (1980). *Symbolic Interactionism: A Social Structural View*. Reading, MA: Benjamin Cummings.

Tannen, D. (1990). *You Just Don't Understand: Women and Men in Conversation*. New York: Ballantine.

Tannen, D. (1993). The relativity of linguistic strategies: Rethinking power and solidarity in gender and dominance. In Tannen, D. (Ed.), *Gender and Conversational Interaction* (pp. 165–188). New York: Oxford University Press.

Thompson, L. F., & Lynch, B. J. (2003). Web-based instruction: Who is inclined to resist and why? *Journal of Educational Computing Research*, *29*(3), 375–385. doi:10.2190/3VQ2-XTRH-08QV-CAEL

Tu, C.-H., & McIsaac, M. (2002). The Relationship of Social Presence and Interaction in Online Classes. *American Journal of Distance Education*, *16*(3), 131–150. doi:10.1207/S15389286AJDE1603_2

Vygotsky, L. S. (1978). *Mind in Society*. Cambridge, MA: Harvard University Press.

Whitley, B. E. (1997). Gender Differences in Computer-Related Attitudes and Behavior: A Meta-Analysis. *Computers in Human Behavior*, *13*(1), 1–22. doi:10.1016/S0747-5632(96)00026-X

Wirt, J., & Livingston, A. (2004). *Condition of Education 2002 in Brief (Rep. No. 2002011)*. Washington, DC: National Center for Education Statistics.

Wood, W., & Rhodes, N. D. (1992). Sex differences in interaction style in task groups. In Ridgeway, C. (Ed.), *Gender, Interaction, and Inequality* (pp. 97–121). New York: Springer-Verlag.

Yoo, Y., & Alavi, M. (2001). Media and group cohesion: Relative influences on social presence, task participation and group consensus. *Management Information Systems Quarterly*, *25*(3), 371–390. doi:10.2307/3250922

ENDNOTE

[1] Discussions with the last GA found that other than first week assistance in setting up WebCT by a few students (<20), students chose to communicate via course email for assistance.

APPENDIX

Research Constructs and Scale Items

The following scales were assessed using a 7-point Likert-type scale with anchors of strongly disagree and strongly agree.

Social Presence

For social presence, participants were asked to evaluate the characteristics of the class environment as delivered through WebCT. Each question used the stem, "The environment is:" Higher numbers represent more presence and lower numbers represent less presence. The anchors for each item were:

1. Impersonal…Personal
2. Unsociable…Sociable
3. Insensitive…Sensitive
4. Cold…Warm
5. Passive…Active

Satisfaction

1. I am satisfied with the clarity with which the class assignments were communicated
2. I am satisfied with the degree to which the types of instructional techniques that were used to teach the class helped me gain a better understanding of the class material.
3. I am satisfied with the extent to which the instructor made the students feel that they were part of the class and "belonged".
4. I am satisfied with the instructor's communication skills.
5. I am satisfied with the accessibility of the instructor outside of class.
6. I am satisfied with the present means of material exchange between you and the course instructor.
7. I am satisfied with the accessibility of the graduate assistants.

Course Instrumentality

1. I feel more confident in expressing ideas related to Information Technology.
2. I improved my ability to critically think about Information Technology.
3. I improved my ability to integrate facts and develop generalizations from the course material.
4. I increased my ability to critically analyze issues.
5. I learned to interrelate the important issues in the course material.
6. I learned to value other points of view

This construct was assessed with a 10 point scale ranging from 0 (Cannot Do) to 100 (Totally Confident) in increments of 10.

General Computer Self-Efficacy

1. I believe I have the ability to unpack and set up a new computer.
2. I believe I have the ability to describe how a computer works.
3. I believe I have the ability to install new software applications on a computer.
4. I believe I have the ability to identify and correct common operational problems with a computer.
5. I believe I have the ability to remove information from a computer that I no longer need.
6. I believe I have the ability to understand common operational problems with a computer.
7. I believe I have the ability to use a computer to display or present information in a desired manner.

This work was previously published in the Journal of Organizational and End User Computing, Volume 23, Issue 1, edited by Mo Adam Mahmood, pp. 79-92, copyright 2011 by IGI Publishing (an imprint of IGI Global).

Chapter 11
Participation in ICT–Enabled Meetings

Katherine M. Chudoba
Utah State University, USA

Mary Beth Watson-Manheim
University of Illinois, Chicago, USA

Kevin Crowston
Syracuse University, USA

Chei Sian Lee
Nanyang Technological University, Singapore

ABSTRACT

Meetings are a common occurrence in contemporary organizations. The authors' exploratory study at Intel, an innovative global technology company, suggests that meetings are evolving beyond their familiar definition as the pervasive use of information and communication technologies (ICTs) changes work practices associated with meetings. Drawing on data gathered from interviews prompted by entries in the employees' electronic calendar system, the authors examine the multiple ways in which meetings build and reflect work in the organization and derive propositions to guide future research. Specifically, the authors identify four aspects of meetings that reflect work in the 21st century: meetings are integral to work in team-centered organizations, tension between group and personal objectives, discontinuities, and ICT support for fragmented work environment.

INTRODUCTION

Meetings are "a focused interaction of cognitive attention, planned or chance, where people agree to come together for a common purpose, whether at the same time and same place, or at different times in different places" (Romano & Nunamaker, 2001, p. 1). Meetings are pervasive in contemporary work life, serving as important forums where social relationships are created and changed in organizations (Schwartzman, 1989; Weick & Meader, 1993). Time spent in meetings

DOI: 10.4018/978-1-4666-2059-9.ch011

appears to be increasing (Romano & Nunamaker, 2001; Stephens & Davis, 2009) and meetings are prevalent across all levels of workers (National Research Council, 1999) as a mechanism for collaboration, coordination, information sharing, and decision-making (Tropman, 1996).

Though meetings would appear to be a stable and mundane feature of organizational life, we suggest that the practice of meetings is in fact changing dramatically with the pervasive use of information and communication technologies (ICTs). We present an exploratory study conducted at Intel Corporation where participation in meetings has evolved beyond an activity that is perceived as peripheral to work to one where it is an integral part of the work. The taken-for-granted image of individuals sitting around a table engaging in verbal discussion is no longer the norm. At Intel (and in many other organizations), meetings are frequently enabled by ICT, specifically voice and audioconferencing, screen sharing (such as Net-Meeting®) and document sharing (e.g., through a Web site or shared file storage), to enable the participation of geographically distributed individuals. This technological change is reflected in evolving work practices, which include new expectations about who participates, what participation looks like, and more generally, how work gets done when people meet. Understanding the role of ICT in changing the way meetings are enacted provides insight into the conditions under which employees work, as well as the infrastructure that is necessary to support work.

We begin with a review of the relevant literature, including the practice lens and prior research on traditional meetings and GSS-supported meetings. Then we describe our methods and the findings from our qualitative analyses, which are reflected in the form of propositions derived from our data. We conclude with a discussion of our findings, including implications for research and practice.

BACKGROUND

Practice View

Because work and use of ICT can be highly intertwined, it is important to have a theoretical perspective that helps make sense of their relation. In this paper, we adapted Orlikowski's (2000) application of a practice lens to analyze the interaction between ICT use and meetings. A practice lens focuses on human agency and the open-ended set of structures or work practices that arise through recurrent human activities. The approach uses people's everyday activities as the unit of analysis, and examines the structural and interpersonal elements that create and are created by these activities (Schultze & Orlikowski, 2001). While offering a broad perspective on which to base an investigation of meetings, a practice lens encourages a focus on specific work practices and the structures and norms associated with them. It provides guidance about what factors should serve as the focus of an investigation, rather than providing a predictive framework of cause and effect relationships that are examined. In our adaptation of the practice lens view, we focus on practices as organizational members report them and use the insights that emerge from these practices to generate propositions to guide future research.

Figure 1, adapted from Orlikowski (2000), shows the relationship between agency and work practices and their constituents. Work practices of social systems are enacted through recurrent human activities and are mediated through settings, norms, and interpretive schemes that guide human action. Settings include the context that supports meeting activities such as participants' physical location and technology; norms, the codes of conduct and etiquette that guide and regulate the activities; and interpretive schemes, the categories and assumptions that give meaning to the activities. By examining meetings as a collection of work practices, we can study the effects of ICT-enabled distributed meetings. To the extent technology is used in different ways, different practices emerge, thus leading to a shift in the way people accomplish their work.

Research on Face-to-Face Meetings

As an introduction to our empirical study, we first briefly analyze prior research on the role of face-to-face (FTF) meetings. We identify the settings, norms and interpretative schema explicitly or implicitly assumed in this research, and the related structures and practices.

Settings. Traditionally, meetings are described as FTF events held in a conference or meeting room. Participants use technologies such as whiteboards and computer projection equipment

to facilitate information sharing. For example, Romano and Nunamaker (2001) reviewed research on meeting analysis over the prior 15 years. While their definition of a meeting is broad enough to include distributed participants, in fact all the literature they reviewed assumes that participants in meetings are collocated. The influence on the effectiveness of the meeting's location on the meeting itself, e.g., on-site vs. off-site, is discussed but no literature is cited where meeting participants are in different locations. Similarly, Luong and Rogelberg (2005) surveyed 37 people in a university setting who attended at least three meetings per week, but meetings were defined as events held in conference rooms with all participants collocated. Sonnentag (2001) studied 60 software professionals from 10 software projects, where teams members were collocated in various

Figure 1. Practice perspective template

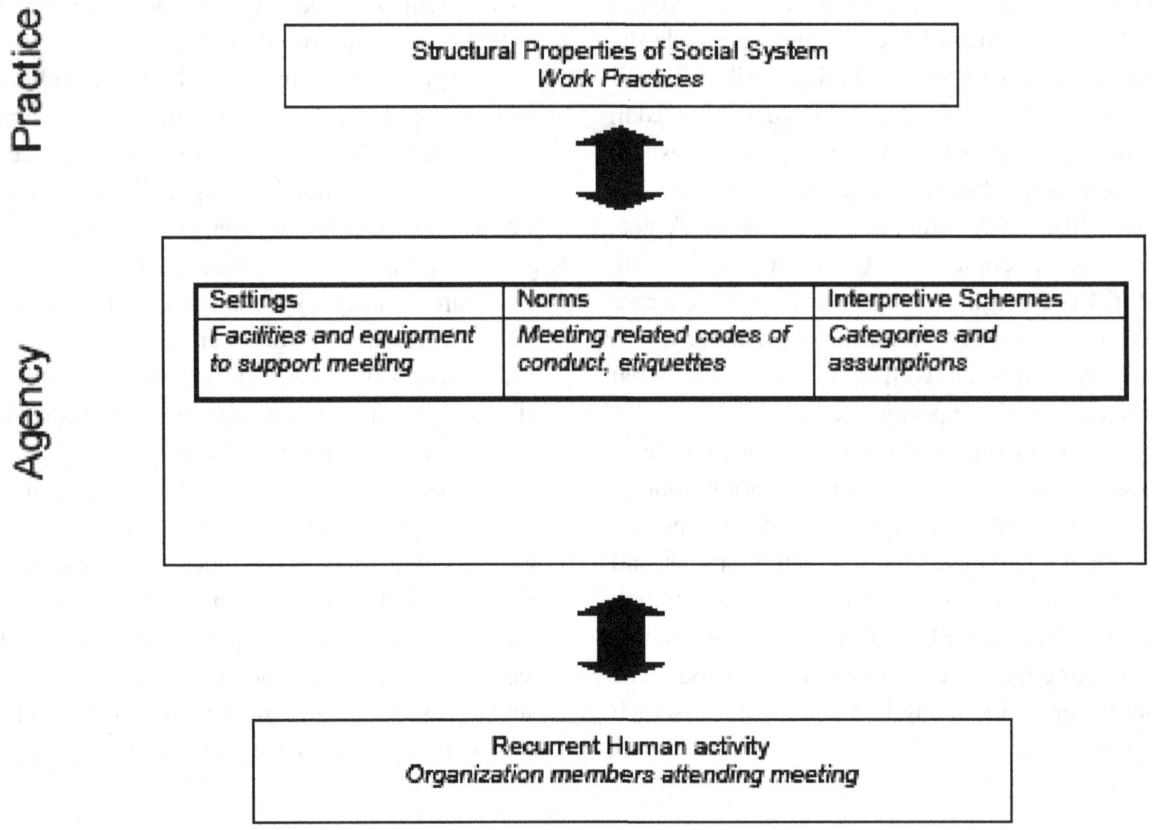

parts of Germany. Thus, historically research on meetings has focused on FTF gatherings and not distributed settings.

Norms and rules. Prescriptive rules for improving meetings have long been a topic of practical interest, as evidenced by books such as *How to Make Meetings Work* (Doyle & Strauss, 1976) and, almost 30 years later, a *Harvard Business Review* article, "Stop Wasting Valuable Time" (Mankins, 2004). Both publications suggest adding regulating processes (e.g., agendas, minutes) to make meetings more efficient and effective. In the research literature, meeting success is proposed to be largely dependent on the existence of group processes that regulate activities (Nunamaker, Dennis, Valacich, Vogel, & George, 1991; Sonnentag, 2001), though there is often little evidence of meeting preparation that includes these regulating processes (Romano & Nunamaker, 2001).

Interpretive schema. Interpretive schemas are the meanings that participants give to meetings, such as its purpose. Sometimes meetings stress individual accountability (Luong & Rogelberg, 2005) and sometimes group accountability (Sonnentag, 2001). A second aspect of the interpretive schema is the way meetings fit with other work. Rogelberg and colleagues conducted research on individuals' perceptions of meeting effectiveness and the link between perceived effectiveness and employee well being (Luong & Rogelberg, 2005; Rogelberg, Leach, Warr, & Burnfield, 2006). Participants in these studies described meetings as peripheral activities that were a disruption to their primary responsibilities. As a result, increased frequency of meetings led to feelings of fatigue and increased workload because the completion of primary tasks was delayed. A final aspect of interpretive schema is the likely effectiveness of meetings. In the literature, meetings are often perceived as costly and ineffective, wasting money and time, and leading to decreased morale and productivity (Sonnentag, 2001).

Work practices. In summary, the work performed by individuals involved in FTF meetings may be individual or collaborative, and meetings are more successful when there are group-regulating processes. There is a general feeling that meetings are often costly and ineffective. Individuals believe meetings are an interruption to primary work activities, but are also necessary activities to generate some discrete outcome, e.g., a decision is made.

Research on GSS-Supported Meetings

We next review the literature to examine how the use of group support systems (GSS) has been documented to change the structures and practices described above. GSS for meeting support has been the subject of information systems research for more than 20 years (Fulk & Collins-Jarvis, 2001).

Settings. The most obvious difference between FTF and GSS-supported meetings research is the technology available to support the meetings. Research has focused on the development of tools and techniques to support and structure group interaction, e.g., by supporting group activities such as brainstorming and voting (Nunamaker et al., 1991) or strategic planning (Adkins, Burgoon, & Nunamaker, 2003). Almost all of the published research has examined the use of GSS when all participants meet FTF in a specially equipped meeting room. Some researchers have suggested that the technology can support people participating in the meeting from different locations for specific tasks such as requirements negotiation (cf., Boehm, Grunbacher, & Briggs, 2007) or collaborative engineering (Briggs, Vreede, & Nunamaker, 2003), but we are aware of no published research that has specifically examined the use of a GSS to support synchronous meetings with distributed participants.

Norms and rules. Consistent with the prescriptive literature on FTF meetings, the role of GSS is to provide increased structure for meetings, which is in turn hypothesized to lead to more successful group meetings. A basic tenet of GSS is that enhancing group outcomes depends on maximizing process gains and minimizing process losses. Increased use of technology support for meetings should lead to increased effectiveness by supporting group processes and providing structure (Romano & Nunamaker, 2001). In this study, we will consider how norms surrounding the use of other types of ICT are associated with meeting effectiveness.

The support for group processes and structure become instantiated as norms and rules to the extent that participants use GSS technology as intended by its designers and faithfully appropriate the GSS (DeSanctis & Poole, 1994). The information technology infrastructure facilitates communication among meeting participants through parallel communication, anonymity, and collective memory (Nunamaker et al., 1991) or by enhancing workspace awareness (Haines & Cooper, 2008). To the extent that meeting participants faithfully appropriate a GSS they have adopted the norms and rules generally associated with use of a GSS. In other words, faithful appropriation means 1) participants use the GSS for parallel communication instead of engaging in extensive verbal communication that requires sequential turn-taking, 2) participants evaluate ideas independently without knowing who submitted them (e.g., they appropriate the anonymity feature of a GSS), and 3) participants use the textual record of a GSS session to document ideas and decisions (cf., Fjermestad & Hiltz, 1999; Jackson & Poole, 2003).

Interpretive schema. Generally speaking, the interpretive schema for GSS-supported meetings research appears to be the same as for traditional FTF meetings, with two exceptions. First, norms and rules are assumed to be derived from use of the technology, to the extent that it is faithfully appropriated. Second, an additional assumption focuses specifically on the role of anonymity. Traditional FTF meetings are believed to suffer from participants' inhibitions against contributing publicly, and so the technology is believed to help

by providing an anonymous channel, although most of the research on anonymity is lab- not field-based.

Work practices. In both traditional and GSS-supported meetings, there is an underlying assumption that meetings are held to accomplish specific purposes outside of normal work activities (cf., Briggs et al., 1999). Much of the GSS research is experimental and focuses primarily on enhancing the group process, rather than investigating its use in field settings. In other words, while the settings are different for traditional and GSS-supported meetings, the assumptions about appropriate rules and norms and individual interpretive schemes have changed very little, with the exception of anonymity.

Changes in Work Practices

From our review, we note that the missing element in the study of meetings is how meetings with geographically distributed participants are supported with ICT and the effects on work practices. Our interest differs from studies of virtual teams, (Powell, Piccoli, & Ives, 2004), which have mostly examined asynchronous interactions using electronic mail (Ahuja & Carley, 1999; Cramton, 2001) or computer conferencing (Sarker & Sahay, 2002). Instead, we are interested in how ICT enables synchronous interactions. Some teams may rarely or never meet FTF and yet still form effective teams (Crowston, Howison, Masango, & Eseryel, 2005; Orlikowski, 2002). The goal of our empirical study is to provide evidence for the relationship between ICT use for meeting support and changes in work practices.

METHODS

Research Setting

The study was conducted at Intel Corporation, a Fortune 100 company in the information technology industry. (The field site has not been anonymized at the request of our sponsor in the company.) Company headquarters is in the United States, and R&D, manufacturing, and sales operations are located in multiple sites within the U.S. and around the world. Intel is recognized as an innovative company in a fast-paced, global environment. Employees routinely work and collaborate with colleagues across the globe. For example, it is common to find employees who work in the same project team to be distributed in different worldwide sites or to find employees who report to managers who are not collocated (up to 25% of employees in support areas).

Data Collection

Data were collected from semi-structured telephone interviews, guided by a protocol developed according to recommendations for conducting qualitative research (Miles & Huberman, 1994; Strauss & Corbin, 1990). We interviewed 30 employees for whom communication with colleagues was an important and significant part of their work responsibilities. A snowballing technique was used to identify respondents. We were initially given names of two mid-level knowledge workers. At the end of each interview, employees were asked to refer us to other employees. As the interviews progressed, researchers asked respondents to identify subjects located outside the U.S. and across different functional areas in order to achieve a stratified sample (Eisenhardt, 1989). Data in Table 1 show the demographic profile of the respondents. Each interview lasted approximately one hour with two members of the research team (one interview only had a single researcher). Interviews were not tape-recorded, although both interviewers took copious notes. Having two researchers allowed one to focus on listening and note taking while the other led the questioning (Mason, 1996).

A semi-structured interview protocol was used to ensure that comparable data was collected from each interviewee. The interview protocol was grounded by reference to entries in the respondent's electronic calendar for the most recent typical week. Everyone in the company used the same electronic calendar application and kept their schedules online and accessible to others. Respondents were asked to discuss all entries on their electronic calendar from the most recent, typical workweek. We asked a number of questions to characterize each meeting including its purpose and frequency (e.g., one time or recurring); what happened during the course of the meeting, including interactions among participants; participants and their location, first language, and function and organizational affiliation; and the technologies used and how they were used. These questions were aimed at uncovering the settings, norms and interpretive schemes that illustrate agency, as well as the structural properties of the overarching social systems (Mason, 1996).

Data Analysis

After each interview, notes were transcribed and compared across the researchers who conducted the interview. Transcriptions were entered into qualitative analysis software. Two members of the team inductively and iteratively studied the data, and independently identified issues and themes. These were compared and differences resolved. The synthesized coding was repeatedly reviewed with the rest of the research team (Strauss & Corbin, 1990). Analysis continued as we sought to understand respondents' perspectives about meetings, ICTs, and roles they played at Intel.

We now present our findings and research propositions derived from them. The practices associated with meetings attended by knowledge workers at Intel allowed us to move toward an understanding of participation practices in ICT-enabled meetings.

Table 1. Demographic Profile of the Interview Respondents (N=30)

Function	Number	Percent
Sales & Marketing	10	33
Information Technology[1]	7	23
Engineering	6	20
Human Resource	2	7
Others	5	17
Location		
United States of America[2]	21	70
Asia	3	9
Europe/Middle East	6	21
Years with organization		
1 to 5	9	30
6 to 10	8	27
11 to 14	3	10
More than 15	9	30
Unknown	1	3
Number of meetings in a week		
1 to 5	2	7
6 to 10	2	7
11 to 15	6	20
16 to 20	7	23
21 to 25	6	20
26 to 30	6	20
More than 30	1	3
Number of different teams in a week		
1	1	3
2 to 3	5	17
4 to 5	15	50
6 to 7	6	20
8 to 9	3	10
>10	0	0
Meetings with peripheral[3] membership in a week		
<20%	16	53
20-40%	11	37
41-60%	2	7
61-80%	1	3
>80%	0	0

continued on following page

Table 1. Continued

Function	Number	Percent
Meetings with integral membership in a week		
<20%	0	0
20-40%	1	3
41-60%	2	7
61-80%	11	37
>80%	16	53

1. Respondents in the IT function supported operations in other functional areas (e.g., manufacturing, sales) as well as internal IT operations (e.g., strategy).

2. Respondents were in 5 states located across 3 time zones.

[3] Peripheral membership refers to those who only attended a portion of the meeting, "partial participation" (e.g., respondent reported significant time spent multi-tasking during meeting), or reviewed agenda and/or minutes and then decided not to attend the meeting. Integral membership refers to full participation during most of the meeting.

TOWARD AN UNDERSTANDING OF PARTICIPATION IN ICT-ENABLED MEETINGS

In this section, we first summarize our research findings. We next elaborate the findings and offer a set of propositions to further examine the relationship between ICT use for meeting support and changes in work practices.

Research Results

Our data clearly show that meetings are a defining aspect of work life for knowledge workers at Intel. Data in Table 1 describe the demographic characteristics of our respondents and their reported meeting participation. Sixty-six percent of respondents reported having more than 15 meetings scheduled on their calendars during the week being covered in the interview. Almost two-thirds of the respondents reported 20 or more hours a week in meetings during the week, with more than a quarter of them spending 30 or more hours in meetings. Strikingly, 80% of respondents reported meeting with more than 4 different teams each week.

Our 30 respondents provided data covering a total of 524 meetings. Table 2 presents data about these meetings. One-third of meetings in our sample involved two participants, while another 30% had 3-9 participants, and 20% had 10 or more participants. Ninety percent of meetings were scheduled regular meetings, occurring on a recurrent and planned basis. Only 10% were ad hoc meetings, occurring on a one-time, as-needed basis, though this small proportion may be due in part to our sampling strategy, which was based on respondents' calendars—it may be that some ad hoc meetings did not get recorded, and were not recalled by the respondent.

We asked respondents to discuss the purpose of each meeting that appeared on their calendars. The overwhelming majority of meetings reported were for team collaboration (76%). Meetings for collaboration purposes included meetings for status updates, brainstorming, problem solving, and coordination. Almost all of these meetings were recurring, as members of teams met periodically to address the team's charge and objectives. Our analysis of the data revealed no significant difference in work practices in meetings that involved these different collaboration objec-

Table 2. Meeting characteristics

ICT used during meetings (N=524)	Count (%)
• Audio bridge	221 (42)
• NetMeeting or similar application	179 (34)
• Telephone	102 (19)
• E-mail	60 (11)
• Shared workspace	61 (12)
• Instant messaging	22 (4)
Meeting size (N=524)	
• 2 people	174 (33)
• 3-5 people	77 (15)
• 6-9 people	77 (15)
• 10-14 people	50 (10)
• 15 or more people	54 (10)
• Unknown	92 (18)
Extent of FTF interaction (N=524)	
• No FTF (all participants distributed)	305 (58)
• Some FTF (combination FTF & distributed)	63 (12)
• All FTF (all participants in same room)	101 (19)
• Unknown	55 (11)
Purpose of Meetings (N=524)	
• Collaboration	399 (76)
• Management	79 (15)
• Information dissemination	41 (8)
• Social	5 (1)
Meeting Frequency (N=524)	
• Recurring	471 (90)
• Ad hoc (one of)	53 (10)
Meeting Attendance (N=524)	
• Attended	502 (96)
• Did not attend	22 (4)
Discontinuities in meetings attended (N=502)	
• Time (2 or more time zones)	92 (18)
• Function (2 or more)	47 (9)
• Organization (2 or more)	37 (7)
• Nationality (2 or more)	122 (24)
• Technology (at least 1 participant did not have equal access to ICT)	5 (1)
• Language (2 or more "first" languages)	70 (14)

tives. Fifteen percent of meetings were between a manager and employees reporting directly to that manager. Eight percent of the meetings were classified as information dissemination. In these meetings the direction of communication is one-way (e.g., announcement, presentation, training). Finally, respondents described a small number of meetings (1%) that were held for a purely social purpose.

A number of different ICT applications were instrumental in the meetings described to us by participants, with many in heavy use. Most prevalent was the audio bridge (42% of meetings), an internal teleconferencing facility. The audio bridge had to be requested prior to the meeting and information sent to meeting participants to establish the connection. Employees differentiated telephone use from audio bridge use, with the telephone used for meetings with two or three participants. Team collaboration tools that allowed shared view of documents and collaborative editing were often used in conjunction with teleconferencing (34% of meetings). Some respondents reported use of shared web repositories (12%) to store project documents. Email was used to disseminate information before, during, and after the actual meeting. On the other hand, instant messaging (IM) was reported used in only 4% of meetings. IM, and sometimes email, were used for impromptu communication during meetings although this number is likely understated. IM technology was just being introduced into Intel at the time of this research, and its use did not surface until relatively late in our interviews, so it may be that our numbers reflect an early stage of adoption. Employees also reported using IM and email for unplanned communication throughout the day because they considered them less intrusive than an unplanned telephone call to a colleague.

The pervasive use of ICTs in meetings reflects a form and structure of meetings that is different from the traditional setting of meetings with participants seated around a conference room table. Participants no longer have to be in the same location in order to attend a meeting. Only 19% of the meetings reflected the traditional scenario of a meeting as a FTF gathering with all participants in the same room. In fact, 58% of the meetings involved no collocated participants. An additional 12% were partially collocated, meaning some participants were in the same room and some were in different locations.

Employees described an organizational culture emphasizing individual productivity influenced by the rapid pace of change in Intel's product line and the intense global competitive environment. The ease of including people in meetings, no matter where they were physically located, was reflected in a proliferation of scheduled meetings on individuals' calendars. The flexibility to attend a meeting from whatever location was most convenient—be it home, office, or conference room—meant time saved from reduced travel, but concurrently, savings in time dissipated because of an ever-increasing number of meetings to which knowledge workers were invited. Because they did not have to physically attend a meeting, employees felt there was an expectation they would be available to attend meetings whenever they were scheduled. This was especially notable for employees who lived in places that were not in the same time zone as the majority of those with whom they worked, which meant frequent early morning or late evening meetings for them.

Research Propositions

While Mintzberg (1973) documented the extensive amount of time executives spend in meetings, prior research has not indicated that knowledge workers can spend 20% or more of their work week in meetings, as they do at Intel. The sheer number of meetings meant our respondents frequently had to make decisions about the extent of their participation in a given meeting because of conflicting demands on their time. Three levels of participation in meetings emerged from our data.

- **Non-participation:** people who are on a distribution list for a group and who have been invited to participate in the meeting, but after reviewing the agenda, decide not to attend this particular meeting.
- **Partial participation:** those who attend only a portion of the meeting or who "listen with one ear" without being fully engaged in the meeting, most often because they are multi-tasking and working on other things such as checking email during the meeting.
- **Full participation:** participants who are actively engaged in the meeting.

Table 1 shows that about half our respondents participated only partially in some of their meetings, and 10% of the respondents partially participated in half of their scheduled meetings. The employees with whom we talked explicitly noted that different degrees of participation were not only accepted, but were expected and deemed necessary in order to manage one's various responsibilities. Employees felt they were empowered to decide for themselves whether a given meeting called for their integral membership, or full participation during most of the meeting, versus partial participation in which they would only attend a portion of the meeting or peripherally participate by multi-tasking during the meeting. Therefore, we propose:

P1: As the number of ICT-enabled meetings increases, participants will be more likely to adapt interpretive schemas and media practices to include different modes of participation.

There was a general belief among respondents that increased structure and full participation, especially among core team members, led to enhanced meeting effectiveness. Norms were evidenced by a number of different strategies employed by meeting leaders to enhance effectiveness. For example, one norm was the regular preparation of meeting agendas and minutes. These documents were either distributed as email attachments or posted in team repositories. Published meeting agendas and minutes served as screening mechanisms and provided meeting participants greater flexibility in deciding the extent and nature of their participation for any given meeting. With open blocks of time increasingly scarce on one's calendar, people reviewed agendas or minutes to help them decide the importance of attending a given meeting, and more specifically, how engaged they needed to be in the communicative interactions that took place during the meeting.

Agendas were sometimes structured to facilitate partial participation or participation during only a portion of the meeting. One respondent noted about his group: *"They begin with product related updates, delivery schedules, customer's requirements, etc. immediately during the first 10 to 20 minutes, then people can leave if they're not interested in the rest of the items on the agenda."* Another respondent observed, *"Agendas are sent out before the meeting so some people review the agenda and choose not to attend the meeting if they think that their item(s) may not get discussed."* Some of our respondents chose to remain on a meeting list even if they rarely attended the group's meetings, because by doing so, they received the agenda and minutes of the meetings and could monitor project progress and activities. This was especially important in teams with distributed membership. At other times, respondents chose not to attend a meeting based on its agenda in order to have time to accomplish other, more pressing responsibilities. This suggests that artifacts that have previously been viewed as structuring meetings and providing documentation can also be used as boundary objects (Star, 1989). In other words, agendas and minutes provide information that is recognized by multiple categories of participants, but each participant may interpret the information differently, based on her or his interpretive schema. Based on this evidence, we propose:

P2: As the number of ICT-enabled meetings increases, artifacts such as meeting agendas and minutes will serve as boundary objects that influence the extent of an individual's participation in meetings based on the individual's interpretive schema.

Some respondents chose to attend a meeting remotely by using the audio bridge, even when FTF participation was an option. By attending meetings from their desks, meeting participants were physically isolated from the rest of the group, which made it easier for them to multi-task during the course of the meeting. Multi-tasking behavior took different forms in these situations. ICT, specifically IM and sometimes email, were used to create side conversations during meetings, the electronic version of whispering or passing notes. At other times, *"side conversations are a problem in these meetings. If the meeting really disintegrates, then people will stop talking about the task"*, which increased the isolation of remote attendees. While side conversations happen at traditional meetings, participation is limited by physical proximity, e.g., talking to the person sitting next to you or passing a note to someone close. Such conversations are also limited by timing, e.g., conversations take place during breaks in formal meetings. Also, side conversations are usually visible in traditional meetings (although they may not be noticed). But when meeting participants are in different locations, these interactions are not visible to other participants in a meeting. One respondent noted that she did not attend a meeting that was on her calendar because of a schedule conflict but while in the other meeting, received an IM from a participant in that the first meeting requesting information.

Respondents expressed understanding of and support for meeting activities that encouraged full participation in order to enhance group effectiveness (e.g., limited or no multi-tasking). At the same time, they also understood the importance of, even necessity of, ensuring their own personal productivity to meet performance objectives. This conflict often led to tension in their decisions as to meeting behavior. For example, respondents understood the importance of a 'level playing field' for meeting participation. Intel Human Resources even recommended meeting etiquette norms for this purpose, e.g., if all participants could not meet in the same room, everyone should join the meeting over the audio bridge. While this etiquette rule was understood as a way to insure participation equality, it had the unintended effect of actually enabling partial participation, albeit for different reasons. A number of interviewees saw the ability to attend meetings from their desk as necessary to enhance individual productivity since it helped them balance their responsibly to multiple teams. As one respondent explained: *"In terms of productivity, using phone allows more flexibility but the risk is I may miss out a couple of questions during the meeting [if multitasking]. The good side is I can do more activities (e.g.: check email) than in a FTF meeting. Particularly, for FTF meeting, it is rude to type on a computer especially when people travel all the way to [location name] to meet up. Thus, FTF meeting will probably result in around 60% productivity lost for Intel."* Based on the preceding discussion, we propose:

P3: As the number of ICT-enabled meetings increases, participants who expect to be only partially engaged in a meeting are less likely to attend a meeting FTF if they can attend using ICT.

Meetings with non-collocated others meant a change in expectations about meeting participation at Intel such that "full participation" by everyone who attended a meeting was the exception. While daydreaming and side conversations invariably happens in traditional FTF meetings, the use of ICT led to new forms of participation when people did not gather in the same room for a meeting. Team collaboration tools made it possible for people in different locations to view the same document, making active participation by everyone possible. Yet, being in a different location and out of visual sight of coworkers made it easier for meeting participants to reduce their participation by multi-tasking, such as checking email or using IM to respond to a question from someone attending another meeting. Partial participation was also evident in the use of published meeting agendas and minutes as screening mechanisms, as discussed earlier. Use of agendas and minutes have been used in the past to help a person decide whether or not to attend a meeting, but they allow a more nuanced decision in distributed meetings since attendance no longer means one must at least give the pretence of full participation as it does in a FTF meeting.

Use of the telephone audio bridge allowed employees to attend a meeting from anywhere, which provided a positive sense of freedom and allowed the team members to accommodate individual priorities. On the other hand, its use did sometimes leave respondents feeling left out of interactions. This effect was especially evident when a subset of meeting participants was located together in a conference room, with remote participants joining the meeting over the audio bridge. One respondent observed: *"You can sense the energy and activity in [location] and that shuts down activity for those on the telephone ... balancing creative energy with total team participation [is difficult]. You don't want to squash that [creative energy] for those in the conference room by doing a more structured thing like asking people on the bridge 'what is your input?' It's hard to balance."*

Recent research on geographically-distributed teams indicates that the configuration of the team, i.e., how many different sites team member are located in and how many members at each site, influences the norms of behavior and ultimately productivity of a distributed team (O'Leary & Mortensen, in press). Imbalanced teams can lead to the formation of subgroups with negative effects on team dynamics and performance. Alternatively, one isolated member can lead to

higher team performance as members adjust their interactions to accommodate this person. Although we did not measure team performance, our data provides additional support for the changes in the norms of behavior and assumptions about the meeting effectiveness that occur in different team configurations.

Thus, we propose:

P4: As the number of ICT-enabled meetings increases, the configuration of attendees and their interpretive schemes will be more likely to lead to different levels of participation.

While employees agreed meetings were important to getting work done, they also felt that time spent in meetings was sometimes unproductive. Use of ICT enabled more meetings to take place, however, employees felt that technology was not always used as effectively as possible in this global environment. As one employee explained, there is *"lots of non-productive time in meetings ...people have not changed their behavior to deal with asynchronous work...everything doesn't have to be done together in a meeting ... they need to save synchronous time for discussion."*

Because meetings at Intel frequently crossed multiple, far-flung time zones, participants, especially those outside of headquarters location, were expected to attend meetings outside of their traditional work hours. While most respondents accepted the need for these meetings, as noted above, respondents devised interesting mechanisms to manage how the workday intruded on personal time. For example, respondent 6 noted that his wife also worked at Intel so they made use of each other's calendars to schedule time commitments for family responsibilities. Before accepting a late evening meeting, he might check his wife's calendar to see if she was available to pick up the children, and if so, add that commitment to her schedule so he could attend the evening meeting.

Meeting scheduling typically revolved around the same time zone as Intel's headquarters in the western U.S., understandable since the majority of employees worked in several locations within this time zone. This meant that "minority" participants residing in other time zones often set their clocks to "headquarters' time." These employees accepted evening meetings (for those in Europe/ Middle East) and pre-dawn meetings (for those in Asia) as "just the way it was." One respondent in Europe/Middle East blocked the 6-8 p.m. slot every day on his calendar so no meetings would be scheduled then to keep this time slot to travel home, have dinner, and put his children to bed. After 8 p.m. local time, he then joined meetings via audio bridge, and when only his partial participation was required, spent the late evening with a phone to his ear and the local football game muted on the television.

Managers and other meeting conveners at Intel were aware of multi-tasking techniques used by meeting participants because they also used the techniques when attending meetings that they were not leading. The presence of laptops and/or use of ICT led meeting conveners to employ strategies to reduce multi-tasking such as cold-calls to assure that participants were paying attention or directing "laptops down" in FTF meetings so that no one was tempted to use a computer for a purpose not directly associated with the meeting. Managers reported that they did this when they believed that full attention and participation by members was necessary for team productivity on a specific task.

However, many participants with whom we talked revealed the self-oriented strategy of blocking time on one's calendar for individual work time. This meant that anyone else trying to schedule a meeting during the block of time would see it as unavailable, and so would be more likely to look for another time to hold the meeting. Respondents told us this was the only way they could assure themselves of quiet work time to prepare analyses or presentations. Interestingly, a number of Intel employees considered unplanned telephone calls intrusive. One respondent noted, *"The norm is to set people's expectations ahead of time [before a telephone meeting]. You schedule time on their calendar for a phone call so they can get the materials ready."* Instead of impromptu telephone calls, Intel employees used email or IM, technologies that were perceived as less intrusive.

Based on these observations, we propose:

P5: As the number of ICT-enabled meetings increases, the tension individuals experience between maximizing personal objectives and maximizing group objectives will increase.

Our respondents reported multiple team memberships, with 80% reporting concurrent membership on four or more teams. The frequency of multi-teaming is an important insight into the reality of networked organizations that is only now being recognized (Mortensen, Woolley, & O'Leary, 2007). Although probably not a new phenomenon (Watson-Manheim & Belanger, 2002), multi-teaming is likely more prevalent today, partially because ICT makes it easier to connect across distance, and partly because of the emergence of networked organizational forms. Multi-teaming makes it easier for people to follow new trends within the organization and expand their social networks (Griffith, Sawyer, & Neale, 2003; Orlikowski, 2002). It also promotes knowledge sharing across teams, a common knowledge management objective (Alavi & Leidner, 2001), but the benefit comes with the cost of time spent in additional meetings. In general, however, multi-teaming likely benefits both the organization and individual even as it brings complications of coordinating across multiple responsibilities and possible cognitive overload from persistent boundary spanning activities (Carlile, 2002).

The vast majority of meetings we examined at Intel were recurring, usually on a weekly, monthly, or quarterly basis. The predictability and repetition from recurring meetings establishes and reinforces norms about the meeting process, including which ICTs will be used and how they are used (Maznevski & Chudoba, 2000). This enhances the social context for non-collocated participants, which reduces the information that must be conveyed at any one time since much of it resides in the minds of meeting participants. Indeed, another study at Intel identified the lack of shared practices as a bigger detriment to effective team performance than distance (Chudoba, Watson-Manheim, Lee, & Crowston, 2005).

HR at Intel tried to promote practices such as meeting agendas and minutes to increase individual productivity, although not all teams implemented these practices in the same way. For example, some teams distributed agendas and minutes as email attachments to meeting notices, while other teams posted such documents to a team repository. Teams also used different ICT during meetings (e.g., a Microsoft meeting support tool versus a locally developed meeting support application). Respondents noted the frustration of trying to remember the set of norms associated with a given team, which was especially challenging when they had back-to-back meetings on their schedules and had limited time to "shift gears" from one team's work to another.

An employee's work life consists of concurrent membership on multiple teams, along with their attendant administration and reporting, so an emergent, unstated responsibility people have is to manage across the entire bundle of teams, each with its own documents, deadlines and deliverables. Work, therefore, requires a perpetual state of polychronic activity or multi-tasking (Ancona, Goodman, Lawrence, & Tushman, 2001; Lee & Liebenau, 1999). Work behavior becomes fragmented so that people cannot as readily organize their responsibilities across meetings and teams or "see" the overlapping activities and deliverables for which they are responsible. Thus we propose:

P6: As multi-teaming increases for an individual, decreased productivity is likely to be associated with ICT-enabled meeting structures and norms that are inconsistent across teams.

DISCUSSION

We have seen that meetings enabled by ICT are an occasion to enact new work practices around participation in meetings. Drawing on data gathered from interviews that used entries in the employees' electronic calendar system, we found that knowledge workers at Intel attend a large number of meetings, with two-thirds spending more than half their workweek in meetings. The vast majority of these meetings included non-collocated participants, enabled by extensive use of ICT. When meetings are held with people joining the event from separate workspaces, it is easier for participants to emphasize their individual objectives (e.g., to multitask) than when everyone meets in the same room. ICT provides a degree of anonymity in the sense that work on other tasks is less visible to non-collocated participants. GSS research emphasized use of technology to mask a person's identity, whereas today's ICT-enabled distributed meetings mask a person's work habits. To the extent that use of ICT enables different participation practices to emerge, the nature of work in organizations shifts.

Our research suggests that the conventional view of meetings as a unitary phenomenon differs from what was experienced by these workers. A more complex picture emerged. While a relatively comparable repertoire of ICTs was used across all 524 meetings, they were used differently depending on whether the objective of the meeting participant was to advance group needs or to protect individual productivity. Similarly, use of an ICT broke barriers through its use in one situation, and created barriers in another. ICT simultaneously enabled a flexible organizational structure by linking people independent of their geography and reduced flexibility as additional structured meetings were added to electronic calendars to compensate for the lack of heretofore commonplace, impromptu interactions. ICT enabled individuals to participate in meetings without being physically co-located and provided them more control over the extent of their participation in a meeting. There is evidence in the literature that the total time spent in meetings has increased concurrent with the deployment of these technologies (Majchrzak & Malhotra, 2004), which is consistent with our finding that 2/3 of those we interviewed spent 20 or more hours a week in meetings. Our data point to new ways of participating in meetings as reflected in evolving work practices. Four aspects of meetings emerged from our data that highlight these changes.

Meetings are Integral to Work in Team-Centered Organizations

Meetings continue to be integral to work in that they are a central means of communication and sharing of information across the organization. This is especially important when work must be accomplished by far-flung organizational members who work together in geographically dispersed teams. Unlike prior research on meetings that portrays them as an artifact of work that is peripheral (Luong & Rogelberg, 2005; Rogelberg et al., 2006), meetings at Intel were the venue where work got done to the extent that they were a central means of sharing information across the organization. Our data suggest this is the case for all types of meetings, whether all participants met FTF or participants were distributed. It was not just managers, the subject of Mintzberg's (1973) ethnography, who spent a large part of their work-day in meetings, but also knowledge workers. This trend is in line with the move toward flatter, networked organizational forms. As the ranks of middle managers decrease, those who remain in

the organization assume some of their boundary spanning responsibilities as represented by frequent attendance in meetings. These meetings provided an opportunity for sense making from the perspective of both individuals and the organization (Schwartzman, 1986, 1989). They also appear to be especially important to effectiveness in teams with non-collated members (Maznevski & Chudoba, 2000).

Tension between Group and Self-Oriented Objectives

The proliferation of meetings to which knowledge workers are invited increases the difficulty of balancing between group and personal objectives, which is exacerbated by concurrent membership on multiple teams. Employees must reconcile conflicting organizational norms that encourage their participation in multiple meetings every day at the same time they must complete work requiring individual attention. When meetings are no longer held with collocated participants, individuals are isolated physically from the group. Tension between group and individual objectives may emerge as the different form of meeting facilitates a change in practices. Traditionally, meetings include someone who facilitates discussion and participants who are, at least theoretically, fully engaged in the meeting. In fact, much of the early research and development of GSS was centered on the need to provide process structure to meetings in order to enhance full participation and encourage input from all meeting participants (Nunamaker et al., 1991).

Partial participation has emerged as an acceptable practice at Intel, in part because of the widespread structure of self-managed work teams and the results-oriented culture that valued self-reliance. Employees had the autonomy to decide how best to manage the myriad demands on their time. In such an environment, self-oriented practices were often beneficial to the organization, even if they might conflict with the group-oriented practices enacted by a meeting convener whose objective was to encourage active participation. Participating in meetings by listening "with one ear" was necessary since it facilitated knowledge sharing and served as a mechanism for keeping track of developments throughout the organization as what was important moved beyond the physical confines of the work location of the individual. "One of the key battlegrounds in the future knowledge war will be the management of attention: understanding how it is allocated by individuals and organizations ..." (Davenport & Volpel, 2001, p. 218).

Discontinuities

Two especially interesting changes in work practices that we observed were the presence of multiple discontinuities in meetings and multi-teaming. The notion of boundaries or discontinuities (e.g., geographical, functional, temporal, organizational, and technology) has often been used as a conceptual anchor to help clarify the challenges and opportunities encountered in the distributed environment (cf., Espinosa, Cummings, Wilson, & Pearce, 2003; Orlikowski, 2002; Watson-Manheim, Chudoba, & Crowston, 2002). However, most of this research has investigated geographically distributed workers (Powell et al., 2004; Watson-Manheim et al., 2002). Our research at Intel highlights the complexity of this environment, as workers were members of multiple teams that crossed multiple boundaries. Moreover, the concurrent membership on multiple teams highlights the potential for negative impacts on personal productivity when different teams have different norms for conducting their work, sharing information, and using ICT. Finally, over 80% of the meetings described to us included one or more non-collocated participants, suggesting an additional level of complexity to a distributed

meeting that has rarely been examined in prior research (O'Leary & Mortensen, in press).

ICT Support for Fragmented Work Environment

Although some form of ICT was used in almost all of the 524 meetings we analyzed, our data did not suggest that meetings were uniformly easier or more difficult. For example, while use of ICT made it easier for people to participate in meetings irrespective of their geographic location, spontaneous interactions became more difficult, adding structure to one's work routine. Even telephone calls were scheduled in advance and noted on one's electronic calendar. Thus, it was not harder to have scheduled meetings at Intel; it was more difficult to have unscheduled meetings. Perhaps because of the number of meetings that filled their calendars, people appeared covetous of unscheduled time and tried to maintain control of it as much as possible.

In their review of research on technology-supported meetings, Fulk and Collins-Jarvis (2001) found that most of the research focused on enhancing social presence (Short, Williams, & Christie, 1976) and the information carrying capacity of the ICT (Daft & Lengel, 1986). One would have expected, then, to see use of ICT such as video conferencing during meetings to replicate collocation, especially since the research was conducted in a technology company known for being innovative. Instead, we found that most meeting participants chose the least rich interaction possible in order to make it easier for them to engage in self-oriented practices such as multi-tasking and balance the many demands on their time.

As communication channels were added within meetings, they were more likely to be used to visually share meeting artifacts rather than increase the social presence among meeting participants through the use of video cams.

Meeting participants sometimes had the option of attending a meeting from their desk or traveling to a conference room that had been reserved for the collated invitees to a meeting. However, rather than choosing collocation or attempting to replicate it with a laptop camera, the more important objective was to maximize one's control over how much non-verbal information was conveyed to other meeting participants in order to maintain the flexibility to multi-task.

Indeed, our data suggest it is more important that ICT be used to facilitate people's need to monitor work across teams and meetings, with multiple means of communication available. This can be even harder with computer files than the traditional reliance on piles of paper files of documents related to a topic. Now files are buried in trees within the electronic filing system. The coherence principle includes the ability to see at a glance one's personal world of work at least in some graphic form so that it is peripherally "present" rather than buried in a hierarchy.

Visibility is a critical component of work. In collocated meetings, people watch a presentation, or use whiteboards and flipcharts at the front of the room for diagrams and to-do lists so that everyone can see and comment on issues. Available tools for distributed meetings offer some degree of visibility in the form of screen-sharing applications and in some cases, video connections. However, people tend to focus on individual elements and specific tasks. Screen sharing applications are widely used at Intel, but people concurrently check e-mail, run Internet searches and perform other tasks while viewing foils on a shared screen. The content is there but it is limited and often not sufficiently compelling to be the sole focus of attention. More integrated ICT tools would allow a more coherent form of multi-tasking and possibly concentrate some of this peripheral activity back to the team at hand, e.g. through online chat with other team members, opening other team-related documents and allowing people to peruse team-mates' current

situated environments of either content, physical conditions or cultural surroundings.

Implications for Practice

Our research suggests several implications for practice. Meeting artifacts such as agendas and minutes serve to make meetings more effective not only because they provide structure and documentation, but also because they make it easier for someone to manage her or his own productivity. To the extent that employees have some autonomy in directing their own work activities, these artifacts can make it easier for them to make decisions about how best to spend their time at work and accomplish their performance objectives. If a meeting convener needs full participation from meeting participants rather than having them participate with "one ear", then agendas and minutes need to convey this in a way that participants will choose to devote their full attention during a meeting. While a meeting convener may be able to mandate attendance at a meeting, he or she must explicitly "earn" participation by helping participants understand why it is in their interest to remain engaged in meeting discourse.

Meeting conveners should also recognize that active participation may not be required of everyone who attends a meeting, and that organizational objectives may be best met with different levels of participation. For example, a meeting convener could require "laptops down" for those who are gathered in a conference room as a way to remove the temptation to multitask from collocated meeting participants by not allowing participants to use ICT during the meeting. Doing so, however, means meeting participants do not have access to information from a colleague who is not attending a meeting. As noted earlier, the ability to use ICT to request and receive information from someone who is not in attendance allows the meeting to continue and decisions to be made based on information received from the non-participating person, which enhances meeting processes and improves decision-making. These productivity gains may outweigh losses from multi-tasking and so we recommend that meeting conveners be judicious in their restrictions on ICT use during meetings.

Organizational norms play a strong role in whether and how ICT are used in meetings (Stephens & Davis, 2009), and meeting conveners must learn to take advantage of organizational norms regarding ICT use in order to meet their objectives for a given meeting. For example, to the extent that organizations value multi-tasking, conveners may choose to invite people to meetings even when their input is required for only a portion of the meeting's objectives. The participant can provide information and insights during the portion of the meeting that concerns her and focus on other tasks during the portion of the meeting that does not need her input, maximizing both personal and group productivity objectives. In addition, as organizations increasingly rely on far-flung teams, frequent meetings are important to keep everyone on the same page and to replace impromptu meetings around the water cooler that FTF team members have relied on in the past. Meetings can serve as a heartbeat or predictable rhythm for team members, as Maznevski and Chudoba (2000) noted in their paper, which is especially important for far flung team effectiveness. Organizations should recognize this important role of meetings, and not be overly aggressive in reducing the number of meetings that are held for members of distributed teams.

Organizational policies that seek to create a level-playing field for all participants in a meeting (e.g., if everyone cannot meet FTF, then all participants will join the meeting using ICT) may unintentionally encourage partial participation in meetings because distributed participation makes it easier for team members to multi-task during the meeting. However, as discussed above, this unintended consequence may be acceptable in organizations that value multi-tasking. Organizations must also weigh the demands they make on employees to work outside normal hours with the need for employees balance their work and personal lives. To the extent that some degree of participation (e.g., partial participation) is better than no participation, meeting conveners should make it as easy as possible for those who attend a meeting outside of normal work hours.

Limitations

Our evidence paints a rich picture of the evolution of meetings as a form of organizing enabled by the use of ICT. Of course, it is important to acknowledge the limitations of our work and possible boundaries to our conclusions. The primary limitation of this research is that we only studied meetings among knowledge workers at Intel Corporation. Intel is obviously a meeting-intensive organization, with a proclivity toward distributed work. As such, our findings may not reflect work in other organizations today. At the same time, we suggest that work at Intel may be a bellwether for the evolution of work practices in the 21st century. By considering our findings, practitioners can be alerted to these new ways of working and make appropriate adaptations within their own organizations.

By examining the electronic calendars of the interviewees, our study focused more on the scheduled interactions than the unscheduled interactions. Specifically, we might have missed out on ad-hoc or unscheduled meetings that might have taken place but were forgotten by our respondents because they were not recorded in their electronic calendars. However, such ad-hoc and unscheduled meetings that were not recorded in the calendars were not common in Intel Corporation. Nonetheless, caution should still be exercised when generalizing the finding to organizations that do not use the electronic calendars as actively.

CONCLUSION

Meetings are such a common occurrence in contemporary organizations that almost everyone shares an understanding of what a meeting is and what participation in a meeting looks like. Even upon first joining an organization, people bring a set of assumptions about meetings that likely includes individuals sitting around a table, engaging in verbal discussion about one or more topics. Yet our exploratory study at Intel, an innovative global technology company, suggests that the structural role of meetings and how they are enacted is evolving beyond this familiar perspective. Whereas twenty years ago the use of technology such as group support systems to support meeting processes was optional, in the distributed work environment, ICT is necessary for the meetings to occur. By examining meetings as a collection of work practices, we have highlighted these changes in taking a first step toward understanding new ways to participate in ICT-enabled meetings. Specifically, we identified four aspects of meetings that reflect work in the 21st century: meetings are integral to work in team-centered organizations, tension between group and personal objectives, discontinuities, and ICT support for a fragmented work environment. Together, these point to new ways of working in the distributed work environment as seen in the practice of meetings.

ACKNOWLEDGMENT

This research was funded in part by a grant from the IT R&D Council at Intel. It was initiated and supported by Eleanor Wynn, our corporate sponsor and a member of Intel's IT Innovation Group, who suggested that interesting insights about the distributed work environment could be gleaned from examining entries on employees' electronic calendars. We also appreciate the support of Intel employees. Ferdi Eruysal, a PhD student at the University of Illinois-Chicago helped with the data analysis. Members of the MIS Department Research Colloquium at Florida State University provided insightful feedback on the research.

REFERENCES

Adkins, M., Burgoon, M., & Nunamaker, J. F. Jr. (2003). Using group support systems for strategic planning with the United States Air Force. *Decision Support Systems*, *34*(3), 315–337. doi:10.1016/S0167-9236(02)00124-0

Ahuja, M., & Carley, K. (1999). Network Structure in Virtual Organizations. *Organization Science*, *10*(6), 741–757. doi:10.1287/orsc.10.6.741

Alavi, M., & Leidner, D. (2001). Review: Knowledge management and knowledge management systems: Conceptual foundations and research issues. *Management Information Systems Quarterly*, *25*(1), 107–136. doi:10.2307/3250961

Ancona, D. G., Goodman, P. S., Lawrence, B. S., & Tushman, M. L. (2001). Time: A new research lens. *Academy of Management Review*, *26*(4), 645–663. doi:10.2307/3560246

Boehm, B., Grunbacher, P., & Briggs, R. (2007). Developing groupware for requirements negotiation: Lessons learned. In Selby, R. (Ed.), *Software Engineering* (pp. 301–314). Hoboken, NJ: Wiley.

Briggs, R. O., Adkins, M., Mittleman, D., Kruse, J., Miller, S., & Nunamaker, J. F. Jr. (1999). A technology transition model derived from field investigation of GSS use aboard the U.S.S. Coronado. *Journal of Management Information Systems*, *15*(3), 151–195.

Briggs, R. O., Vreede, G.-J. D., & Nunamaker, J. F. Jr. (2003). Collaboration Engineering with ThinkLets to Pursue Sustained Success with Group Support Systems. *Journal of Management Information Systems*, *19*(4), 31–64.

Carlile, P. R. (2002). A pragmatic view of knowledge and boundaries: Boundary objects in new product development. *Organization Science, 13*(4), 442–455. doi:10.1287/orsc.13.4.442.2953

Chudoba, K. M., Watson-Manheim, M. B., Lee, C. S., & Crowston, K. (2005). *Meet Me in Cyberspace: Meetings in the Distributed Work Environment*. Paper presented at the Academy of Management Conference.

Cramton, C. D. (2001). The mutual knowledge problem and its consequences for dispersed collaboration. *Organization Science, 12*(3), 346–371. doi:10.1287/orsc.12.3.346.10098

Crowston, K., Howison, J., Masango, C., & Eseryel, U. Y. (2005). *Face-to-face interactions in self-organizing distributed teams*. Paper presented at the Academy of Management Conference.

Daft, R. L., & Lengel, R. H. (1986). Organizational information requirements: Media richness and structural design. *Management Science, 32*(5), 554–571. doi:10.1287/mnsc.32.5.554

Davenport, T., & Volpel, S. (2001). The rise of knowledge towards attention management. *Journal of Knowledge Management, 5*(3), 212–221. doi:10.1108/13673270110400816

DeSanctis, G., & Poole, M. S. (1994). Capturing the complexity in advanced technology use: Adaptive structuration theory. *Organization Science, 5*(2), 121–147. doi:10.1287/orsc.5.2.121

Doyle, M., & Strauss, D. (1976). *How to make meetings work*. Chicago, IL: Playboy Press.

Eisenhardt, K. M. (1989). Building theory from case study research. *Academy of Management Review, 14*(4), 532–550. doi:10.2307/258557

Espinosa, J. A., Cummings, J. N., Wilson, J. M., & Pearce, B. M. (2003). Team boundary issues across multiple global firms. *Journal of Management Information Systems, 19*(4), 157–190.

Fjermestad, J., & Hiltz, S. R. (1999). An assessment of group support systems experiment research: Methodology and results. *Journal of Management Information Systems, 15*(3), 7–149.

Fulk, J., & Collins-Jarvis, L. (2001). Wired meetings: Technological mediation of organizational gatherings. In Jablin, F. M., & Putnam, L. L. (Eds.), *The New Handbook of Organizational Communication: Advances in Theory* (pp. 624–663). Research, and Methods.

Griffith, T. L., Sawyer, J. E., & Neale, M. A. (2003). Virtualness and knowledge in teams: Managing the love triangle of organizations, individuals and teams. *Management Information Systems Quarterly, 27*(3), 265–287.

Haines, R., & Cooper, R. (2008). The Influence of Workspace Awareness on Group Intellective Decision Effectiveness. *European Journal of Information Systems, 17*(6), 631–648. doi:10.1057/ejis.2008.51

Jackson, M., & Poole, M. S. (2003). Idea generation in naturally-occurring contexts: Complex appropriation of a simple procedure. *Human Communication Research, 29*, 560–591. doi:10.1093/hcr/29.4.560

Lee, H., & Liebenau, J. (1999). Time in organizational studies: Towards a new research direction. *Organization Studies, 20*(6), 1035–1058. doi:10.1177/0170840699206006

Luong, A., & Rogelberg, S. G. (2005). Meetings and more meetings: The relationship between meeting load and the daily well-being of employees. *Group Dynamics, 9*(1), 58–67. doi:10.1037/1089-2699.9.1.58

Majchrzak, A., & Malhotra, A. (2004). *Virtual Workspace Technology Use and Knowledge-Sharing Effectiveness in Distributed Teams: The Influence of a Team's Transactive Memory*. Los Angeles, CA: Marshall School of Business, University of Southern California.

Mankins, M. (2004). Stop wasting valuable time. *Harvard Business Review, 82*(9), 58.

Mason, J. (1996). *Qualitative Researching*. London, UK: Sage.

Maznevski, M. L., & Chudoba, K. M. (2000). Bridging space over time: Global virtual team dynamics and effectiveness. *Organization Science, 11*(5), 473–492. doi:10.1287/orsc.11.5.473.15200

Miles, M. B., & Huberman, A. M. (1994). *Qualitative Data Analysis: An Expanded Sourcebook* (2nd ed.). Thousand Oaks, CA: Sage.

Mintzberg, H. (1973). *The Nature of Managerial Work*. New York, NY: Harper & Row.

Mortensen, M., Woolley, A. W., & O'Leary, M. B. (2007). Conditions Enabling Effective Multiple Team Membership. In K. Crowston, S. Sieber, & E. Wynn (Eds.), *Proceedings of the IFIP Working Group 8.2 Working Conference on Virtuality and Virtualization,* Portland OR (Vol. 236, pp. 215–228). Berlin, Germany: Springer.

National Research Council. (1999). *The Changing Nature of Work: Implications for Occupational Analysis*. Washington, DC: National Academy Press.

Nunamaker, J. F., Dennis, A. R., Valacich, J. S., Vogel, D. R., & George, J. F. (1991). Electronic meeting systems to support group work. *Communications of the ACM, 34*(7), 40–61. doi:10.1145/105783.105793

O'Leary, M. B., & Mortensen, M. (2010). Go (Con) figure: Subgroups, Imbalance, and Isolates in Geographically Dispersed Teams. *Organization Science, 21*(1), 115–131.

Orlikowski, W. J. (2000). Using technology and constituting structures: A practice lens for studying technology in organizations. *Organization Science, 11*(4), 404–428. doi:10.1287/orsc.11.4.404.14600

Orlikowski, W. J. (2002). Knowing in practice: Enacting a collective capability in distributed organizing. *Organization Science, 13*(3), 249–273. doi:10.1287/orsc.13.3.249.2776

Powell, A., Piccoli, G., & Ives, B. (2004). Virtual teams: A review of current literature and directions for future research. *The Data Base for Advances in Information Systems, 35*(1), 6–36.

Rogelberg, S. G., Leach, D. J., Warr, P. B., & Burnfield, J. L. (2006). "Not another meeting!"? Are meeting time demands related to employee well-being? *The Journal of Applied Psychology, 1*, 86–96.

Romano, N. C., Jr., & Nunamaker, J. F., Jr. (2001). Meeting Analysis: Findings from Research and Practice. In *Proceedings of the 34th Hawaii International Conference on System Sciences.*

Sarker, S., & Sahay, S. (2002). Information Systems Development by US-Norwegian Virtual Teams: Implications of Time and Space. In *Proceedings of the 35th Annual Hawaii International Conference on System Sciences* (pp. 1–10).

Schultze, U., & Orlikowski, W. J. (2001). Metaphors of virtuality: Shaping an emergent reality. *Information and Organization, 11*, 45–77. doi:10.1016/S1471-7727(00)00003-8

Schwartzman, H. (1986). The meeting as a neglected social form in organizational studies. In Staw, B., & Cummings, L. (Eds.), *Research in organizational behavior* (*Vol. 9*, pp. 233–258). Greenwich, CT: JAI.

Schwartzman, H. (1989). *The meeting: Gatherings in organizations and communities*. New York, NY: Plenum.

Short, J., Williams, E., & Christie, B. (1976). *The social psychology of telecommunications*. New York, NY: John Wiley & Sons.

Sonnentag, S. (2001). High performance and meeting participation: An observational study in software design teams. *Group Dynamics*, *5*(1), 3–18. doi:10.1037/1089-2699.5.1.3

Star, S. L. (1989). The structure of ill-structured solutions: Boundary objects and heterogeneous distributed problem solving. In Gasser, L., & Huhns, M. N. (Eds.), *Distributed Artificial Intelligence* (*Vol. 2*, pp. 37–54). San Francisco, CA: Morgan Kaufmann.

Stephens, K., & Davis, J. (2009). The social influences on electronic multitasking in organizational meetings. *Management Communication Quarterly*, *23*(1), 63–83. doi:10.1177/0893318909335417

Strauss, A., & Corbin, J. (1990). *Basics of qualitative research: Grounded theory procedures and techniques*. Newbury Park, CA: Sage.

Tropman, J. (1996). *Effective meetings: Improving group decision making*. Thousand Oaks, CA: Sage.

Watson-Manheim, M. B., & Belanger, F. (2002). Exploring Communication-Based Work Processes in Virtual Work Environments. In *Proceedings of the 35th Hawaii International Conference on System Sciences (HICSS-35)*.

Watson-Manheim, M. B., Chudoba, K. M., & Crowston, K. (2002). Discontinuities and continuities: A new way to understand virtual work. *Information Technology & People*, *15*(3), 191–209. doi:10.1108/09593840210444746

Weick, K. E., & Meader, D. K. (1993). Sensemaking and group support systems. In Jessup, L., & Valacich, J. (Eds.), *Group Support Systems: New Perspectives* (pp. 230–252). New York, NY: Macmillan.

This work was previously published in the Journal of Organizational and End User Computing, Volume 23, Issue 2, edited by Mo Adam Mahmood, pp. 15-36, copyright 2011 by IGI Publishing (an imprint of IGI Global).

Chapter 12

Understanding User Dissatisfaction:
Exploring the Role of Fairness in IT-Enabled Change

Tim Klaus
Texas A&M University – Corpus Christi, USA

ABSTRACT

This paper examines the role of fairness and how it shapes a user's view in IT-enabled change. Drawing from several fairness theories, components of fairness are identified and examined in two studies. The first study examines the role of fairness through user interviews and finds that all five components of fairness are considered by users in enterprise system implementations. The second study operationalizes and analyzes the components of fairness through a questionnaire distributed to users. This second study finds that fairness is comprised of all five components that were proposed and a significant relationship exists with user dissatisfaction. The two studies lead to a new theoretical perspective and provide practical implications regarding the role of fairness in IT-enabled change and their strategic implications.

INTRODUCTION

An Enterprise System (ES) is a commercial software package which can integrate business processes and data throughout an organization (Markus, Axline, Petrie, & Tanis, 2003). They include organizational-wide software such as Enterprise Resource Planning (ERP) systems, scheduling, customer relationship management, product configuration, and sales force automation (Markus & Tanis, 2000). The ES market will be an estimated $47.7 billion in 2011 (Jacobson, Shepherd, D'Aquila, & Carter, 2007) as there are many technical and business reasons for why

DOI: 10.4018/978-1-4666-2059-9.ch012

organizations choose to implement ESs to enable organizational change (Burns, Jung, & Hoffman, 2009; Klaus, Wingreen, & Blanton, 2010; Markus & Tanis, 2000). The dynamic business environment requires most organizations to change to stay competitive, and managers have often facilitated major organizational changes through an ES implementation. However, there are many articles that report costly ES implementation failures which do not meet the expected return on investment (Bingi, Sharma, & Godla, 1999; Robey, Ross, & Boudreau, 2002; Stein, 1999) and others that discuss outright failure (Hill, 2003; Krasner, 2000; Maurer, 2002).

In an attempt to minimize failures, there have been quite a few studies that have addressed critical success factors to ES implementations (Akkermans & Van Helden, 2002; Gupta, 2000; Nah & Lau, 2001; Rao, 2000; Sternad, Bobek, Dezelak, & Lampret, 2009; Stratman & Roth, 2002; Willcocks & Sykes, 2000). These studies have described critical success factors that are related to the technology, commitment of management, process changes, project management and other areas. Despite these recommendations, however, there continue to be ES implementations which outright fail or do not reap the projected benefits. One particular issue that has been described as "the root of many enterprise software project failures" (Hill, 2003, p. 1) is user resistance. Other studies also describe the significant impact user resistance has on ES implementation failures (Barker & Frolick, 2003; Robey et al., 2002; Umble & Umble, 2002). User dissatisfaction with various implementation issues is primarily the reason for user resistance (Hirschheim & Newman, 1988; Klaus & Blanton, 2010). This study explores concepts which may help to explain user dissatisfaction, an antecedent of user resistance.

Previous studies have examined a variety of issues that have led to user dissatisfaction. However, previous research has not examined the role of various aspects of fairness in an implementation. Fairness is definitely not the only predictor of user attitudes, but this study proposes that it may be an important factor in explaining the level of user resistance based on several reasons: 1) An ES implementation affects employees in many different ways with some employees bearing the brunt of the change due to a workload redistribution; 2) Employees' face unmet expectations as their jobs are changed and tighter management monitoring is added; and 3) Organizational procedures and reward structures are changed, affecting required job skills and causing a redistribution of workload or role expansion. Folger (1993) points out that change increases employees' sensitivity to the level of fairness, and since an ES implementation necessitates change, employees are likely to be more sensitive to fairness based on the way they are treated.

Although previous research has not tested the role of fairness in an implementation, the idea that employees' perceived fairness may affect dissatisfaction is supported by previous studies, primarily from the organizational change literature. For example, Folger and Skarlicki (1999) and Cobb et al. (1995) argue that perceived fairness is important in explaining resistance to change. Other studies have found that when employees perceive a lack of fairness, they desire revenge and socially withdraw (Bies & Tripp, 1996), place blame and desire retribution (Sheppard, Lewicki, & Minton, 1992), are more likely to steal from the organization (Greenberg, 1990), and resist the change (Shapiro & Kirkman, 1999). Cohen-Charash and Spector (2001) suggests that an employee's perception of fairness significantly affects attitudes and behaviors. Although an IT study has looked at equity (Joshi, 1990), no study was found which examines the multiple components of fairness.

As previous research has not focused on the individual users' perceptions of fairness, this paper proposes that examining the equity, justice, and psychological contract literature can provide a sound theoretical basis for understanding user attitudes in ES implementations. Previous literature has shown that if there is even one aspect of unfairness (i.e., inequity), it can cause negative attitudes (Jex, 2002). However, there also are sometimes interactive effects between multiple types of fairness (De Cremer, 2005) and thus fairness overall may be better to focus on than only individual aspects of fairness. Furthermore, this paper explores and tests the idea that fairness is an important antecedent to user attitudes in the implementation of an ES. The fairness theories used in this paper provide important, yet unused lenses by which to examine user attitudes in ES implementations. This study extends the ES implementation research by examining the fairness literature which provides an applicable, yet relatively understudied explanation to the system implementation context. This paper first discusses the relevance of several components of fairness, as identified through equity theory, justice theory, and the psychological contract literature. Using a mixed-methods design, the impact of fairness on user attitudes is examined and then tested. Two studies were conducted; the first study uses semi-structured interviews to assess the applicability of fairness in an ES implementation and to understand the perspective of ES users. The second study assesses the individual components of fairness through a questionnaire distributed to ES users. Theoretical and practical implications are described from each of the studies.

There are several contributions for this paper: One contribution is to bring a new theoretical perspective to the examination of user attitudes through the examination of fairness. This is beneficial for both researchers and practitioners as fairness is an issue not studied extensively in the strategic implementation literature yet impacts the response of users. A second contribution is integrating and examining specific components of fairness in the ES implementation context. This study examines and integrates multiple fairness components not examined together in previous studies to determine the impact of fairness on user dissatisfaction. Furthermore, this study brings a broader application of fairness models into the ES

implementation context. A third contribution is a comparison of how perceptions of equity, justice, and the psychological contract affect a user's attitudes. It is important for a manager making strategic decisions about the implementation to know which types of fairness most impact user attitudes. This contribution also further develops the body of knowledge regarding the impact of various fairness components.

LITERATURE REVIEW

Rather than developing one construct to measure fairness, this study views fairness as a multidimensional concept due to various dimensions described by fairness theories. Fairness is defined in this paper as employees having equal probabilities of obtaining outcomes. Through a thorough examination of fairness-related literature, components of fairness are identified. The first component addressed in this paper is from equity theory, which essentially is the fairness of obtaining outcomes based on the inputs an employee provides to the organization. The second set of components are from justice theory, which is the fairness of the means of the outcomes. This includes procedural justice (fairness of procedures), distributive justice (fairness of reward structures), and interactional justice (fairness of the manner of how an employee is treated). The third component is the psychological contract, which essentially is the fairness of the organization fulfilling the expectations of an employee. This study analyzes these fairness components that have been examined in previous studies to examine their effect on user attitudes. Each component is discussed individually in the following paragraphs, starting with the organizational behavior literature, then describing how each component has been examined in the information systems context. First, perceived fairness based on equity theory will be examined, followed by justice theory and the psychological contract literature.

Equity Theory

The first component of fairness that will be described is perceived equity. Research examining the level of fairness in organizations began with Adams's early work on equity theory (Adams, 1963). There are two parts to equity theory: the inputs employees put into an organization and the outputs they receive from the organization. Inputs include job-related skills, prior experience, and effort. Outputs include monetary compensation, job security, stock ownership, and feelings of accomplishment (O'Reilly & Pfeffer, 2000). Based upon the input/output ratio and its comparison to the ratio of others, employees decide if there is equity, or fairness. If employees feel that there is inequity, adjustments usually are made, such as: 1) reducing inputs; 2) work harder to potentially increase organizational outcomes; 3) changing the perception of inputs or outputs; 4) compare the input/output ratio to a different person; or 5) leave the field (Jex, 2002).

Equity theory is a part of the broader social exchange theory which revolves around the nature and level of what employees and organizations exchange (Makin, Cooper, & Cox, 1996). The outputs that employees receive, based on their inputs, help explain *how* employees determine fairness in social exchanges (Jex, 2002). For an ES implementation, employees will compare their inputs (i.e., workload) with the outputs (i.e., compensation), and make a judgment that will affect their attitude. The perception of equity is subjective, based on the perceptions of the employee.

There have been several IS studies that have used equity theory. For example, Joshi (1990) used equity theory in examining how the level of equity affects user information satisfaction, Joshi (1991) used equity theory in a equity-implementation model, and Joshi (1992) used equity theory in examining the allocation of IS resources. Recent IS studies also have used this theory (Au, Ngai, & Cheng, 2008; Carr, 2007). Furthermore, there are system implementation studies that have discussed

issues affecting equity, such as employees' loss of status and power (Hussain & Hussain, 1984), economic insecurity (Hussain & Hussain, 1984), parochial self-interest (Shang & Su, 2004), and increased efforts without additional compensation (Shang & Su, 2004). However, equity has not been examined in conjunction with other forms of fairness in order to examine the effect on user attitudes.

Justice Theory

While equity focuses on the fairness of a particular outcome, justice theory focuses on the means by which the outcome was reached. Justice theory assumes an individual's perceptions of the organizational practices affect the level of perceived justice, which, in turn, affects the individual's attitudes and behaviors. Some of the attitudes and behaviors affected by perceived justice are job performance, extra-role behavior (i.e., performing extra tasks not required by the job description), counterproductive behavior (i.e., spreading rumors, destroying equipment), emotions (i.e., anger, guilt, happiness) (Cohen-Charash & Spector, 2001), attitudes (i.e., job satisfaction, organizational commitment), and behavioral outcomes (i.e., organizational citizenship behavior) (Moorman, 1991). Figure 1 shows a general framework of the empirical work of justice theory related to organizations. There have been a multitude of studies that have examined at least one of the relationships shown in Figure 1, with most studies finding significant results. However, few, if any, studies have examined the role of justice in system implementations to explain negative user attitudes.

On one hand, justice research has shown that when employees perceive that they are being treated fair, attitudes and behaviors are cultivated which are needed for successful change (Cobb et al., 1995). On the other hand, Folger and Skarlicki (1999) argues that a rational person tends to not want to cooperate with a change in which something is lost that is valued. Part of the reason for

Figure 1. Framework of justice research

this is that employees "restore" justice by resisting the change (Knights & Vurdubakis, 1994). One study points out that an employee's attempts to restore justice may be indirect (Skarlicki & Folger, 1997) while another suggests indirect attempts may take the form of withdrawing citizenship behaviors or increasing the level of resistance behaviors (Knights & Vurdubakis, 1994). Although resistance to the change may be perceived as deviant behavior, it is possible that some managers act unfairly to users which makes the resistance to be perceived as legitimate by the users. Davidson (1994) discusses that resistance is a reaction to management's actions and that resistance provides a way for employees to "settle the score" for mistreatment. In other words, the actions such as resistance are justified in the minds of employees because of the perceived injustice.

Researchers examining organizations have identified three types of justice: distributive, procedural, and interactional (Cohen-Charash & Spector, 2001). Procedural justice theory suggests employees' perception of fairness is based on appropriate procedures, such as procedures which determine pay. Distributive justice theory suggests employees' perception of fairness is based on reward structure fairness. Interactional justice theory suggests employees' perceptions of fairness are based on the relationship between themselves and their supervisors and how they are treated (Cohen-Charash & Spector, 2001). Folger and Skarlicki (1999) discusses these three types of justice and argues that focusing on any one of the three types may lead to a failure in addressing resistance adequately. Thus in the following paragraphs, each of the three justices are described as to why they can affect user attitudes.

For procedural justice, Cobb et al. (1995) found that when the procedures are perceived to be fair, individuals exert citizenship behaviors and are more likely to consent to the change. On the other hand, if the procedures are perceived to be unfair, resentment is likely to develop (Cropanzano & Folger, 1989). In a study examining a change requiring employees to work in teams, it was found that a large number of procedural-related comments regarding the change-related impediments dealt with fairness-related issues. Some of the comments from the employees included "How will I be appraised? … How can someone at another location know what I am doing? … [Will I] be appraised fairly for work performed?" (Kirkman, Shapiro, Novelli, & Brett, 1996, p. 56). One IS study which used procedural justice found that user attitudes and participation in voluntary system use was affected by procedural justice (Hunton, 1996). Another systems implementation study found that users were frustrated with the new procedures because there was no communication channel to address disagreements with superiors during the implementation (Marakas & Hornik, 1996).

In regards to distributive justice, the ES implementation often affects the reward structures, the work schedule, workload, and job responsibilities. Thus ES users are likely to be affected by the way rewards are distributed to employees for their work with the ES. Some of the distributive justice concerns of employees that were faced with a change include "[Will I] do more work than others for the same money? … Will I be fairly recognized for my contributions? … Everyone [might not] do their share and work as hard as the others" (Kirkman et al., 1996, p. 56). An information systems study which looked at user resistance found that the redistribution of resources was a reason for user resistance during the implementation (Hirschheim & Newman, 1988). During ES implementations, there is usually some redistribution of resources, which can definitely impact the perceived distributive justice of users.

For interactional justice, the way supervisors and other managers treat the users throughout the change will likely affect their attitudes and behaviors. For example, managers which adequately explain situations were found to enhance the perceived fairness of employees (Shapiro, Buttner, & Barry, 1994). On the other hand, if a low level of interactional justice exists, users may not feel respected or may perceive that their opinions are not valued by management. This can cause users to less readily express their opinions or engage in the organizational change. Some of the interactional justice concerns faced by employees encountering change include "Will management change back after some short period of time? ... How supportive [will] upper management be ... [Will there be] respect for others" (Kirkman et al., 1996, p. 57). One ES study identified the process changes as one of the reasons for user resistance (O'Leary, 2000) and another implementation study found that interpersonal relationships were altered (Hussain & Hussain, 1984). The process changes occurring in ES implementations generally affect the way that people interact with each other and in some cases it likely for users to feel that they are not being respected, impacting the perceived interactional justice of the users.

Psychological Contract

Another component of fairness is the psychological contract. "Psychological contract" appears to have first been coined by Argyris (1960) and is traced back to organizational psychology literature (Schein, 1980). Psychological contract has been defined as "beliefs that individuals hold regarding promises made, accepted, and relied on between themselves and another" within an organiza-

tion (Rousseau, 1995, p. 9). In regards to an ES implementation, the end-users' perspectives are of importance; thus for this paper the psychological contract is defined as a user's beliefs regarding the employer's promises that were made and implied. The equity and justice components of fairness entails a comparison with another, but the psychological contract is only focused on the perceptions of unfulfilled perceived obligations to the individual.

There are numerous information sources from which an employee accumulates thoughts about the psychological contract – during the initial interview with the company, from managers, from co-workers, news articles about the company, the company mission statement, the employee manual, training, job reviews, gossip, and seeing the job tracks of other workers. Based on these sources, employees perceive a psychological contract in areas such as feedback, job security, appropriate training/development, promotion, nature of job, management of change, job responsibilities and appropriate compensation (Rousseau, 1995, p. 116). Studies have found that an employee's psychological contract in these areas affects employee attitudes and actions (McDonald & Makin, 2000; Porter, Pearce, Tripoli, & Lewis, 1998).

It is inevitable that employees develop psychological contracts with their employer. However, a change, such as that brought forth from an ES, disrupts the psychological contract. Change often causes employees to perform different tasks than they are accustomed to, work in a different environment than accustomed, and perform more tasks than those performed in the past (Cummings & Worley, 2005). During an organizational change, the psychological contract often is unilaterally changed by the organization, which causes the employee to have a perceived psychological contract breach (Rousseau, 1996). A psychological contract breach occurs when employees do not perceive that their employer is meeting their promises/obligations, whether or not the employer knowingly failed to fulfill the perceived promises

(Morrison & Robinson, 1997). Employees that have experienced a psychological contract breach have been shown to have lower organizational commitment, in-role and extra-role performance (Coyle-Shapiro, 2002), job performance (Robinson, 1996), job satisfaction and employee trust (Coyle-Shapiro, 2002; Robinson & Rousseau, 1994), organizational citizenship behavior (Robinson & Morrison, 1995), an increased level of turnover intentions (Coyle-Shapiro, 2002; Robinson, 1996), withdrawal behaviors (Conway & Briner, 2002), and feelings of deception, anger, injustice, or betrayal (Morrison & Robinson, 1997).

As ES users are faced with change, the old psychological contract is upset and replaced by a new one. Since the status quo is familiar to employees, this serves as the referent for new expectations. The idea that employees have some form of cognitive anchor or psychological contract based on the status quo is consistent with the findings on psychological contract violations (Rousseau, 1995). For example, Meckler, Drake, and Levinson (2003) addresses resistance and the psychological contract, and discusses how dealing with resistance to change affects the psychological contract and the development of a new contract.

Despite the use of the psychological contract in non-IS literature, there is only a limited examination of the psychological contract in IS research, such as examining how a psychological contract breach in virtual teams affects trust (Piccoli & Ives, 2003) or examining IT outsourcing (Koh, Ang, & Straub, 2004). However, some systems implementation research has indicated issues which are likely that the change will trigger a psychological contract breach for many users. For example, one study addressed the change in job content that occurs (Hussain & Hussain, 1984) while two other studies found that managing the users' expectations are important (Akkermans & Van Helden, 2002; Ginzberg, 1981) and one study discussed the values, work habits and dilemmas that carry over to the new system (Alvarez & Urla, 2002). This breach in the employee-employer relationship will affect the users' attitudes. As a psychological contract breach in other areas of a job causes employees to respond negatively, an ES implementation causing a breach of a psychological contract also should cause employees to respond negatively towards the system. Particularly, in mandatory, role transforming systems, such as an ES, an employee will likely consider a lack of fairness as a breach in the employee's psychological contract, which may manifest itself in resistance to the system.

Based on the five components of fairness described in the literature review, the following two research questions will be examined:

1. In what areas does an ES implementation affect perceived fairness?
2. How do the components of perceived fairness affect user attitudes in an ES implementation?

STUDY 1: USER INTERVIEWS

Study 1 was conducted to address the first research question by exploring equity, justice, and psychological contract in an ES implementation. A large university in the southern U.S. with approximately 10,000 employees was contacted which recently implemented an ES. Employees were contacted based on three criteria: 1) represent different positions; 2) represent different departments; and 3) regularly use the system. Although all interviewed employees used the system for their respective jobs, some of the employees admitted they had resisted the system implementation in various ways and also had stories of others in their department who had resisted and/or were dissatisfied with the implementation. There were 22 people interviewed: 5 clerical staff, 4 office managers, 4 middle management, 3 trainers, 2 IT professionals, 2 top management, 1 accountant, and 1 purchaser. Seven of these interviewees were superusers – employees that take a part-time or full-time role with the implementation, have more system skills and knowledge of business processes, and are often where a group of users will go to for support.

The interviews were semi-structured based on four categories of questions, starting with a basic question in a category and then asking more detailed questions based on how the user responded to the first question. Rather than asking specific questions from a script, the interviewer asked a variety of questions related to the four categories based on the users' responses. The first category of questions related to the experiences of the users throughout and after the ES implementation (Example: Please describe your involvement in the Enterprise System implementation, the amount of time you were involved in the project, and the name/type of system). The second category of questions related to the changes experienced by the users (Example: What degree of change has the enterprise system had on your job?). The third category of questions related to the user's response (attitude and behavioral) to the changes (Example: How did you respond to the system implementation?). The fourth category of questions related to the users' perceptions of fairness throughout the implementation (Example: Describe your perceptions of fairness throughout the system implementation). All interviews were recorded lasting an average of 47 minutes and then transcribed by a third-party transcription service yielding 242 pages of single-spaced transcripts (135,200 words).

The transcripts of the interviews were analyzed independently by three researchers to identify the components of fairness that existed. Each of the researchers was given a coding scheme that described equity, procedural justice, distributive justice, interactional justice, and the psychological contract. The three researchers then were assigned to read through the transcripts and indicate any comment that appeared to be related to these five types of fairness as well as any other type of fairness other than the five that were on the coding

scheme. Next, the researchers met and discussed several discrepancies that arose among the identification of fairness components. A discussion on each of these discrepancies led to an eventual resolution regarding whether or not the interviewee's comment was a specific component of fairness already identified in the literature review or if this was a new component of fairness not found in the literature review. However, every fairness-related comment identified in the transcript fit into one of the five fairness components identified in the literature review.

Although several users thought that all aspects of the ES implementation were fair, others pointed out examples of unfairness. Following are sample quotes from users which demonstrate the various components of perceived fairness.

Equity

User 12: I would say the people that aren't comfortable with change and didn't like the system would say no [it's not fair]. And the people that get better information out of it and like it would say yes. You could almost predict that question by how they like the system.

User 20: All the groups are disadvantaged... it seems like everybody had to kind of step up and assume, at least initially, more of a work load.

User 8: I have a 40-hour a week job for the department that pays me and now you want me to do this system work as well.

User 9: I would be very surprised if someone is able to demonstrate how this is saving anybody any time.

Procedural Justice

User 10: Those units or individuals responsible for doing central billing operations or billing accounts receivable type stuff – it may have been harder for them because they had to develop new systems. In addition to learning the system, they had to develop new systems for feeding information and so it was probably less fair for them.

User 8: *If you were a small department you didn't get much say in what was happening... Some of the modifications that were made suit their needs, but my auxiliary that's less than $100,000 a year, it doesn't fit my needs at all, so even though it was brought up, I mean I'm not the only small auxiliary, but you know it's like you guys are such a small piece you're just going to have to adjust to the information that's going to be provided.*

User 2: *[Was there much management support?] Well, in the academic side – no. In some of the accounting sections – yes.*

Distributive Justice

User 14: *I've seen people in general accounting work just like seven days a week with no compensation ... those same people are now still putting in days and days of time what they did before, plus the complexity that this was added one time and I don't think they were adequately compensated.*

User 8: *We weren't paid by the [implementation] team; we were still paid for by the departments we worked for even though we weren't doing work in that department.*

User 15: *Volunteered time. There's no special pay and no promotion because of that. It's just basically you're doing two jobs for the price of one because you have your regular duties and your [implementation] duties and both do require quite a bit of time on your part. It wasn't until July 1ˢᵗ ... that I was able finally to present a solution to my director in trying to cover me on the purchasing side of it because it was killing me. I was missing a lot of my kids, coming home at 10 o'clock, helping them with their homework, so it was more on the*

personal side. Something has to give after a year and a half. Something has to give, my personal life, what means a lot to me, which is my family, but I also realize that with my job I can't take care of them either.*

Interactional Justice

There was no indication by any of the interviewees that a manager failed to treat employees with respect or was inconsiderate. However, around half of the interviewees voiced comments regarding a lack of communication from top management and management failing to consider the opinions of users.

User 8: *We have a directors group that are mid-management and there were suggestions presented to them and either they didn't understand the system, they didn't understand what the people in the trenches went through that their choices have a ripple effect, so some of the problems we have now is because the directors.*

User 3: *Communication is very bad here ... it gets filtered down person by person.*

Psychological Contract

User 9: *Is it fair to that employee to be fired when they were able to be performing the job they were hired for, but their job has changed and to me that isn't fair. Whether or not if a person has a degree doesn't determine their level of understanding or capability, but you know now I'm hearing people not wanting to hire people for these support positions unless they have a degree in accounting and things like that.*

User 19: *[Expectations] had to be changed in a major way and in ways that they weren't familiar with. So we had, you know I don't want to sound like I'm age discriminating or whatever, but we have these middle aged folks who are used to* *their baskets, their forms come in their baskets in this way and this is what they do [rather than obtaining it through the system].*

User 11: *It's not fair to ask me to change what I know how to do, you are asking me to become more computer literate.*

DISCUSSION OF STUDY 1

Qualitative analysis of the transcripts has revealed that users perceive components of fairness in the ES implementation which influenced user attitudes. In fact, all of the components of fairness described in the literature review were identified by the users. Equity was an issue due to employees facing the ES change in different ways, affecting the jobs of some employees much more than others. For example, some employees experienced a much heavier workload while some employees were only minimally affected. Procedural justice was an issue since some implementation procedures were perceived as not being fair. For example, the new procedures required for the ES was very useful for the jobs of some employees, but others did not reap any benefits and only experienced frustrations due to issues such as dealing with the complexity of the system and being required to go through long training sessions. Distributive justice was an issue since the reward structure was slightly altered. For example, some employees received a small bonus to compensate them for the extra work that was required while most others did not receive any benefit despite their expanded job requirements. Interactional justice also was an issue, as indicated next to the quotations, as close to half of the interviewees voiced comments regarding a lack of communication from top management and management failing to consider the opinions of users. Finally, the psychological contract was an issue as employees felt that the implied promises of their employer and their expectations were not fulfilled due to the changes in areas such as

job skill requirements. All of these areas related to fairness definitely affect user perceptions and it is likely that users experiencing a low level of fairness in the system implementation also tend to be dissatisfied with the system.

A few interesting observations can be gleaned from the transcripts. First of all, to further understand the components of fairness, the transcripts were analyzed to see if there was any relationship among the components. However, no comments were made by any of the users that indicate that one component of fairness affects the perceptions of the user towards another component of fairness. This suggests that although multiple dimensions of fairness exist, each component of fairness is distinct from the others. Although perceived fairness from multiple components may influence users to a greater degree, the interviews suggest that the components are distinct.

Another observation is that even though all users were involved in the same implementation since they work for the same organization, each user still has different perceptions regarding the fairness of the implementation. From a manager's perspective, it is definitely difficult to implement the ES because there are such varying perceptions from the users. Furthermore, it was found that managers have very different viewpoints on the ES than the clerical staff. The quote below is from a manager who reaped the benefits of the system implementation:

User 19: *[the ES] leads to better decision-making; it certainly greatly reduces the likelihood that you're going to overspend on your budget...It would allow anybody, anytime to go online and look and see where a transaction was in terms of being processed... I think there are some reduced labor costs... More access, more real time access to information is available now to everybody ... [there is] the ability to ride or craft your own reports to get the kind of management information back out of the system that might not have been* *available to us with the legacy application that we moved away from.*

This is a very different viewpoint from the staff, which expressed:

User 2: *So from a management's perspective, it was a plus but the headaches of the individual users weren't accounted for.*

User 14: *I don't think the upper management really understands what it means to use the system and they don't care.*

One issue faced by management involved in the implementation is what to communicate to the users. In trying to prevent a psychological contract violation, top management must be aware of the temptation to oversell the benefits of a system and not mention the undesirable effects; this can also cause a psychological contract violation and may increase user dissatisfaction. An alternative is that management may prefer communicating nothing rather than communicating information that turns out to be incorrect. However, in dealing with this dilemma, one study found that employees that received a preview of the change that is to occur resulted in less dysfunctional outcomes (Schweiger & Denisi, 1991).

Another issue faced by management is how to change the psychological contract. Rosseau (1996) discusses interventions that can be performed in order to transition employees from old psychological contracts to new ones. Four stages are suggested: 1) Challenge the old contract; 2) Prepare employees for change; 3) Contract generation; 4) Live out the new contract (Rousseau, 1996). These four stages are applicable to an ES implementation. For stage 1, the old contract can be challenged through educating the future users of the system why the system is needed for the organization, which includes explaining why it is necessary to change the old contract. For stage 2, users can be prepared for the system change

by involving and informing them through team involvement, obtaining their input in the process reengineering, and through information meetings about the ES-enabled change. For stage 3, a new contract can be generated by involving employees in discussions on the "new workplace" and the "new expectations" that must occur because of the ES and how the company will adapt to better serve its employees. Finally, for stage 4, living out the new contract would entail following through with the new workplace expectations that were set.

STUDY 2: THE EFFECTS OF FAIRNESS

As shown through study 1, users perceive various components of fairness in an ES implementation. The purpose of study 2 is to address the second research question, which focuses on the components of perceived fairness that affect user attitudes. For a measure of user attitudes, user dissatisfaction has been selected since dissatisfied users are less likely to use the system for its appropriate uses and may resist the change. Considering the mandated and role-transforming nature of ES, system use cannot be used as a measure in this context since even people who may covertly resist the system will be using the system. Based on study 1, which found that all the components of fairness existed in a systems implementation, and the literature review, it is proposed here that a lesser degree of fairness in the ES change will increase the degree of user dissatisfaction.

Fairness is a complex concept comprised of multiple components which may or may not interact with each other. Thus, to best examine the effects of fairness on user dissatisfaction, a higher level of abstraction is used. This study represents fairness as a second-order construct and uses a second-order molar model to examine the components of fairness as well as the relationship between fairness and user dissatisfaction. Following are the hypotheses which address these relationships (Figure 2).

Hypothesis 1: Perceived fairness has a negative relationship with user dissatisfaction.

Hypothesis 2: The components of fairness have a positive relationship with perceived fairness.

H2a: Perceived equity has a positive relationship with perceived fairness.

H2b: Procedural justice has a positive relationship with perceived fairness.

H2c: Distributive justice has a positive relationship with perceived fairness.

H2d: Interactional justice has a positive relationship with perceived fairness.

H2e: Psychological contract fulfillment has a positive relationship with perceived fairness.

STUDY 2 METHODOLOGY

A questionnaire was developed based on the literature review. Previous studies with organizational fairness questionnaire items were examined. Since the purpose of this study is to examine fairness in an ES implementation, the questionnaire items were slightly modified to apply to fairness in the ES context rather than just overall organizational fairness. Questions were brought in from four

studies (Joshi, 1990; Kickul, Neuman, Parker, & Finkl, 2001; Niehoff & Moorman, 1993; Shapiro & Kirkman, 1999). Five questions were used for each of the variables, as shown in the Appendix. Questions were also developed to obtain demographic information. The questionnaire instructed respondents to indicate their level of agreement to the questions in regards to the system implementation. Each item was measured on a 6-point Likert scale ranging from "strongly disagree" to "strongly agree". Feedback from this questionnaire was first obtained from two researchers not involved in the study which led to minor revisions to some of the questions. Next, a pilot test was conducted to examine the reliability of the questionnaire items. 35 ES users filled out the pilot questionnaire. Cronbach's alpha was >.7 for all of the variables.

The second step was the main data collection, using a convenience sample from ES users in multiple organizations. Through conducting web searches on large ERP vendors, information was found on various user groups and listservs. 317 ES user groups were emailed directly and an email was sent to three user group listservs to solicit involvement with the study. The emails requested a person who is willing to participate in the study through distributing the questionnaire to 15-20 users within their user group. Only organizations

Figure 2. Study 2 model: perceived fairness as a multidimensional aggregate construct – formative second-order, reflective first-order

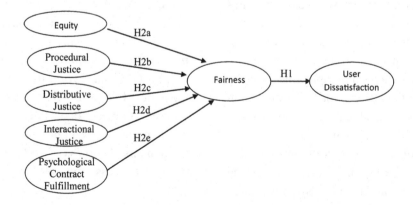

that had rolled out a system less than three years ago were included in the data collection. 30 members from the user groups responded to the emails and agreed to distribute the questionnaires. Each of these questionnaire distributors was mailed a packet of 20 questionnaires and business reply envelopes along with instructions on distributing the questionnaires (Except for two members, who received the questionnaire via email). Several weeks after mailing the packets, a follow-up email was sent to each distributor. It was found that a total of 494 questionnaires were actually distributed to ES users. 196 of these questionnaires were returned, which shows a 39.7% response rate from ES users who actually received the questionnaire. However, since 600 questionnaires were sent out to user group members to distribute the questionnaires, there was a 32.7% response rate to questionnaires that were sent out. Since some of the returned questionnaires were incomplete, only 179 of these questionnaires were actually used in the data analysis.

Partial Least Squares (PLS), a method for estimating path models with latent constructs measured by multiple indicators (Lohmoller, 1988), was used for the analysis based on several reasons. First, PLS is robust for small and moderate sample sizes (Cassel, Hackl, & Westlund, 2000). Second, PLS is suggested to be used for causal-predictive analysis (Chin, 1998; Joreskog & Wold, 1982). Third, PLS enables the construction and use of unobservable variables measured by indicators (Chin, 1998). Fourth, since fairness is proposed as a second-order molar model, PLS is suitable to test such a model (Chin & Gopal, 1995).

STUDY 2 RESULTS AND DISCUSSION

Several questions were asked to obtain the demographic information of the respondents. As shown in Table 1, a variety of ages, education levels, and positions were represented. Table 2 shows descriptive information about all the indicators.

It is recommended that bootstrapping is used for hypothesis testing (Chin, 1998), reestimating path coefficients by a large number of random samples. Thus both the parameter estimates and standard errors are computed based on the samples. Following common practice (Brown & Chin, 2004; Mathieson, Peacock, & Chin, 2001), the alpha coefficients, the composite reliabilites, the average variance extracted for the constructs, the construct to item correlations, and the inter-construct correlations are examined, shown in Table 3. This table shows that both Cronbach's Alpha and the composite reliability of each construct is greater than the recommended 0.70 level (Nunnally & Bernstein, 1994).

Convergent and discriminant validity was supported in two ways. First, the average variance extracted exceeds the square of the correlations (Chin, 1998); as shown on Table 3, the diagonal elements are greater than the off-diagonal elements. Also, the average variance extracted exceeds.50 for each construct (Chin, 1998). Second, all items on the factor loadings, shown in Table 4, loaded on their own construct rather than others and are above the 0.5 level that Chin (1998) states is acceptable with most indicators above the 0.7 level recommended by Nunnally (1978). This supports the idea that the components of fairness identified in study 1 are distinct dimensions of fairness.

In order to assess construct validity, the factor loadings and weights were calculated for each of the indicators that comprised a latent variable. As shown in Table 5, all of the factor loadings are greater than 0.70, which represents a substantial correlation between the indicator and the latent variable (Chin, 1998).

Figure 3 displays the path loadings among the latent variables. For Hypothesis 1, the relationship between fairness and user dissatisfaction turned out to be significant. R-squared for user dissatisfaction in this model is 35.5%. For Hypothesis 2, the components of fairness all have a significant relationship with fairness. As Chin et al. points out, "…at an aggregate level, the relative weightings used to construct the molar attitude can be

Table 1. Survey respondent demographics

Gender	Male	63	35.2%
	Female	115	64.2%
	No Response	1	0.6%
Age	Under 25	1	0.6%
	26-35	22	12.3%
	36-45	42	23.5%
	46-55	45	25.1%
	Above 55	22	12.3%
	No Response	47	26.3%
Highest Completed Education	High School	37	20.7%
	Associate's Degree	20	11.2%
	Bachelor's Degree	73	40.8%
	Master's Degree	44	24.6%
	Doctorate Degree	3	1.7%
	No Response	2	1.1%
Position	Clerical/Data Entry	8	4.5%
	Support Staff	39	21.8%
	IT Staff	11	6.1%
	Supervisor	14	7.8%
	Mid-level manager	56	31.3%
	Top management	3	1.7%
	No Response/Other	48	26.8%
System Vendor	SAP	4	2.2%
	Peoplesoft/Oracle	156	87.2%
	J.D. Edwards	9	5.0%
	Don't Know	3	1.7%
	Other	7	4.0%

used to indicate relative importance" (Chin & Gopal, 1995, p. 49). Thus the weightings on Figure 3 show the relative importance of the various components of fairness.

As a post-hoc analysis, Harman's one-factor test has been performed as a test for common method variance. Following Podsakoff and Organ (1986), an unrotated factor solution of a factor analysis was used to determine the number of factors necessary for the variance in the variables. Table 6 shows the results, which indicates a total of six factors with eigenvalues greater than 1.00. This test indicates that common method variance

may not be a problem for this data as one factor does not account for most of the covariance in the variables (Podsakoff & Organ, 1986; Schriesheim, 1979).

Study 2 results show that fairness significantly explains 35.5% of the variance in user dissatisfaction. Although a considerable amount of variance is explained, fairness is not the only antecedent of user dissatisfaction. As described in the literature review section, there are quite a few other factors that can influence user attitudes in ES implementations. Previous literature also points out a reason for why the amount of variance

Table 2. Descriptives

	MIN	MAX	AVG	STD DEV
ATT1	1	6	2.31	1.45
ATT2	1	6	2.17	1.41
ATT3	1	6	3.05	1.62
ATT4	1	6	3.15	1.63
ATT5	1	6	2.32	1.49
PRO1	1	6	3.53	1.42
PRO2	1	6	3.69	1.56
PRO3	1	6	3.50	1.39
PRO4	1	6	3.09	1.43
PRO5	1	6	3.26	1.50
DIST1	1	6	3.40	1.55
DIST2	1	6	3.42	1.53
DIST3	1	6	3.43	1.56
DIST4	1	6	3.45	1.54
DIST5	1	6	3.45	1.50
INT1	1	6	4.91	1.26
INT2	1	6	4.84	1.15
INT3	1	6	4.35	1.38
INT4	1	6	4.32	1.41
INT5	1	6	4.92	1.28
PSY1	1	6	3.99	1.41
PSY2	1	6	3.97	1.42
PSY3	1	6	4.18	1.34
PSY4	1	6	4.06	1.35
PSY5	1	6	4.02	1.30
EQU1	1	6	4.00	1.50
EQU2	1	6	4.06	1.48
EQU3	1	6	4.08	1.48
EQU4	1	6	4.24	1.49
EQU5	1	6	4.07	1.55
	MIN	**MAX**	**AVG**	**STD DEV**
ATT1	1	6	2.31	1.45
ATT2	1	6	2.17	1.41
ATT3	1	6	3.05	1.62
ATT4	1	6	3.15	1.63
ATT5	1	6	2.32	1.49
PRO1	1	6	3.53	1.42
PRO2	1	6	3.69	1.56
PRO3	1	6	3.50	1.39

continued on following page

Table 2. Continued

	MIN	MAX	AVG	STD DEV
PRO4	1	6	3.09	1.43
PRO5	1	6	3.26	1.50
DIST1	1	6	3.40	1.55
DIST2	1	6	3.42	1.53
DIST3	1	6	3.43	1.56
DIST4	1	6	3.45	1.54
DIST5	1	6	3.45	1.50
INT1	1	6	4.91	1.26
INT2	1	6	4.84	1.15
INT3	1	6	4.35	1.38
INT4	1	6	4.32	1.41
INT5	1	6	4.92	1.28
PSY1	1	6	3.99	1.41
PSY2	1	6	3.97	1.42
PSY3	1	6	4.18	1.34
PSY4	1	6	4.06	1.35
PSY5	1	6	4.02	1.30
EQU1	1	6	4.00	1.50
EQU2	1	6	4.06	1.48
EQU3	1	6	4.08	1.48
EQU4	1	6	4.24	1.49
EQU5	1	6	4.07	1.55

explained is not a higher amount. As Woodroof and Burg (2003) suggests, it is possible that some of the users are predisposed towards their satisfaction level rather than being affected by the level of fairness. Thus, despite a lack of fairness in the ES implementation, some users continue to express their same satisfaction level. This idea is consistent with one of the comments of a study 1 inter-

*Table 3. Intercorrelation and internal consistencies of constructs * Significant at p<0.001*

	N	Cronbach's Alpha	Composite Reliability	Equ	Pro Jus	Dis Jus	Int Jus	Psy Con	User Dis
Equity	179	0.961	0.965	**.87**					
Procedural Justice	179	0.827	0.877	.35	**.59**				
Distributive Justice	179	0.961	0.970	.39	.39	**.87**			
Interactional Justice	179	0.892	0.915	.28	.37	.33	**.70**		
Psychological Contract Fulfillment	179	0.906	0.930	.49	.41	.66	.48	**.73**	
User Dissatisfaction	179	0.915	0.938	-.68	-.33	-.35	-.24	-.43	**.75**

* The numbers on the diagonal is the square root of the variance between the constructs and their measures. Off-diagonal elements are correlations among the latent constructs.

Table 4. Factor loadings (varimax rotation)

	1	2	3	4	5	6
ATT1	0.12	0.38	**0.78**	-0.03	0.11	0.19
ATT2	0.10	0.25	**0.74**	0.05	0.19	0.21
ATT3	0.13	0.28	**0.79**	0.07	0.21	0.10
ATT4	0.09	0.26	**0.80**	0.08	0.04	0.01
ATT5	0.12	0.29	**0.82**	0.13	-0.02	0.05
PRO1	0.09	0.06	0.23	0.31	**0.56**	0.13
PRO2	0.10	-0.03	0.10	0.09	**0.75**	0.12
PRO3	0.15	0.04	0.14	0.09	**0.75**	0.18
PRO4	0.12	0.19	0.03	0.14	**0.80**	0.00
PRO5	0.17	0.26	0.01	0.04	**0.76**	0.01
DIST1	**0.81**	0.17	0.00	0.10	0.07	0.19
DIST2	**0.87**	0.15	0.20	0.10	0.15	0.20
DIST3	**0.87**	0.11	0.13	0.14	0.19	0.24
DIST4	**0.90**	0.12	0.17	0.11	0.17	0.19
DIST5	**0.87**	0.13	0.11	0.14	0.17	0.26
INT1	0.15	-0.04	0.06	**0.86**	0.06	0.12
INT2	0.16	0.05	-0.01	**0.87**	0.06	0.06
INT3	0.07	0.16	0.08	**0.73**	0.18	0.23
INT4	0.05	0.22	0.12	**0.67**	0.22	0.29
INT5	0.08	0.02	0.05	**0.87**	0.10	-0.01
PSY1	0.25	0.30	0.27	0.35	0.19	**0.61**
PSY2	0.47	0.02	0.21	0.24	0.04	**0.67**
PSY3	0.38	0.13	0.12	0.26	0.09	**0.77**
PSY4	0.32	0.13	0.17	0.08	0.12	**0.81**
PSY5	0.23	0.40	0.02	0.09	0.22	**0.64**
EQU1	0.20	**0.82**	0.31	0.07	0.12	0.19
EQU2	0.09	**0.86**	0.31	0.06	0.16	0.09
EQU3	0.13	**0.85**	0.34	0.08	0.12	0.16
EQU4	0.16	**0.84**	0.34	0.09	0.12	0.14
EQU5	0.16	**0.74**	0.41	0.13	0.07	0.11

viewee, who indicated that there were definitely aspects of the implementation that were unfair, but she was not going to let that affect her attitude towards the implementation.

One other issue that should be addressed is that a dissatisfied user does not necessarily hurt an ES implementation; rather, there may be dissatisfied users because an organization is implementing a system or changing the business processes in a way that is detrimental to the organization. For example, just as conflict can constructively affect change, dissatisfied users expressing their dissatisfaction can lead to improved implementation plans. As shown through the interviews, even within one organization, multiple views of fairness can be expressed. The way ES are perceived as well as

Table 5. Weights and loadings

Variable	Weight	Loading
ATT1	0.2505	0.9008
ATT2	0.2433	0.8520
ATT3	0.2453	0.8882
ATT4	0.1941	0.8217
ATT5	0.2186	0.8681
EQU1	0.2242	0.9279
EQU2	0.2037	0.9335
EQU3	0.2185	0.9516
EQU4	0.2183	0.9537
EQU5	0.2082	0.8923
PRO1	0.2786	0.7142
PRO2	0.2236	0.7420
PRO3	0.2750	0.7909
PRO4	0.2627	0.8199
PRO5	0.2604	0.7767
DIST1	0.1879	0.8490
DIST2	0.2206	0.9428
DIST3	0.2204	0.9504
DIST4	0.2191	0.9618
DIST5	0.2235	0.9497
INT1	0.2171	0.8283
INT2	0.2223	0.8510
INT3	0.2699	0.8572
INT4	0.2888	0.8283
INT5	0.1963	0.8228
PSY1	0.2593	0.8542
PSY2	0.2265	0.8440
PSY3	0.2417	0.9193
PSY4	0.2236	0.8763
PSY5	0.2198	0.7700

their impact will likely differ from one sub-culture to another within the organization. The needs of the sub-cultures differ as well as the processes and functions that embody them and should be considered by management. ES success does not just occur from a one-time implementation, but rather through on-going improvements (Kraemmergaard & Rose, 2002).

CONCLUSION AND LIMITATIONS

The results of study 1 and 2 have practical implications for the strategic information systems and fairness literature as well as for managers which make strategic decisions in an organization. There are several contributions of these studies. The first contribution is that both study 1 and 2 find that fairness is an issue in ES implementations.

Figure 3. Results of Study 2 Model

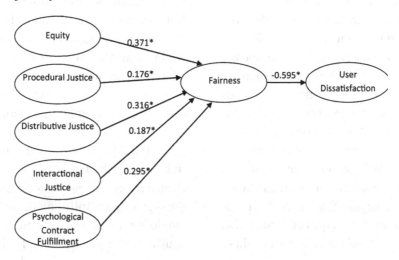

For researchers, this suggests an area that may be fruitful to consider for better understanding user responses to implementation strategies. For practitioners, this is an area that needs to be considered when developing implementation strategies. Effective managers of ES implementations should create strategies which will be perceived as fair to users. However, if managers know that the implementation strategy will not be perceived as fair, then a judgment call needs to be made whether the change is valuable enough to the organization to justify the likelihood of increased user dissatisfaction and ultimately user resistance. Fairness is important to address in the strategic information systems literature since it affects the response of users to the strategies developed by the implementation team. Managers creating

the implementation strategy must understand the lifecycle of an implementation and how user views change throughout the lifecycle. Markus et al. (2003) expands on Markus and Tanis (2000), both of which discuss problems and successes in various phases of ES implementations. Understanding the lifecycle of an ES would be useful in developing effective management strategies and affecting user attitudes. In addition, user views prior to the implementation have not been considered in this article, but by examining user views throughout the implementation, employee perceptions on fairness prior to the implementation can be compared to the effect the implementation has on perceptions of fairness.

The second contribution is that study 1 identifies that multiple components of fairness are

Table 6. Unrotated factor solution

Component	Eigenvalues	% of Variance	Cumulative %
1	11.746	39.153	39.2
2	3.873	12.910	52.1
3	2.730	9.100	61.2
4	2.113	7.042	68.2
5	1.439	4.798	73.0
6	1.253	4.175	77.2

considered by users and study 2 identifies quantitatively the impact of fairness on user dissatisfaction, yielding significant results. Employees' views on equity, the three types of justice, and the psychological contract have been shown to influence perceived fairness. Since the components of fairness have not been examined in previous literature, this finding leads to some interesting conclusions and implications for both researchers and practitioners. The five components of fairness clearly are distinct from one another, as shown through the factor analysis. Since users have different perceptions about the types of fairness that exist, future studies should consider all of these components of fairness since they are distinct from each other. For practitioners, this indicates various aspects of fairness which need to be considered and integrated into their strategic implementation plans. This suggests that there are specific areas of implementation plans that should be analyzed and possibly altered to better facilitate these types of fairness. For example, plans can be made more equitable through the use of incentives or rewards. For procedural justice, analyzing the procedures by which an ES is rolled out in an organization can lead to identification of problem areas which, when resolved, can increase the level of perceived procedural justice. In regards to distributive justice, strategies to share the additional workload required by the ES implementation may be helpful in increasing the level of perceived distributive justice. For interactional justice, making users feel valued and treated well, such as holding townhall meetings and obtaining user input, can increase the level of perceived interactional justice. In regards to psychological contract fulfillment, some of the ideas suggested in the study 1 discussion to facilitate a psychological contract change effectively can be useful in increasing the level of perceived psychological contract fulfillment.

A third contribution is that equity has the most important influence on fairness in an ES implementation. For researchers, examining why equity has a more important influence on fairness than the other fairness components may bring interesting results and a better understanding of user perspectives. Perhaps this is because an implementation causes most employees to experience either a relative advantage or disadvantage. Thus these additional inputs or outputs can affect each employee's fairness perception more than changes in the fairness of procedures (procedural justice), reward structures (distributive justice), the manner of how an employee is treated (interactional justice), and the fairness of the organization fulfilling the expectations of an employee (psychological contract). In regards to equity, if a change significantly increases the input of employees (i.e., additional workload) without increasing the output (i.e., employee bonus), then management must decide whether the benefits of implementing this policy justifies the likely problems that will arise from user dissatisfaction.

There are several limitations to this study. First, a potential limitation for both studies is based on the bias of interviewees and questionnaire respondents, which were reflecting on their own ES experiences. One aspect of this bias results from some respondents responding to the questionnaire regarding an experience they had perhaps a year earlier. Even though respondents may be trying to provide accurate information, they may have a skewed view concerning what actually happened. Another bias is social desirability, which may have occurred in the interviews and may have affected the responses of some of the interviewees. For example, interviewees may not have discussed their own resistance to the system in order to present a certain image about themselves. This impact of this limitation was minimized through the use of interviewing multiple people within the same organization as well as distributing questionnaires to multiple users within the same organization.

A second limitation is that a convenience sample was used for study 2. Packets of questionnaires and business reply envelopes were distributed to members of various user groups. The users which received the questionnaire from these user group members were not randomly selected among the general population; rather, they were people known to the user group members who may have filled it out as a favor to the user group member. The result is that certain groups may have been underrepresented, such as small businesses that do not have a user group member. Thus, although the respondents represented many different positions within many different organizations, it may not be representative of the overall population. To minimize the impact of this limitation, user groups from various ES vendors were selected in order to represent a wide variety of businesses that implement an ES.

There are some ways that future research could expand on this study. First, a better psychological understanding of users' attitudes may be helpful in predicting the resulting behaviors. For example, Eagly and Chaiken (1995) discuss attitude strength, attitude structure, and resistance to change. For a user, there may be dissatisfaction towards the ES and the change; however, the attitude strength and structure has not been examined. It is possible that if the users are dissatisfied, but the dissatisfaction attitude is not strong enough, resistant behaviors will not exist. On the other hand, users greatly dissatisfied have a strong attitude and may exhibit a greater degree of resistance. Second, examining user attitudes and the behaviors of users also could be important in determining which user behaviors are impacted by user dissatisfaction in ES implementations. The impact of fairness constructs on behavioral constructs, such as organizational citizenship behavior, would be useful in understanding how fairness impacts the user. Generally, a successful change requires more than only satisfied users; some employees will need to perform jobs beyond their assigned duties, demonstrating organizational citizenship behavior. Thus, examining how fairness affects both attitudinal as well as behavioral responses may bring forth interesting results.

REFERENCES

Adams, J. S. (1963). Toward an Understanding of Inequity. *Journal of Abnormal Psychology*, *67*(5), 422–436. doi:10.1037/h0040968

Akkermans, H., & Van Helden, K. (2002). Vicious and Virtuous Cycles in ERP Implementation: A Case Study of Interrelations between Critical Success Factors. *European Journal of Information Systems*, *11*, 35–46. doi:10.1057/palgrave/ejis/3000418

Alvarez, R., & Urla, J. (2002). Tell me a Good Story: Using Narrative Analysis to Examine Information Requirements Interviews during an ERP Implementation. *The Data Base for Advances in Information Systems*, *33*(1), 38–52.

Argyris, C. (1960). *Understanding Organizational Behavior*. Homewood, IL: Dorsey Press.

Au, N., Ngai, E. W. T., & Cheng, T. C. E. (2008). Extending the Understanding of End User Information Systems Satisfaction Formation: An Equitable Needs Fulfillment Model Approach. *Management Information Systems Quarterly*, *32*(1), 43–66.

Barker, T., & Frolick, M. N. (2003). ERP Implementation Failure: A Case Study. *Information Systems Management*, *20*(4), 43–49. doi:10.1201/1078/43647.20.4.20030901/77292.7

Bies, R. J., & Tripp, T. M. (1996). Beyond Distrust: 'Getting Even' and the Need for Revenge. In Kramer, R. M., & Tyler, T. R. (Eds.), *Trust in Organizations: Frontiers of Theory and Research* (pp. 246–260). Thousand Oaks, CA: Sage.

Bingi, P., Sharma, M. K., & Godla, J. K. (1999). Critical Issues Affecting an ERP Implementation. *Information Systems Management*, *16*(3), 7–14. doi:10.1201/1078/43197.16.3.19990601/31310.2

Brown, S. P., & Chin, W. W. (2004). Satisfying and Retaining Customers through Independent Service Representative. *Decision Sciences*, *35*(1), 527–550. doi:10.1111/j.0011-7315.2004.02534.x

Burns, J. R., Jung, D. G., & Hoffman, J. J. (2009). Capturing and Comprehending the Behavioral/Dynamical Interactions within an ERP Implementation. *Journal of Organizational and End User Computing*, *21*(2), 67–90. doi:10.4018/joeuc.2009040104

Carr, C. L. (2007). The FAIRSERV Model: Consumer Reactions to Services Based on a Multidimensional Evaluation of Service Fairness. *Decision Sciences*, *38*(1), 107–130. doi:10.1111/j.1540-5915.2007.00150.x

Cassel, C. M., Hackl, P., & Westlund, A. H. (2000). On Measurement of Intangible Assets: A Study of Robustness of Partial Least Squares. *Total Quality Management*, *11*(7), 897–907. doi:10.1080/09544120050135443

Chin, W. W. (1998). The Partial Least Squares Approach to Structural Equation Modeling. In Marcoulides, G. A. (Ed.), *Modern Methods for Business Research* (pp. 295–336). Mahwah, NJ: Lawrence Erlbaum Associates.

Chin, W. W., & Gopal, A. (1995). Adoption Intention in GSS: Relative Importance of Beliefs. *The Data Base for Advances in Information Systems*, *26*(2-3), 42–64.

Cobb, A. T., Wooten, K. C., & Folger, R. (1995). Justice in the Making: Toward Understanding the Theory and Practice of Justice in Organizational Change and Development. In Pasmore, W. A., & Woodman, R. W. (Eds.), *Research in Organizational Change and Development* (pp. 243–295). London, UK: Jai Press.

Cohen-Charash, Y., & Spector, P. E. (2001). The Role of Justice in Organizations: A Meta-Analysis. *Organizational Behavior and Human Decision Processes*, *86*(2), 278–321. doi:10.1006/obhd.2001.2958

Conway, N., & Briner, R. B. (2002). A Daily Diary Study of Affective Responses to Psychological Contract Breach and Exceeded Promises. *Journal of Organizational Behavior*, *23*(3), 287–302. doi:10.1002/job.139

Coyle-Shapiro, J. A.-M. (2002). A Psychological Contract Perspective on Organizational Citizenship Behavior. *Journal of Organizational Behavior*, *23*(8), 927–946. doi:10.1002/job.173

Cropanzano, R., & Folger, R. (1989). Referent Cognitions and Task Decision Autonomy: Beyond Equity Theory. *The Journal of Applied Psychology*, *74*(2), 293–299. doi:10.1037/0021-9010.74.2.293

Cummings, T. G., & Worley, C. G. (2005). *Organization Development and Change* (8th ed.). Mason, OH: Thomson South-Western.

Davidson, J. O. C. (1994). The Sources and Limits of Resistance in a Privatized Utility. In Jermier, J. M., Knights, D., & Nord, W. R. (Eds.), *Resistance and Power in Organizations* (pp. 69–101). London, UK: Routledge.

De Cremer, D. (2005). Procedural and Distributive Justice Effects Moderated by Organizational Identification. *Journal of Managerial Psychology*, *20*(1), 4–13. doi:10.1108/02683940510571603

Eagly, A. H., & Chaiken, S. (1995). Attitude Strength, Attitude Structure, and Resistance to Change. In Petty, R. E., & Krosnick, J. A. (Eds.), *Attitude Strength: Antecedents and Consequences* (pp. 413–432). Mahwah, NJ: Lawrence Erlbaum Associates.

Folger, R. (1993). Reactions to Mistreatment at Work. In Murnighan, J. K. (Ed.), *Social Psychology in Organizations: Advances in Theory and Research* (pp. 161–183). Englewood Cliffs, NJ: Prentice-Hall.

Folger, R., & Skarlicki, D. P. (1999). Unfairness and Resistance to Change: Hardship as Mistreatment. *Journal of Organizational Change Management, 12*(1), 35–50. doi:10.1108/09534819910255306

Ginzberg, M. J. (1981). Early Diagnosis of MIS Implementation Failure: Promising Results and Unanswered Questions. *Management Science, 27*(4), 459–478. doi:10.1287/mnsc.27.4.459

Greenberg, J. (1990). Employee Theft as a Reaction to Underpayment Inequity: The Hidden Costs of Pay Cuts. *The Journal of Applied Psychology, 75*(5), 561–568. doi:10.1037/0021-9010.75.5.561

Gupta, A. (2000). Enterprise Resource Planning: The Emerging Organizational Value Systems. *Industrial Management & Data Systems, 100*(3), 114–118. doi:10.1108/02635570010286131

Hill, K. (2003). *System Designs that Start at the End (User)*. CRM Daily.

Hirschheim, R., & Newman, M. (1988). Information Systems and User Resistance: Theory and Practice. *The Computer Journal, 31*(5), 398–408. doi:10.1093/comjnl/31.5.398

Hunton, J. E. (1996). Involving Information System Users in Defining System Requirements: The Influence of Procedural Justice Perceptions on User Attitudes and Performance. *Decision Sciences, 27*(4), 647–671. doi:10.1111/j.1540-5915.1996.tb01830.x

Hussain, D., & Hussain, K. M. (1984). *Information Resource Management*. Homewood, IL: Richard D. Irwin.

Jacobson, S., Shepherd, J., D'Aquila, M., & Carter, K. (2007). *The ERP Market Sizing Report, 2006-2011*. Stamford, CT: AMR Research.

Jex, S. M. (2002). *Organizational Psychology*. New York, NY: John Wiley & Sons.

Joreskog, K. G., & Wold, H. (1982). *Systems under Indirect Observation*. Amsterdam, The Netherlands: North Holland.

Joshi, K. (1990). An Investigation of Equity as a Determinant of User Information Satisfaction. *Decision Sciences, 21*(4), 786–807. doi:10.1111/j.1540-5915.1990.tb01250.x

Joshi, K. (1991). A Model of Users' Perspective on Change: The Case of Information Systems Technology Implementation. *Management Information Systems Quarterly, 15*(2), 229–242. doi:10.2307/249384

Joshi, K. (1992). A Causal Path Model of the Overall User Attitudes Toward the MIS Function: The Case of User Information Satisfaction. *Information & Management, 22*, 77–88. doi:10.1016/0378-7206(92)90063-L

Kickul, J. R., Neuman, G., Parker, C., & Finkl, J. (2001). Settling the Score: The Role of Organizational Justice in the Relationship between Psychological Contract Breach and Anticitizenship Behavior. *Employee Responsibilities and Rights Journal, 13*(2), 77–93. doi:10.1023/A:1014586225406

Kirkman, B. L., Shapiro, D. L., Novelli, L. Jr, & Brett, J. M. (1996). Employee Concerns Regarding Self-Managing Work Teams: A Multidimensional Perspective. *Social Justice Research, 9*(1), 47–67. doi:10.1007/BF02197656

Klaus, T., & Blanton, J. E. (2010). User Resistance Determinants and the Psychological Contract in Enterprise System Implementations. *European Journal of Information Systems, 19*, 625–636.

Klaus, T., Wingreen, S. C., & Blanton, J. E. (2010). Resistant Groups in Enterprise System Implementations: A Q-methodology Examination. *Journal of Information Technology, 25*, 91–106.

Knights, D., & Vurdubakis, T. (1994). Foucault, Power, Resistance, and All That. In Jermier, J. M., Knights, D., & Nord, W. R. (Eds.), *Resistance and Power in Organizations* (pp. 167–198). London, UK: Routledge.

Koh, C., Ang, S., & Straub, D. W. (2004). IT Outsourcing Success: A Psychological Contract Perspective. *Information Systems Research, 15*(4), 356–373. doi:10.1287/isre.1040.0035

Kraemmergaard, P., & Rose, J. (2002). Managerial Competences for ERP Journeys. *Information Systems Frontiers, 4*(2), 199–211. doi:10.1023/A:1016054904008

Krasner, H. (2000). Ensuring E-Business Success by Learning from ERP Failures. *IT Professional, 2*(1), 22–27. doi:10.1109/6294.819935

Lohmoller, J.-B. (1988). The PLS Program System: Latent Variables Path Analysis with Partial Least Squares Estimation. *Multivariate Behavioral Research, 23*(1), 125–127. doi:10.1207/s15327906mbr2301_7

Makin, P. J., Cooper, C. L., & Cox, C. J. (1996). *Organizations and the Psychological Contract: Managing People at Work*. London, UK: Wiley-Blackwell.

Marakas, G., & Hornik, S. (1996). Passive Resistance Misuse: Overt Support and Covert Recalcitrance in IS Implementation. *European Journal of Information Systems, 5*, 208–219. doi:10.1057/ejis.1996.26

Markus, M. L., Axline, S., Petrie, D., & Tanis, C. (2003). Learning from Adopters' Experiences with ERP: Problems Encountered and Success Achieved. In Shanks, G., Seddon, P. B., & Willcocks, L. P. (Eds.), *Second-Wave Enterprise Resource Planning Systems*. Cambridge, UK: Cambridge University Press.

Markus, M. L., & Tanis, C. (2000). The Enterprise System Experience - From Adoption to Success. In Zmud, R. W. (Ed.), *Framing the Domains of IT Management* (pp. 173–208). Cincinnati, OH: Pinnaflex Educational Resources.

Mathieson, K., Peacock, E., & Chin, W. W. (2001). Extending the Technology Acceptance Model: The Influence of Perceived User Resources. *The Data Base for Advances in Information Systems, 32*(3), 86–112.

Maurer, R. (2002). *Plan for the Human Part of ERP*. Workforce Online.

McDonald, D. J., & Makin, P. J. (2000). The Psychological Contract, Organisational Commitment and Job Satisfaction of Temporary Staff. *Leadership and Organization Development Journal, 21*, 84–91. doi:10.1108/01437730010318174

Meckler, M., Drake, B. H., & Levinson, H. (2003). Putting Psychology Back into Psychological Contracts. *Journal of Management Inquiry, 12*(3), 217–228. doi:10.1177/1056492603256338

Moorman, R. H. (1991). Relationship between Organizational Justice and Organizational Citizenship Behaviors: Do Fairness Perceptions Influence Employee Citizenship? *The Journal of Applied Psychology, 76*(6), 845–855. doi:10.1037/0021-9010.76.6.845

Morrison, E. W., & Robinson, S. L. (1997). When Employees Feel Betrayed: A Model of How Psychological Contract Violation Develops. *Academy of Management Review, 22*(1), 226–256.

Nah, F. F.-H., & Lau, J. L.-S. (2001). Critical Factors for Successful Implementation of Enterprise Systems. *Business Process Management Journal, 7*(3), 285–296. doi:10.1108/14637150110392782

Niehoff, B. P., & Moorman, R. H. (1993). Justice as a Mediator of the Relationship between Methods of Monitoring and Organizational Citizenship Behavior. *Academy of Management Journal, 36*(3), 527–556. doi:10.2307/256591

Nunnally, J. C. (1978). *Psychometric Theory.* New York, NY: McGraw-Hill.

Nunnally, J. C., & Bernstein, I. (1994). *Psychometric Theory.* New York, NY: McGraw-Hill.

O'Leary, D. E. (2000). *Enterprise Resource Planning Systems: Systems, Life Cycle, Electronic Commerce, and Risk.* Cambridge, UK: Cambridge University Press.

O'Reilly, C. A., & Pfeffer, J. (2000). *Hidden Value: How Great Companies Achieve Extraordinary Results with Ordinary People.* Cambridge, MA: Harvard Business School Press.

Piccoli, G., & Ives, B. (2003). Trust and the Unintended Effects of Behavior Control in Virtual Teams. *Management Information Systems Quarterly, 27*(3), 365–395.

Podsakoff, P. M., & Organ, D. W. (1986). Self-Reports in Organizational Research: Problems and Prospects. *Journal of Management, 12*(4), 531–544. doi:10.1177/014920638601200408

Porter, L. W., Pearce, J. L., Tripoli, A. M., & Lewis, K. M. (1998). Differential Perceptions of Employers' Inducements: Implications for Psychological Contracts. *Journal of Organizational Behavior, 19*(S1), 769–782. doi:10.1002/(SICI)1099-1379(1998)19:1+<769::AID-JOB968>3.0.CO;2-1

Rao, S. S. (2000). Enterprise Resource Planning: Business Needs and Technologies. *Industrial Management & Data Systems, 100*(2), 81–88. doi:10.1108/02635570010286078

Robey, D., Ross, J. W., & Boudreau, M.-C. (2002). Learning to Implement Enterprise Systems: An Exploratory Study of the Dialectics of Change. *Journal of Management Information Systems, 19*(1), 17–46.

Robinson, S. L. (1996). Trust and Breach of the Psychological Contract. *Administrative Science Quarterly, 41*(4), 574–599. doi:10.2307/2393868

Robinson, S. L., & Morrison, E. W. (1995). Psychological Contracts and OCB: The Effect of Unfulfilled Obligations on Civic Virtue Behavior. *Journal of Organizational Behavior, 16*, 289–298. doi:10.1002/job.4030160309

Robinson, S. L., & Rousseau, D. M. (1994). Violating the Psychological Contract: Not the Exception but the Norm. *Journal of Organizational Behavior, 15*, 245–259. doi:10.1002/job.4030150306

Rousseau, D. M. (1995). *Psychological Contracts in Organizations: Understanding Written and Unwritten Agreements.* Thousand Oaks, CA: Sage.

Rousseau, D. M. (1996). Changing the Deal While Keeping the People. *The Academy of Management Executive, 10*(1), 51–61. doi:10.5465/AME.1996.9603293198

Schein, E. H. (1980). *Organizational Psychology* (3rd ed.). Englewood Cliffs, NJ: Prentice-Hall.

Schriesheim, C. A. (1979). The Similarity of Individual Directed and Group Directed Leader Behavior Descriptions. *Academy of Management Journal, 22*(2), 345–355. doi:10.2307/255594

Schweiger, D. M., & Denisi, A. S. (1991). Communication with Employees following a Merger: A Longitudinal Field Experiment. *Academy of Management Journal, 34*(1), 110–135. doi:10.2307/256304

Shang, S. S. C., & Su, T. C. C. (2004, August). *Managing User Resistance in Enterprise Systems Implementation*. Paper presented at the Americas Conference on Information Systems, New York.

Shapiro, D. L., Buttner, E. H., & Barry, B. (1994). Explanations: What Factors Enhance Their Perceived Adequacy. *Organizational Behavior and Human Decision Processes*, *58*, 346–368. doi:10.1006/obhd.1994.1041

Shapiro, D. L., & Kirkman, B. L. (1999). Employees' Reaction to the Change to Work Teams: The Influence of "Anticipatory" Injustice. *Journal of Organizational Change Management*, *12*, 51–66. doi:10.1108/09534819910255315

Sheppard, B. H., Lewicki, R. J., & Minton, J. W. (1992). *Organizational Justice: The Search for Fairness in the Workplace*. New York, NY: Lexington Books.

Skarlicki, D. P., & Folger, R. (1997). Retaliation in the Workplace: The Roles of Distributive, Procedural, and Interactional Justice. *The Journal of Applied Psychology*, *82*(3), 434–443. doi:10.1037/0021-9010.82.3.434

Stein, T. (1999). Making ERP Add Up. *InformationWeek*, 59-68.

Sternad, S., Bobek, S., Dezelak, Z., & Lampret, A. (2009). Critical Success Factors (CSFs) for Enterprise Resource Planning (ERP) Solution Implementation in SMEs: What Does Matter for Business Integration. *International Journal of Enterprise Information Systems*, *5*(3), 27–46. doi:10.4018/jeis.2009070103

Stratman, J. K., & Roth, A. V. (2002). Enterprise Resource Planning (ERP) Competence Constructs: Two-Stage Multi-Item Scale Development. *Decision Sciences*, *33*(4), 601–628. doi:10.1111/j.1540-5915.2002.tb01658.x

Umble, E. J., & Umble, M. M. (2002). Avoiding ERP Implementation Failure. *Industrial Management (Des Plaines)*, *44*(1), 25–33.

Willcocks, L. P., & Sykes, R. (2000). The Role of the CIO and IT Function in ERP. *Communications of the ACM*, *43*(4), 32–38. doi:10.1145/332051.332065

Woodroof, J., & Burg, W. (2003). Satisfaction/Dissatisfaction: Are Users Predisposed? *Information & Management*, *40*, 317–324. doi:10.1016/S0378-7206(02)00013-7

APPENDIX

Table A1. User dissatisfaction

| I am upset with the system |
| I feel animosity towards the system |
| I am frustrated with the system |
| I am not satisfied with the system |
| I have a negative attitude towards the system |

Table A2. Procedural justice

| The decision makers made decisions in an unbiased manner |
| The decisions that were made influence employees uniformly |
| Job decisions were applied consistently across all affected employees |
| The job changes affect employees evenly |
| The procedures that were put in place affect employees equally |

Table A3. Distributive justice

| My level of compensation is fair, considering the stress that I encountered with the system |
| Based on the responsibilities I undertook with the system, the rewards I receive are fair |
| The rewards I receive are fair, considering my performance with the system |
| Based on the amount of effort I put forth with the system, the rewards I receive are fair |
| My level of compensation is fair, considering the job changes I endured during the implementation |
| **Interactional Justice** |
| My supervisor treated me with kindness and consideration throughout the system implementation |
| My supervisor dealt with me in a truthful manner throughout the implementation |
| My supervisor offered adequate justification for the decisions that were made |
| My supervisor offered explanations that made sense to me |
| My supervisor treated me with respect and dignity throughout the implementation |
| **Psychological Contract Fulfillment** |
| Even through the changes caused by the system, management has fulfilled what was originally promised to me about my job |
| My organization has rewarded me to the degree that they promised me |
| Even though my job may have changed, my organization still fulfills their pledges to me |
| Even through the changes caused by the system, my organization has fulfilled what they promised me |
| The job changes I have been faced with are still in line with what my organization originally told me |
| **Equity** |
| The benefits of the system outweigh the additional effort I had to put forth |
| I am at an advantage because of the system |
| Even though I may have put forth additional effort to learn the system, the benefits of the system made up for this effort |
| Overall, I benefited more than I lost from the system |
| I prefer my job more after the system than before the system |

Chapter 13
Antecedents of Improvisation in IT-Enabled Engineering Work

William J. Doll
The University of Toledo, USA

Xiaodong Deng
Oakland University, USA

ABSTRACT

The success of engineering work depends on the ability of individuals to improvise in response to emerging challenges and opportunities (Kappel & Rubenstein, 1999). Building on experiential learning theory (Eisenhardt & Tabrizi 1995; Kolb, 1984) and improvisation theory (Miner, Bassoff, & Moorman, 2001), this authors argue that information systems facilitate the generation of new product and process design ideas by providing richer feedback, creating shorter learning cycles, and enabling engineers to try a variety of new ideas more easily. An empirical research model of the antecedents of improvisation in IT-enabled engineering work is proposed. This model is examined using a sample of 208 individuals engaged in computer-intensive engineering design work. The multiple regression results suggest that software capability, autonomy, problem solving/decision support usage, system use for work planning, and length of use explain the extent of new product and process ideas that are generated. The practical and theoretical implications of these findings are discussed.

INTRODUCTION

Eisenhardt and Tabrizi (1995) describe two strategies for engineering design work. If all requirements are known in advance and stable, the design engineer can pre-plan all work to save time. Where design requirements are not fully known or subject to change, design engineers may have to engage in experiential learning (Kolb, 1984). Using this experiential strategy, engineers approach problems with some general, often vague, idea of what they want to accomplish and some learned routines for interacting with the computer. They repetitively interact with the computer, assess the substantive

DOI: 10.4018/978-1-4666-2059-9.ch013

results using analysis and intuition, and then, sometimes, they extemporaneously generate new ideas (i.e., improvisations) that they incorporate in their intellectual work product or process. As the design process unfolds, design engineers must improvise to adapt to emerging task requirements (Brown & Duguid, 1991; Orr, 1990b).

Kappel and Rubenstein (1999) describe how the introduction of information technology into the engineering design process can enhance creativity in design. Improvisation in design work is the interactive use of an information system application by an engineer to extemporaneously design, execute, and assess novel ideas. The novel ideas may pertain to the design of the individual's product (end result) or the design of the individual's work processes (activities). Thus, improvisation is the impromptu generation and trying out of new ideas. It is planning spontaneously based upon substantive feedback and the temporal convergence of planning and execution (Miner, Bassoff, & Moorman, 2001). This generation and trying out of novel ideas helps engineers respond to emerging challenges and opportunities in their work. Novel ideas enhance learning and productivity in engineering work.

The organizational literature on improvisation emphasizes its importance for organizational learning and renewal (Miner et al., 2001). This literature has clarified the concept of improvisation, suggesting novelty, substantive convergence, temporal convergence, and intentionality as the defining characteristics of improvisation. It distinguishes between planned ideas (innovations) and emergent ideas (improvisations).

Little is known about the antecedents of on-the-spot new idea generation in engineering design work. Improvisations have referents (e.g., antecedents) that may stimulate or constrain novel productions (Miner et al., 2001). Improvisation theory uses jazz as a metaphor and assumes that the real time interactions among musicians or between individual musicians and their audience stimulate

improvisations (e.g., new melodies). Or, product development team members interact and, thus, trigger the creation of new ideas. In IT-enabled engineering design work, there is an interaction among an active human agent, an emergent work process, and an interpretively flexible technology that may stimulate the on-line creation of impromptu new ideas (Orlikowski, 1992, 2000).

While the information systems literature has devoted considerable attention to the design of information systems that support knowledge work (Alavi & Leidner, 2001; Majchrzak, Rice, Malhotra, King, & Ba, 2000), emergent knowledge processes (Markus, Majchrzak, & Gasser, 2002), and the situated and emergent nature of information technology usage (Orlikowski, 1992, 2000), it has not specifically focused on the antecedents that stimulate improvisation in engineering design work. There has been an increased interest in application design to support experimentation (Terry & Mynatt, 2002) and creativity (Greene, 2002) in knowledge work. More recently, the information systems literature has focused on related phenomena such as personal innovativeness in the use of the internet (Larsen & Sørebø, 2005), the need for flexible systems where users themselves will be able (and increasingly enabled by technology) to satisfy their changing needs (Kanellis & Paul, 2005), and the need for users to take personal initiative with respect to experimenting in their use of information technology (Spitler, 2005). However, the information systems literature has not specifically focused on the role of improvisation in engineering work, and the subtle and often unrecognized ways in which information technology may transform the nature of engineering work by facilitating the experiential approach.

In the next section, this paper first describes how computers make engineering design work more improvisable. Then, in Section 2.1, we review the literature on improvisation. We contrast improvisation in the context of emergencies with improvisation in an engineering work context.

Section 2.2 describes engineering work as an experiential learning process where IT-enabled new ideas contribute to knowledge creation. Section 2.3 presents a model of the antecedents of IT-enabled improvisations and develops five hypotheses. Section 3 describes the research methods followed by the results (Section 4), discussion (Section 5), and conclusions (Section 6).

IMPROVISATION IN IT-ENABLED ENGINEERING WORK

We view engineering design work as an emergent process of deliberations characterized by a rapidly changing task environment with unexpected opportunities and problems, dynamic interaction with a community of knowing, uncertain work processes, and an unknowable intellectual end product. In this environment, design engineers learn experientially. They are given considerable autonomy to craft their own work process, share perspectives with members of their community, and, thus, shape the intellectual product and process of their work (Malone, 1997; Wrzesniewski & Dutton, 2001). In doing so, engineers often use information technology to facilitate their generation and trying out of new ideas.

The use of information systems supports the experiential strategy for conducting engineering design work. Information technology enables short experiential learning cycles, enhances substantive feedback, supports perspective making and perspective taking in distributed cognition, and facilitates the reproducibility of work by enabling the storing and sharing of alternative product and process designs in detailed digital form. The detailed product and process plans can be easily changed, permitting the trying out of new work product or process innovations. This technology enhanced context also improves human problem solving by rapidly building knowledge and intuitive judgment through iterative experiential cycles (Eisenhardt & Tabrizi, 1995). Engineering work

processes are naturally transformed in ways that help individuals extemporaneously generate and try out new ideas (i.e., to improvise).

Improvisation Theory

The defining characteristics of improvisation help explain this natural linkage between an experiential strategy for engineering work and improvisation. Miner et al. (2001) identify temporal convergence, substantive convergence, novelty, and deliberateness as the defining characteristics of improvisation. In their early work, Moorman and Miner (1998b) argue that one good measure of improvisation is the narrowness of the time gap between design and execution (i.e., temporal convergence). Temporal convergence means that the time between composition and execution is compressed.

Improvisation is an extemporaneous process leading to impromptu or "spur of the moment" action (Ciborra, 1999, p. 78). It is not the product of a slow, judicious, prior planning process. Improvisation occurs in settings where planning and execution are intertwined in an iterative process of thought and action. In such settings, it is difficult to clearly distinguish between planning and execution as they occur almost simultaneously. Improvisation is an emergent phenomenon grounded in the here and now. Crossan, Cunha, Vera, and Cunha (2005) suggest that, in improvisation, time is relative. Clock-event-time management (speed) should focus on manipulative flexibility (e.g., responsiveness to changes in the environment so as to shape that environment).

Later, Miner et al. (2001) clarify that substantive convergence rather than time convergence is a better indicator of improvisation. Substantive convergence occurs when design not only informs execution, but execution also informs design as well. What is learned by execution is fed back to inform planning. Feedback arrives and is applied before the decision is firmed up. This pre-empts the deployment of any pre-planned procedure or solution. In improvisation, implementation is a form of instantaneous exploration. Implementation becomes a source of discovery (i.e., idea generation) so that problem setting and solving feed continuously upon each other through anticipatory feedback (Ciborra, 1999).

Planning is assumed to inform work "doing". In substantive convergence, work "doing" also informs work planning in important, material, and novel ways. In improvisation, real-time experience informs novel action at the same time that the action is being taken (during the act). Miner et al. (2001) consider the real-time impact of experience on action as the defining characteristic of improvisation.

In improvisation, novelty means new ideas for doing work that are generated and executed in real-time. Miner et al. (2001) identify three forms of improvisation – behavioral productions (new processes or sequences of behavior), artifactual productions (new physical structures or tools), and interpretive productions (new interpretive frameworks). Novelty is the extent of deviation from prior routines, structures, or interpretations. Improvisations involve "an existential moment of vision" of a new way to solve a surprising problem or seize an unexpected opportunity (Ciborra, 1999, p. 79). The level of novelty (newness) in action can vary from relatively unoriginal to exceedingly unusual, but a threshold degree of deviation from prior routines is necessary for an improvisation.

Improvisation requires some degree of deliberateness or purpose. Improvisations are not random deviations from prior routines; they are goal directed actions (Hatch, 1997; Miner et al., 2001; Weick, 1998). Intentionality grounds the improvisational act in the experience and goal directed behavior of the actor. The improviser is aware that his/her novel actions may deviate from the controls of previous routines without abandoning the purposes those routines seek to achieve. Thus, the deliberate two-way interaction between work planning and work doing and the novel ideas it creates is the hallmark of improvisation.

Miner et al. (2001) distinguish between improvisation on one hand and adaptation (Campbell, 1969), creativity (Amabile, 1988), innovation, trial-and-error learning, and experimentation on the other in that, while these other constructs may also produce a novel product, they do not, like improvisation, always require the deliberate on-line fusion of design and execution that is inherent in the experiential learning cycle.

Improvisations have referents to facilitate the generation and editing of improvised behavior (Miner et al., 2001; Pressing, 1984). You cannot improvise on nothing; you have to improvise on something (Kernfeld, 1995). Situated and emerging task requirements, existing work plans and routines, and the capabilities/features of the software packages being used both infuse meaning into improvisational actions and guide/constrain the generation of novel productions. Markus et al. (2002) describe how emergent knowledge processing software both constrains and guides (facilitates) an emerging process of deliberations.

Organizational researchers have analyzed improvisation in fast-moving and uncertain organizational settings such as new product development (Eisenhardt & Tabrizi, 1995; Moorman & Miner, 1998a, 1998b), emergencies such as a strike (Preston, 1987), a failed navigational system (Hutchins, 1991), and a firestorm (Weick, 1993b). The importance of improvisation is often readily apparent in emergency or fast moving competitive business situations where expedient action is required. Improvisation enables individuals or organizations to quickly modify their behavior to changing circumstances. In rapidly changing environments, a plan developed some time ago may not fit the current situation, that is, lack coherence (Rumelt, 1995).

In emergencies, improvisation is often viewed as the opposite of planning (Weick, 1993a, 1998). In the work context, planning and improvisation may be complementary rather than opposite or competitive activities (Chelariu, Johnston, & Young, 2002). Work planning (i.e., the development of explicit objectives, criteria for decision making, and action plans) and how these plans are embedded into the design of business processes provide a useful point of departure for studying improvisation in work (Ciborra, 1999). Work contexts range from the stable to the highly unstructured. Improvisational actions help the individual or group adapt when exceptions occur that were not anticipated in the plans. While some scholars prefer to maintain a clear distinction between planning and improvisation (Miner et al., 2001), improvisation may also be viewed as a type of planning, that is, on the spot ad hoc planning and execution of new ideas generated by feedback from implementation (Ciborra, 1999).

Engineering design work is an emerging process of deliberations. Pre-planned methods, processes, or standard operating procedures (SOP) seldom provide a detailed map of how design work should be done. Unexpected problems and surprise opportunities occur as the emerging design process unfolds. This happens more frequently where the environment is dynamic or uncertain, but it occurs in all jobs. The map of how work should be done, though potentially useful, by itself provides little insight into how ad hoc decisions presented by changing conditions can be resolved. The actual journey is characterized by surprise occurrences (e.g., detours, accidents, etc.), new opportunities, equivocal interpretations between abstract instructions and situated demands, and inaccuracies in the original instructions (Ciborra, 1999).

Contrary to much conventional wisdom, people continuously learn and improvise while working (Brown & Duguid, 1991). Ethnographic studies of workplace practices indicate that the ways people actually work differ frequently, and often fundamentally, from the ways organizations describe that work in manuals, training programs, organizational charts, and job descriptions (Orr, 1987a, 1987b, 1990a, 1990b). To deal with these exceptions, workers need to improvise. For example, Massa and Testa (2005) describe three case studies where IT professionals improvise in their

use of data warehouse tools to match situational requirements and exceptions, enacting different technologies-in-practice in each organization. Success will depend upon the coherence or fit between the emergent situation and the improvisational action.

Novelty, speed, and coherence determine the success of improvisations (Chelariu et al., 2002; Moorman & Miner, 1998b). The degree of novelty varies. Some problems require more novel solutions than others. Speed is how quickly improvisation enables individuals to seize opportunities or avoid coherence problems caused by changing environmental conditions (Moorman & Miner, 1998a) or task demands. IT-enabled improvisation can often improve internal and external coherence by enabling the rapid incorporation of new ideas in product or process designs.

Internal coherence is the degree to which an action displays internal fit. When the individual is the unit of analysis, internal fit may mean that the individual's product and process design components work well together. External coherence is the degree to which an action displays external fit. External fit means that product and process design components interact to achieve external performance criteria (e.g., market success or integration of a focal part design with the overall product design). Execution via computer simulation models or virtual prototypes enhances how quickly novel ideas can be tried out and how thoroughly they can be assessed for internal and external coherence. Thus, in engineering design work, improvisations are often assumed to be effective.

Over time, individuals, groups, and organizations may develop competencies in improvisation through the intensity, type, and time span of their experiences. By shaping and directing the improvisations that take place, such competencies may enable them to capture the benefits of improvisation while limiting the detrimental effects or unintended consequences (Baker, Miner, & Eesley, 2003; Miner et al., 2001). For example, product designers develop considerable expertise in checking out virtual product designs for internal and external coherence. Such design expertise is an improvisational competency because it enhances the speed and coherence, and thus the effectiveness, of improvisations.

What we know about improvisation is largely based on ethnographic studies at the organizational level as illustrated by the work of Miner et al. (2001). Ethnographic methods are used to identify an organization's original plan and assess whether new ideas qualify as improvisations by confirming the presence of intentionality, substantive convergence, and temporal convergence. These ethnographic studies indicate that improvisation is a type of experiential learning. The experiential learning model can be contrasted with a planning model that assumes a well-known and rational process.

In this study, the individual level of analysis and the experiential nature of the engineering design work ensure intentionality, substantive convergence, and temporal convergence. The novel ideas that emerge extemporaneously from reflective thought about substantive feedback can be viewed as improvisations. These on-the-spot new ideas are the direct outcomes of an experiential learning process that includes the deliberate and substantive fusion of IT-enabled design and execution.

The Experiential Process in Engineering Design Work

To better understand the iterative and cyclical nature of experiential work, Eisenhardt and Tabrizi (1995) contrast a rational (planned) strategy with an experiential learning strategy for engineering work in the product development context. The rational strategy assumes a predictable or certain process, one that can be planned out as a series of discrete steps. Under conditions of certainty, some aspects of engineering work might be entirely preplanned, i.e., conducted in a single cycle of

work planning and execution where there is no opportunity or need to obtain/assess feedback and incorporate additional ideas. Such a certain process can then be optimized by pre-planning e.g., shortening the time for each step, overlapping the execution of steps, and rewarding individuals for attaining the compressed schedule. In this predictable process, a single cycle of planning and execution can optimize performance.

The experiential learning strategy assumes an uncertain process and an, as yet, indeterminate final outcome or intellectual product. It relies on real-time experience, analytical and intuitive skills, improvisation, and flexibility. Flexibility requires individual autonomy in designing work processes and interpreting whether, or how, technology is used. Accelerating the experiential cycles rapidly builds experience, intuition, and flexible options that can help individuals adapt to an unclear and dynamically changing task environment. Here the short iterative cycles of work planning and execution enable the individual worker to obtain substantive feedback, learn, and generate and incorporate novel ideas for improving their intellectual product or process. In this emergent work process, the multiple cycles of planning and execution enhance the production of novel ideas, provide flexibility for adapting to dynamically changing task requirements, and, thus, improve performance.

This experiential learning strategy is based on Kolb's (1984) work on learning from experience. Kolb defines learning as a process whereby knowledge is created by the transformation of experience. As he describes experiential learning, a person continuously cycles through a process of having a concrete experience, making observations and reflections on that experience, forming abstract concepts and generalizations based on those reflections, and testing those ideas in a new situation, which leads to another concrete experience.

Kim (1993) describes this experiential learning process in terms of an observe-assess-design-implement (OADI) cycle. The OADI model preserves the salient features of Kolb's experiential learning model, yet the terms have clearer connections to the activities of individuals engaged in experiential work. Figure 1 is an adaptation of Kim's (1993) OADI model to experiential work. In experiential work, individuals experience a dynamically changing task environment and actively observe what is happening (e.g., obtain feedback). They deliberately assess (consciously or subconsciously) this experience by reflecting on their observations and then, sometimes, develop a novel idea that seems to be an appropriate response to the assessment. Figure 1 describes the context in which individual engineers work experientially, using the software to help them create new ideas or improvisations.

In each experiential cycle, the engineers could enact preexisting ideas for the product and/or process design, enact preexisting routines in their usual patterns without making changes, or they can improvise. The dashed circle around the novel idea in Figure 1 indicates that every experiential learning cycle does not necessarily result in a novel idea even if substantive and temporal convergence is inherent in the process. Engineers incorporate the idea(s) into the design of their work product or process and implement it via

computer, leading to new observations and triggering another cycle.

In developing the novel ideas, engineers may incorporate perspectives they take from others in their community of knowing (Boland & Tenkasi, 1995). If they feel their novel idea has value to the community, they may also share it (perspective making) with others. Perspective taking and perspective making are an important aspect of how information technology supports distributed cognition, but they may or may not occur in a particular cycle. Here, taken perspectives are not novel ideas. They are one input to an individual engineer's ideation process.

The multiple short-cycles greatly enhance substantive feedback and temporal convergence of planning and execution. Thus, they facilitate the generation of new ideas (i.e., improvisations). The short cycles enable individuals to try out new ideas for product or process designs and evaluate their internal and external coherence.

Since external and internal coherence can be more easily assessed, improvisations that are adopted are likely to be successful. Storing product and process data on the computer enhances the reproducibility of work and facilitates knowledge sharing. This reproducibility of work also facilitates new idea generation. It enables the

Figure 1. An experiential process for IT-enabled new ideas in engineering design work

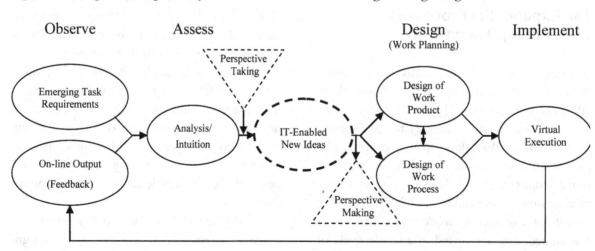

easy comparison of current designs with previous designs, building intuitive judgment.

The learning process can be facilitated and/or constrained by the capabilities of the software. Users can not execute functions that are prohibited by the software. Embedded features in the software can also guide/facilitate the problem solving process.

The Research Model

The interaction among an active (motivated) human agent, an emergent work process, and an interpretively flexible technology may stimulate the real time creation of on-the-spot new ideas (Orlikowski, 1992, 2000). Since engineering work and experiential learning require the generation and trying out of new ideas, the exploratory model presented in Figure 2 explains IT-enabled improvisations in terms of the empowerment of engineers and their building of improvisational competencies through application usage experience. This model contends that, to take advantage of the improvisational opportunities inherent in engineering design work, individuals need to be empowered. Empowering workers requires (1) software capability, and (2) sufficient autonomy to design their own work processes, and, thus, actively shape the intellectual product and processes of their work (Malone, 1997). Software capability refers to the entirety of the software's

functions, features, and interpretive flexibility. The opportunity itself to design their own work processes motivates workers to create new ideas (Wrzesniewski & Dutton, 2001). Software capability and an active human agent (e.g., autonomy) may actually be preconditions for the occurrence of extemporaneous new idea generation.

Individuals also need to build improvisational competencies over time by short learning cycles that rapidly build intuitive judgment (Eisenhardt & Tabrizi, 1995). Experience is essential to building improvisational competencies. Dimensions of experience might include the extent of use for problem solving/decision support, the type of use (e.g., work planning), and the length of use.

Miner et al. (2001) observed that the firms they studied, as they gained experience in improvisation, appeared to develop distinctive competencies related to improvisation itself. Organizations appeared to have learned how to limit the risks of improvisational learning and/or to facilitate fruitful higher-level improvisational learning.

We contend that engineers can also build improvisational competencies over time by using computers in their work. As suggested by Eisenhardt and Tabrizi (1995), engineers can learn to redesign their work processes to emphasize short-cycle experiential learning rather than pre-planning everything before work is started. In their problem solving/decision support activities,

Figure 2. The antecedents of IT-enabled new ideas

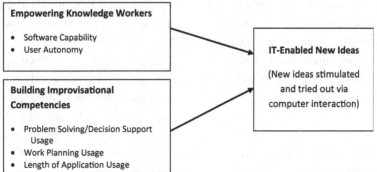

engineers can learn to use software to check the internal and external coherence of their work so that a part design can be successfully integrated into a virtual product design. By checking internal and external coherence, they greatly reduce risks and enhance the probability that the new idea will be successful.

Over time, engineers can reconfigure their software or modify their usage patterns to take advantage of the reproducibility of work and the short experiential learning cycles provided by computers. This greatly enhances the engineers' ability to try out new ideas under different parameters or contexts or to share ideas within a community of knowing. This ensures that the new ideas are properly vetted before they are actually incorporated into work processes or the design of the intellectual work product.

Empowering Engineers

Markus et al. (2002) describe the software capabilities required for emergent knowledge processing (EKP) in terms of whether the software is interpretively flexible (e.g., self-deploying, self-evolving, and actionable). They argue that such powerful software capabilities are necessary to support an emerging process of deliberations. For example, powerful software such as CATIA can reduce experiential cycle time enabling engineers to try out more new ideas on the spot. Some of these new ideas may improve the reliability, performance, or aesthetics of the designs while confirming conformance to design requirements. We contend that software capabilities support IT-enabled novel idea generation by enriching substantive interactive feedback, enabling individuals to explore more alternatives, reducing OADI cycle time, and enabling communities of knowing to share ideas. Thus:

H1: The greater the software's capability, the more the engineers will use the software to generate new ideas.

Nonaka (1994) argues that individual autonomy widens the possibility that individuals will motivate themselves to create new knowledge. A professional with higher autonomy will perceive that he/she has more choice or freedom to initiate and regulate application usage (Spreitzer, 1995, 1996). Many users, especially professionals such as engineers, are able to craft their own work processes (Wrzesniewski & Dutton, 2001). Amabile (1988) indicates that having freedom to decide what to do and how to do one's work enhances the capability for creative behavior.

Autonomy provides individual engineers with increased opportunity and motivation to spontaneously create and tryout new ideas in their work. User autonomy is a powerful variable because discretion in how the computer is used in work has implications for the work process and work product innovations. Conceptualized as self-determination, autonomy is positively related to innovation (Spreitzer, 1995; Spreitzer, De Janasz, & Quinn, 1999). Thus, we contend:

H2: The greater the autonomy, the more the engineers use information technology to generate new ideas.

Building Improvisational Competencies Thru Experience

Kolb (1984) argues that learning is a process whereby knowledge (e.g., new ideas) is created by the transformation of experience. Prior experience is thought to facilitate improvisation by providing individuals and groups with larger and more appropriate sets of existing routines, knowledge, and skills (Hatch, 1997). To the degree that individuals use the computer intensively in their problem solving and decision support activities, they build task knowledge, computer skills, and intuitive judgment that enhance their ability to generate new ideas. For example, information technology may enable engineers to solve problems or make design decisions by trying out different parameters or data sets. This experiential problem solving process may trigger new design ideas. Thus, we contend:

H3: The more information technology is used for problem solving/decision support, the more the engineers generate new ideas.

In the past, some scholars preferred to maintain a clear distinction between planning and improvisation (Moorman & Minor, 1998b). More recently, improvisation is viewed as a type of planning, that is, on-the-spot or ad hoc planning (Miner et al., 2001). Ciborra (1999) argues that work planning (i.e., the development of explicit objectives, criteria for decision making, and action plans) and how these plans are embedded into the design of business processes provide a useful point of departure for studying improvisation in work. Improvisational actions help the individual or group adapt when exceptions occur that were not anticipated in the plans.

Where engineers use information technology extensively for work planning (e.g., to plan the intellectual product and process of their work), the greater detail and reproducibility of the IT-enabled work planning makes it easier for engineers to modify product or process designs and try out new ideas. Planning enhances improvisation by providing a base line for evaluating improvement. Cunha, Kamoche, and Cunha (2003) imply that using software for work planning provides flexibility and, thus, enables improvisation to thrive.

H4: The more information technology is used for work planning, the more the engineers use the technology to generate new ideas.

Length of application usage can also lead to learning and improvisation. Over time, using the application for a process helps human agents refine their frameworks and routines, i.e., improve their mental models. The longer an individual uses a software package for a work process, the more

learning cycles the individual experiences and the greater the likelihood he/she will be able to improvise as task requirements change (Eisenhardt & Tabrizi, 1995; Kolb, 1984).

As engineers tackle a variety of problems over the years, they build knowledge, skills, and intuitive judgment that help them improvise in similar situations. Thus, the longer the software has been used for the process, the more one might expect the engineer to develop deeper insights that help generate novel ideas on how to solve problems that emerge in this work. Thus, we contend:

H5: The longer the individual has used the IT application, the more the engineers use the technology to generate new ideas.

This list of antecedent variables should not be viewed as exhaustive or all-inclusive. For the reasons argued above, these five factors are judged by the researchers to have a direct influence on IT-enabled generation of extemporaneous new ideas.

RESEARCH METHODS

This study uses a survey to explore possible antecedents of IT-enabled new idea generation in engineering work. A large sample gathered across several engineering organizations was chosen because it would enable us to test the hypotheses above and have greater confidence in the external validity of our findings. Since individuals may use several software packages with different levels of experience, software capabilities, and autonomy, we define the unit of analysis as an individual who uses a specific software package for a specific engineering process.

Sample

The survey was administered in five firms doing engineering design work in the greater Detroit area. The firms are asked to identify the engineers, software packages, and processes that will be included in the survey. Management asked 743 engineers doing highly analytical engineering design work to respond to the survey concerning their use of a specific software package for a specific process/task. The 208 responses represent a 28% response rate. All respondents were engaged in designing products, parts, or manufacturing processes for the automotive industry. One of the firms was a large original equipment manufacturer (OEM). The other firms were smaller engineering service firms who did engineering work for OEMs.

The processes under study are: computer-aided design, computer-aided engineering, or computer-aided manufacturing (79%); project management (17%); and manufacturability analysis (4%). The major software packages used for these processes included: AutoCAD and Eagle Point for computer aided design; ANSA, HYMESH, CATIA, MSC for computer aided engineering; Simul8, Arena, Witness, and ProModel for simulation; and Lotus Notes for project management.

The work of these engineers consists largely of using information systems in product development to generate ideas for possible solutions to specific design or project management problems that emerge. The information systems are also used to make explicit the reasons for the decisions, thus, supporting a decision process to get the users or their colleagues/superiors, as necessary, to accept and implement a solution. This is an interactive process with problem solving (i.e., sense making and idea generation) inextricably intertwined with decision support.

Most of the users are moderate to heavy users who have used the application for some time. Forty percent say they use the software "a great deal"; twenty seven percent use it "much"; and twenty percent use the software "moderately". By length of use, nineteen percent have used the software for more than five years; fifty two percent between one year and five years; twenty three percent between one month and one year; and six percent for several weeks, but less than a month.

The sample consists of twelve percent top or middle level engineering management, fourteen percent first-level engineering supervisor, sixty two percent professional engineer without supervisory responsibility, and twelve percent technical personnel. The sample is highly educated with thirteen percent having a Ph.D. degree, forty one percent having a master's degree, thirty percent having a bachelor's degree, six percent having an associate degree, and ten percent having only a high school diploma.

The respondents were asked to rate their level of knowledge/skill in using the software for their process compared to users who can make full use of the software in their work. In indicating their level of knowledge/skill, forty eight percent rated their knowledge/skill as 80% or more of that required for full use, twenty four percent indicated 60-79% of full use, fifteen percent indicated 40-59% of full use, eight percent indicated 20-39% of full use, and five percent indicated less than 20% of full use.

Respondents were also asked to rate their software's capabilities/features compared to software that would enable one to make full use of the software in their work. For this question, forty percent rated their software as having 80% or more of the capabilities required to make full use of the software, forty two percent indicated 60-79% of full use capabilities, eleven percent indicated 40-59% of full use capabilities, two percent indicated 20-39% of full use capabilities, and five percent indicated less than 20% of full use capabilities.

To test for non-response bias (Bailey, 1978), the researchers compare early and late respondents (Hu, Chau, Sheng, & Tam, 1999). Early respondents are defined as those who complete and return the questionnaire before a reminder is sent out. Approximately two thirds are early respondents and one third are late respondents. Using a chi-square test, the researchers assess differences between early and late responses by gender, educational level, and position of respondent categories (Sabherwal & King, 1995). No significant differences are found. Using a t-test, the researchers also assess whether there are significant differences between early and late respondents in mean scores for the six variables in the research model (Figure 2). All six t-tests reveal no significant differences between early and late respondents. Thus, the results suggest that the threat of non-response bias is not serious.

Instruments

The software package and process relevant for each individual are displayed at the beginning of the questionnaire. Respondents are asked to answer all questions in relationship to the specific software package and engineering process that

they use. Thus, all the measures described below are application specific.

IT-enabled engineering work is an emerging process of deliberations characterized by intentionality, substantive feedback, and the temporal convergence of planning and execution. IT-enabled new ideas are outcomes of this process. To measure this outcome, we use a three-item scale (Table 1) developed by Torkzadeh and Doll (1999) and cross-validated by Torkzadeh, Koufteros, and Doll (2005). The three-item scale demonstrates excellent psychometric properties in both prior samples with Cronbach (1951) alpha >.90. IT-enabled new ideas uses a five point scale ranging from 1= "not at all" to 5 = "a great deal".

This three-item scale is chosen because the items capture the essence of whether the applica-

Table 1. Survey instruments

Items	Item Descriptions
IT-Enabled New Ideas	
CFI1	This application helps me come up with new ideas.
CFI2	This application helps me create new ideas.
CFI3	This application helps me try out innovative ideas.
Software Capability	
SFT	How would you rate the capabilities/ features of the software compared to a software package that has all the capabilities/ features necessary in your job?
Autonomy	
AUT1	I have considerable opportunity for independence in how I use the software for this process.
AUT2	I have significant autonomy in determining how I use the software for this process.
AUT3	I have a say in how I use this software for this process.
Problem Solving/ Decision Support Usage	
PSDS1	I use this application to improve the efficiency of the decision process.
PSDS2	I use this application to help me make explicit the reasons for my decisions.
PSDS3	I use this application to make sense out of data.
PSDS4	I use this application to analyze why problems occur.
Work Planning Usage	
WPL1	I use this application to help me manage my work.
WPL2	I use this application to monitor my own performance.
WPL3	I use this application to plan my work.
Length of Usage	
LGU	Please indicate how long you have been using the software for the process.

tion facilitates the generation and trying out of new ideas (improvisational productions). The scale was originally identified as a measure of task innovation. When applied in the experiential work context where substantive convergence, temporal convergence, and intentionality are implicit in the iterative and cyclical nature of the extemporaneous design, execution, and assessment process, the items provide a general measure of IT-enabled new idea generation. Each item is prefixed with "This application helps me…" to capture the new ideas that are facilitated by the computer application rather than by other means. The term application describes a software package that is used for a specific work process.

Respondents were also asked to rate their software's capabilities/features compared to software that would enable one to make full use of the software in their work. For this single item, the scale choices were less than 20% of full use capabilities, 20-39% of full use capabilities, 40-59% of full use capabilities, 60-79% of full use capabilities, and 80% or more of the capabilities required to make full use of the software. This scale captures a user's perspective rather than a designer's perspective of the software's capability. A user's perspective provides a relative measure of software capability, relative to the software capability they need for their work.

Autonomy is measured using the three-item scale illustrated in Table 1. A five-point scale ranging from 1 = "strongly disagree" to 5 = "strongly agree" is used. The items were adapted from Spreitzer's (1995) measures of employee autonomy in a work setting. The questions were modified for the unit of analysis in this study, i.e., an individual who uses a specific software package for a specific work process. Each of the items makes reference to the individual's software and the process. Thus, the scale provides a measure of the degree of choice available to individuals in how they use the software for their work process.

The items measuring problem solving/decision-support usage are illustrated in Table 1. The

development of this four-item scale is reported in Doll and Torkzadeh (1998). Problem solving/decision support uses a five point scale ranging from 1= "not at all" to 5 = "a great deal". The scale provides an indication of the extent of use for problem solving/decision support.

The items measuring work planning usage are illustrated in Table 1. The development of this three-item scale is reported in Doll and Torkzadeh (1998). A five point scale is used, ranging from 1= "not at all" to 5 = "a great deal". The scale provides an indication of a specific type of use, i.e., the degree that individuals use the computer to plan or manage their own work activities.

To measure the length of use, respondents are asked how long they have been using the software for their work process (Table 1). The response categories are: (1) for several weeks, but less than a month, (2) between one month and one year, (3) between one year and five years, (4) more than five years.

RESULTS

As suggested by Ghiselli, Campbell, and Zedeck (1981), Table 2 reports the mean, variance, skewness, and kurtosis for each of the six variables. The means for the five point scales range between 2.65 for work planning usage and 4.12 for software capabilities. Each variable has adequate variance ranging from a standard deviation of 0.79 for length of use to 1.15 for IT-enabled new ideas. Each variable had reasonable skewness (less than 2) and kurtosis (less than 5).

Reliability (Cronbach's alpha) and average variance extracted (AVE) scores for the multi-item scales are on the diagonals in Table 3. The reliabilities are 0.83, 0.91, 0.84, and 0.90 for autonomy, problem solving/decision support usage, work planning usage, and IT-enabled new ideas, respectively. Thus, all four multi-item scales have acceptable reliability (Nunnally, 1978). Average variance extracted (AVE) is used to assess the

Table 2. Descriptive statistics

	Empowering		Building Improvisational Competence			IT-Enabled New Ideas
	Autonomy	**Software Capability**	**Problem Solving/ Decision Support Usage**	**Work Planning Usage**	**Length of Use**	
Mean	3.66	4.12	3.30	2.65	2.85	3.16
Standard Deviation	0.80	1.01	1.14	1.14	0.79	1.15
Skewness	-0.76	-1.52	-0.55	0.13	-0.43	-0.34
Kurtosis	1.14	2.33	-0.61	-0.92	-0.10	-0.75
						AVE = 0.74

amount of variance that is captured by each construct in relation to the amount of variance due to measurement error (Fornell & Larcker, 1981). For convergent validity, the AVE of each factor should be greater than 0.50. AVE values for all four variables are all greater than or equal to 0.63, indicating adequate convergent validity. With the exception of work planning usage and length of use variables, all correlations are significant at $p < 0.01$. Work planning has a moderate correlation ($r = 0.159$) with autonomy that is significant at the $p < 0.05$ level. Work planning also has a moderate correlation ($r = 0.168$) with length of use (significant at $p < 0.05$).

Problem solving/decision support usage has the highest correlation ($r = 0.689$) with IT-enabled new ideas. This suggests that the intensity of individual's application usage for problem solving may be the most immediate antecedent of new idea generation. Work planning usage ($r = 0.499$), autonomy ($r = 0.490$), and software capability ($r = 0.439$) also appear to be strong referents or facilitating conditions for IT-enabled new ideas. Length of use has a lower but significant ($p < 0.01$) correlation with IT-enabled new ideas. Intercorrelations between pairs of antecedents are significant (Table 3). Thus, in the step-wise mul-

tiple regression analysis below we examine whether the antecedents explain the same or unique variance in IT-enabled new ideas.

A chi-square test described by Segars (1997) is used to assess discriminant validity between pairs of latent factors. The chi-square value indicates whether a unidimensional rather than a two-dimensional model can account for the correlations among the observed items in each pair. As further evidence of discriminant validity, the value of AVE for each variable should be higher than the squared correlation between the focal factor and other factors (Segars, 1997).

Table 4 summarizes the results of a chi-square test of discriminant validity between pairs of variables. For the 15 pair-wise comparisons, the chi-square value must be equal to or greater than 11.58 for significance at $p < 0.01$ (Cohen & Cohen, 1983). Findings reported in Table 4 indicate that all chi-square differences are significant at the 0.01 level, indicating discriminant validity between each pair of constructs. The fact that the AVE value of each multi-item scale is greater than the squared correlations between the focal variable and the other variables (Table 3) provides further evidence of the discriminant validity.

Table 3. Reliability, average variance extracted, and correlations

	Empowering		Building Improvisational Competence			IT-Enabled New Ideas
	Autonomy	Software Capability	Problem Solving/ Decision Support Usage	Work Planning Usage	Length of Use	
Autonomy	α = 0.83 AVE = 0.63					
Software Capabilities	0.442 (p = 0.000)	--				
Problem Solving/ Decision Support Usage	0.449 (0.000)	0.334 (0.000)	α = 0.91 AVE = 0.72			
Work Planning Usage	0.159 (0.022)	0.191 (0.006)	0.514 (0.000)	α = 0.84 AVE = 0.64		
Length of Use	0.271 (0.000)	0.237 (0.001)	0.316 (0.000)	0.168 (0.015)	--	
Computer-Facilitated Improvisational Productions	0.490 (0.000)	0.439 (0.000)	0.689 (0.000)	0.499 (0.000)	0.355 (0.000)	α = 0.90

Table 4. Chi-square test of discriminant validity

Construct Pair	Δ Chi-Square	Δ Degree of Freedom	Discriminant Validity *
Autonomy – Software Capability	183.79	1	Yes
Autonomy – Problem Solving/Decision Support Usage	184.55	1	Yes
Autonomy – Work Planning Usage	231.68	1	Yes
Autonomy – Length of Use	218.45	1	Yes
Autonomy – IT-Enabled New Ideas	171.56	1	Yes
Software Capability – Problem Solving/Decision Support Usage	596.26	1	Yes
Software Capability – Work Planning Usage	247.16	1	Yes
Software Capability – Length of Use	--	--	--
Software Capability – IT-Enabled New Ideas	299.14	1	Yes
Problem Solving/Decision Support Usage – Work Planning Usage	180.23	1	Yes
Problem Solving/Decision Support Usage – Length of Use	599.37	1	Yes
Problem Solving/Decision Support Usage – IT-Enabled New Ideas	190.95	1	Yes
Work Planning Usage – Length of Use	245.56	1	Yes
Work Planning Usage – IT-Enabled New Ideas	189.80	1	Yes
Length of Use – IT-Enabled New Ideas	309.52	1	Yes

* Indicates significance at p < 0.01, based on the Δ Chi-square greater than 11.58 for the fifteen comparisons conducted.

Since the variables have significant inter-correlations, hypotheses 1 through 5 are tested using step-wise multiple regression (Neter, Kutner, Nachtsheim, & Wasserman, 1996). Step-wise multiple regression enters one variable at a time in the regression equation, starting with the one that explains the most variance and adding additional variables that explain the most additional variance in IT-enabled new ideas. Thus, where you have significant inter-correlations among independent variables, it helps identify those variables that explain some significant unique portion of the variance in the dependent variable.

The step-wise multiple regression results are depicted in Table 5. Column 1 contains the sequential model number, column 2 contains the variables entered in the model, columns 3 and 4 contain the unstandardized B and its standard error, column 5 contains the standardized Beta coefficient, column 6 depicts the t-values, column 7 depicts the significance level, and column 8 contains the R square value of the model. The results are described below in the order that the variables entered the regression equation rather than the sequential number of the hypothesis (i.e., H1 thru H5).

Consistent with the correlation analysis, problem solving/decision support usage has the highest association with IT-enabled new ideas and enters the regression equation in Model 1 with a standardized Beta coefficient of 0.69 and a significance level of 0.0000. Thus, hypothesis H3 is supported, i.e., the greater the extent that the engineer uses information technology for problem solving/decision support, the more technology is used by the engineer to generate new ideas. Model 1 explains forty seven percent of the variance in IT-enabled new ideas.

Since work planning and autonomy have high correlations (0.514 and 0.449, respectively) with problem solving/decision support usage, the second variable to enter the regression equation is software capability. In Model 2, software capability has a standardized Beta coefficient of 0.24 and a significance level of 0.0000. Thus, hypothesis H1 is supported, i.e., the greater the software's capability, the more information technology is used by the engineer to generate new ideas. Model 2 with problem solving/decision support usage and software capability explains fifty two percent of the variance in IT-enabled new ideas.

The third variable to enter the regression equation is work planning usage. In Model 3, work planning usage has a standardized Beta coefficient of 0.19 and a significance level of 0.0006. Thus, hypothesis H4 is supported, i.e., the more the engineer uses information technology for work planning, the more technology is used by the engineer to generate new ideas. Model 3 explains fifty five percent of the variation in IT-enabled new ideas.

The fourth variable to enter the regression equation is autonomy. In Model 4, autonomy has a standardized Beta coefficient of 0.18 and a significance level of 0.0011. Thus, hypothesis H2 is supported, i.e., the greater the engineering worker's autonomy, the more information technology is used by the engineer to generate new ideas. Model 4 explains fifty seven percent of the variation in IT-enabled new ideas. These results suggest that autonomy motivates IT-enabled new ideas. The engineer's choice is a prerequisite for novel ideas.

The fifth variable to enter the regression equation is length of use. In Model 5, length of use has a standardized Beta coefficient of 0.10 and a significance level of 0.0347. Thus, hypothesis H5 is supported, i.e., the longer the individual has used the IT application, the more technology is used by the engineer to generate new ideas. Model 5 explains fifty eight percent of the variation in IT-enabled new ideas. These results indicate that, even after the extent of use for problem solving/decision support, software capability, type of use, and autonomy are entered into the regression equation, the length of time that an engineer uses the application still has a significant influence on

Table 5. Step-wise multiple regression results

Model	Variables Entered	Unstandardized Coefficients		Standardized Coefficients	t-Value	Sig.	R Square
		B	Std. Error	Beta			
1	(Constant)	0.87	0.18		4.92	0.0000	0.4746
	Problem Solving/Decision Support Usage	0.69	0.05	0.69	13.64	0.0000	
2	(Constant)	0.03	0.25		0.12	0.9079	0.5238
	Problem Solving/Decision Support Usage	0.62	0.05	0.61	11.93	0.0000	
	Software Capability	0.27	0.06	0.24	4.60	0.0000	
3	(Constant)	-0.14	0.25		-0.57	0.5713	0.5504
	Problem Solving/Decision Support Usage	0.52	0.06	0.51	9.01	0.0000	
	Software Capability	0.26	0.06	0.23	4.64	0.0000	
	Work Planning Usage	0.19	0.06	0.19	3.47	0.0006	
4	(Constant)	-0.63	0.28		-2.23	0.0271	0.5735
	Problem Solving/Decision Support Usage	0.45	0.06	0.44	7.40	0.0000	
	Software Capability	0.19	0.06	0.17	3.29	0.0012	
	Work Planning Usage	0.21	0.05	0.21	3.90	0.0001	
	Autonomy	0.26	0.08	0.18	3.32	0.0011	
5	(Constant)	-0.86	0.30		-2.86	0.0047	0.5829
	Problem Solving/Decision Support Usage	0.42	0.06	0.42	6.98	0.0000	
	Software Capability	0.18	0.06	0.16	3.09	0.0023	
	Work Planning Usage	0.21	0.05	0.21	3.90	0.0001	
	Autonomy	0.24	0.08	0.17	3.09	0.0023	
	Length of Use	0.15	0.07	0.10	2.13	0.0347	
	Dependent Variable: **IT-Enabled New Ideas**						

IT-enabled new ideas. This suggests that building improvisation competencies may take a long time (greater than five years) to fully develop.

In summary, all five independent variables enter the step-wise regression equation and, thus, all five hypotheses H1 thru H5 are supported. The regression equation with all five variables entered explains approximately 58% of the variance in IT-enabled new ideas.

DISCUSSION

This experiential approach to IT enabled design is well recognized in the engineering field and described in some detail by Eisenhardt and Tabrizi (1995). This study has identified five antecedents of improvisations created by experiential learning cycles and provided some empirical evidence of their efficacy.

While engaged in an emergent work process of deliberations, engineers use information technology to help them generate and try out new ideas. Such IT-enabled new ideas are a type of short-term or experiential individual learning. They may or may not be retained by the engineer or shared with the community of knowing (Miner et al., 2001). Thus, one of the limitations of this survey study is the inability to follow the ideas generated to see if they are shared with others and retained by the community. Other limitations include the use of single item measures of software capabilities and length of use. Further research is needed to confirm whether these results can be generalized outside of the engineering setting.

IT-enabled new idea generation is viewed as having two parts – process components (deliberate, substantive feedback, temporal convergence) and an outcome component (novel idea). An experiential work process with intentionality, substantive convergence, and temporal convergence does not guarantee an improvisational outcome. In each experiential cycle, the workers could enact preexisting ideas for the product and/or process design, enact preexisting routines in their usual patterns without making changes, or improvise. Since all experiential work has the process characteristics of substantial convergence, temporal convergence, and intentionality, it seems appropriate to measure improvisation in terms of whether a novel outcome is produced. This approach is conceptually consistent with Miner et al.'s (2001) definition of improvisation, but differs in how improvisation is operationalized.

This does not imply that the process components are not essential to IT-enabled new idea generation. Rather, experiential work is a process that meets the process characteristics essential to improvisation. Experiential engineering design work is a situated and emergent process of deliberations. It is deliberate and provides the substantive and temporal convergence essential to novel productions. Thus, new ideas that emerge from this process can justifiably be called improvisations.

5.1 Theoretical Implications

While Eisenhardt and Tabrizi (1995) focus on the distinction between rational (pre-planned) versus experiential work and Miner et al. (2001) focus on describing the defining characteristics of improvisation, this research merges these two contributions to explain how information technology changes the nature of experiential engineering work, i.e., making it more improvisable. The extant information systems literature has largely ignored the role of information technology in facilitating improvisation, and the subtle and often unrecognized ways in which technology transforms the nature of engineering design work itself by making it more improvisable.

In the product engineering context, Eisenhardt and Tabrizi (1995) describe how engineering development time can be reduced by an experientially based process of discovery and learning. Information technology can support experiential strategies for engineering work by providing substantive and temporal convergence. This experiential experience in turn improves human problem solving by rapidly building knowledge and intuitive judgment through iterative experiential cycles. Short cycles of planning and execution can stimulate novel ideas. Computers make it easier to generate and "try out" these new ideas and thus, create new knowledge.

Markus et al. (2002) provide guidelines for the design of information systems that support emergent knowledge processes such as engineering design or strategic planning. They argue that emergent knowledge processing systems must be interpretively flexible (i.e., self-deploying, self-evolving, and actionable). This research suggests that software capabilities such as interpretive flexibility play an important role in stimulating new ideas. Interpretively flexible systems enable knowledge workers to create a dynamic fit between their software and their task as new task requirements emerge. Further research is necessary to explore and confirm whether and how interpretively flexible software enhances the generation of IT-enabled new ideas.

We argue that information technology enables a transformation of engineering design work from a pre-planned or rational work process to an experiential process characterized by short cycles and enhanced substantive feedback. We are not arguing for a complete transition, but rather, for a balanced approach of both planning and improvising. Information technology facilitates this balanced approach by enabling an engineer to use the same medium for both work planning and experiential learning cycles.

Spontaneous idea creation often works best when the improvisational outcome builds upon an initial plan or vision of what the intellectual product should look like or what type of work process is necessary to produce this product. In experiential engineering design work, planning and improvisation are complementary rather than competitive modes of work. Further research is needed on how to best achieve this balanced approach.

While the phenomena of experiential learning cycles and the on-the-spot generation of new product or process ideas (improvisations) are reported in the management literature, the information systems literature has made little use of these two related theoretical constructs to explain how business users create new ideas (improvise) in their IT enabled work processes. These theoretical constructs (experiential learning and improvisation) fit the context of IT enabled work well, providing a sound theoretical base. These linked phenomena of experiential learning and IT enabled new ideas (improvisation) by end users (Figure 1) are not explored adequately in the extant information systems literature.

The business information systems community and IS academics can benefit from what empowered users have learned in engineering – shorter experiential learning cycles enhance improvisations if the users have power software capabilities, user autonomy, and have built improvisation com-

petencies over time by problem solving/decision support usage, work planning usage, and length of application usage. What users have experienced in engineering may also apply to other business users. Future research might focus on whether, and to what extent, information technology makes some business knowledge work processes more improvisational.

Practical Implications

Information technology has enabled engineers to improve a design's functional specifications and aesthetic appeal through faster design feedback (experiential learning cycles with substantive convergence) leading, at times, to new ideas for product or process design. For example, in one interview with an automotive OEM, the management pointed out that CATIA could generate the stresses on a surface thirty times faster than other CAE software packages, enabling more ideas to be generated and tried out. Used in an experiential mode, software packages like CATIA have made engineering work more improvisational.

If information technology makes engineering design work more improvisable by facilitating experiential learning processes, what conditions encourage this transformation? This paper identifies some antecedents of improvisation grouped into two categories (1) empowering workers through powerful software capability and autonomy, and (2) building IT-enabled improvisational competencies through the extent of usage for problem solving/decision support, type of use, and length of application usage.

Empowering workers through powerful and interpretively flexible software and autonomy is a prerequisite for experiential engineering design work and learning. Unlike business process re-engineering efforts that may require intentional efforts by management to redesign the work process, individual designers craft their own work processes as their situated and emergent task requirements reveal themselves. In this situation where there is no best structure, autonomy and powerful software packages seem to be two essential ingredients that encourage IT-enabled new idea generation.

Engineers build improvisational competencies through the extent of use for problem solving/decision support, type of use, and length of application usage. Problem solving/decision support usage is the antecedent most closely associated with IT-enabled new idea generation. Other antecedents such as individual differences may indirectly influence IT-enabled new ideas through their impact on the extent of computer usage for problem solving/decision support.

CONCLUSION

The information systems literature has focused largely on automating and informating work. The role of computers in facilitating improvisation in experiential work has been largely ignored. This paper describes why computers make engineering work more improvisable. It helps reveal the subtle and often unrecognized ways in which information technology transforms the nature of engineering design work itself, making it more improvisable.

This initial exploratory study identifies a set of antecedents that facilitate improvisation in the engineering design work. The benefits of improvisation in an individual's experiential work are more effective designs of the intellectual work product, more effective designs of work processes, enhanced ability to take advantage of emerging opportunities, and improved ability to adapt to emerging problems.

Future research might focus on developing differentiated measures of improvisational outcomes rather than a general measure. Measures of behavioral, artifactual, and interpretive improvisations might be developed. Different antecedents may be associated with different improvisational outcomes. A general measure of improvisational outcomes (i.e., generating and trying out new

ideas) may not enable a finer distinction concerning which antecedents drive which types of improvisations.

REFERENCES

Alavi, M., & Leidner, D. (2001). Review: Knowledge Management and Knowledge Management Systems: Conceptual Foundations and Research Issues. *Management Information Systems Quarterly*, 25(1), 107–136. doi:10.2307/3250961

Amabile, T. M. (1988). A Model of Creativity and Innovation in Organizations. In Straw, B. M., & Cummings, L. L. (Eds.), *Research in Organizational Behavior* (*Vol. 10*, pp. 123–167). Greenwich, CT: JAI Press.

Bailey, K. D. (1978). *Methods of Social Research*. New York, NY: Free Press.

Baker, T., Miner, A. S., & Eesley, D. T. (2003). Improvising firms: Bricolage, account giving and improvisational competencies in the founding process. *Research Policy*, 31(2), 255–276. doi:10.1016/S0048-7333(02)00099-9

Boland, R. J. Jr, & Tenkasi, R. V. (1995). Perspective making and perspective taking in communities of knowing. *Organization Science*, 6(4), 350–372. doi:10.1287/orsc.6.4.350

Brown, J. S., & Duguid, P. (1991). Organizational Learning and Communities-Of-Practice: Toward a Unified View of Working, Learning, and Innovation. *Organization Science*, 2(1), 40–57. doi:10.1287/orsc.2.1.40

Campbell, D. T. (1969). Variation and Selective Retention in Socio-cultural Evolution. *General Systems*, 16, 69–85.

Chelariu, C., Johnston, W. J., & Young, L. (2002). Learning to Improvise, Improvising to Learn: A Process of Responding to Complex Environments. *Journal of Business Research, 55*(2), 141–147. doi:10.1016/S0148-2963(00)00149-1

Ciborra, C. U. (1999). Notes on Improvisation and Time in Organizations. *Accounting Management and Information Technologies, 9*(2), 77–94. doi:10.1016/S0959-8022(99)00002-8

Cohen, J., & Cohen, P. (1983). *Applied Multiple Regression/Correlation Analysis for the Behavioral Sciences* (2nd ed.). Hillsdale, NJ: Lawrence Erlbaum Associates.

Cronbach, L. J. (1951). Coefficient Alpha and the Internal Structure of Tests. *Psychometrika, 16*, 297–334. doi:10.1007/BF02310555

Crossan, M., Cunha, M. P. E., Vera, D., & Cunha, J. (2005). Time and Organizational Improvisation. *Academy of Management Review, 30*(1), 129–145. doi:10.5465/AMR.2005.15281441

Cunha, M. P. E., Kamoche, K., & Cunha, R. C. E. (2003). Organizational improvisation and leadership: A field study in two computer-mediated settings. *International Studies of Management & Organization, 33*(1), 34–57.

Doll, W. J., & Torkzadeh, G. (1998). Developing a Multidimensional Measure of System-Use in an Organizational Context. *Information & Management, 33*(4), 171–185. doi:10.1016/S0378-7206(98)00028-7

Eisenhardt, K. M., & Tabrizi, B. N. (1995). Accelerating Adaptive Processes: Product Innovation in the Global Computer Industry. *Administrative Science Quarterly, 40*(1), 84–110. doi:10.2307/2393701

Fornell, C., & Larcker, D. F. (1981). Evaluating Structural Equation Models with Unobservable Variables and Measurement Error. *JMR, Journal of Marketing Research, 18*(1), 39–50. doi:10.2307/3151312

Ghiselli, E. E., Campbell, J. P., & Zedeck, J. P. (1981). *Measurement Theory for the Behavioral Sciences*. San Francisco, CA: Freeman.

Greene, S. L. (2002). Characteristics of applications that support creativity. *Communications of the ACM, 45*(10), 100–104. doi:10.1145/570907.570941

Hatch, M. J. (1997). Jazzing up the theory of organizational improvisatione. In Walsh, J. P., & Huff, A. S. (Eds.), *Advances in Strategic Management* (pp. 181–191). Greenwich, CT: JAI Press.

Hu, P. J., Chau, P. Y. K., Sheng, O. R. L., & Tam, K. Y. (1999). Examining the Technology Acceptance Model Using Physician Acceptance of Telemedicine Technology. *Journal of Management Information Systems, 16*(2), 91–112.

Hutchins, E. (1991). Organizing Work by Adaptation. *Organization Science, 2*(1), 14–39. doi:10.1287/orsc.2.1.14

Kanellis, P., & Paul, R. J. (2005). User Behaving Badly: Phenomena and Paradoxes from an Investigation into Information Systems Misfit. *Journal of Organizational and End User Computing, 17*(2), 64–91. doi:10.4018/joeuc.2005040104

Kappel, T. A., & Rubenstein, A. H. (1999). Creativity in Design: The Contribution of Information Technology. *IEEE Transactions on Engineering Management, 46*(2), 132–143. doi:10.1109/17.759140

Kernfeld, B. (1995). *What to Listen for in Jazz*. New Haven, CT: Yale University Press.

Kim, D. H. (1993). The Link between Individual and Organizational Learning. *Sloan Management Review, 35*(1), 37–50.

Kolb, D. A. (1984). *Experiential Learning: Experience as the Source of Learning and Deployment.* Englewood Cliffs, NJ: Prentice-Hall.

Larsen, T. J., & Sørebø, Ø. (2005). Impact of Personal Innovativeness on the Use of the Internet among Employees at Work. *Journal of Organizational and End User Computing, 17*(2), 43–63. doi:10.4018/joeuc.2005040103

Majchrzak, A., Rice, R. E., Malhotra, A., King, N., & Ba, S. (2000). Technology Adaptation: The Case of a Computer-Supported Inter-Organizational Virtual Team. *Management Information Systems Quarterly, 24*(4), 569–600. doi:10.2307/3250948

Malone, T. W. (1997). Is Empowerment Just a Fad? Control, Decision Making, and IT. *Sloan Management Review, 38*(2), 23–35.

Markus, M. L., Majchrzak, A., & Gasser, L. (2002). A Design Theory for Systems that Support Emergent Knowledge Processes. *Management Information Systems Quarterly, 26*(3), 179–212.

Massa, S., & Testa, S. (2005). Data warehouse-in-practice: Exploring the function of expectations in organizational outcomes. *Information & Management, 42*(5), 709–718. doi:10.1016/j.im.2004.06.002

Miner, A. S., Bassoff, P., & Moorman, C. (2001). Organizational Improvisation and Learning: A Field Study. *Administrative Science Quarterly, 46*(2), 304–337. doi:10.2307/2667089

Moorman, C., & Miner, A. S. (1998a). Organizational Improvisation and Organizational Memory. *Academy of Management Review, 23*(4), 698–723.

Moorman, C., & Miner, A. S. (1998b). The Convergence of Planning and Execution: Improvisation in New Product Development. *Journal of Marketing, 62*(3), 1–20. doi:10.2307/1251740

Neter, J., Kutner, M., Nachtsheim, C., & Wasserman, W. (1996). *Applied Linear Statistical Models* (4th ed.). Chicago, IL: Irwin.

Nonaka, I. (1994). A Dynamic Theory of Organizational Knowledge Creation. *Organization Science, 5*(1), 14–38. doi:10.1287/orsc.5.1.14

Nunnally, J. C. (1978). *Psychometric Theory.* New York, NY: McGraw-Hill.

Orlikowski, W. J. (1992). The Duality of Technology: Rethinking the Concept of Technology in Organizations. *Organization Science, 3*(3), 398–427. doi:10.1287/orsc.3.3.398

Orlikowski, W. J. (2000). Using Technology and Constituting Structures: A Practice Lens for Studying Technology in Organizations. *Organization Science, 11*(4), 404–428. doi:10.1287/orsc.11.4.404.14600

Orr, J. (1987a). Narratives at Work: Story Telling as Cooperative Diagnostic Activity. *Field Service Manager,* 47-60.

Orr, J. (1987b). *Talking About Machines: Social Aspects of Expertise.* Palo Alto, CA: Xerox, Palo Alto Research Center.

Orr, J. (1990a). *Talking About Machines: An Ethnography of a Modern Job.* Unpublished doctoral dissertation, Cornell University, Ithaca, NY.

Orr, J. (1990b). Sharing Knowledge, Celebrating Identity: War Stories and Community Memory in a Service Culture. In Middleton, D. S., & Edwards, D. (Eds.), *Collective Remembering: Memory in Society* (pp. 169–189). Beverley Hills, CA: Sage.

Pressing, J. (1984). Cognitive processes in improvisation. In Crozier, W. R., & Chapman, A. J. (Eds.), *Cognitive Processes in the Perception of Art* (pp. 345–363). Amsterdam, The Netherlands: North-Holland. doi:10.1016/S0166-4115(08)62358-4

Preston, A. (1987). Improvising Order. In Mangham, I. L. (Ed.), *Organization Analysis and Development* (pp. 81–102). New York, NY: John Wiley & Sons.

Rumelt, R. (1995). The Evaluation of Business Strategy. In Mintzberg, H., Quinn, J. B., & Voyer, J. (Eds.), *The Strategy Process* (pp. 73–79). Englewood Cliffs, NJ: Prentice Hall.

Sabherwal, R., & King, W. R. (1995). An Empirical Taxonomy of the Decision-Making Processes Concerning Strategic Applications of Information Systems. *Journal of Management Information Systems, 11*(4), 177–214.

Segars, A. H. (1997). Assessing the Unidimensionality of Measurement: A Paradigm and Illustration within the Context of Information Systems Research. *Omega, 25*(1), 107–121. doi:10.1016/S0305-0483(96)00051-5

Spitler, V. K. (2005). Learning to Use IT in the Workplace: Mechanisms and Masters. *Journal of Organizational and End User Computing, 17*(2), 1–25. doi:10.4018/joeuc.2005040101

Spreitzer, G. M. (1995). Psychological Empowerment in the Workplace: Dimensions, Measurement, and Validation. *Academy of Management Journal, 38*(5), 1442–1465. doi:10.2307/256865

Spreitzer, G. M. (1996). Social Structural Characteristics of Psychological Empowerment. *Academy of Management Journal, 39*(2), 483–504. doi:10.2307/256789

Spreitzer, G. M., De Janasz, S. C., & Quinn, R. E. (1999). Empowered to lead: the role of psychological empowerment in leadership. *Journal of Organizational Behavior, 20*(4), 511–526. doi:10.1002/(SICI)1099-1379(199907)20:4<511::AID-JOB900>3.0.CO;2-L

Terry, M., & Mynatt, E. D. (2002). Supporting experimentation with side-views. *Communications of the ACM, 45*(10), 106–108. doi:10.1145/570907.570942

Torkzadeh, G., & Doll, W. J. (1999). The Development of a Tool for Measuring the Perceived Impact of Information Technology on Work. *Omega, 27*(7), 327–339. doi:10.1016/S0305-0483(98)00049-8

Torkzadeh, G., Koufteros, X., & Doll, W. J. (2005). Confirmatory factor analysis and factorial invariance of the impact of information technology instrument. *Omega, 33*(2), 107–118. doi:10.1016/j.omega.2004.03.009

Weick, K. E. (1993a). Organizational Redesign as Improvisation. In Huber, G. P., & Glick, W. H. (Eds.), *Organizational Change and Redesign* (pp. 346–379). New York, NY: Oxford University Press.

Weick, K. E. (1993b). The Collapse of Sensemaking in Organizations: The Mann Gulch Disaster. *Administrative Science Quarterly, 38*(4), 628–652. doi:10.2307/2393339

Weick, K. E. (1998). Improvisation as a Mindset for Organizational Analysis. *Organization Science, 9*(5), 543–555. doi:10.1287/orsc.9.5.543

Wrzesniewski, A., & Dutton, J. (2001). Crafting A Job: Revisioning Employees as Active Crafters of Their Work. *Academy of Management Review, 28*(2), 179–201.

This work was previously published in the Journal of Organizational and End User Computing, Volume 23, Issue 3, edited by Mo Adam Mahmood, pp. 26-47, copyright 2011 by IGI Publishing (an imprint of IGI Global).

Chapter 14
The Role of Computer Attitudes in Enhancing Computer Competence in Training

James P. Downey
University of Central Arkansas, USA

Lloyd A. Smith
Missouri State University, USA

ABSTRACT

Computer competence is poorly conceptualized and inconsistently measured. This study clarifies computer competence and examines its relationship with anxiety, affect, and pessimism, along with self-efficacy and previous experience. Using a survey of 610 end users, the strengths of anxiety, affect (positive), pessimism, self-efficacy, and previous experience were compared for nine different competency measures in seven different domains, including word processing, email applications, spreadsheets, graphic programs, databases, web design, and overall computing. Results suggest that for most domains, affect and anxiety are significant predictors, as are self-efficacy and previous experience, but pessimism is not. In addition, competence in a domain was found to mediate the relationship between competence and its antecedents. These results suggest that organizations focus not only on skills training, but on ways to enhance computing attitudes during the training process.

INTRODUCTION

Computer competence is critical to organizations in that they rely on employees of all kinds who possess the skills necessary to carry out increasingly technical tasks. In many companies today, IT skills belong not to some cadre of technology experts, but to a majority of employees who are now technology end users (Downey, 2004). Technology provides an inherent strategic and synergistic capability that is necessary for most organizations merely to compete in today's mar-

DOI: 10.4018/978-1-4666-2059-9.ch014

ketplace. It is therefore essential for both organizations and researchers to understand the process by which employees gain these competencies as well as the factors which enable and enhance this learning process. Although the process of gaining IT skills, through education, training, or on the job learning, is relatively well documented, not all factors important in the acquisition of computer skills have been examined. Much of the previous research is situation based, that is, it examines the learning environment, method of instruction, and common individual factors such as previous experience as it relates to learning outcomes. For example, Compeau and Higgins (1995a) found that for spreadsheet applications, subjects who observed others perform the behavior outperformed subjects who only learned through lecture. Many studies examine factors important for enhancing learning outcomes, especially prior experience (Compeau & Higgins, 1995a; Guimaraes & Igbaria, 1997) and self-efficacy (Compeau & Higgins, 1995a, 1995b; Johnson & Marakas, 2000), though there are many others reported in the literature: age and education (Guimaraes & Igbaria, 1997), gender (Hoxmeier, Nie, & Purvis, 2000), health issues or ergonomics (Gattiker & Hlavka, 1992; McMurtrey, McGaughey, & Downey, 2008), and organizational support (Guimaraes & Igbaria, 1997), to name a few. However, there are additional factors which may contribute to successful learning that have been largely untested. In particular, the influence of computer attitudes on competence is largely unexplored. Attitudes toward computing have reported significance in *other* areas, including end user satisfaction (Aladwani, 2002), computer experience (Potosky & Bobko, 2001), IT implementation in small companies (Winston & Dologite, 2002), early adopters of IT (Burkhardt & Brass, 1990), computer confidence (Hoxmeier et al., 2000), and intention to use a computer or actual computer use (Chau, 2001; Compeau & Higgins, 1995b). But, to our knowledge, the link between attitudes and computer competence has not been tested.

As shall be presented, theory suggests that computer attitudes are important in competences, particularly in the training process (Ford & Noe, 1987; Gattiker & Hlavka, 1992; Noe, 1986). Indeed this makes intuitive sense because positive attitudes should provide additional motivation and perseverance during the learning process. In fact, the case must be made that the attitude/competence link even merits a study. We make the case for this study based on two factors, which we believe compels an investigation into the effect attitudes have on learning outcomes (and, in particular, competence).

First, in extant studies the effects of attitudes on learning outcomes has been inconsistent and even at times counter-theoretic. Although no study examined "competence" (see next section), some studies examined subsequent performance and found the relationship between attitudes and performance significant as expected (Jawahar & Elango, 2001; Nickell & Pinto, 1986; Webster & Martocchio, 1993). While this is hardly surprising, what is surprising is that other studies did not. In one study which examined the effect of five different attitudes on student performance (measured by the final grade in a computer class), only one of the five attitudes was significant (attitudes toward computing complexity). Attitudes toward computer productivity, computer health attitudes, computer interest, and attitudes toward computer consequences were not significant, leading to their assertion that "overly positive attitudes towards computers could actually hinder learning performance" (Gattiker & Hlavka, 1992, p. 99). Another study found computer anxiety (a well-researched attitude) was not significantly related to computer usage, computer satisfaction, or productivity (Guimaraes & Igbaria, 1997). The same study found that positive attitudes did have a significant relationship with the same outcomes. Other studies have found similarly puzzling findings (Kernan & Howard, 1990; Szajna & Mackay, 1995). These studies suggest that the relationship between attitudes and competence (or even performance) may not be as intuitive as first thought. One reason for such findings is that the effect of computer attitudes on performance may operate *through* competence. That is, attitudes indirectly influence performance through computer competence. Although this is not directly tested in this study, it is a fruitful area for further work; this study tests only the attitudes-competence relationship.

Second, although enhancing competence (and subsequently performance) is the goal of most IT training, competence itself is not a well understood construct. It has been conceptualized and measured in many ways (Marcolin, Compeau, Munro, & Huff, 2000), creating confusion and making generalization and triangulating results difficult. Unless we understand the nature of competence, it is difficult to draw any conclusions on which factors and processes are important in the learning process. To this end, it is important to agree on the construct of competence, to place it in perspective by distinguishing it from performance, and to examine its relationship with important antecedent factors.

It is the purpose of this study to shed light on our understanding of competence and the attitudes that may affect it. If user attitudes toward computing have a significant influence on user competence, and the literature makes a theoretical case that they should, this would suggest organizations should emphasize enhancing positive attitudes and reducing negative attitudes as part of the training process. Focusing solely on skill development and improvement may not be enough; appropriate attitude reinforcement should be important as well. In fact, practitioner literature suggests that attitude training is important of itself, leading to greater learning, effectiveness and efficiency (Feiertag, 2001). One article declared that some multinational corporations (including Motorola) that invested specifically in attitude training earned a $30 return

on investment for each $1 spent ("Invest In", 1998). Given the amount and cost of technology training today, maximizing its effects is a critical outcome that could be enhanced by focusing on improved attitudes. This study examines competence and its relationship with computer attitudes, while suggesting methods that organizations may use to more effectively conduct IT training.

THE PROBLEM OF COMPUTER COMPETENCE

In spite of its importance, computer competence is not well understood—it is poorly conceptualized, inconsistently measured, and unpredictably defined (Marcolin et al., 2000). There are several factors which have obscured a more precise understanding of computer competence. The first is that any range of computing competence (from high to low) is subjective and, therefore, difficult to generalize across subjects. Computer competence to a software engineer is different than the competence ascribed to an office clerk or an accounting end user. A second factor concerns how computer competence is represented in studies. Because it is measured in so many different ways, establishing consensus is difficult. Besides typologies (Rockart & Flannery, 1983), computer competence and skills have been evaluated by performance-based measures (Marcolin et al., 2000; Munro, Huff, Marcolin, & Compeau, 1997), self-reports of ability (Winter, Chudoba, & Gutek, 1997), and perceptions of ability (Harrison & Rainer, 1992). As Marcolin et al. (2000) convincingly point out, these different measures are not the same and make comparisons difficult, if not impossible, which in turn makes generalization more problematic. In order to draw convincing conclusions across studies, there must be agreement on the theoretical nature of computer competence and an understanding of the different ways of measuring the construct. Because each method has different ramifications, an understanding of strengths and weaknesses of each is important (but beyond the scope of this paper) (Marcolin et al., 2000).

A final factor concerns the computing domain. When a study reports computing competence, to which domain is it referring? The computing domain may be divided in many ways, and we suggest that competence may be present and measured in any of them. These range from "sub-domains" like applications (e.g., spreadsheets, databases, accounting or finance packages, etc.) and environments (Windows XP or Vista, Mac OS, UNIX, etc.) to "mid-domains" like software (such as some or all applications and environments), hardware (drives, memory, routers, switches, etc.), languages, and many more. Competence in the entire computing domain may also be measured. When one speaks of computing competence, the domain of competence *must be clearly delineated*.

Based on these factors, three general rules should be followed in studies concerning competence. First, the population to which competence is ascribed must be established. Competence can be considered relative and depends on the population involved, and derived generalization must be appropriate to the population. In this study, for example, that population is college students in ROTC in fourteen different universities in the U.S. Secondly, the measure of competence must be established. There are many methods of measuring computer competence, each having its own positive and negative qualities (Marcolin et al., 2000). This study includes both written tests and self-reports of competence. Finally, the specific domains must be delineated. In this study, we examine seven domains: six application sub-domains (word processing, email applications, spreadsheets, graphic programs, databases, and web development applications), plus competence in the entire computing domain. These domains were chosen because the respondents were likely to have some experience in most but not all of them, thus including full range of competence. It

was important to include domains with a range of expected competency levels in order to analyze the effect of attitudes on each. We fully expect that these respondents will have more competence in word processing and email applications than for databases and web design, making testing of domains with different competency levels possible. In addition, some of these applications (e.g., word processing, spreadsheets) can be considered building blocks for most software applications, rendering them especially important from a competence perspective.

Any "competence" has three defining characteristics: it is behavioral in nature, it is learned, and it involves some combination of cognitive, perceptual, and motor processes (Adams, 1987). A competence is thus described as the interaction of cognitive and motor processes, learned through training or practice. Computing or computer competence may be defined as a learned behavior involving some combination of cognitive and motor processes, used to maximize any computer-related performance (Gattiker, 1992; Marcolin et al., 2000). This implies one's competence, or ability to do something well, is antecedent to performance. While one's computer performance is based in part on competence, there are other factors (nature of task and its complexity, work environment, alertness of performer, etc.) affecting performance. Because competence is antecedent to performance, and necessary for successful performance of computer tasks, we consider this the more valuable construct for study. Although organizations are clearly interested in performance, the first objective for training is building competence.

The acquisition of competence has been the topic of much research, particularly in the training literature (Gattiker, 1992; Marcolin et al., 2000). Competence is directly enhanced through training, which may be defined as planned learning designed to bring about a permanent change in an individual's knowledge, skills, or attitudes (Noe,

1986). The training literature suggests that there are several antecedent factors and a range of training methodologies, which affect the acquisition of competence. This process is displayed in Figure 1. This study examines some of these antecedents (in italics, Figure 1), in particular the attitudes that may enhance competence.

FACTORS IMPORTANT TO COMPETENCE AND HYPOTHESES

The literature suggests five general areas which influence the acquisition of computer competence (Figure 1): ability, motivation, attitudes, previous experience, and socio-demographic factors. These are individual factors; not included are factors based on the organization, such as the expected role an employee fills, organizational support, frequency of organizational training, etc.

Ability

Some factors which influence computing competence have little to do with computing. One such factor is IQ, or general intelligence. General intelligence influences competence by its effect on learning, remembering, responding to feedback, information processing, etc., particularly in an intellectual task (Brody, 1992). Another factor is perceptual speed, which is the individual's capability to learn through practicing and to quickly automate procedures, by means of repetition (Gattiker, 1992). A final factor is psychomotor ability, which includes motor processes intrinsic to the area of competence. For example, hand-eye coordination is critical for mastering many athletic sports. In the computing realm, psychomotor ability is less important, though there are some facets of competence which require motor skills, such as keyboarding, mouse movement,

Figure 1. Computer competence acquisition process

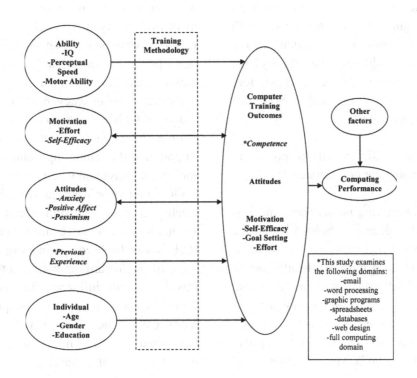

etc. Ergonomics, or the study of human-computer physical interaction, plays a role in competence, particularly with older learners. Older computer learners can learn more quickly if there are no visual impediments (e.g., small buttons) or the environment includes more appropriate interfaces (McMurtrey et al., 2008).

Motivation

An individual's motivation can make a significant difference in skill acquisition. Motivation affects the quality and quantity of attention resources focused on learning (Kanfer & Ackerman, 1989). One of the contributors of motivation is self-efficacy, which is an individual's perception of his or her ability in performing a task (Bandura, 1997). Self-efficacy influences choice of action, effort expended in accomplishing or learning the task, and persistence in the face of adversity (Murphy, Coover, & Owen, 1989). Self-efficacy helps govern and stimulates the motivation necessary to learn and conduct the behavior. In the computing field, self-efficacy may be defined as an individual judgment of one's capability to use a computer (Compeau & Higgins, 1995b). Its relationship with computer skills acquisition is well established in the literature. Those with higher self-efficacy have more skills or perform at a higher level (Martocchio & Webster, 1992; Munro et al., 1997). This leads to our first hypothesis.

H1: Computer self-efficacy will be positively related to computer competence

One of the interesting points about motivation (and attitudes, discussed below) is that the relationship between motivation and competence is reciprocal (indicated by a two-headed arrow in Figure 1). Self-efficacy, for example, is also influenced by competence in the domain. As competence increases, one's perception of ability toward that domain (i.e., self-efficacy) also increases (Bandura, 1997). Motivation and self-

efficacy, therefore, both influence competence and are influenced by that competence. It should be noted that we include self-efficacy (and experience, covered shortly) in this model to serve as controls. Our primary interest is the relationship between attitudes and competence. But because of the strong influence of self-efficacy and experience (Marakas, Yi, & Johnson, 1998), and in order to isolate attitudes, the model should account for the influence of both self-efficacy and previous experience.

Attitudes

Attitudes have long been recognized as predictors of action. In their seminal Theory of Reasoned Action (TRA) (Fishbein, & Ajzen, 1975) and the follow-on Theory of Planned Behavior (TPB) (Ajzen, 1991), the models suggest that behavior is predictable, and based on both positive and negative attitudes toward the behavior, subjective norm (perceived expectations of salient others), and perceived control. TPB extended TRA in that it added perceived control, which is an individual's assessment of the "resources and opportunities" available to carry out that behavior (Ajzen, 1991, p. 183). Ajzen states that perceived control is one's perceived self-efficacy for conducting the behavior and "most compatible" with Bandura's construct of self-efficacy (p. 184). Ajzen suggests that those with higher perceived control display a stronger intention to learn the behavior, which is important in the computing realm. In line with this model, an individual who has a positive attitude (or less anxiety or less negative attitudes) toward a behavior tends to perform or learn that behavior, while one with negative attitudes tends to avoid the behavior. In the computing realm, this suggests that attitudes toward computing, both negative and positive, should influence computer behaviors, including usage, acceptance, adoption, desire to learn, competence, etc. TPB thus incorporates attitudinal as well as motivational constructs in explaining human behavior.

An attitude is defined as "a learned predisposition to respond in a consistently favorable or unfavorable manner" towards a domain (Fishbein & Ajzen, 1975, p. 6). Attitudes are dynamic, domain-specific individual differences that affect the conduct of the individual's activities within the domain (Thatcher & Perrewe, 2002). Computer attitudes influence how an individual reacts in the computing environment. Theory suggests that positive and negative attitudes are important in computer competence (Noe, 1986). Attitudes impact an individual's motivation to learn during the training phase. Individuals who have more positive feelings and less anxiety and pessimism are more interested in learning new skills which subsequently improves job performance and increases feelings of self-worth (Noe, 1986). Attitudes affect how an individual perceives future outcomes, such as career growth, job choice, and performance, which enhance skill acquisition through their usefulness.

As mentioned, studies of computing attitudes on computing competence or performance have had mixed results. Similarly, studies in other fields have also had mixed results. In one study of British global companies, it was found that attitudes toward learning a new (foreign) language, both positive and negative, influenced one's inclination to learn the language and their competence in the new language (Swift & Smith, 1992). However in another study of college students taking statistics, anxiety towards statistics and ultimate performance in learning statistics was significantly influenced by major, suggesting the relationship may be mediated by other factors (Lipscomb, Hotard, Shelley, & Baldwin, 2002). Similar to the computing field, the influence of attitudes on behaviors or competence may be more complex than is at first apparent.

The current study examines the three attitudes of anxiety, affect, and pessimism. These were chosen primarily because they are attitudes toward computers in general (as opposed to attitudes about a particular technology) and because they have a relatively rich history in extant literature.

- **Computer Anxiety:** Computing anxiety is a fear of computers or of computer use (Loyd & Gressard, 1984). It is a domain-specific fear (specific to the computing domain), and is distinguishable from trait anxiety, which is a general feeling of anxiety (Thatcher & Perrewe, 2002). Computer anxiety is influenced by a variety of emotional and environmental factors (Marakas et al., 1998). Like any attitude, anxiety influences choice of behavior, motivation to learn, effort, and persistence. Those with lower anxiety should have higher levels of competence, while those with higher anxiety will have lower levels of competence:

H2: Computer anxiety will have a negative relationship with computer competence

- **Computer Affect:** Computer affect is the feeling of like or dislike towards computing. A person's affect towards computing is an important factor in user acceptance as well as computer usage (Al-Jabri & Al-Khaldi, 1997) and skills acquisition. Individuals tend to pursue activities they like while avoiding disliked activities (Loyd & Gressard, 1984). One who has a liking for computing, therefore, ought to be more persistent in learning and have higher competence. It should be noted that affect (like or dislike) is not the same as anxiety (Kernan & Howard, 1990). An individual could simultaneously dislike computing and have little anxiety towards it. This study examines positive affect. Therefore:

H3: Positive computer affect will have a positive relationship with computer competence

- **Computer Pessimism:** Pessimism is a tendency to take a gloomy view of situations, which contribute to expectations of negative outcomes. In an experiment involving stress and feedback, pessimistic individuals were found to be more affected negatively by feedback and had more stress (Szalma, Hancock, Dember, & Warm, 2006). In the computing realm, one study found that pessimists used computers significantly less than optimists (Arndt, Clevenger, & Meiskey, 1985), while another study found that those with pessimistic attitudes towards computers had significantly less computer self-efficacy (Harrison & Rainer, 1992). Computer pessimism is not the same as *negative* affect, which is a dislike of computing (Brock & Sulsky, 1994). Thus:

H4: Computer pessimism will have a negative relationship with computer competence

Previous Experience

Prior experience in computing is known to influence performance (Brock & Sulsky, 1994; Chan & Storey, 1996; Munro et al., 1997). One reason is that prior experience enables individuals to make mental models which enable them to understand and apply technology and its applications (Bostrom, Olfman, & Sein, 1990). Because competence is a learned behavior enhanced through repetition, previous experience should play a key role in the acquisition of competence. Individuals who have more experience typically have more competence in any computing domain. In this study, we examine the relationship between experience and competence in six different sub-domains, as well as experience in all sub-domains (labeled "overall application experience"). Both domain experience (e.g., spreadsheet experience) and overall application experience should be

related to competence in each sub-domain, the latter because each sub-domain is a constituent part of overall experience. Like hypothesis 1, our fifth hypothesis is included as a control because the relationship between prior experience and competence is established:

H5: Previous computer experience in a sub-domain (word processing, spreadsheet, graphic programs, databases, email, and web design) and overall application experience will have a positive relationship with computer competence in those same sub-domains.

As stated above, this study includes self-efficacy and prior experience primarily as control constructs. Self-efficacy can influence computer attitudes (Harrison & Rainer, 1992); therefore, it is important to distinguish the effects that each has on competence individually. Similarly, one with more prior experience in a domain should have higher competence and better attitudes. To distinguish between the two effects, prior experience was included.

Socio-Demographic Factors

The final factors affecting competence are demographic in nature, such as age, gender, education, and economic status. These factors have had mixed results in extant studies. There have been a number of studies which suggest that older computer users have less competence (Burkhardt & Brass, 1990). The reasons for this may be because age affects learning, in particular processing resources (Gattiker, 1992) or even ergonomic factors, such as mouse movement among the elderly (McMurtrey et al., 2008). Gender has also been a factor in computing related activities in some empirical studies. In general, studies which have found gender significant have found that women have less competence than men (Munro et al., 1997). Other factors have been found to be significant in some studies, including education (Munro et al., 1997), computer access (Evans & Simkin, 1989), and goal setting (Kanfer & Ackerman, 1989).[1]

RESEARCH METHODOLOGY

Participants

The population chosen for this study is Midshipmen in the U.S. Navy's commissioning program (providing officers for both the Navy and Marine Corps), using a survey methodology. This was part of an ongoing study to examine the effectiveness of technology training in universities for newly commissioned officers in the U.S. Navy and Marine Corps. Thirteen universities with Navy ROTC (NROTC) programs were chosen at random (from 57 total), as well as the US Naval Academy (with the largest number of Midshipmen). Each NROTC university was sent 48 surveys, while the Naval Academy received 128. Of the 751 sent out, 630 were returned (86%), of which 610 were usable (81%). The list of universities includes the US Naval Academy, South Florida, Florida, Missouri, Kansas, Minnesota, South Carolina, Penn State, Idaho, Ohio State, Washington, Purdue, Oregon State, and Vanderbilt.

On average, responders were 21.1 years of age (sd = 2.78), 510 were male (83.6%), 98 female (16.1%), with two unmarked. Respondents, on average, had 2.4 years of college (sd =.99). The respondents' majors included liberal arts (30%), engineering (19%), IT/MIS (14%), math or science (12%), business other than MIS (12%), and computer science (6%), with the rest being a mixture of agriculture, education, health/nursing, and others (7%).

Study Measures

Almost all measures used previously reported and validated instruments.

- **Attitudes:** Anxiety and affect (positive) were measured using those subscales of the Computer Attitude Scale developed by Loyd and Gressard (1984) and validated by Al-Jabri and Al-Khaldi (1997). Woodrow (1991) determined that the subscales were reliable enough to be administered separately. The anxiety subscale included eight items; the affect subscale included seven items. Pessimism was measured using the Computer Attitude Scale developed by Nickell and Pinto (1986). It consists of six items and was validated by Harrison and Rainer (1992). All three scales used a seven-point Likert scale, where 1 is "completely disagree" and 7 is "completely agree." The items for all three attitude scales are listed in Appendix A as part of the factor analysis table.

- **Self-Efficacy:** Self-efficacy was measured using the ten-item instrument of Compeau and Higgins (1995b). The instrument asks respondents to judge their ability to complete a job using an unfamiliar software product, given ten different levels of assistance. This instrument measures general computer self-efficacy, or an individual's perception of ability for the entire computing domain.

- **Previous Experience:** Previous experience was measured for six different application domains: word processing, spreadsheets, graphic programs, databases, email programs, and web design. For each domain, respondents indicated the frequency of previous use (from "Almost never" to "Several times a day") and duration (from "Almost never" to "More than 3 hours per day"). The six domain results were summed to provide a measure of overall previous application experience.

- **Computing Competence**: Computer competence was measured two ways. One was a multiple choice declarative knowledge test for the domains of word processing and spreadsheets. The second way was a self-reported competence measure for the same six domains as measured in previous experience. This measure is from Munro et al. (1997), based on number of domain packages used, number of academic or training courses taken in the domain, and thoroughness of current knowledge of the domain (on a scale of 0 = "No Knowledge" and 1 = "Very Limited Knowledge" to 7 = "Complete Knowledge"). To obtain a measure of overall computing competence, these six domain measures were summed, and to this result was added two additional areas, "other" software and hardware (number of courses taken and thoroughness of current knowledge). There were four other software domains, including programming languages, task managers, financial software, and statistics software packages. The hardware domains consisted of PCs, mainframes, network hardware, graphic media hardware, and PDAs.

- **Application Tests:** Two multiple choice tests of fourteen questions each were developed for this study, for the sub-domains of word processing and spreadsheets. Both consisted of declarative knowledge and procedural items in a multiple choice format. The tests were specific to Microsoft's Word and Excel and based on intermediate skill level tasks (Microsoft, 2006). In order to avoid a potential confound for

those respondents who were skilled in a non-Microsoft application, each respondent also indicated which word processing and spreadsheet application they "know best". The test results were only included for those respondents who indicated they knew Word and Excel best. Each respondent received only one of the tests, randomly distributed. The tests were optional, but 67% (408 total, 210 word processing and 198 spreadsheet) of respondents took one of the tests.

RESULTS

Attitudes and Self-Efficacy

Survey responses for self-efficacy, anxiety, affect, and pessimism were each factor analyzed independently. This resulted in two items being dropped due to cross-loads among items; one item was from the self-efficacy scale the other from the affect scale. Following these deletions, each scale was unidimensional with satisfactory internal reliability. In order to test discriminant validity among the three attitude scales, they were factor analyzed simultaneously; discriminant validity is indicated if each item loads highest on its own scale. This proved to be the case, with loadings mostly above .70 and no cross-loads above .33. Appendix A presents the factor analysis.

Table 1 presents means and standard deviations of self-efficacy, anxiety, affect, and pessimism,

Table 1. Descriptive statistics and correlations between antecedent variables

Construct	M	SD	α	1	2	3	4
1. Self-efficacy	7.04	1.74	.93	1.0			
2. Anxiety	1.81	1.00	.92	-.50**	1.0		
3. Affect	4.58	1.21	.88	.57**	-.54**	1.0	
4. Pessimism	3.06	1.32	.84	-.26**	.34**	-.36**	1.0

** p <.01 α is Cronbach's alpha. n = 610

Table 2. Descriptive statistics and correlations between previous experience variables

	M	SD	1	2	3	4	5	6	7
1. Email	5.32	.95	1.0						
2. Word Processing	4.27	1.2	.30**	1.0					
3. Spreadsheet	2.59	1.3	.21**	.43**	1.0				
4. Graphic Programs	2.39	1.4	.30**	.29**	.40**	1.0			
5. Web Design	1.84	1.5	.16**	.16**	.17**	.24**	1.0		
6. Database	1.49	1.0	.16**	.30**	.33**	.31**	.31**	1.0	
7. Overall Experience	2.98	.77	.52**	.65**	.69**	.69**	.58**	.62**	1.0

** p <.01

as well as pair-wise correlations among them. In general, this population had above average self-efficacy (7.04 of 10), relatively low anxiety (1.81 of 7, where 7 indicates high anxiety), above average positive affect (4.58 of 7), and a pessimism mean in the middle. Correlations were all significant and in the assumed direction.

Previous experience was measured for six domains, as well as overall previous application experience. Descriptive statistics and pair-wise correlations are presented in Table 2. Respondents had the most prior experience in email and word processing applications; they had the lowest prior experience in web design and database applications.

There were nine competence measures, six for application domains, one overall computing competency, and two application tests in word processing and spreadsheets. For the six application domains, respondents had the highest competence in email and word processing and the lowest in web design and databases. For the two tests, the mean is derived from the number of correct answers (of 14). Descriptive statistics and pair-wise correlations for competencies are given in Table 3.

Hypothesis Testing

To test which factors significantly influenced computer competence for this population, correla-

Table 3. Descriptive statistics and correlations between competence variables

	M	SD	1	2	3	4	5	6	7	8	9
1. Email	3.87	1.17	1.0								
2. Word Processing (WP)	3.71	.97	.60**	1.0							
3. Graphic Programs	2.74	1.03	.51**	.52**	1.0						
4. Spreadsheet (SS)	2.57	1.03	.42**	.60**	.52**	1.0					
5. Web Design	1.32	1.53	.34**	.41**	.35**	.39**	1.0				
6. Database	1.05	1.19	.23**	.37**	.31**	.45**	.43**	1.0			
7. Overall Competence	38.1	17.0	.59**	.66**	.60**	.62**	.61**	.57**	1.0		
8. WP Test	7.79	2.51	.29**	.34**	.25**	.29**	.25**	.21**	.37**	1.0	
9. SS Test	8.45	2.12	.21**	.22**	.25**	.20**	.21**	.21**	.30**	n/a	1.0

** p <.01 Correlations for all competencies (except the two tests) are based on entire sample (n = 610). Correlations involving the two tests are based only on those that took that particular test.

tions and multiple regression analyses were used. Correlations were used to test whether each factor variable *independently* had a significant relationship with each dependent variable (a pair-wise comparison). This provides the same significance as simple regression. Next, multiple regression was used to test strength of all independent variables simultaneously, in order to determine which variables had the *strongest* relationship. Multiple regression was used instead of Structured Equation Modeling (SEM) in order to use the Cohen and Cohen (1983) formal t-test procedure to determine statistical strength among multiple independent variables. While standardized path strength (of SEM) allows some comparison, this procedure determines whether one path is statistically stronger (p <.05) than another. Nine separate models were created, because there were nine separate competencies (dependent variables), including six domain competencies, one overall computer competence measure, and two application tests in word processing and spreadsheets. The factors of interest depended on the dependent variable. Some were included in all nine models (overall previous experience, self-efficacy, affect, anxiety, and pessimism), because these independent variables were hypothesized to influence all competencies. For the six domain competencies and two tests, previous experience in that domain was also included (e.g., for spreadsheet competence and the spreadsheet performance test, only previous spreadsheet experience was included); but previous spreadsheet experience was not included in database competence (or any other outside competencies).

Results are summarized in Table 4. Correlations show pair-wise significance between individual antecedents and competence measures (column marked correlation or "Corr"). The table also provides multiple regression results that indicate which antecedents were strongest, using standardized betas with associated t-values that can

be directly compared (columns marked "β" and "t" are multiple regression results). Most of the models accounted for a considerable amount of the variation in the dependent variables, with R^2 values ranging from .45 (overall computer competence) to between .19 and .30 for the six sub-domain competencies. Only the two application tests had lower R^2 values (.12 for the WP test and .07 for the SS test). With only four exceptions, all factors individually had significant relationships with each dependent variable (see correlation column). Pessimism was not significant for WP and GP competencies or the SS test, and overall computer experience was not significant for the SS test. For six of the eight dependent variables, all factors were significant. This provides evidence of convergent validity for the constructs (except perhaps for pessimism), because the literature suggests such relationships.

When all applicable factors were included in the model using multiple regression, the number of significant variables differed, ranging from one to five. Self-efficacy was the only factor significant in every model. Prior domain experience was significant in all models except for the two performance tests. Affect and anxiety were significant in most of the nine models (affect was not significant for graphic programs and databases; anxiety not significant for spreadsheets and web design; neither were significant for the performance tests). Overall computer experience was significant in five of the models, while pessimism was not significant in any model. These results provide support for hypotheses 1-3 and 5.

While standardized βs provide some measure of comparison, we used the formal t-test procedure for multiple regression models from Cohen and Cohen (1983) to determine which antecedents were statistically strongest. The results are indicated by numbers in parenthesis in the table next to the domain name. Domains with the same number mean that there was *no statistical differ-*

Table 4. Correlations and regression results for variables

WP Competence (R² =.290)			SS Competence (R² =.300)			GP Competence (R² =.210)		
		Multiple Reg.			Multiple Reg.			Multiple Reg.
	Corr.	β	t		Corr.	B	t		Corr.	β	t
SE (1)	.46**	.22	6.63**	SS Exp (1)	.43**	.28	5.88**	SE (1)	.36**	.23	5.09**
Affect (1)	.43**	.17	3.64**	SE (1)	.40**	.23	5.36**	GP Exp (1)	.35**	.23	4.60**
Anxiety (3)	-.36**	-.11	-2.49*	Affect (3)	.37**	.11	2.41*	Anxiety (3)	-.30**	-.13	-2.92**
Ovl. Exp (3)	.32**	.11	2.26*	Ovl. Exp (3)	.40**	.10	2.12*	Ovl. Exp	.30**	ns	ns
WP Exp (3)	.25**	.10	2.12*	Anxiety	-.28**	ns	ns	Affect	.29**	ns	ns
Pessimism	-.13	ns	ns	Pessimism	-.14**	ns	ns	Pessimism	-.09	ns	ns

DB Competence (R² =.245)			Web Competence (R² =.242)			Email Competence (R² =.187)		
		Multiple Reg.			Multiple Reg.			Multiple Reg.
	Corr.	β	t		Corr.	B	t		Corr.	β	t
DB Exp (1)	.41**	.34	7.66**	Web Exp (1)	.35**	.22	5.01**	SE (1)	.38**	.25	5.34**
SE (2)	.32**	.19	4.16**	Affect (1)	.37**	.19	3.97**	Email Exp (2)	.19**	.14	3.24**
Anxiety (3)	-.25**	-.09	-2.05*	SE (3)	.32**	.14	3.18**	Anxiety (2)	-.32**	-.10	-2.26*
Ovl. Exp	.34**	ns	ns	Ovl. Exp (3)	.38**	.14	2.96**	Affect (2)	.34**	.10	2.02*
Affect	.28**	ns	ns	Anxiety	-.24**	ns	ns	Ovl. Exp	.19**	ns	ns
Pessimism	-.11**	ns	ns	Pessimism	-.09*	ns	ns	Pessimism	-.18**	ns	ns

WP Test (R² =.122)			SS Test (R² =.067)			Overall Competence (R² =.449)		
		Multiple Reg.			Multiple Reg.			Multiple Reg.
	Corr.	β	t		Corr.	B	t		Corr.	β	t
Ovl Exp. (1)	.27**	.30	2.44*	SE (1)	.28**	.20	2.17*	Ovl. Exp (1)	.47**	.30	9.27**
SE (2)	.32**	.20	2.35*	Affect	.22**	ns	ns	SE (1)	.54**	.29	7.72**
Affect	.27**	ns	ns	Anxiety	-.20**	ns	ns	Affect (2)	.53**	.20	4.94**
Anxiety	-.23**	ns	ns	SS Exp.	.14*	ns	ns	Anxiety (3)	-.43**	-.13	-3.35**
WP Exp.	.14*	ns	ns	Pessimism	-.10	ns	ns	Pessimism	-.17**	ns	ns
Pessimism	-.09*	ns	ns	Ovl. Exp.	.10	ns	ns				

** p <.01 * p <.05 Corr. are pair-wise Pearson Correlations. n = 610 for all competencies. SE = self-efficacy.
Numbers in parenthesis indicate statistical differences in antecedents for multiple regression models

ence in strength between the two domains in its relationship with competence. In web competence, for example, there was no statistical difference in the strength of the top two antecedents, prior web experience and affect, or in the strength of the next two factors, self-efficacy and overall experience. But prior web experience and affect were statistically stronger then SE and overall experience.

DISCUSSION

The goals of this study were to examine the competence construct and its relationship with attitudes. We offered three suggestions to aid in purifying the competence construct: the population to which competence is ascribed must be established, the strengths and weaknesses of the methodology used measure competence must be understood and used appropriately, and the specific domain(s) or applications must be clearly delineated.

Concerning the relationship between attitudes and competence, results indicate that while most of the independent variables (self-efficacy, prior experience, and anxiety, affect and pessimism) significantly influenced all nine competence measures, there were differences in strength which depended in part on the application domain. This section discusses important findings.

First, almost all antecedents independently had a significant relationship with each form of competence. Except for pessimism (hypothesis 4), this provided support for all hypotheses. In support of the literature, one's previous experience, attitudes (anxiety and affect), and self-efficacy all influence one's competence in the seven computing domains of word processing, email, graphic programs, spreadsheets, web design, databases, and overall computing competence. The only exception was pessimism, not significant in three of the domains.

After controlling for self-efficacy and previous experience, results suggest that attitudes play a key factor in competence for most domains. Affect, or computer liking, had a significant independent relationship with all nine competency measures and was significant in four domains in the multiple regression models (word processing, spreadsheets, email, and overall competence). Anxiety was similar; it had a significant independent relationship with all nine competency measures and was significant in five of the domains when tested with all antecedents (word processing, graphic programs, databases, email, and overall competence). These findings suggest that how an individual feels about the computing domain, measured by positive affect and anxiety, are significant and important influences on competence. If a person is anxious, it has a negative influence on competence; similarly, how much a person likes computing influences competence.

The implication is clear: *organizations should emphasize enhancing attitudes during and outside of actual training.* An individual's computer anxiety level and affect (computer liking) influences competence in application domains in several ways, such as training perseverance, more enjoyment of the computing learning process (and therefore higher skills attained), and a willingness to succeed (Marakas et al., 1998). Organizations should foster these attitudes, first through actual skills training (formal training), in any or all important organizational applications (e.g., office software, accounting packages, ERP software, etc.). Consideration should be given to short and frequent technical training, refresher courses, and recognized certifications. Skills training enhance both competence and attitudes; in fact, the resulting improved attitudes provide an improved foundation for the next round of training or application use. Secondly, organizations should enhance attitudes in ways *other than skills training.* Anxiety can be dispelled and liking increased by non-skill-based computer education, including computer use in non-application environments. Another avenue is through vicarious experience, that is, observing others using the application(s), which can enhance attitudes (Bandura, 1997; Compeau & Higgins, 1995a). Other possibilities for improving attitudes may include Internet use and playing computer games (particularly for novices). An organization that actively promotes more positive attitudes, instead of waiting for it to occur through formal training, should see increased competence and subsequent performance in computer end users.

Not all attitudes are important, however. Pessimism, or one's feeling of gloom (concerning computers) or its dehumanization effect, was not significant in any of the multiple regression models. When examined pair-wise with each competence (using correlations), it was usually significantly related to competence. However, even these significant relationships were weak. Pessimism towards computing was only a weak predictor at best of computing competence. Including training that dispels feelings of gloom toward computing does not appear warranted.

The antecedents of self-efficacy and prior experience were included as controls, because the literature strongly supports their relationships with competence (Compeau & Higgins, 1995a; Marakas et al., 1998; Munro et al., 1997). This study was no exception. When all antecedents were tested simultaneously using multiple regression, self-efficacy and previous experience in a domain usually had the strongest relationship with that domain's competency. Self-efficacy was the strongest antecedent (standardized β) for competency in three applications (word processing, graphic programs, and email) and the spreadsheet test. Previous experience was the strongest for the other three applications (spreadsheets, databases, and web design). Overall previous experience (the sum of previous experience in the six application domains) was the strongest predictor for the word processing test and overall computing competence, as well as significant for three of the application domains (word processing, spreadsheets, and web design). How an individual perceives their ability in a domain (i.e., their self-efficacy), remains a strong predictor of competence. That previous experience in the domain is also a strong predictor is also not surprising.

One of the important findings was the way the strengths of antecedents changed as domain competence increased. For the domains where respondents had the most competence (email and word processing), self-efficacy was significantly the strongest predictor, using the t-test procedure of Cohen and Cohen (1983) (though the difference between self-efficacy and affect, for word processing, was not statistically significant). For the domains with the lowest competencies (database and web design), previous experience was strongest, though again, the difference between previous experience and affect was not statistically significant for web design. In the domains of graphic programs and spreadsheet applications, where respondents had moderate competency, the predictive power of self-efficacy and previous experience was similar (and not significantly different). This suggests that as mastery occurs, the influence of previous experience diminishes

and self-efficacy becomes the stronger predictor. This finding does not support the contention of some cognitive theorists (e.g., Eden & Kinnar, 1991), who suggest that self-efficacy (in particular, general self-efficacy, as used in this study) loses predictive power as mastery occurs. As competence in a domain increases, it makes intuitive sense that the amount of previous experience should have less of an impact on that competence. One proficient in word processing, for example, rarely experiences new application features that would lead to greater competence. Likewise, a person's high self-efficacy (in particular general self-efficacy which is perceived ability for the entire computing domain), has been more significantly influenced by those applications in which there is high competence. Further research is required to further investigate the dynamics of this relationship.

The two application tests in word processing and spreadsheets present a puzzling interpretation problem. Although almost all antecedents were significant when analysed independently of other variables; when they were combined in a multiple regression model, the amount of variance explained was low (especially for the spreadsheet test) and the lack of previous experience in the domain as a significant predictor was unexpected. For word processing, the only two significant antecedents were overall previous experience and self-efficacy. For spreadsheets, only self-efficacy was significant. Clearly self-efficacy was an important antecedent for these tests, but not much else. There are a number of possible explanations for the lack of other significant antecedents. The measure of competence was different than the self-reported competence measures used elsewhere. This may suggest that the tests did not accurately measure an individual's knowledge in the domain. Each test had fourteen items, based on an intermediate level of knowledge (Microsoft, 2006). But other items could have been included. It is apparent that there are other factors important in how an individual performs on such a test, factors which were not captured. One possibility is survey fatigue; the optional test was placed at the end of this relatively long survey. Another possibility is that the tests actually measured performance, rather than competence. This could explain the lack of significant attitudes, as they may indirectly influence performance (through competence). An individual with a positive attitude and low anxiety towards computing may have high word processing or spreadsheet competence, but actual performance (in word processing or spreadsheets) is a function of more than just competence.

Limitations

As with any cross-sectional instrument, common method bias and other related limitations arise. Although reliabilities were high, any time all measures are collected simultaneously validity may suffer. However this is routinely practiced in academic settings, particularly when the variables of interest must be self-reported (such as self-efficacy, which is a personal judgment of ability). Attempts were made to reduce the influence of these biases and limitations. For example, most instruments included multiple items for each measure to increase reliability and validity (Netemeyer, Bearden, & Sharma, 2003). The use of knowledge tests provides an objective outcome that adds validity to this study in which most other variables are self-reported. A related limitation is the inability to draw causal results; this study is based on cross-sectional data and therefore only correlational analyses may be legitimately performed. Although a theoretical case was presented that suggests causal (and reciprocal) relationships, it cannot be statistically concluded that the attitudes of anxiety and affect influenced competence.

Generalizing to a wide population must be approached with caution. This population is one in a Navy commissioning program and may be different from other populations at large or even a university student population. However there is some indication that this population is no different from many other student groups in terms of academic ability and goals (Carlson & Grabowki, 1992; Neiberg, 2000). There was also a gender difference in this study, as only 14% were female. That factor is somewhat mitigated by the total number of respondents; there were still almost 100 females surveyed. We tested to determine any gender differences among variables and found no differences in age, college class, college attended, major, self-efficacy, or any of the three attitudes. We did find that there was a significant gender difference in competence in spreadsheet (self-reported, not the test), database, and overall competence (in each case males reported higher competence), but not in any of the other six competencies. The amount of variance explained by gender in the three competencies was less than 1%, suggesting that the gender effect was minimal.

CONCLUSION AND DIRECTIONS FOR FURTHER RESEARCH

This study examined the construct of computer competence and significant antecedents to computing competence in seven different computing domains, six applications and a combined software/application domain (which we labeled overall competence).

In testing antecedents, results demonstrate that both prior experience and self-efficacy are significantly related to competence in all domains. But results also demonstrate that computing attitudes, in particular anxiety and positive affect, are also significantly related to competence, even after controlling for both prior experience and self-efficacy. Affect, in particular, was statistically as strong as the strongest predictor for the domains of word processing and web design. Computer pessimism, on the other hand, has a weak and sporadic relationship with competence. Further research is necessary in a number of areas. Theory suggests that the relationship between attitudes (and self-efficacy) and competence is reciprocal. For example, over time positive affect should enhance competency, which in turn should increase positive affect. A longitudinal study should examine and confirm this relationship. One study which examined the relationship between attitudes and computer use suggested a causality effect in that negative affect influenced use which in turn influenced positive affect (Brock & Sulsky, 1994). Could something similar be observed for computer competence? Another interesting issue is *how* attitudes affect competence. This study examined a direct effect, but attitudes should also indirectly affect competence through the learning process, effort expended in learning, etc. We have posited that attitudes may affect performance through competence. This clearly needs evaluation to confirm such a phenomenon. This study examined six applications and one summed measure, but clearly there are other domains to examine. In addition, we only looked at operational or basic competencies and did not consider strategic level competencies (leveraging operational competencies for sophisticated analysis and decision making purposes). Though strategic level competencies may be considered "performance" rather than competencies, this is still a fruitful area of study. This study also found differences in the significant antecedents depending on an individual's competence in the domain. For individuals with low competence in a domain, previous experience was the strongest antecedent. When the domain was mastered, self-efficacy was stronger. Further research is needed to investigate this phenomenon.

IT training is an expensive and time-consuming endeavor in most companies and determining which factors are most important in building competence is important. This study adds to this understanding by studying three attitudes as antecedents to competence. Two of these, anxiety and affect, had a significant relationship with competence for the six application domains in-cluded. This suggests that organizational training strategies should include efforts to allay anxiety and enhance computer liking. Individuals who are less anxious and have greater affect will likely be more competent and presumably will have higher subsequent performance. Computer competence is a critical prerequisite for successful organizations, particularly in today's high-technology marketplace. Competence impacts organizational readiness, ability to compete, and efficiency. Improved competence should be one of the main objectives for organizational training programs, and understanding this process is of great importance.

REFERENCES

Adams, J. A. (1987). Historical review and appraisal of research on the learning, retention, and transfer of human motor skills. *Psychological Bulletin*, *101*(1), 41–74. doi:10.1037/0033-2909.101.1.41

Ajzen, I. (1991). The theory of planned behavior. *Organizational Behavior and Human Decision Processes*, *50*, 179–211. doi:10.1016/0749-5978(91)90020-T

Al-Jabri, I. M., & Al-Khaldi, M. A. (1997). Effects of user characteristics on computer attitudes among undergraduate business students. *Journal of End User Computing*, *9*(2), 16–22.

Aladwani, A. M. (2002). Organizational actions, computer attitudes, and end-user satisfaction in public organizations: An empirical study. *Journal of Organizational and End User Computing*, *14*(1), 42–49. doi:10.4018/joeuc.2002010104

Arndt, S., Clevenger, J., & Meiskey, L. (1985). Students' attitudes toward computers. *Computers and the Social Sciences*, *14*(3-4), 181–190.

Bandura, A. (1997). *Self-efficacy: The Exercise of Control*. New York, NY: W.H. Freeman and Company.

Bostrom, R. P., Olfman, L., & Sein, M. K. (1990). The importance of learning style in end-user training. *Management Information Systems Quarterly*, *14*(1), 101–119. doi:10.2307/249313

Brock, D. B., & Sulsky, L. M. (1994). Attitudes toward computers: construct validation and relations to computer use. *Journal of Organizational Behavior*, *15*, 17–35. doi:10.1002/job.4030150104

Brody, N. (1992). *Intelligence*. San Diego, CA: Academic Press.

Burkhardt, M. E., & Brass, D. J. (1990). Changing patterns or patterns of change: the effects of a change in technology on social network structure and power. *Administrative Science Quarterly*, *35*(1), 104–127. doi:10.2307/2393552

Carlson, R., & Grabowski, B. (1992). The effects of computer self-efficacy on direction-following behavior in computer-assisted instruction. *Journal of Computer-Based Instruction*, *19*(1), 6–11.

Chan, Y. E., & Storey, V. C. (1996). The use of spreadsheet in organizations: determinants and consequences. *Information & Management*, *31*, 119–134. doi:10.1016/S0378-7206(96)00008-0

Chau, P. Y. (2001). Influence of computer attitude and self-efficacy on IT usage behavior. *Journal of Organizational and End User Computing*, *13*(1), 26–33. doi:10.4018/joeuc.2001010103

Cohen, J., & Cohen, P. (1983). *Applied Multiple Regression/Correlation Analysis for the Behavioral Sciences*. Hillsdale, NJ: Lawrence Erlbaum Associates.

Compeau, D. R., & Higgins, C. A. (1995a). Application of social cognitive theory to training for computer skills. *Information Systems Research*, *6*(2), 118–143. doi:10.1287/isre.6.2.118

Compeau, D. R., & Higgins, C. A. (1995b). Computer self-efficacy: development of a measure and initial test. *Management Information Systems Quarterly*, *19*(2), 189–202. doi:10.2307/249688

Downey, J. (2004). Toward a comprehensive framework: EUC research issues and trends (1990-2000). *Journal of Organizational and End User Computing*, *16*(4), 1–16. doi:10.4018/joeuc.2004100101

Eden, D., & Kinnar, J. (1991). Modeling Galatea: boosting self-efficacy to increase volunteering. *The Journal of Applied Psychology*, *76*(6), 770–780. doi:10.1037/0021-9010.76.6.770

Evans, G. E., & Simkin, M. G. (1989). What best predicts computer proficiency? *Communications of the ACM*, *32*(11), 1322–1327. doi:10.1145/68814.68817

Feiertag, H. (2001). Front office attitude training ensures business in tough times. *Hotel and Motel Management*, *216*(20), 22.

Fishbein, M., & Ajzen, I. (1975). *Belief, Attitude, Intention and Behavior: An Introduction to Theory and Research*. Reading, MA: Addison-Wesley.

Ford, J. K., & Noe, R. A. (1987). Self-assessed training needs: the effects of attitudes toward training, managerial level, and function. *Personnel Psychology*, *40*, 39–53. doi:10.1111/j.1744-6570.1987.tb02376.x

Gattiker, U. E. (1992). Computer skills acquisition: a review and future directions for research. *Journal of Management*, *18*(3), 547–574. doi:10.1177/014920639201800307

Gattiker, U. E., & Hlavka, A. (1992). Computer attitudes and learning performance: Issues for management education and training. *Journal of Organizational Behavior*, *13*(1), 89–101. doi:10.1002/job.4030130109

Guimaraes, T., & Igbaria, M. (1997). Assessing user computing effectiveness: An integrated model. *Journal of End User Computing, 9*(2), 3–14.

Harrison, A. W., & Rainer, R. K. (1992). The influence of individual differences on skill in end-user computing. *Journal of Management Information Systems, 9*(1), 93–111.

Hoxmeier, J. A., Nie, W., & Purvis, G. T. (2000). The impact of gender and experience on user confidence in electronic mail. *Journal of Organizational and End User Computing, 12*(4), 11–20. doi:10.4018/joeuc.2000100102

Invest in Developing Employee Attitude. (1998, August 14). *Business Line*, p. 1.

Jawahar, I. M., & Elango, B. (2001). The effect of attitudes, goal setting, and self-efficacy on end-user performance. *Journal of Organizational and End User Computing, 13*(2), 40–45. doi:10.4018/joeuc.2001040104

Johnson, R. D., & Marakas, G. M. (2000). Research report: the role of behavioral modeling and computer skills acquisition-toward refinement of the model. *Information Systems Research, 11*(4), 402–417. doi:10.1287/isre.11.4.402.11869

Kanfer, R., & Ackerman, P. L. (1989). Motivation and cognitive abilities: An integrative/aptitude-treatment interaction approach to skill acquisition. *The Journal of Applied Psychology, 74*(4), 657–690. doi:10.1037/0021-9010.74.4.657

Kernan, M., & Howard, G. S. (1990). Computer anxiety and computer attitudes: An investigation of construct and predictive validity issues. *Educational and Psychological Measurement, 50*, 681–690. doi:10.1177/0013164490503026

Lipscomb, T., Hotard, D., Shelley, K., & Baldwin, Y. (2002). Business students' attitudes toward statistics: A preliminary investigation. *Proceedings of the Academy of Educational Leadership, 7*(1), 47–50.

Loyd, B. H., & Gressard, C. (1984). The effects of sex, age, and computer experience on computer attitudes. *AEDS Journal*, 67–77.

Marakas, G. M., Yi, M. Y., & Johnson, R. D. (1998). The multilevel and multifaceted character of computer self-efficacy: Toward clarification of the construct and an integrative framework for research. *Information Systems Research, 9*(2), 126–163. doi:10.1287/isre.9.2.126

Marcolin, B. L., Compeau, D. R., Munro, M. C., & Huff, S. L. (2000). Assessing user competence: Conceptualization and measurement. *Information Systems Research, 11*(1), 37–60. doi:10.1287/isre.11.1.37.11782

Martocchio, J. J., & Webster, J. (1992). Effects of feedback and cognitive playfulness on performance in microcomputer software training. *Personnel Psychology, 45*, 553–578. doi:10.1111/j.1744-6570.1992.tb00860.x

McMurtrey, M., McGaughey, R., & Downey, J. (2008). Seniors and information technology: Are we shrinking the digital divide? *Journal of International Technology and Information Management, 17*(2), 121–136.

Microsoft. (2006). *Certifications*. Retrieved April 12, 2006, from http://office.microsoft.com/marketplace/default.aspx?

Munro, M. C., Huff, S. L., Marcolin, B. L., & Compeau, D. R. (1997). Understanding and measuring user competence. *Information & Management, 33*, 45–57. doi:10.1016/S0378-7206(97)00035-9

Murphy, C. A., Coover, D., & Owen, S. V. (1989). Development and validation of the computer self-efficacy scale. *Educational and Psychological Measurement, 49*(2), 893–899. doi:10.1177/001316448904900412

Neiberg, M. S. (2000). *Making Citizen Soldiers: ROTC*. Cambridge, MA: Harvard University Press.

Netemeyer, R. G., Bearden, W. O., & Sharma, S. (2003). *Scaling Procedures Issues and Applications*. Thousand Oaks, CA: Sage.

Nickell, G. S., & Pinto, J. N. (1986). The computer attitude scale. *Computers in Human Behavior*, *2*, 301–306. doi:10.1016/0747-5632(86)90010-5

Noe, R. A. (1986). Trainees' attributes and attitudes: neglected influences on training effectiveness. *Academy of Management Review*, *11*(4), 736–749.

Potosky, D., & Bobko, P. (2001). A model for predicting computer experience from attitudes toward computers. *Journal of Business and Psychology*, *15*(3), 391–403. doi:10.1023/A:1007866532318

References

Rockart, J. F., & Flannery, L. S. (1983). The management of end user computing. *Communications of the ACM*, *26*(10), 776–784. doi:10.1145/358413.358429

Swift, J., & Smith, A. (1992). Attitudes to language learning. *Journal of European Industrial Training*, *17*(7), 7–15.

Szajna, B., & Mackay, J. M. (1995). Predictors of learning performance in a computer-user training environment: A path analytic study. *International Journal of Human-Computer Interaction*, *7*(2), 167–185. doi:10.1080/10447319509526118

Szalma, J. L., Hancock, P. A., Dember, W. N., & Warm, J. S. (2006). Training for vigilance: The effect of knowledge of results format and dispositional optimism and pessimism on performance and stress. *The British Journal of Psychology*, *97*, 115–135. doi:10.1348/000712605X62768

Thatcher, J. B., & Perrewe, P. L. (2002). An empirical examination of individual traits as antecedents to computer anxiety and computer self-efficacy. *Management Information Systems Quarterly*, *26*(4), 381–396. doi:10.2307/4132314

Webster, J., & Martocchio, J. J. (1993). Microcomputer playfulness: Development of a measure with workplace implications. *Management Information Systems Quarterly*, *16*(2), 201–226. doi:10.2307/249576

Winston, E. R., & Dologite, D. (2002). How does attitude impact IT implementation: A study of small business owners. *Journal of Organizational and End User Computing*, *14*(2), 16–29. doi:10.4018/joeuc.2002040102

Winter, S. J., Chudoba, K. M., & Gutek, B. A. (1997). Misplaced resources? Factors associated with computer literacy among end-users. *Information & Management*, *32*, 29–42. doi:10.1016/S0378-7206(96)01086-5

Woodrow, J. E. (1991). A comparison of four computer attitude scales. *Journal of Educational Computing Research*, *7*(2), 165–187. doi:10.2190/WLAM-P42V-12A3-4LLQ

ENDNOTE

[1] This study conducted post-hoc testing of demographics and found few that were significant (three age related, three gender related); those that were significant were weak, accounting for less than 1% of the variance in the respective competence.

APPENDIX

Table A1. Factor analysis

Simultaneous Factor Analysis of Anxiety, Affect, and Pessimism Scales	1	2	3
Anx1: Computers do not scare me at all	.74		
Anx2: Working with a computer makes me nervous	.77		
Anx3: I do not feel threatened when others talk about computers	.73		
Anx4: Computers make me feel uncomfortable	.85		
Anx5: I would feel at ease in a computer class	.70	-.33	
Anx6: I get a sinking feeling when I think of trying to use a computer	.79		
Anx7: I would feel comfortable working with a computer	.81		
Anx8: Computers make me feel uneasy and confused	.80		
Lik1: The challenge of solving problems with computers does not appeal to me			.71
Lik2: Working with computers is enjoyable and stimulating			.72
Lik3: Figuring out computer problems does not appeal to me	-.32		.80
Lik4: When I can't solve a computer problem, I stick with it until I have the answer			.83
Lik5: I don't understand how some people can spend so much time working with computers and seem to enjoy it			.74
Lik6: Once I start to work with the computer, I find it hard to stop			.58
Pess1: Computers turn people into just another number		.73	
Pess2: Computers are lessening the importance of too many job done by humans		.76	
Pess3: People are becoming slaves to computers		.83	
Pess4: Computers are becoming dehumanizing to society		.65	
Pess5: The overuse of computers may be harmful and damaging to humans		.74	
Pess6: Soon our world will be completely run by computers		.74	

Only values greater than .30 presented. Some items are reverse scored.

Chapter 15
Development of a Mesh Generation Code with a Graphical Front-End:
A Case Study

Jeffrey Carver
University of Alabama, USA

ABSTRACT

Scientists and engineers are increasingly developing software to enable them to do their work. A number of characteristics differentiate the software development environment in which a scientist or engineer works from the development environment in which a more traditional business/IT software developer works. This paper describes a case study, specifically about the development of a mesh-generation code. The goal of this case study was to understand the process for developing the code and identify some lessons learned that can be of use to other similar teams. Specifically, the paper reports on lessons learned concerning: requirements evolution, programming language choice, methods of communication among teammates, and code structure.

INTRODUCTION AND BACKGROUND

The development of software applications for use in Computational Science and Engineering (CSE) is an important and growing segment of the software development landscape. The development of CSE software differs from the development of the typical types of software that software engineering researchers study (Carver, Kendall, Squires, & Post, 2007; Carver, 2009; Kelly, 2007; Segal, 2008; Wilson, 2006). These differences mean that

DOI: 10.4018/978-1-4666-2059-9.ch015

many existing software engineering practices are not directly applicable within the CSE domain. Some of these characteristics are:

- The requirements discovery and gathering process is often difficult due to the large number of unknowns at the beginning of a project;
- Achieving the correct scientific or engineering output is more important to the developers than using correct or appropriate software engineering techniques; and
- Even when they are aware of traditional software engineering processes, which is not common, developers tend to be averse to using those processes.

Most software engineering research has been conducted in a business-oriented environment rather than a research-oriented environment. It is not clear whether the results of this research and the lessons learned about the usefulness of various techniques will translate to the CSE environment. To better understand the impact of the CSE environment and to document successful CSE software development practices, the author, along with other colleagues, as part of the DARPA High Productivity Computing Systems (HPCS) project (http://www.highproductivity.org), have conducted a series of case studies of successful CSE projects. To protect the anonymity of the projects, we assign each one a bird pseudonym. This paper describes the seventh case study in the series after Falcon (Post, Kendall, & Whitney, 2005), Hawk (Kendall et al., 2005), Condor (Kendall, Mark, Post, Squires, & Halverson, 2005), Eagle (Kendall, Post, Squires, & Carver, 2006), Nene (Kendall, Post, & Mark, January, 2007), and Osprey (Kendall et al., 2008) which will be referred to as the HARRIER project.

The common objectives for all of the case studies in the series are to (Carver et al., 2007):

- provide feedback to software vendors about issues that should be addressed to improve productivity;
- develop a body of case studies of CSE projects that can serve as a guide to the community; and
- document lessons learned about CSE software development to provide a resource for other similar teams.

The case study approach, as defined by Yin, is a well-understood method for gathering this type of information (Yin, 2002). The case studies, including the one reported in this paper, all followed a similar format that has proved successful. To provide context for this paper, a brief summary of the methodology is provided. The complete methodology is described in a previous publication (Carver et al., 2007). After identifying a suitable project, members of the code team complete a survey to provide background information about the project. Using this information, the researcher develops a set of questions that will be used to explore interesting topics further during an on-site interview. The researcher conducts the on-site interview with the team to gather more detailed information about the software development processes. Finally, using information from the survey and interview, the researcher writes a report that is circulated with the development team to ensure its accuracy.

Specifically for HARRIER, the project was identified through collaborations of the author. After identifying the project, the project manager completed the background survey. Based on the responses, the on-site interview guide was developed to elicit more details about some important topics. Table 1 lists the topics covered by the pre-interview survey and the on-site interview. The on-site interview lasted approximately two hours. The interviewees included: The project manager, all members of the core development team, and

Table 1. Topics covered by pre-interview survey and on-site interview

Pre-Interview Survey	On-site Interview
Details about the goals, requirements and deliverables	How knowledge was captured and shared
Project characteristics, structure, organization and risks	Use of programming languages
Code characteristics and structure	Project lifecycle
Staffing	Validation and Verification
Project workflow	Success Measures
Verification, Validation and Testing	
Success measures	
Lessons Learned	

two other developers who were tangentially involved in the project.

The remainder of this paper is organized as follows. First, I describe the context of the HARRIER project and the characteristics of the code. Next, I describe the development team and environment. The next section discusses the software lifecyle of the HARRIER project. Next, I discuss some insight into the types of defects that must be addressed during development. Then, I describe lessons learned from the practices employed by the HARRIER team that helped them to be successful. The final section provides a summary of the paper.

CODE CHARACTERISTICS

HARRIER's main goal is to assist computational scientists in the task of generating and manipulating geometric meshes that will serve as a basis for computational simulations. As such, it has two main goals:

1. Provide a high-quality implementation of standard mesh generation algorithms.
2. Provide users with a graphical front-end to ease their task of interacting with and generating meshes.

Mesh Generation

As Post (2009) describes computational engineers are relying more often on computational simulations to speed-up their engineering processes. Computational simulation of product designs is significantly cheaper and quicker than building a physical prototype of each design. The basic computational engineering process is shown in Figure 1. Engineers must first develop a mathematical model or geometry for their simulation. The model is then converted into a mesh (the focus of HARRIER). This mesh serves as the basis for a simulation. Finally some post-processing of the simulation results is performed.

The ability to adequately generate a mesh that accurately conforms to the underlying geometry is an important step in obtaining a valid and useful simulation result. *Mesh generation* (Shih, Yu, & Soni, 1994; Soni, Thompson, Stokes, & Shih, 1992; Thompson, Warsi, & Mastin, 1985) is a research discipline that studies the mathematical methods that generate coordinate values of discrete points in a continuous space. It is closely associated with mesh-based computational technologies, such as Computational Fluid Dynamics (CFD) and Computational Structural Mechanics (CSM). The distribution and arrangement of these points affect the "quality" of the mesh, which is measured geometrically in many ways, such as aspect ratio

Figure 1. Computational engineering process

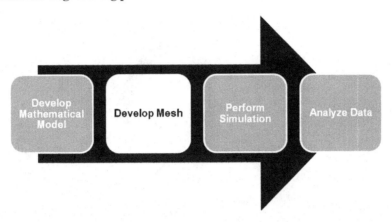

and skew angles. It can be very challenging to produce high quality meshes for complex real-world applications, such as highly detailed aircraft with control surfaces or car engine blocks. The mesh-generation challenges arise because the underlying geometry may not be complete, i.e., it may have holes or missing elements thus requiring a lot of user manipulations to remove defects. Or, the underlying geometry can be so complex that the distribution of the discrete surface points is not correct in some regions, preventing the volume elements from being properly generated. These challenges require a user to have extensive experience and knowledge about the application to be able to generate a quality mesh. There is a heavy burden on the mesh generation tools to provide sufficient features to enable the user to perform the mesh generation process. These tools must be robust, interactive and general. However, the more general the tool is, the more complicated it is to learn and use. Therefore, it is very difficult for mesh generation software developers to strike the right balance in the software design. Existing mesh generation codes from other developers had many limitations in terms of features allowed and grid size. They also could not do many of the required complex calculations properly. For example, some existing mesh generation codes failed to generate proper meshes for certain types of geometries. The codes are not defective.

They just do not handle some special cases. HARRIER has some unique features to allow it to handle these challenging cases. By addressing these concerns, HARRIER tackles a vital portion of the mesh generation process. As an example, Figure 2 shows a notional wind turbine mesh. A wind turbine is a device that generates energy from moving air, like a windmill. The mesh in Figure 2 is a generic example provided for illustration and does not represent a real wind turbine.

HARRIER's Origin and Evolution

HARRIER began in 2002 and had its origins in a geometry and mesh generation code which had been in existence since the 1980s. The mathematics of the core algorithms, such as how to generate a Spline Surface, upon which the code is based are well-known. There are various versions of the algorithms in existence. Depending on the data structures used and the particular implementation choices made, these versions provide different efficiency and quality for different types of applications. Based on requests from outside groups for their mesh generation code, the HARRIER team believes that they have a good implementation of the algorithms.

Like most CSE codes, the goal of HARRIER is to reduce the *time-to-solution* for the end user. Time-to-solution is a measure of how long it

Figure 2. Wind turbine mesh example

takes an end-user to obtain a quality mesh for the geometry of interest. This measure includes both the time the user spends generating the proper input, e.g., geometric modes from CAD software or input from 3D scanners, and the time the computer spends executing algorithms to process that input. But, unlike many CSE codes, which find their bottleneck in the computer processing time, the major time-to-solution bottleneck in the development of the geometry and mesh is the time spent by the end-users developing the mesh. The process of developing a geometry and associated mesh is not straightforward and simple. To generate a suitable mesh, the user must perform numerous tedious interactions to prepare the correct geometry and topology. In some cases, a user may spend six months or more developing the required geometry and mesh (much more time than it takes the mesh generation algorithm to actually generate the computational mesh!). By providing an interactive, graphical front-end, the developers of HARRIER seek to reduce the amount of time the end-user must invest in creating meshes and significantly reduce the overall time-to-solution.

To improve the user experience with HARRIER and reduce the amount of time required to obtain a mesh, the team has spent a large amount of effort since 2002 refactoring the mesh generation code from FORTRAN to C++, updating the data structures, removing redundancies in the code and redesigning the front-end GUI. After porting the code to C++, much of the work has focused on improving the mesh generation algorithms and updating the GUI. Ultimately, the success of HARRIER is largely dependent upon robustness and user friendliness. Robustness can be judged by measuring the mathematical quality of the meshes and the frequency of successfully producing a quality mesh from beginning to end without any code modifications. User friendliness is much more vaguely defined. The time required to generate mesh is heavily dependent upon user experience and knowledge.

Current State of HARRIER

HARRIER's user base is made up of two or three local students and many external users worldwide. The team views the student users as beta testers. Unlike some of the previous CSE codes studied, HARRIER's users do not contribute any code to the code base. There is a scripting facility built into the system, but most of the users do not take advantage of it. One limitation on the size of the user base is the level of robustness of the code. As robustness is one of the main success criteria for HARRIER, the development team is not yet ready for wide dissemination during its research phase. They have decided that it is better to wait until they have completed the necessary development of functionality and improved the robustness of the code rather than foul the water with potential users before the code is really ready. The development team is planning to develop a tutorial about HARRIER to recruit additional users in the near future.

The two main parts of HARRIER, the graphical front end and the mesh generation algorithm, total approximately 50K SLOC. The graphical front end is written mostly in Python (wxPython), while the underlying mesh generation algorithms are written in C++. HARRIER began as a small GUI code written in C++ that sat on top of an existing mesh generation code written in FORTRAN by engineers. The original code was written in FORTRAN because it was efficient and widely used in the engineering community, but it was difficult to maintain. The HARRIER team has spent a lot of effort to reimplement the mesh generation code in C++ to facilitate portability, compatibility and maintainability. In addition, the original GUI code was rewritten in Python for two reasons: 1) to allow for more user control and scripting and 2) to allow the GUI code to be maintained more easily and more quickly compared with C++. Currently HARRIER does not contain any parallelism. The software is labor-intensive rather than computationally intensive initially (to develop the surface mesh). After developing the surface mesh, it becomes computationally intensive for volume mesh generation. The team has considered parallelizing some of the computationally intensive code, but has yet to do so.

CODE TEAM AND ENVIRONMENT

HARRIER's core team consists of three Mechanical Engineering (ME) professors, with the most senior one acting as the project manager/director. In addition, there are two primary developers, and several students. The two primary developers are computer scientists, although one does have a B.S. in ME and a M.S. in computer science. For the most part, the developers work independently. Each developer has his own niche and respects the niche areas of others. The students perform minor, peripheral tasks not directly related to the core of the code base. They primarily develop GUI code using a version control mechanism, which is tightly controlled by the primary developers. This arrangement prevents developers from accidentally "stepping on each other's toes" and overwriting earlier changes to the code. The team composition has turned out to be quite successful, although there was no advance planning that went into the team composition. The project manager attributes the successful team composition to the professionalism and the interpersonal harmony of the team.

The preferred development environment for HARRIER is Emacs. Some specific reasons why the developers prefer the Emacs environment are: 1) they are familiar with it, 2) it does what they want it to, and 3) they have spent a lot of time customizing it for their own working styles and needs. The developers have considered using Integrated Development Environments (IDEs) such as Eclipse or Visual Studio, but could find no compelling reason to switch from Emacs. One developer tried Visual Studio and found that it slowed down his work process. Because each developer has customized Emacs to do everything he needs it to, there is little reason to move to a more complex IDE. The developers do admit that there may be some potential benefits gained from using an IDE such as Eclipse, but they do not want to invest the extra time required to become proficient in its use.

Because the development team is relatively small, they have developed a communication framework that has worked to successfully keep everyone updated and prevent them from stepping on each other's work. There is a lot of daily communication among the primary developers. For everyday questions, the developers prefer to walk down the hall and communicate face-to-face. Email is used to keep the team updated on what each person is working on and to ensure that work is not overlapping. In addition, any commits to the code repository generate an email to the whole development team. The team also has a weekly group meeting, primarily for the project manager to understand and identify high-level problems.

CODE LIFECYCLE

Most of the choices related to the development approach for HARRIER are the result of working in a context with constantly changing requirements, as is true for many CSE projects. Because the application domain and needs of the end-users are constantly evolving, HARRIER's requirements are ever changing. The typical requirements engineering process is as follows:

1. The HARRIER team meets quarterly with the project sponsors, from a government agency.
2. During this meeting, the team receives some high-level requirements based on the sponsor's current goals.
3. The HARRIER team then begins to implement those requirements.
4. Meanwhile, the goals of the sponsor shift.
5. By the next quarterly meeting, some of requirements from the previous meeting are no longer relevant.
6. The sponsor gives the HARRIER team some new requirements.

The constantly evolving requirements affect four aspects of HARRIER's development process: the choice of software lifecycle, the testing process, the release schedule and documentation of the code. Each of these issues is discussed in more detail as follows.

First, to address the constant requirements evolution, the HARRIER team has employed various agile software development practices. Specifically, they have found the rapid prototyping approach to work well within their context. Rapid prototyping allows the team to quickly produce software that can serve as a basis for discussion with the sponsors. It also helps the team get a better idea of whether a particular function can work as planned before investing a lot of effort into its development. Table A1 in the Appendix summarizes the extent to which the HARRIER team uses various agile (http://www.agilealliance.com) (Cockburn, 2002) and plan-driven development practices (Paulk, Weber, Curtis, & Chrissis, 1995). As suggested by Boehm and Turner, these practices should be balanced based on the specific context of the application (Boehm & Turner, 2004). According to the Appendix, HARRIER makes use, at least partially, of a number of agile practices and few plan-driven practices. These practices are discussed in relevant sections of the paper. The typical situation is that when the sponsor requests a new feature, they do not yet know all of the requirements. The development team is relatively small, so, there is no official 'design team' or formal design process. Design is done as the project progresses. The developers know the general direction of the project, but they have to address the changing requirements. It is not a straight path! As a result, there is no overall design for HARRIER. When a new feature is required, the developers discuss the feature and agree upon the best implementation approach (partially following the Feature-Driven Development practice). One of the HARRIER developers does create his own formal design documents and use cases for the functionality he is implementing. These design documents are stored on his local machine rather than in the project repository. He has decided that it is often easier to refer to the UML than to read the code. His basic approach is that if the implementation is straightforward, he proceeds to develop code. If it is complicated, then he does some design work to get an idea of how to implement before beginning to write code. The typical development process is to begin implementation of a new feature, then test and refactor the code. The main advantage of this approach is that it produces code quickly. The main disadvantage is that it has the potential to introduce defects.

The second issue for HARRIER is testing. Because it is a relatively small project, HARRIER does not have a dedicated test team. In addition, because the requirements are constantly changing it is difficult to create thorough test plans or conduct formalized, systematic regression testing. The correctness of the underlying algorithms can be systematically tested. But, because there are several ways to achieve the same goal, the user interaction is more difficult to test. Furthermore, because the sponsor does not require testing results, the team has trouble justifying the investment of a large amount of resources for testing. As a result, the team has to be efficient with the small amount of effort that can be invested in testing, especially by the expert users. The team has determined that emphasizing testing of the correctness of the core algorithms is a better use of effort than testing of the usability of the code. In addition, the team has not been able to locate good tools that are appropriate to their context to aid in usability testing and subsequently reduce the required effort. The HARRIER team has developed a testing process that works successfully under these constraints. HARRIER is primarily tested in two ways. First, mechanical engineering and computer science undergraduate and graduate students work on specific assigned problems, i.e., developing a particular geometry. As they work on these assigned problems, the students test various aspects of the code. The goal of this type of testing is to evaluate both the correctness and the usability of the code. Because they have low experience in the domain, the testing performed by the students usually focuses on the usability issues in the interface, e.g., the placement of GUI widgets, rather than focusing on the more complicated algorithms in the code. This student testing also helps to evaluate how useful HARRIER is for general cases. The testing performed by the students is the primary evaluation of the usability of the code. In addition, because each student is working only on a small isolated problem, there is no full benchmark test of the code. On the positive side, because each student is continually working on the same problem, they provide a kind of informal regression testing as new features are integrated into HARRIER. The second type of testing is conducted by the end-users of the code. Again, this testing is not systematic. It is similar to beta-testing as the end-users execute the code to do their work. As opposed to the user interface problems found by the students, these end-users tend to find deeper problems with the core algorithms.

The third important development issue related to changing requirements is the release schedule. While the HARRIER team realizes the need for a specified release schedule, they find it difficult to create one due to the ever evolving requirements and the lack of a formalized project schedule. A release is typically made when the development team examines the implemented features and decides that they have enough to make a release. Because of this informal, *ad hoc* process, when one developer is ready to release a feature, other developers are often in the middle of developing their feature. Developers seldom complete development of their features at the same time. This situation makes it difficult to find a good time to release a stable product. Therefore, the team often releases incomplete features. Typically there are two official releases per year with a number of unofficial, incremental releases occurring in between.

Table 2. Harrier code and team characteristics

Application Domain	Mesh Generation for Engineering Applications
Duration	~8 years
# Releases	~2/year
Staffing	5 primary + students
Customers	10s
Code Size	50,000 LOC
Primary Language	C++ (50%); Python (50%)
Target Hardware	Linux, Windows

The final issue caused by changing requirements is how to best create project documentation. Due to the heavy workload and the lack of concrete, stable requirements from the sponsor, the team finds it difficult to generate detailed project documentation. HARRIER has avoided the communication problems that the lack of documentation typically causes in projects by having different developers become experts on portions of the code and, in a sense, "owning" that code. This arrangement prevents a developer who lacks detailed knowledge from making a change that could break the code. Previous attempts to create detailed documentation have struggled because of limited resources and lack of time allocated in the project schedule. In addition, because of the rapid requirements evolution, the code changes radically in six months. The team cannot maintain documentation that frequently, so they view documentation as useless. One of the team members said that it was "better to have no documentation than outdated documentation." The development team does realize the drawbacks that result from the lack of formal documentation. The most glaring problem they see is the amount of time required for someone new to adequately understand the code to be able to contribute to the project. Typically a new developer is given a specific task to work on. That developer spends time with a more experienced developer to learn the code. The knowledge transfer process happens tacitly from developer to developer rather than explicitly through formal documentation. Table 2 summarizes the characteristics of HARRIER using the same list of characteristics as in the previous case studies.

DEFECTS

It is important to analyze the types of defects that occur in a software system in order to improve the software development process. When a development team understands the types of defects that occur, they can take active steps to fix those defects and to prevent similar defects from occurring in the future. To provide this type of insight, the HARRIER team discussed their defect identification and tracking process.

An overall observation is that the two main testing groups, students and end-users, tend to find different types of defects. The student testers tend to identify user interface defects. These defects are difficult to fix because students often do not provide enough detail in their defect reports, including the steps required to reproduce the defect. Because these defects often relate to the user's workflow through the system and there is no set procedure that everyone has to follow when generating a mesh, the specific order of commands may vary from user to user. If the sequence of steps followed is not clearly documented, then it is difficult to reproduce the fault. Conversely, the engineer end-users tend to report more complex defects with the underlying algorithms. One issue with the defects reported by the engineers is that what they report are not always defects. They may identify some aspect of the system that is confusing or may identify a new user workflow that the developers had not previously considered. These situations are generally not considered to be defects.

Regarding the effort required to repair the defects, the end-users' defects tend to be easier to fix than the students' defects. The developers believe that if the students reported defects with the same level of detail as the end-users do, then the difficulty of fixing those defects would be greatly reduced. Furthermore, the length of time to identify a problem did not necessarily correspond to the difficulty of fixing the problem. Algorithmic defects tend to take a long time to identify, but are easy to fix. Conversely, a problem with the user interface, which may be simple to find, could require a restructuring of the code.

The HARRIER team uses *Bugzilla* to record their defects. Information about defect repairs, features and enhancements is not tracked in Bugzilla. Typically, a student who is a good programmer examines the defects recorded in Bugzilla and fixes those that he can. Then the members of the core development team examine the defects and assign each one to the developer most knowledgeable about that portion of the code. Interface defects are assigned to graduate students. To determine which defect fixes should be included in each release, the developers examine the list of defects and decide what needs to be done. This process is *ad hoc* rather than formalized. The developers do not have time to examine the hundreds of reported defects in detail. Each individual developer looks through his list of assigned defects, finds something broken and fixes it. If developing a new feature is deemed to be more important, then the defects are not repaired until that feature is completed.

LESSONS LEARNED

Based on the discussion in the previous section, the HARRIER team has made choices in four areas to successfully deal with the constraints imposed by their context. These lessons learned should be instructive to other CSE projects in similar contexts.

Requirements Evolution

The HARRIER team used two practices to address the challenge of constantly changing requirements. First, as described in an earlier section, the team uses rapid prototyping, an agile software development lifecycle, to enable them to adjust to the sponsors' changing needs. Second, the team explicitly addressed the requirements evolution problem with the sponsor and as a result developed a workable solution. At each quarterly meeting, the sponsor now provides the development team with detailed requirements for the current quarter along with the general direction of the work for future quarters. This additional information helps the development team better plan for and work on future requirements. In addition, many of the sponsor's requirements are not unique to that sponsor. These requirements improve the quality and functionality of the user interface and mesh generation code. Even when the sponsor's goals do change, the effort spent developing this code is not wasted effort. The HARRIER team still views this newly developed code as being useful for other potential customers.

To make the quarterly meetings effective for both the development team and the sponsor, they follow this process:

1. The development team restates their understanding of the requirements to the sponsor to ensure they are not deviating from the sponsor's needs;

2. The development team demonstrates their progress since the last quarterly meeting and describes which features can be developed in the next quarter; and

3. The sponsor provides feedback on the project and, if necessary, modifies the plan for the next quarter.

Due to the success of these meetings, the development team is working to increase their frequency. This increased frequency would allow for better interaction between the development team and the sponsor.

Choice of Programming Languages

A fairly unique characteristic of HARRIER, at least among the CSE codes that were the subject of the previous case studies, is the absence of any FORTRAN code. The two main parts of the code, the user interface and the mesh generation algorithms, are written in Python and C++, respectively. In both cases, the development team made a conscious, reasoned choice to re-implement the code in a different language than the one in which it was originally developed. These changes are one of the important success factors for HARRIER. One difference between HARRIER and the codes studied previously which may have affected the language choice is the lack of parallelism within HARRIER. It is possible that the move away from FORTRAN may have been more complex if there was significant parallelism in the code. It will be interesting to observe any impacts from the choice of languages if/when the HARRIER team choose to add parallelism to the code.

The legacy code that implemented the mesh generation algorithms, which was the forerunner to HARRIER, was written in FORTAN. Because the members of the code development team all have a computer science background, they decided to rewrite this core computational code in C++, with the goal of making it easier to maintain and port. The user interface, which was built on top of the legacy code, was originally written in C++. It has now been rewritten in Python. This language choice allows students, who can more easily learn Python than C++, to write the less critical and less complex user interface code. The core development team can then focus their time on implementing the more complex computational algorithms.

Methods of Communication

The development team has created a communication structure that works well for their small team. The two important communication goals are: 1) to ensure everyone on the team is informed of project progress, and 2) to prevent collisions among the work of different developers. Three specific practices that have helped with preventing collisions are: 1) each section of code is "owned" by one developer who becomes the expert in that code, 2) developers primarily fix defects in their own code and 3) if a developer finds a defect in code that another developer owns, they ask the owner of the code to fix that defect.

When the expertise of more than one developer is needed to implement a portion of the code, the team employs pair programming, where two developers sit at the same machine and write code collaboratively (Beck, 2000). The HARRIER team typically uses pair programming a couple of times per week for about an hour each time. For example, one developer is an expert in the mesh generation algorithms, while another developer is an expert in the development of the geometry (the step before mesh generation). Whenever there is a problem in the interface between these two portions of code, the two developers use pair programming to work together to solve the problem. Because the developers work together and exchange tacit knowledge about the code, the use of pair programming reduces the need for more formal documentation.

As a result of developers owning portions of the code and using pair programming, the HARRIER team is able to be successful with less documentation than otherwise might be necessary. This approach works primarily because the environment facilitates close communication and the small size of the core development team. Because there is relatively little turnover among the core development team, long-term maintenance, especially of the core modules, is not an issue. Generally, the developer who wrote the code is there to maintain the code. The lack of documentation does make it more difficult to help new team members, primarily students who are working on the team for a short time, become familiar with the code. In these cases, the new teammate is first given a small task to work on. Then she sits with a more experienced developer to learn a portion of the code. Therefore, knowledge transfer is tacit, i.e., person to person, rather than explicit, i.e., through documentation.

Code Structure

The HARRIER team made three important decisions relative to the structure of the code to allow HARRIER to successfully evolve and grow. The three decisions, each of which is described in more detail as follows, are: use of modular code, use of open source software, and use of a coding standard.

First, the decision to use a modular code structure has allowed HARRIER to successfully handle the constantly changing requirements provided by its sponsor. HARRIER was initially going to be a relatively simple three month project that would add a user interface to an existing mesh generation algorithm. Since then, not only has the user interface evolved, but the underlying mesh algorithm was rewritten. The modularity of the code allowed it to evolve to effectively meet additional needs. Specifically, the modularity has facilitated three activities.

- The developers have been able to successfully re-factor the code numerous times to improve its quality.
- The user interface and the underlying mesh generation code can be changed independently, allowing them to be developed by different developers.
- The developers can work somewhat independently in their own sections of the code.

Second, the HARRIER team has used open source software to support the development of the code and ensure its modularity. For example, they have used wxPython (http://www.wxpython.org/what.php) to help build the GUI. Because the development team does not have the "not developed here" syndrome, common to many teams, they are very willing to use open source tools that have a compatible license and perform required tasks. The frequent use of open source software makes HARRIER different from many of the other CSE codes studied in prior case studies.

Finally, the HARRIER team has developed and follows a coding standard to help ensure coherency across different parts of the code. The coding standard describes naming conventions and the documentation that must be provided. The use of a coding standard is especially important because of the large number of transient students who work on the code. While each developer can use their own programming style, there are some conventions that remain constant. Minor deviations from the coding standard may occur, but not in code that another developer is going to see. The use of the coding standard has supported the constant personnel churn among that occurs in an academic environment with students working on a project for a couple of years at a time (note this personnel churn is not present within the core development team, only among the students).

SUMMARY

This paper has described the code characteristics, code team, and project lifecycle of a mesh generation code. Similar to many projects within the CSE domain, this project has faced the challenge of constant requirements evolution. The HARRIER team has employed several practices that have allowed them to be successful in achieving their development goals. This paper described the motivation for each decision along with its concrete impacts on the code development process. The Introduction posed three objectives for conducting case studies in this area. This paper has provided results relative to each of those objectives.

Objective 1: *Provide feedback to software vendors about issues that should be addressed to improve productivity* – This study highlighted that the HARRIER team, and likely other similar CSE projects, would benefit from better support for usability and regression testing in a domain with complex software that requires high domain knowledge for its end-users to be effective.

Objective 2: *Develop a body of case studies of CSE projects that can serve as a guide to the community* – This code project that was the subject of this study makes an interesting addition to the set of existing case studies. The heavy use of developers with computer science backgrounds on the main team makes this case different from the previous case studies. Even so, many of the challenges faced by HARRIER were consistent with those faced by CSE teams that were subjects of previous case studies.

Objective 3: *Document the lessons learned about CSE software development to provide a resource for other similar teams* – The HARRIER team employed a number of approaches that made them successful from which other teams could learn. Specifically:

- To deal with the problems caused by constantly evolving requirements, the HARRIER team developed a specific method of interacting with the sponsor that included rapid prototyping and a specific process for the quarterly meetings. They have also created a modular code structure to prevent changes from one section of the code from affecting the entire code base.
- The move from FORTRAN to C++ for the implementation of the core algorithms and from C++ to Python for the user interface has increased the portability and maintainability of the algorithm code and has increased the development speed of the user interface.
- The use of code ownership allows the main developers to work independently until there is an issue at the interface of code modules, in which case pair-programming is employed.

Future work in this area will be to investigate the usability issues of mesh generation software in more detail. By better understanding the process of how engineers build and validate a mesh, the software engineering community and software vendors can provide better support and facilitate more productivity among the engineer end-users.

ACKNOWLEDGMENT

I would like to thank Christine Halverson from IBM for her assistance during the interview process. I also would like to thank Lorin Hochstein for helpful comments and feedback. Finally, I would like to thank the HARRIER team for their time and openness in participating in the interview and follow-up.

REFERENCES

Beck, K. (2000). *Extreme programming explained: Embrace change*. Reading, MA: Addison-Wesley.

Boehm, B., & Turner, R. (2004). *Balancing agility and discipline: A guide for the perplexed*. Reading, MA: Addison-Wesley.

Carver, J. C. (2009). First international workshop on software engineering for computational science & engineering. *Computing in Science & Engineering, 11*(2), 7–11. doi:10.1109/MCSE.2009.30

Carver, J. C., Kendall, R. P., Squires, S., & Post, D. (2007). Software development environments for scientific and engineering software: A series of case studies. In *Proceedings of the 29th International Conference on Software Engineering*, Minneapolis, MN (pp. 550-559).

Cockburn, A. (2002). *Agile software development*. Reading, MA: Addison-Wesley.

Kelly, D. F. (2007). A software chasm: Software engineering and scientific computing. *IEEE Software, 24*(6), 120–119. doi:10.1109/MS.2007.155

Kendall, R. P., Carver, J., Mark, A., Post, D., Squires, S., & Shaffer, D. (2005). *Case study of the hawk code project* (Tech. Rep. No. LA-UR-05-9011). Los Alamos, CA: Los Alamos National Laboratories.

Kendall, R. P., Carver, J. C., Fisher, D., Henderson, D., Mark, A., & Post, D. (2008). Development of a weather forecasting code: A case study. *IEEE Software, 25*(4), 59–65. doi:10.1109/MS.2008.86

Kendall, R. P., Mark, A., Post, D., Squires, S., & Halverson, C. (2005). *Case study of the condor code project* (Tech. Rep. No. LA-UR-05-9291). Los Alamos, CA: Los Alamos National Laboratories.

Kendall, R. P., Post, D., & Mark, A. (2007). *Case study of the NENE code project* (Tech. Rep. No. CMUI/SEI-2006-TN-044). Pittsburgh, PA: Carnegie Mellon University.

Kendall, R. P., Post, D., Squires, S., & Carver, J. (2006). *Case study of the eagle code project* (Tech. Rep. No. LA-UR-06-1092). Los Alamos, CA: Los Alamos National Laboratories.

Paulk, M. C., Weber, C., Curtis, B., & Chrissis, M. (1995). *The capability maturity model: Guidelines for improving the software process*. Reading, MA: Addison-Wesley.

Post, D. E. (2009). The promise of science-based computational engineering. *Computing in Science & Engineering, 11*(3), 3–4. doi:10.1109/MCSE.2009.60

Post, D. E., Kendall, R. P., & Whitney, E. (2005). Case study of the falcon project. In *Proceedings of the 2nd International Workshop on Software Engineering for High Performance Computing Systems Applications*, St. Louis, MO (pp. 22-26).

Segal, J. A. (2008). Models of scientific software. In *Proceedings of the 1st International Workshop on Software Engineering for Computational Science and Engineering*.

Shih, M. H., Yu, T. Y., & Soni, B. K. (1994). Interactive grid generation and NURBS applications. *Journal of Applied Mathematics and Computation, 65*(1-3), 279–294. doi:10.1016/0096-3003(94)90183-X

Soni, B. K., Thompson, J., Stokes, M. L., & Shih, M. H. (1992). GENIE++, EAGLEView and TIGER: General and special purpose graphically interactive grid systems. In *Proceedings of the 30th AIAA Aerospace Science Meeting*, Reno, NV.

Thompson, J. F., Warsi, Z. U. A., & Mastin, C. W. (1985). *Numerical grid generation: Foundations and applications*. New York, NY: Elsevier North-Holland.

Wilson, G. (2006). Where's the real bottleneck in scientific computing. *American Scientist, 94*(1), 5–6.

Yin, R. K. (2002). *Case study research: Design and methods (applied social research methods)* (3rd ed.). Thousand Oaks, CA: Sage.

APPENDIX

Table A1. HARRIER's use of agile and plan-driven development practices

Practice	Description	Followed	Comments
Collective ownership	Allow anyone to change any code anywhere in the system at any time	No	Code typically modified by the original developer
Configuration management	Establish and maintain the integrity of work products using configuration identification and control.	Yes	Code is stored in a repository. Developers are emailed when commits are made
Continuous integration	Integrate and build the system many times a day, each time a task is completed.	No	Code is integrate periodically, but not necessarily daily
Feature-driven development	Establish an overall architecture and feature list, then design by feature and build by feature.	Partially	Feature request from users are designed and developed incrementally
Frequent delivery/ small releases	Have many releases with short time spans; implement the highest-priority functions first.	Partially	Large number of informal, partial releases as features are completed
Onsite customer	Include a real, live user on the team who is available full time to answer questions.	No	Frequent customer interaction, but not onsite
Organizational process definition	Follow an organization-wide process	No	Not used
Organizational training	Develop team members' skills and knowledge so that they can perform their roles effectively.	Partially	Several team members are working on advanced degrees
Pair programming	Work side by side with another programmer at one computer, collaborating on the design, algorithm, and code.	Yes	Weekly, when working on portions of the code that cover expertise of two developers
Planning game	Quickly determine the scope of the next release with business priorities and technical estimates.	No	Not used
Peer reviews	Review peers' software artifacts (requirements, design, code) to improve quality.	No	Not used
Process and product quality assurance	Objectively evaluate adherence to process descriptions and resolve noncompliance.	No	Not used
Project monitoring and control	Provide an understanding of the project's progress so that appropriate corrective actions can be taken if progress deviates from plan.	Partially	No formal process. Weekly team meetings and periodic meetings with customers used to track progress.
Project planning	Establish and maintain plans that define project activities.	Partially	Roadmap and milestones established with sponsors
Refactoring	Restructure software to remove duplication, improve communication, simplify, or add flexibility.	Partially	No formal process, but developer refactor as needed
Requirements development	Produce, analyze, and verify customer, project, and product requirements.	Partially	Requirements and features are established with sponsors
Requirements management	Manage the project's requirements and identify inconsistencies with the project plan.	Yes	Through use of project-management software
Retrospective	Perform a postiteration review of the effectiveness of the work performed, methods used, and estimates.	Partially	Through quarterly meetings and demo sessions with sponsors
Risk management	Identify potential problems and adequately handle them.	No	Not used
Simple design	Design only what is being developed, with little planning for the future.	No	Not used
Tacit knowledge	Maintain and update project knowledge in participants' heads rather than in documents.	Partially	Due to the lack of documentation, tacit knowledge is important
Test-driven development	Write module or method tests before and during coding.	No	Not Used

This work was previously published in the Journal of Organizational and End User Computing, Volume 23, Issue 4, edited by Mo Adam Mahmood, pp. 1-16, copyright 2011 by IGI Publishing (an imprint of IGI Global).

Chapter 16

Characterizing Data Discovery and End-User Computing Needs in Clinical Translational Science

Parmit K. Chilana
University of Washington, USA

Peter Tarczy-Hornoch
University of Washington, USA

Elishema Fishman
University of Washington, USA

Fredric M. Wolf
University of Washington, USA

Estella M. Geraghty
University of California, Davis, USA

Nick R. Anderson
University of Washington, USA

ABSTRACT

In this paper, the authors present the results of a qualitative case-study seeking to characterize data discovery needs and barriers of principal investigators and research support staff in clinical transla-tional science. Several implications for designing and implementing translational research systems have emerged through the authors' analysis. The results also illustrate the benefits of forming early partner-ships with scientists to better understand their workflow processes and end-user computing practices in accessing data for research. The authors use this user-centered, iterative development approach to guide the implementation and extension of i2b2, a system they have adapted to support cross-institutional ag-gregate anonymized clinical data querying. With ongoing evaluation, the goal is to maximize the utility and extension of this system and develop an interface that appropriately fits the swiftly evolving needs of clinical translational scientists.

INTRODUCTION

Rapid advances in information technology are opening up new avenues for conducting research in biomedicine. The application of new technologies has enabled greater ability to generate, capture and analyze biological data for basic research, but there remain significant challenges in integrating and translating these data with clinical data into forms that can be used to improve clinical out-comes. Clinical translational science is an emerg-ing interdisciplinary field that seeks to facilitate the translation of biomedical research advances from the laboratory to improve clinical and public health outcomes and vice-versa (Zerhouni, 2007).

DOI: 10.4018/978-1-4666-2059-9.ch016

Institutions that are part of the 46 member Clinical Translational Science Award (CTSA) consortium, sponsored by the National Institutes of Health (NIH) are beginning to develop, implement and support research on improving human health, both in local environments as well as collaboratively across sites. Individual CTSA sites are composed of researchers with expertise across the clinical research workflow, including biostatistics, informatics, bioethics, as well as community outreach and clinical translational research scientists. Within this novel consortia, there are challenges to consider in developing and incorporating information tools and methods that can catalyze and advance research both into as well as across the rapidly evolving and increasingly heterogeneous clinical research environments.

To better characterize the data discovery needs and end-user computing practices of clinical translational scientists and to steer the development of a novel cross-institutional clinical data discovery project, we conducted a pilot study involving semi-structured interviews with twelve principal investigators and research support staff working on a range of clinical research projects within the University of Washington's Medical Center, and affiliated with the UW CTSA site, the Institute of Translational Health Sciences (ITHS). We adopted the widely-used techniques of user and task analysis (Hackos & Redish, 1999) and focused on understanding workflows in translational research, data gathering methods, and previous experience in developing or customizing data discovery tools.

Our findings shed light on the diversity of user needs and expertise within clinical translational science and the potential barriers scientists face in accessing clinical data and using existing querying systems. The diversity exists not only in terms of the technical prowess of the scientists, but also in the range of their research questions, which still typically rely on and require customized data discovery and analysis features for domain specific work. The implications of these results are useful for supporting and enhancing our on-going project implementation as well as to other clinical data-discovery systems that seek to integrate and catalyze collaboration across complex and heterogeneous data domains.

At a higher level, we have found it critical to understand the context of use and the querying practices of clinical translational scientists early in the development process of our system, as the resulting computational tools become necessarily integrated components of the research enterprise, though in novel ways that test existing expectations of the end-user researchers. Developing query access to clinical data systems for research is particularly unique and challenging, as in addition to managing patient data privacy and data security, establishing methods to extract, analyze and compare highly heterogeneous clinical data challenges normative assumptions of how researchers understand and interact with operational clinical health environments. This challenge is enhanced when multiple institutions seek to build collaborative services of this form, and is demanding new roles and expertise to support characterization, deployment and support of these new systems. We plan to continue our partnership with end-user translational scientists by employing an iterative development process to refine use-cases and to better understand user perceptions, end-user computing needs and usability issues to maximize the overall utility of clinical research systems.

BACKGROUND AND MOTIVATION

We have implemented and are extending Informatics for Integrating Biology and the Bedside (*i2b2*), an interoperable open source software architecture that for our project facilitates the discovery of anonymized aggregate clinical data of patients who may meet eligibility criteria for clinical trial recruitment (e.g., counts). This Cross-Institutional Clinical Research Project (CICTR) effort is a generalization of the Harvard implementation of i2b2 and when fully deployed will provide query-level access to de-identified clinical data through a web interface across three different institutions: University of Washington (UW), University of California, Davis (UC Davis) and University of California, San Francisco (UCSF), with an estimated patient population in excess of 4 million individuals. By constructing a common technical interface and knowledge representation across these three sites, researchers will have the ability to create simultaneous federated queries to discover if sufficient study subjects for clinical trials are available either locally or more broadly within the consortium. The tool will further allow translational scientists to create queries to explore retrospective clinical data characteristics and modify and reuse query criteria to reflect ongoing research questions.

Although technical, semantic and governance issues are major challenges in building this collaborative information exchange environment, ensuring the overall usability and utility of the system to end-user researchers is crucial. Since clinical translational science is a newly emerging field, little prior work has specifically focused on the needs of scientists or software development within this domain.

Informatics and human-computer interaction (HCI) studies related to biomedical research have largely focused on biologists in lab settings. Some have focused on understanding information tasks and workflows (MacMullen & Denn, 2005; Tran et al., 2004; Bartlett & Toms, 2005), while others have investigated larger socio-technical issues of biomedical research environments (Anderson et al., 2007; Ash et al., 2008). Other works have focused on facilitating end-user programming needs in bioinformatics (Massar et al., 2005; Letondal, 2005) and understanding software development practices (Umarji & Seaman, 2008; Chilana et al., 2009). Such studies have not considered the issues that scientists face in clinical translational science where the focus is on patient data rather than molecular data.

On the other hand, studies related to patient data have largely revolved around clinicians and use of patient information systems. For example, some studies have outlined barriers and solutions in the use of electronic medical records (EMRs) (Miller & Sim, 2004), while others have focused specifically on usability issues (Rose et al., 2005). Others have focused on understanding errors in patient information systems (Ash et al., 2004). Research on doing formal design, verification and prototyping of patient information systems has also been explored (Mathe et al., 2007). Although these works are clearly important for clinical practice, they have not looked at how scientists query and interpret patient data for secondary research purposes, which has been the primary goal of our study.

The insights gained from our study are useful not only for developing and extending *i2b2*, but also serve as a case study of data discovery and end-user computing needs in clinical translational science. In this regard, this paper also complements other case studies of scientific software development (Carver et al., 2007; Segal, 2005, 2009) but sheds new light on needs and barriers that are unique to clinical translational science.

STUDY DESIGN

We used a semi-structured interview technique for informant data collection, since it allowed for an open-ended discussion that could capture the situational aspects of data use and provide us an approach to finding consistencies among responses (Strauss & Corbin, 1998). We developed a list of structured interview questions, focusing on user and task analysis techniques (Hackos & Redish, 1999) to understand the work environments and tasks of participants. We used this instrument to initiate the interview and probed into interesting responses by asking unstructured follow-up questions. Each interview lasted approximately one hour. Where possible we interviewed participants

Table 1. Profile of interview participants

	Role	Research Area
P01	Clinical Data Support	Basic/bench science & retrospective clinical
P02	Research Scientist	Clinical trials, health quality
P03	Clinical Data Support	Various
P04	Clinical Data Support	Various
P05	Principal Investigator	Pre- and Post-tests, drug effects, healthcare guidelines
P06	Research Scientist/ Clinical Data Support	Health services, quality metrics
P07	Principal Investigator	Health services, clinical trials
P08	Principal Investigator	Clinical trials, cardiovascular health
P09	Principal Investigator	Epidemiology, observational studies
P010	Principal Investigator	Clinical Trials
P011	Clinical Data Support	Various
P012	Clinical Data Support	Various

in their work settings to establish context and facilitate recall.

Our participants were identified through word-of-mouth, email and snowball sampling (where our current participants helped identify other participants). There were 12 participants in total (see summary in Table 1).

The interviews began with a focus on understanding how scientists used clinical and EMR data for research purposes. Based on the critical incident technique, we asked participants to show us the steps they recently carried out in exploring or analyzing such data. Responses included explanations of the tools they used, the types of queries they formulated and descriptions of how they formatted and visualized the output, where applicable. In particular, we were interested in learning about the types of technical and/or non-technical challenges or barriers they faced in using clinical data for secondary research purposes. Lastly, we sought to understand what scientists expected or wished they could do with a federated clinical data querying system that would provide access to anonymized EMR data across a larger pool of institutions.

Two of the authors transcribed the interviews independently. We analyzed the transcripts induc-tively, by following an iterative process of open coding and axial coding to discover relationships among emerging concepts in our data (Strauss & Corbin, 1998). This was followed by selective coding where all the results of axial coding were integrated. We identified recurring themes in the interviews by following this inductive analysis approach.

RESULTS

Our key findings confirm the existence of different types of users of clinical systems, the varying range in their clinical data discovery and analysis needs, and the barriers scientists face in accessing clinical data for translational research purposes.

Different Needs for Different Users

Our interviewees represented two different user groups based on the tasks they described: (1) researchers who collected and analyzed clinical data on their own and (2) intermediaries or "research IT" (coined by Bernstam et al., 2009) staff members who worked closely with investigators to access and distill complex data requests.

Within the group of researchers, participants had a wide range of technical skills. For example, half of our researcher participants were accustomed to extracting and manipulating clinical data on their own and could write sophisticated SQL queries. One researcher explained that technical skills were necessary to have control over the data:

I would want to do my own analysis – totally depends on what the question is...I want more control over the data, I would want to do subsets... my experience with other reporting software is that they are very limited...it's just a simple report, you can't do back-end processing...

In contrast, one researcher confessed to having very limited experience using any kind of EMR system or other tools. Most other researchers fell somewhere in-between—they had basic knowledge about database design and how EMR systems function in general but could not extract or manipulate data on their own. They consulted with research IT personnel as needed.

Researchers also pointed out that sometimes they had no choice but to go through specialized research IT staff because they did not have permission to access certain patient databases. For example, one researcher described a situation where she needed patient data housed in the hospital billing database:

...they [billing dept] don't have people sitting around willing to help us...there [are] long delays to when we make a take request to when we actually get data back...have to be considerate of their work load... it would be good to get things that may be harder for them but valuable for us....

Research IT participants typically had expertise in database design and programming, but little or no formal training in biomedicine. They acquired the relevant domain knowledge through on the job experience. The type of help they provided to researchers ranged from basic data transfor-

mations to complex queries involving multiple tables and joins from disparate data sources that could sometimes take days to run. For instance, one research IT participant described the intricacies of providing the requested data to a group of clinical researchers as follows:

A lot of what I do is pulling raw data and then putting [it] into tables and then handing off tables to researchers. I don't have statistical background. For some researchers, sometimes they want us to do basic analytics (like counts) – but a lot of the researchers want raw data so they can do high-level analysis.

He further explained that researchers were sometimes not aware of the level of detail required in formulating queries and data transformations to seemingly "simple" research questions. As a result, often there were unrealistic expectations about timelines.

Another research IT participant pointed out that working with researchers required an ongoing partnership approach as the researchers did not always know what to do with raw data:

...they [researchers] had the notion of pulling the data in batch, but once they see the data in tabular form, they have no idea what to do with it...they don't want data, they want the question answered...it's an iterative process...the data has to be cleaned or managed.

Thus, we saw that not only researchers and research IT had different needs and experiences in accessing and making use of clinical data repositories, but that there existed individual differences even within these subgroups. As an example, principal investigators often possessed domain-specific statistical and study design skills – or could gain assistance with this – but were not necessarily those who were directly involved in resolving this experience with the practice of directly accessing the patient data. It is clear that developing high-utility access to clinical data for research purposes will require tools and expertise that cover both the technical skills as well as the ability to transform formal research questions into structured queries that meet necessary study design criteria.

Range of Clinical Data Discovery and Analysis Needs

Another theme that emerged was the variability and range in participants' responses on the use of clinical data for secondary research purposes.

There was general agreement among our participants about the utility of certain types of clinical data for pursuing translational research questions: patient demographics, laboratory values, diagnosis data and medication data. However, participants also described instances where they had more specific data needs for answering larger or more complex questions and there was wide variability within each research domain (i.e., clinical science, epidemiology, quality of care). For example, participants listed sources such as:

- Problem lists
- Nursing notes
- Therapeutic data
- More detailed orders
- Diagnostic test data
- Socio-economic data
- Procedure data (i.e., surgical procedures)

For some researchers, the use of data was more exploratory in that they wanted to find interesting patterns in the data to formulate relevant research questions. For other researchers, it was more important to have access to defined data points to answer various domain-specific research questions.

For instance, a common use of clinical data was for pre- and post-analysis of various groups of patients with a specific condition, such as blood clots. Another use was to determine if an interven-

tion had the expected outcome for which it was designed. Some participants were also engaged in research involving quality of care and were interested in data, such as patient admission and re-admission rates, in a given timeframe. Table 2 summarizes the types of translational research questions researchers were pursuing by making secondary use of patient data.

Barriers to Accessing Clinical Data

Our findings also revealed a number of common barriers to accessing and using clinical data for translational research. These were from the perspective of both researchers and research IT staff. Some of the barriers were related to general medical practices and conventions, while others were more specific to experiences using commercial, open-source and/or locally developed software tools housing clinical data.

- **Accessing data in disparate sources:** Participants agreed that one of the biggest challenges in doing translational research was obtaining access to patient data housed in different systems and formats suitable for secondary research.

For example, one researcher described that the challenge was in dealing with additional cost and time overhead for gathering clinical data from disconnected sources:

We mostly [had] registries of patients – type in their number and go look at their record...nursing notes were more detailed on what actually happened, but that's a separate database and had to be integrated - very costly and time consuming... about an hour per patient and we had a couple of thousand patients...

Another researcher explained that gathering data for certain retrospective studies could be challenging—sometimes requiring 10 year old re-

Table 2. Examples of secondary use of patient data

- identifying target patient cohort(s) for clinical trials
- observational studies of drug benefits and adverse effects
- clinical quality improvement for patients (e.g., for diabetes)
- monitoring errors (e.g., in the emergency room)
- analyzing rates of use of medication to prevent certain conditions (e.g., blood clots)
- monitoring the type of lab test requests

cords that were available only through paper-based records or microfilms. Having access to recent EMR data was crucial, but not always sufficient:

You may have 10 variables you want to satisfy [that] you don't have for whatever reason...you may not have all 10 variables in a database, but you may have in your chart..need photocopy of the medical record, ...[through the] hardcopy, you may have [access to] some things you may not know about [in the EMR]...

Thus, the lack of data integration, different formats, and inefficiency in accessing non-EMR data was a major barrier in pursuing many types of translational research questions in a timely manner.

- **Naming and classification conventions:** Participants also discussed the difficulties of working with EMR data because of inconsistent naming conventions and non-intuitive classification schemes.

For example, one research IT participant described problems he faced when merging data based on attributes that were defined inconsistently between systems:

A data dictionary is very important/necessary with this system, but also there's a need for deep semantics of the data itself. For instance, deceased is a data attribute in this system, but it only applies if the patient dies at [this hospital], and if the patient dies somewhere else it's not marked/included in the software accurately...also not easy to interpret.

Although proper supporting documentation could be used to prevent such misinterpretations, this documentation was often incomplete or unavailable:

..not enough documentation of the data...it's hard to find the person who knows what a particular data field is there or why it's used ...list of codes don't mean anything...not an easy way to just look it up...not enough consistent documentation or availability of expertise.

Ongoing work in biomedical ontologies and terminologies will resolve some of these issues with naming conventions in the near future, but our data suggests that these ontologies should be developed with a user-centered approach to minimize misinterpretations by researchers in the long run.

- **Inaccurate or missing information:** All participants also cited examples of difficulty they faced in working with incomplete and inaccurate data in the clinical repositories they queried.

This problem often appeared, due to data entry errors influencing quality and requiring additional work:

There's always missing data, some of the data is wrong...[we know] after checking multiple sources – in one area it's different than another area...it's hard to determine..

The other problem was that due to the exploratory nature of research, scientists could not predict ahead of time or enforce what patient data should be collected or protected over time:

One of the issues with our data systems is that certain tables are at the patient level and certain ones are the visit level – if someone changes their address...then I don't get to see it changed...like was the patient homeless 3 years ago? As the data goes in, it would be helpful [to know]...how [data] looked on a particular day... what did it look like at that time, instead of keeping messy audit trails...

The researcher participants agreed that an improvement in the design and use of EMRs in clinical settings would be helpful to prevent errors during translational research, but such a change would require long-term institutional changes and could be difficult in practice.

Limitation of systems interfaces: When asked to show example queries in existing systems, many of our participants pointed out the other barriers they faced.

For example, one research IT participant explained that although he could create complex queries with the current system he was using, he could not handoff the system to researchers if they wanted more control over their data and analysis:

...functionally [the system] does pretty much what I ever wanted it to...users are limited... can't customize query [as] they need to know how backend of a database works...I don't know anyone [researcher] who would use it... a lot of things are hard-coded in, have to change all the way back to the underlying C++ code to create custom queries..

In addition to lacking intuitive querying and end-user programming facilities that are usually required to create customized queries, current systems altogether lacked the facility for answering certain research questions, as explained by one researcher:

A lot of the data would get better info if we were able to have natural language processing to extract some free text...some of the coded data in EMR is easier to work with and there are also algorithms that can only access coded data. However, for some of the diagnoses, important data resides as free text...our analytic algorithms can't look at it because we don't have any automatic way of looking at it...

In summary, the barriers discussed above show how researchers were sometimes forced to extend timelines or invest additional resources to obtain and transform the appropriate data. These results further stimulate enthusiasm for tools that would allow and simplify the process of using clinical data for secondary research and better streamline clinical research workflows.

DISCUSSION

As CTSAs and other research organizations seek to develop and deploy tools and processes to meet the needs of the clinical translation research environment, access to high quality clinical data remains a key issue. Although there are significant technical, semantic and policy-level issues underlying the development and implementation of clinical research systems, our study shows that for improving the long-term adoption and success of these emerging systems, there is much value to be gained by understanding the variation in needs of end-user scientists in the context of their translational research workflows.

A number of the themes that have emerged in our data have direct implications for supporting clinical data discovery and designing systems and processes that accommodate the varied needs of different users (i.e., researchers versus research IT staff). As suggested by our results, for some researchers who possess aptitude for development-stage end-user computing, completeness of interfaces is less of a barrier and they may be willing to experiment with tools and formulate complex queries on their own. But, for others, the role of the research IT and collaborative research design and data extraction is critical in getting assistance in effectively exploring data. Although our current focus has been on the *i2b2* system within the CICTR project, we believe the emergent themes apply to other similar clinical data querying tools.

Providing Control Over the Expressiveness of Data Queries

In our results we found the existence of two end-user groups (researchers and research IT) and varying levels of technical expertise and clinical domain knowledge. This finding suggests that different levels of control are needed to tailor queries to research questions. Thus, systems should (1) offer a facility to create custom queries and in this, (2) address the balance between providing simple query interfaces for data characterization (e.g., counts and summary data) and advanced interfaces for formulating complex questions (e.g., trends, limited data sets, visualizations). This is consistent with other findings on EMR-based query construction (Murphy et al., 2003).

Making Sense of Query Results

Our results showed that clinical scientists had different preferences and expectations for processing the results of the data access queries that they carried out themselves or through research IT. Thus, a data discovery system should provide multiple options for export and data delivery (e.g., raw tabular view and visual overview), and support evolving national data representation standards. There may be value in further exploring options for integrating statistical tools and sophisticated visualizations, such as recent work on temporal

patterns in numerical and categorical EMR data (Plaisant et al., 2008; Wang et al., 2008).

Improving Usability of Interfaces

A consistent theme in our results was the need to make input and output behavior consistent to encourage reuse and customization, both in view of the researchers and the research IT participants who used different clinical data querying tools. Thus, users should be able to leverage experience from using other data discovery systems where appropriate and not face an arduous learning curve in customizing new interfaces since that takes away time and resources from the main goal of scientific data discovery. Ongoing usability testing and iterative design (Hackos & Redish, 1999) and training approaches could be valuable for creating user-centered clinical data querying systems.

Data Transformation in *i2b2*

To better support end-user needs, the CICTR project has been building rapid iteration data transformation workflows to allow for each of the three university-based i2b2 nodes to develop common end-user data environments. At the heart of these are tools which allow for raw data to be imported easily into i2b2, including a "Universal" Extraction, Transformation and Load (UETL) tool a ontology tool that allows for customizable "mapping" scripts to leverage web-based terminology servers to generate rich end-user query taxonomies in the existing i2b2 interface. These tools have allowed the researchers and evaluators to separate the mechanics of Extraction, Transformation and Load (ETL) and terminology alignment from the end-user domain-focused experience. As a result, the current i2b2 workbench (Figure 1) remains identical for all end-users within CICTR, though the abilities and data representations contained within the interface are becoming more flexible as the project engages with different domain experts.

Managing Cultural and Social-Technical Issues

Apart from the implications for design, our interviews shed new light on the diversity of data discovery needs affecting translational research and the barriers that still need to be resolved beyond the system level. We believe the key to successful development and integration of translational research systems rests on how well we address the higher level barriers. Our results show that cultural and social-technical issues pose major challenges to clinical translational scientists in obtaining the appropriate data. For example, the barrier of accessing data in disparate sources involves knowledge of who has control and who can grant access or securely "hand off" data between clinical and research environments. Provenance issues, due to data-entry mistakes or data corruption errors, can limit the types of research questions that can be credibly pursued as well as the utility of the data. There are clear comparable risks in translational research to clinical practice – particularly when the potential impact of translational research can be on entire populations, but the present emphasis of this research is on the secondary use of clinical data, in forms that are highly regulated and de-identified as to mediate risk to individual patient groups, researchers or institutions. Based in part on this project, though also in the context of the larger CTSA informatics community, researchers are increasingly interested in applying their unique domain-focused requirements to informing how upstream clinical source systems capture and semantically store data, which shifts the expected role of a researcher from an end-user to one of a collaborator or participant in the development process.

Ongoing Evaluation

We see the need for evaluation of the utility of this end-user clinical-translational query interface to be on-going. Given the diversity of research needs and the rapid advancement in the quantity of electronic clinical and biological data available, the ability to measure the relative utility of this form of service interface within the context of researchers design and implementation workflows

Figure 1. Screen shot from i2b2 web client within CICTR

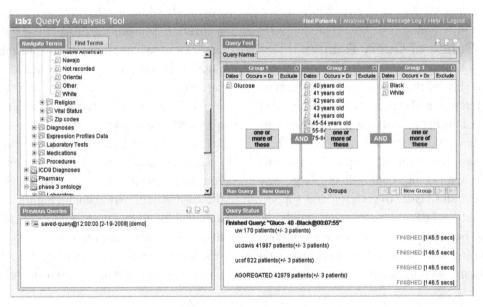

will provide valuable requirements and insights to advancing data-driven collaborations.

While our current study focused on discovery of generalized end-user computing needs for clinical translational scientists, our future evaluation plans are more specific to the i2b2 workbench environment as a method to address our users' needs. Multi-institutional survey tools will be used to determine user response to either a video demonstration of the tool or after a 'hands-on' training session to see how the reality of the i2b2 workbench meshes with their anticipated needs. Next, in depth usability testing, via think aloud sessions with pre-defined use cases, will be done to examine how researchers build queries and interact with the tool. In each phase of the evaluation, a report will be returned to the development team highlighting findings with recommendations for improvement.

Finally, although this study is limited by using data from interviews at a single academic site, it is unique because it establishes an understanding of the needs of scientists in data-rich translational research workflow environments. In-depth studies illuminating other aspects of collaboration, impact, and work practices in translational research will serve as a useful supplement to statistical accounts of user needs and system use.

CONCLUSION

Our study is one of the first of its kind to characterize the data discovery and end-user computing needs of clinical translational scientists. As suggested by our results, there remain major challenges for clinical translational scientists to gain access to and subsequently integrate clinical and laboratory data that is located within and across multiple institutions. Some of these challenges can be met with sustaining and information-needs focused partnerships with end-user scientists and their research IT staff throughout the development and deployment process of clinical translational data discovery tools.

The CICTR project has used the themes of this study to inform the refinement of the data discovery environment as to begin to meet the different end user needs. With the increase in data available in the network (in excess of 4 million patients as of early 2010), and richness of data sources (now including data on medications and laboratory tests) there have been corresponding on-going challenges to introduce these capabilities and limitations of such an interventional architecture. We are now focusing on establishing knowledge development workflows that allow us to capture and then map the different data sources such that we can leverage the input of end-user scientists' perspectives on utility, richness and quality against the data-level transformation requirements that is necessary for our clinical terminologists and software developers to implement. These complementary and overlapping workflows are allowing us to capture and operationalize information representation needs more efficiently, and are in turn enhancing scientists' ability to ask challenging research questions.

Our next steps are to iteratively address further issues raised by users in the current prototype and expand to a multi-site evaluation method that employs other forms of summative and formative evaluation as described above. We hope that this user-centered approach to designing our collaborative information environment will help us better understand and serve the needs of scientists as they address novel clinical translational research challenges.

ACKNOWLEDGMENT

This work has been supported by NCRR grants UL1 RR025014 and UL1 RR024146, and NCRR, NIH and DHHS under Contract No. HHSN268200700031C.

REFERENCES

Anderson, N. R., Lee, E. S., Brockenbrough, J. S., Minie, M. E., Fuller, S., Brinkley, J., & Tarczy-Hornoch, P. (2007). Issues in biomedical research data management and analysis: Needs and barriers. *Journal of the American Medical Informatics Association, 14*(4), 478–488. doi:10.1197/jamia.M2114

Ash, J. S., Anderson, N., & Tarczy-Hornoch, P. (2008). People and organizational issues in research systems implementation. *Journal of the American Medical Informatics Association, 15*(3), 283–289. doi:10.1197/jamia.M2582

Ash, J. S., Berg, M., & Coiera, E. (2004). Some unintended consequences of information technology in health care: The nature of patient care information system-related errors. *Journal of the American Medical Informatics Association, 11*(2), 104–112. doi:10.1197/jamia.M1471

Bartlett, J. C., & Toms, E. G. (2005). Developing a protocol for bioinformatics analysis: An integrated information behavior and task analysis approach. *Journal of the American Society for Information Science and Technology, 56*(5), 469–482. doi:10.1002/asi.20136

Bernstam, E. V., Hersh, W. R., Johnson, S. B., Chute, C. G., Nguyen, H., & Nagarajan, R. (2009). Synergies and distinctions between computational disciplines in biomedical research: Perspective from the Clinical and Translational Science Award programs. *Academic Medicine, 84*(7), 964–970. doi:10.1097/ACM.0b013e3181a8144d

Carver, J. C., Kendall, R. P., Squires, S. E., & Post, D. E. (2007). Software development environments for scientific and engineering software: A series of case studies. In *Proceedings of the International Conference on Software Engineering* (pp. 550-559).

Chilana, P. K., Palmer, C. L., & Ko, A. J. (2009). Comparing bioinformatics software development by computer scientists and biologists: An exploratory study. In *Proceedings of the ICSE Workshop on Software Engineering for Computational Science and Engineering* (pp. 72-79).

Hackos, J., & Redish, J. (1998). *User and task analysis for interface design.* New York, NY: John Wiley & Sons.

Letondal, C. (2005). Participatory programming: Developing programmable bioinformatics tools for end users. In Lieberman, H., Paterno, F., & Wulf, V. (Eds.), *End-user development* (pp. 207–242). Dordrecht, The Netherlands: Springer-Verlag.

MacMullen, W. J., & Denn, S. O. (2005). Information problems in molecular biology and bioinformatics. *Journal of the American Society for Information Science and Technology, 56*(5), 447–456. doi:10.1002/asi.20134

Massar, J., Travers, M., Elhai, J., & Shrager, J. (2005). BioLingua: A programmable knowledge environment for biologists. *Bioinformatics (Oxford, England), 21*(2), 199–207. doi:10.1093/bioinformatics/bth465

Mathe, J., Duncavage, S., Werner, J., Malin, B., Ledeczi, A., & Sztipanovits, J. (2007). Implementing a model-based design environment for clinical information systems. In *Proceedings of the ACM/IEEE International Workshop on Model-Based Trustworthy Health Information Systems* (pp. 399-408).

Miller, R. H., & Sim, I. (2004). Physicians' use of electronic medical records: barriers and solutions. *Health Affairs, 23*(2), 116–126. doi:10.1377/hlthaff.23.2.116

Murphy, S., Gainer, V., & Chueh, H. (2003). A visual interface designed for novice users to find research patient cohorts in a large biomedical database. In *Proceedings of the AMIA Annual Symposium* (pp. 489-493).

Plaisant, C., Lam, S., Shneiderman, B., Smith, M., Roseman, D., Marchand, G., et al. (2008). Searching electronic health records for temporal patterns in patient histories: A case study with microsoft Amalga. In *Proceedings of the AMIA Annual Symposium* (pp. 601-605).

Rose, A. F., Schnipper, J. L., Park, E. R., Poon, E. G., Li, Q., & Middleton, B. (2005). Using qualitative studies to improve the usability of an EMR. *Journal of Biomedical Informatics, 38*(1), 51–60. doi:10.1016/j.jbi.2004.11.006

Segal, J. (2005). When software engineers met research scientists: A case study. *Empirical Software Engineering, 10*(4), 517–536. doi:10.1007/s10664-005-3865-y

Segal, J. (2009). Software development cultures and cooperation problems: A field study of the early stages of development of software for a scientific community. *Computer Supported Cooperative Work, 18*(5-6), 581–606. doi:10.1007/s10606-009-9096-9

Strauss, A., & Corbin, J. (1998). *Basics of qualitative research: Techniques and procedures for developing grounded theory*. Newbury Park, CA: Sage.

Tran, D., Dubay, C., Gorman, P., & Hersh, W. (2004). Applying task analysis to describe and facilitate bioinformatics tasks. *Medinfo, 11*(2), 818–822.

Umarji, M., & Seaman, C. (2008). Informing design of a search tool for bioinformatics. In *Proceedings of the ICSE Workshop on Software Engineering for Computational Science and Engineering*.

Wang, T. D., Plaisant, C., Quinn, A. J., Stanchak, R., Murphy, S., & Shneiderman, B. (2008). Aligning temporal data by sentinel events: discovering patterns in electronic health records. In *Proceeding of the SIGCHI Conference on Human Factors in Computing Systems* (pp. 457-466).

Zerhouni, E. (2007). Translational research: Moving discovery to practice. *Clinical Pharmacology and Therapeutics, 81*, 126–128. doi:10.1038/sj.clpt.6100029

This work was previously published in the Journal of Organizational and End User Computing, Volume 23, Issue 4, edited by Mo Adam Mahmood, pp. 17-30, copyright 2011 by IGI Publishing (an imprint of IGI Global).

Chapter 17
Computational Engineering in the Cloud:
Benefits and Challenges

Lorin Hochstein
USC Information Sciences Institute, USA

Brian Schott
Nimbis Services, USA

Robert B. Graybill
USC Information Sciences Institute, USA

ABSTRACT

Cloud computing services, which allow users to lease time on remote computer systems, must be particularly attractive to smaller engineering organizations that use engineering simulation software. Such organizations have occasional need for substantial computing power but may lack the budget and in-house expertise to purchase and maintain such resources locally. The case study presented in this paper examines the potential benefits and practical challenges that a medium-sized manufacturing firm faced when attempting to leverage computing resources in a cloud computing environment to do model-based simulation. Results show substantial reductions in execution time for the problem of interest, but several socio-technical barriers exist that may hinder more widespread adoption of cloud computing within engineering.

INTRODUCTION

Cloud computing has recently emerged as a service model where users obtain short-term access to large-scale computational resources, potentially at lower cost than purchasing and administering computing hardware (Armbrust et al., 2009). Most of the initial drive and interest in cloud computing has been in the IT community. More recently, there has been growing interest on using cloud computing as a platform for computational science and engineering (Shainer et al., 2010)

The term *computational science* refers to the application of computers for solving scientific problems, particularly the use of computer simulations to predict physical phenomena (Post

DOI: 10.4018/978-1-4666-2059-9.ch017

& Votta, 2005). The term *computational engineering*, analogously, refers to the application of computers in solving engineering problems (Post, 2009). While there are many similarities between computational science and computational engineering, important differences also exist. Computational engineers use computers to do *virtual prototyping*, analyzing the behavior and failure modes of proposed designs through model-based simulations. Virtual prototyping has the potential to reduce both engineering development time and cost by reducing the amount of physical prototyping required to do a design validation, as well as opening up possibilities for design optimizations. Commercially available engineering packages put these simulation techniques within reach of the end-user engineer, although a high degree of domain expertise is required to set up and interpret the results of such simulations.

Engineering simulations are extremely computationally intensive, with simulations taking anywhere from hours to days or weeks, depending on the type of simulation required. Many of these commercial engineering packages have support for running on high-performance computing (HPC) systems, and a survey of larger engineering firms indicates that such firms take advantage of HPC (Joseph et al., 2004). However, such systems are expensive to maintain and require additional IT expertise, rendering them inaccessible to many smaller engineering firms.

In this paper, we describe a feasibility study undertaken by the authors to help determine whether the use of remote HPC resources for modeling-based simulation would have a positive return on investment for a small-to-medium-sized manufacturing company. This paper describes our experiences, including the benefits of reduced processing time, as well as practical challenges that we faced while supporting computational engineers in using remote HPC resources.

RELATED WORK

Armbrust et al. (2009) provide a broad overview of the costs, benefits, and challenges of cloud computing. Although they do not focus specifically on scientific and engineering applications, they discuss several issues that appear in this study, such as batch processing of parallel processing applications, compute-intensive desktop applications, data transfer bottlenecks, and data licensing issues.

Cloud computing for computational science and engineering is a very young but increasingly active area, as evidenced by new workshops emerging in 2010 such as the first Workshop on Science Cloud Computing (ScienceCloud) (http://dsl.cs.uchicago.edu/ScienceCloud2010/) and Cloud Futures 2010: Advancing Research with Cloud Computing Workshop (Faculty Connection, 2007). Some early experience reports have begun to emerge. Hoffa et al. (2008) explored the use of cloud computing for executing a scientific workflow in the field of astronomy. Lauret and Keahy used cloud computing resources to quickly perform a preliminary analysis of a nuclear physics experiment in time to submit a conference paper (Heavy, 2009).

The MapReduce programming model has dominated much of the early interest in cloud computing applications. Dean and Ghemawat (2004) introduced the MapReduce model of processing datasets on large clusters, which has been implemented by the open-source Hadoop project (Bialecki et al., 2007). Much of the current cloud computing research focuses on data-intensive applications that map well to this model, such as indexing of spatial databases (Cary et al., 2009), processing very large graphs (Zhao et al., 2009; Cohen, 2009), and indexing of very large text corpora for information retrieval (Callan & Kulkarni, 2009)

There are also ongoing research projects to develop software infrastructure to support the creation of private clouds for scientific use. These efforts include Eucalyptus (Nurmi et al., 2009), OpenNebula (Sotomayor et al., 2008), Nimbus (Keahy et al., 2009), and Cumulus (Wang et al., 2008).

THE COMPUTATIONAL ENGINEER AS END USER

There is substantial overlap between computational scientists and computational engineers. Both of them model physical phenomena and simulate these models using computers to estimate the physical behavior of interest. Many of the underlying physical and mathematical computational machinery are the same (e.g., finite element methods, sparse linear solvers). However, the goals and contexts of these two groups differ.

Case studies of computational science projects to date suggest that most such code is "research code", written in a research lab such as a university or government-sponsored research environment. Examples include Hochstein and Basili's (2008) case studies of five university-based projects funded by the U.S. Department of Energy, Carver et al.'s (2007) retrospective case studies of five projects sponsored by various U.S. government agencies, and Easterbrook and Johns' (2009) ethnographic study of software development practices in a large U.K. government-funded climate research lab. Computational scientists will typically either write their own research code, or use someone else's, but it is rare to find commercially available computational science software that does model-based simulation (*Gaussian*, a commercial software package for computational chemistry, is a notable exception). By contrast, in computational engineering there is a significant market for commercial software that performs model-based simulation, such as ANSYS, FLUENT, MD NASTRAN, NEI NAS-

TRAN, ABAQUS, STAR-CD, LS-DYNA and COMSOL Multiphysics. These tools enjoy wide commercial usage, and many are now able to take advantage of HPC systems. The effective result is that the computational engineer is much more of a traditional end-user, seldom called upon to write software directly.

Because commercial software packages must support a wide range of simulations, and because significant domain knowledge is required to understand the different simulation options, they are very complex pieces of software. A computational engineer must devote considerable time and effort to master the use of one of these commercial software packages. While many packages have a modern graphical user environment, they also have extensive command languages to support playback and batch processing. For example, the ANSYS software package implements a scripting language called the ANSYS Parametric Design Language (APDL), which has a FORTRAN-like syntax. In effect, these commercial tools can be thought of as domain-specific, graphical programming environments. Figure 1 shows some ANSYS commands, based on a tutorial by Quon and Bhaskaran (2002).

The end-goals of the activities of computational scientists are described in the case studies performed by Hochstein and Basili (2008), Carver at el. (2008) and Easterbrook and Johns' (2009), as well as by Segal (2005) and Basili et al. (2008). Based on our observations, the goals of computational engineers differ. In general, a computational scientist is interested in learning about particular physical phenomena for its own sake. The computational engineer is interested in validating or optimizing the performance of a particular product to be manufactured and sold. These differences in goals can affect the time constraints that they work under (e.g., product life cycles vs. publication deadlines). In addition, being in a corporate environment impacts the exchange of information with others outside of the organization. Even information exchange with

Figure 1. Example ANSYS commands for defining a bicycle crank

```
RESUME, crank,db,
/PREP7
! Add the Mesh Facet 200 element
ET,1,MESH200
! Add the Brick 8node 45 element
ET,2,SOLID45
! For Mesh200, we use 3D quadrilateral with 4 nodes
! K1: QUAD 4-NODE)
KEYOPT,1,1,6
KEYOPT,1,2,0
! Specify structural, linear, elastic, isotropic material
! Young's modulus (EX)=2.8E7, Poisson's Ratio (PRXY)=0.3
MPTEMP,,,,,,,,
MPTEMP,1,0
MPDATA,EX,1,,2.8E7
MPDATA,PRXY,1,,0.3
! Rectangle
BLC4,-3.3465,0,3.3465,1.299
! Big circle
CYL4,-3.3465,.6495,.6495
! Smaller circle
CYL4,-3.3465,.6495,.25
```

customers and suppliers about computer-based models can be limited due to concerns about trade secrets.

Another significant difference between computational science and computational engineering is the role of verification and validation. In many computational science applications, the motivation for using computer simulation is that running physical experiments as an alternative are either prohibitively expensive, or practically impossible. By contrast, in computational engineering, computer simulation is part of a larger process that will involve the construction and testing of physical prototypes. The computational scientist may never discover how well his or her simulations approximate reality. For better or for worse, the computational engineer will eventually find out.

STUDY CONTEXT

Study Motivation

The case study described here is one of a series of studies funded by the Defense Advanced Research Projects Agency (DARPA) to determine whether manufacturing companies in the U.S. Department of Defense's supply chain can benefit from using high-performance computing for model-based simulation. There is increasing concern among policymakers in the United States about the decline of the U.S. manufacturing industry over the past several decades (National Academies, 2007). This is of particular concern to the U.S. Department of Defense because of the potential national security risks involved in becoming dependent on foreign manufacturers.

Because of the structure of the global economy, American manufacturers cannot compete with foreign manufacturers on labor costs. Therefore, American manufacturers are required to innovate to remain competitive in a global marketplace (Helper, 2009). The application of model-based simulation in engineering design has been proposed as one such innovation to improve the competitiveness of the manufacturing sector (Glotzer, 2009), with higher-fidelity simulations performed using high-performance computing (Council on Competitiveness, 2009). The motivation for this study is to understand the practical benefits and obstacles that a manufacturing company would face when trying to leverage high-performance computing resources.

The Organization

To conduct this study, we partnered with an engineering company (hereafter referred to as "Company X") that supplies fuel injection systems for commercial and military applications. The partnership was facilitated by a previous relationship we had with one of their customers. A medium-sized business, Company X employs about 200 people, with three computational engineers on staff. These computational engineers take the CAD models from the design engineers, construct finite element models, and then use computational tools to predict whether the part will fail under various conditions it may be subjected to during operation. Before beginning the study, Company X had no previous plans to pursue HPC.

Their primary software tool is ANSYS Mechanical, a commercial finite-element analysis solver that can simulate structural and thermal stresses. For more sophisticated simulations, the engineers also use CD-adapco's STAR-CD, a computational fluid dynamics solver. Using STAR-CD can increase the accuracy of the simulation outputs under certain scenarios (e.g., to obtain more accurate estimates of heat flows due to convection).

Cloud Computing Resources

To identify a suitable cloud computing provider, we had two considerations: performance, and an environment that complies with the U.S. International Traffic in Arms Regulations (ITAR). Most modern HPC systems are *clusters*: independent computers (or *nodes*) that are connected together via a network. In that sense, any cloud computing resource has the ability to provide users with a cluster by requesting several *nodes* on demand. However, for many model-based simulation problems, the speed of the network interconnect is a critical part of the performance of software that will run in parallel on the system. Because of the need for a high-speed interconnect, it is unclear how well performance will be on cloud computing providers such as Amazon's Elastic Compute Cloud (Amazon, 2009), because there is no guarantee that multiple nodes requested will be on the same local network.

ITAR is a set of United States regulations, enforced by the Department of State, which restricts certain U.S. exports to protect national security. Computer models of designs for military vehicles typically fall under ITAR, which restricts who can access data. In particular, only those the State Department classifies as "U.S. persons" are permitted to handle such data. This severely restricts the use of cloud computing resources for processing of ITAR data, because such organizations would have to certify that no "foreign persons" have electronic or physical access to the data. Since Company X manufactured components for both commercial and military vehicles, support for ITAR processing was an important attribute for selecting a resource provider.

After examining several cloud computing providers, we chose IBM's Computing on Demand (Cod) service because they were the only provider able to provide us with both a high-speed interconnect and an ITAR-compliant environment. We rented a 14-node dual-processor, dual-core, Intel-based Linux cluster, connected via a high-performance Infiniband network between the nodes.

DEMONSTRATION PROBLEM

The computational engineers selected one of their fuel nozzle designs to serve as the driver problem for the feasibility study. The finite element model for the nozzle study has approximately six million degrees of freedom (MDOF), which is a relatively large model for Company X. The corresponding ANSYS database file takes up about 1.7GB of space on the file system. They chose this particular design because it was one they happen to be working on at the time of the study, and because the structure was more complex than their typical models, which made it a good candidate for doing the simulation on an HPC system.

Transient Thermal Structural Response

The main simulation of interest was a transient thermal structural response problem, also known as a thermal shock problem, and is shown in Figure 2. As the nozzle operates, fluids at changing temperatures move through it. These moving fluids result in convective heat transfer that causes changes in temperature in the nozzle. The purpose of the simulation is to estimate the structural stresses that the fuel nozzle undergoes over an interval of time because of these temperature changes. Using information from the customer, the computational engineer estimates the thermal surface loads at several time points within the time interval of interest, which serve as boundary conditions for the simulation.

The complete simulation is a two-stage process. The first stage is a transient thermal simulation, where the ANSYS software calculates how the temperature distribution of the nozzle varies over time given the boundary conditions. The second stage is a static structural simulation, where the ANSYS software calculates the structural stress-es in the nozzle at a particular point in time using the temperature distribution at that time point from the thermal analysis. The engineers chose 81 time points in the interval for doing structural analyses. Note that the structural analysis at each time point is independent: once the temperature distribution is known, no other information from previous time points is required.

The workflow for this problem is shown in Figure 3. Each time ANSYS is invoked, it takes as input a database file containing a finite element model (in this case, nozzle.db for all invocations), and a command file that specifies what analyses are to be done. By convention, we put the word "driver" in the file name for these command files. In addition to the database file and command file, there may be other inputs depending on the problem.

In depicting the thermal analysis, we have separated it out into two stages to show the flow of the various data files. In practice, the bc-driver.txt and thermal-driver.txt file can be concatenated and this can be run as a single ANSYS invocation.

The thermal boundary conditions files contain information about the expected thermal loads of the fuel nozzle over time, based on information from the customer. For this simulation, the engineer selected 40 time points over the interval of interest

Figure 2. Transient thermal structural response

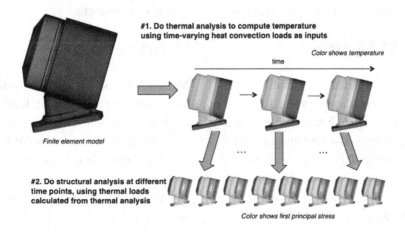

#1. Do thermal analysis to compute temperature using time-varying heat convection loads as inputs

time

Color shows temperature

Finite element model

#2. Do structural analysis at different time points, using thermal loads calculated from thermal analysis

Color shows first principal stress

Figure 3. Workflow for transient thermal structural response

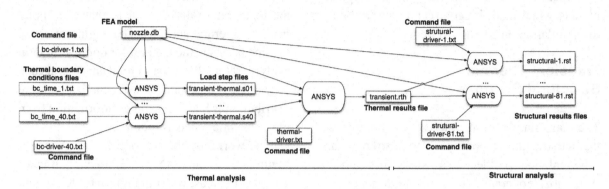

for doing the transient thermal analysis. Within the model, the engineer assigns variable names to areas of interest (e.g., ZONE1), and the boundary condition files specify temperature and heat loading conditions to these regions. Before ANSYS can run this simulation, it must translate those constraints from named regions to constraints on the individual elements within the model. While the thermal boundary condition files are quite small (40-70 LOC, or 1.1KB), the corresponding load files are much larger (170 – 300 KLOC, or 135 MB). Most of the potential advantage of HPC comes from the structural analyses, which can be executed independently. Generating the load step files can also be done in parallel, although this step is not as time consuming as the structural analysis. Since ANSYS has HPC support, there is also potential benefit of running the thermal analysis on an HPC system. In addition, it is also possible to generate the load step files in parallel.

Dynamic Stress Analysis

The engineers also selected a dynamic stress analysis scenario as a challenge problem for the study. This is a frequency analysis of the nozzle at operational temperatures, also known as a model analysis. The engineer specifies a frequency range, and the ANSYS software subjects the model to structural vibrations at frequencies within the range. In such analyses, the engineers wish to verify that the natural frequencies, or modes, of the system, do not fall within the expected vibrations that the nozzle will be subjected to during its operation. These types of simulations are extremely computational intensive, and Company X would not typically attempt this type of simulation using such a large model.

As shown in Figure 4, the workflow of the dynamic stress analysis is much simpler: a single database file and command file, generating a single results file.

Challenges of Running in a Cloud Environment

The previous section demonstrates the potential benefits of running simulations on HPC systems. In this section, we discuss the challenges related to running HPC simulations within a cloud computing environment.

Licensing Issues

Most users have encountered a software license agreement, typically in the form of an end-user license agreement: a series of legalistic text that describes how we are permitted to use the software. As end-users, most of us simply click the "Accept" button without a passing glance. However, in corporate environments, software license agreements may be treated just like any other contract,

Figure 4. Workflow for dynamic stress analysis

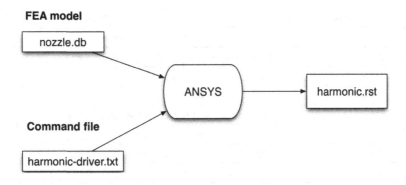

requiring approval and possible modification of license agreements by legal departments.

When we began this study, we were unprepared for the complexity of dealing with license agreements with respect to the ANSYS software. In our case, the cost was not an issue, as ANSYS provided us access to the software without charge for this project. But they had never dealt with the complexity of the situation where the license holder (University of Southern California) was different from the intended users (engineers at company X), which was also different from the site where the software would be installed (IBM Computing on Demand).

Complicating issues even further, the legal department at the University of Southern California desired some modifications to the language of the license agreement. Resolving the license agreement issues to the satisfaction of legal departments at all four organizations (ANSYS, USC, Company X, IBM) took months of document exchanges. Table 1 shows an abbreviated timeline of the events involved in obtaining access to the IBM Computing on Demand resources and the ANSYS software and licenses. Note the span of time between when ANSYS agreed to donate licenses for this project to the time that software was obtained (five months), and the time between establishing what IBM computing services are requested and when the services became available (four months).

Software and Hardware Configuration Issues

IBM's Computing on Demand uses the Infrastructure-as-a-Service (IaaS) model; IBM initializes the nodes with a supported Linux distribution of the user's choice, and then provides the user with root access. It is then up to the user to install and configure the applications they wish to run on the cloud. In our case, that involved installing ANSYS Mechanical, as well as the TORQUE open-source resource manager for launching parallel jobs on the node.

The fourteen IBM compute nodes we rented had both gigabit Ethernet and Infiniband network interfaces. ANSYS can be run in HPC mode to take advantage of multiple processors, implementing parallelism through the use of the Message Passing Interface (MPI) library. The term "MPI" does not refer to a specific library, but rather a standardized library interface (Dongarra et al., 1996) for FORTRAN, C and C++, for which there are multiple implementations. ANSYS ships with two MPI implementations, one by HP (the default) and another by Intel. For ANSYS to take advantage of the high-speed Infiniband interconnect, it requires Infiniband support in both the operating system and the MPI implementation. We initially wanted to install CentOS Linux 5 as the operating system, but the drivers were not compatible with the Infiniband interfaces, so we

Table 1. License negotiation abbreviated timeline

Date	Service	Event
7/15/08	IBM	Initial telecon with IBM to discuss service
7/18/08	IBM	IBM sends USC preliminary service agreement
7/24/08	IBM	IBM sends USC pricing proposal
7/29/08	ANSYS	**ANSYS agrees to donate license**
8/6/08	IBM	**Telecon with IBM to establish service requested**
8/13/08	IBM	Updated pricing info from IBM
9/4/08	IBM	IBM sends USC service agreement
9/9/08	IBM	USC legal requests changes
9/23/08	ANSYS	ANSYS sends license agreement
10/13/08	ANSYS	USC legal requests changes
10/16/08	IBM	IBM responds to USC legal changes
10/21/08	IBM	USC legal requests changes
10/22/08	IBM	IBM legal responds to USC
10/23/08	IBM	USC legal responds to IBM
11/18/08	IBM	USC signs IBM agreement
11/19/08	ANSYS	ANSYS legal responds to USC
11/19/08	IBM	USC requests billing info from IBM
12/1/08	ANSYS	IBM comments on ANSYS letter
12/4/08	IBM	IBM signs agreement, provides VPN access
12/10/08	IBM	IBM sends machine login information
12/17/08	IBM	**IBM nodes become accessible**
12/17/08	ANSYS	ANSYS responds to IBM
12/19/08	ANSYS	USC signs ANSYS agreement
12/26/08	ANSYS	**ANSYS software obtained**

were forced to install CentOS Linux 4, instead. According to the documentation, both the HP and Intel MPI implementations had support for Infiniband. However, in practice, the HP-MPI library was not compatible with the Infiniband interconnects installed on our nodes, which caused ANSYS to crash each time we attempted to run it over Infiniband. Switching to the Intel MPI library eventually resolved the problem.

When we had the software configured and began doing ANSYS runs, we soon discovered errors due to the hard drives on the compute nodes filling up. The default configuration had only 67 GB of space, which was inadequate to store the input files, output files, and intermediate files from our ANSYS runs. We upgraded the storage on the local nodes to support the simulations.

Network and VPN Configuration Issues

To access remote resources in the cloud, the computational engineers must be able to log into the remote machines from inside Company X. However, the company's network configuration was not designed with this type of remote access in mind. In general, the IT personnel work to secure the network to prevent unauthorized machines from joining the network.

Company X's network only allows outbound connections to the web, and only through a web proxy. The IT department had to change the firewall rules to allow outbound SSH connections from certain machines so that computational engineers could log into a test cluster at ISI.

To access IBM Computing on Demand resources, users connect through a virtual private network (VPN). We had installed the ANSYS license server on the head node of the test cluster at ISI, so that we could have an HPC environment for running ANSYS applications when not renting time on IBM. Each node in a cluster that runs an instance of ANSYS needs to be able to connect to an ANSYS license server and check out a token. The implication of this was that the nodes in IBM's cluster had to be able to access the head node of the ISI cluster. Therefore, we had to maintain a VPN connection between the ISI head node and IBM whenever running ANSYS.

Ideally, the VPN connection should be maintained indefinitely. Each time the head node initiated a new connection to the VPN, it was assigned a new IP address on the VPN, which necessitates altering a configuration file on the IBM nodes to point to the right location for the license server.

IBM provides a range of VPN connectivity options depending on the number of concurrent users required and the maximum throughput needed for the application. These options included both "hardware VPNs" (physical machines that are connected to the network) and "software VPNs" (software clients that run on a machine with Internet access). In our circumstance, the overhead of configuring a hardware VPN for a network seemed too high, so we chose a software VPN solution.

IBM offered two software VPN solutions: a software VPN client that works over UDP, and an "SSL VPN" that works over an SSL tunnel, which uses TCP. We began with the UDP-based software VPN client, since a UDP-based VPN tunnel should have better performance. The commercial Linux version of the VPN client that we obtained did not function properly on the license server machine. Fortunately, an open-source implementation of the VPN client was available, and worked properly. However, we discovered that the VPN connection would automatically timeout after about 24 hours. This 24-hour timeout was hard-coded into the VPN server implementation.

We opted instead to switch to an SSL VPN solution. This tunnels the VPN over an SSL connection. This had a lower effective throughput than the UDP-based VPN, but it did not have the timeout issues. In addition, because the network bottleneck was not based on the VPN, the lower performance of the SSL VPN was less of an issue.

We did run into several troubles with the SSL VPN. In particular, the supplied VPN clients did not work properly on Linux. Fortunately, an open-source implementation of a Linux client was available that was compatible with the VPN (Knight, 2007).

Network Performance

The goal of using a cloud computing environment is to save compute time. However, there is overhead associated with transferring data to and from the cloud.

Figure 5 shows the network configuration. As mentioned earlier a customer has a choice of different VPN solutions depending on the network throughput and number of concurrent users required. However, in our case, the performance bottleneck of the network was on Company X's network. The computational engineers were located at a branch office of Company X, and all network traffic bound for the Internet had to be routed through the corporate office, over a T1 line that was shared by the entire branch office.

Our network transfer tests revealed an effective throughput of about 130 KB/s from Company X to IBM CoD. Table 2 shows estimates of the file transfer times. Note that a structural analysis

Figure 5. Network configuration

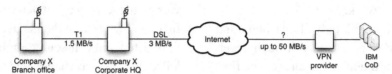

consists of 81 independent structural analyses and therefore generates 81 structural results files, each of which is approximately the same size.

Recall that we saved about 90 hours (about four days) by running the transient analysis in the cloud environment. While the time to upload the input data is reasonable (about four hours), the download times for the output data are clearly prohibitive. At these transfer times, this time saving is swamped by the amount of time required for downloading all of the data (587.9 h), which is roughly three and a half weeks.

The result was that we needed to identify an alternate way for the engineers to be able to access the simulation data for *post-processing*. Typically, when a simulation is complete, the engineers visualize the simulation results through ANSYS, using false color to view the distributions of different types of analysis, such as temperature and various different types of stress measures. This is a highly interactive process, as the engineer manipulates a 3D model, selecting and zooming in on various areas.

We evaluated the feasibility of doing interactive remote post-processing. Since ANSYS on Linux is an X11 application, in principle it can be run remotely. In practice, the bandwidth of the network connection was simply too low to allow the computational engineers to view the simulation results over the network. We evaluated two remote visualization solutions that were designed to provide better performance than X11.

NoMachine's NX accelerates remote X11 applications by using various strategies to compress the X protocol, reducing the amount of round-trip network interactions (Pinzari, 2003). While we did observe better performance using NX, the response time was still much too slow to be usable by the engineers.

One of the drawbacks of NX is that it cannot take advantage of the 3D acceleration provided by modern video cards, and post-processing involves visualization and manipulation of a 3D model. To evaluate whether 3D acceleration would improve performance, we evaluated VirtualGL (Commander, 2009a) to accelerate remote visualization. VirtualGL takes advantage of graphics accelerator hardware located on the remote server. It is designed to be used in conjunction with other remote visualization tools. We evaluated it in conjunction with NX, as well as with TurboVNC (Commander, 2009b), a remote visualization tool

Table 2. File transfer times

		Size	Transfer time
	Input (model) file	1.7 GB	3.8 h
Transient thermal & structural analysis	Thermal results file	46.7 GB	104.6 h
	Single structural results file	3.2 GB	7.2 h
	All (81) structural results files	259.2 GB	580.7 h
	Total	262.4 GB	587.9 h
Harmonic analysis	Harmonic results file	61.1 GB	136.9 h

that implements the Virtual Network Computing protocol (Richardson, 2009). The compute nodes in our cluster at IBM did not contain any graphics accelerator hardware. Fortunately, IBM also maintains a visualization node that is intended specifically for supporting remote visualization and contains graphics accelerators. To evaluate the VirtualGL solution, we obtained access to the visualization node. The responsiveness of the remote application was too poor for it to be usable. Our final solution was to implement a web-based post-processing solution. Another potential solution would be to simply ship an external hard disk via overnight delivery from IBM to Company X.

WEB-BASED REMOTE POST-PROCESSING

To support the engineers in doing post-processing of the data, we had to provide them with information about the results without requiring them to download hundreds of gigabytes of data. Networking issues aside, we also had to provide them with some way of dealing with this large volume of data. In their traditional workflow, the engineers would work sequentially, running a single simulation, examining the results, and then running another one. In this case, we had given them the results from 81 different simulations, and they had no way of dealing with all of this data other than to examine each file sequentially.

We developed the HPC Remote Simulation Portal (RemoteSimPortal) to support remote, parallel post-processing. RemoteSimPortal is a web-based front-end for ANSYS that allows the engineers to explore the results from many simulations at once. Figure 6 shows a screenshot of the initial view of RemoteSimPortal. This view shows the results of the model for several different types of analyses, at several different viewpoints.

Clicking on one of these images will provide a thumbnail view. Assuming the results contain multiple time points (in the case of a transient

analysis) or multiple frequency points (in the case of a harmonic analysis), all of the different points will be visible. The engineer can scan through to identify points of interest, and then click on one for more detail.

The engineer may also select a new viewing angle, as well as selecting a subset of nodes to focus on a region of interest. The engineer brings up the input model in ANSYS, chooses the desired view with the ANSYS interface, and selects the desired nodes. Next, the engineer opens up the ANSYS session editor, which keeps a log of all of the user interface commands, and copies and pastes them into a field in a web form and clicks the "Generate" button. The result is a visualization of the subset of nodes, as well as a text listing of the numerical values of the selected nodes, which can then be copied and pasted into a text file for later analysis with a different program such as Excel. In this way, RemoteSimPortal provides some interactivity without the need for large downloads.

PERFORMANCE ANALYSIS

In this section, we compare the time to execute these simulations on IBM's Computing on Demand resources, as compared to the time to execute the simulations on Company X's desktop. Table 3 shows a summary of the computing hardware used for the comparisons.

Transient Thermal Structural Response

Recall that the main simulation was a transient analysis that involved a thermal and structural analysis over an interval of time (Table 4). In the first phase, a thermal analysis is performed that models the heat flows through the nozzle over the time interval. This analysis will determine how the temperature distribution across the nozzle changes over the time. In the second phase, several time points on the interval are selected and structural

Figure 6. RemoteSimPortal initial view

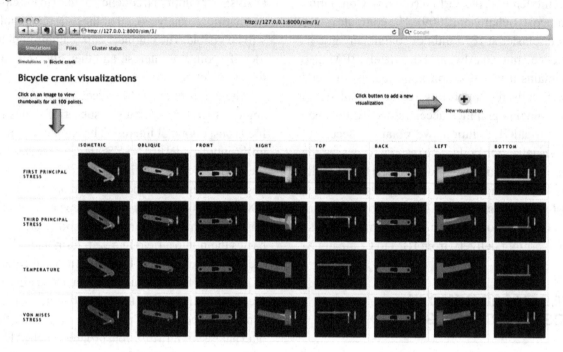

analyses are performed using the temperature distribution. These structural analyses can be executed independently. For this simulation, we ran at 81 different time points.

Company X would not attempt to do such a simulation on the desktop, because of the amount of time it would take (roughly four days of computation).

Note that we did not separate out the time to generate the load files from the time to do the thermal simulation, because we did not have this information separated out in the desktop-based simulations done at Company X. On our cluster, generating the load step files in parallel reduced the load step generation time from 86 minutes to about 9 minutes.

Also note that the structural analysis can potentially scale linearly with the number of cores in an HPC system, because the analyses are all independent. The cluster we used had 56 cores, so in principle we should have been able to reduce the structural execution time even further. Unfortunately, as we only had 10 ANSYS licenses, we could only run 10 concurrent simulations.

Table 3. Machine characteristics

	Desktop (Company X)	**Cluster (IBM CoD)**
# of nodes	1	14
Processor	Intel dual-core Xeon 3.8 GHz	Intel dual-processor dual-core Xeon 3.0 Ghz
RAM	12 GB	16 GB
Interconnect	N/A	Infiniband

Table 4. Transient analysis simulation time

	Desktop (Company X)	Cluster (IBM CoD)
Thermal analysis	1029m (17h 9m)	148 m (2h 28m)
Structural analysis	4852m (80h 52m)	328 m (5h 28m)
Subtotal (computation time)	5881m (98h 1m)	476 m (7h 56m)
Data transfer	0	219 m (3h 39m)
Total (computation+transfer)	5881m (98h 1m)	695 m (11h 35m)
Time saved	5186 m (86h 26m)	

Dynamic Stress Analysis

A harmonic analysis is used to determine how the nozzle will respond to vibration. A physical system such as a fuel nozzle has a set of normal modes, or resonant frequencies. It is important to confirm that the operational behavior of the nozzle does not cause it to be driven at one of its resonant frequencies, or structural failure is more likely. Harmonic analysis is a very compute-intensive task, and Company X would not typically attempt such a simulation on such a large model because of the execution time involved. Table 5 shows the execution time results. The engineers at Company X were not able to run the model to completion on a desktop: even after 4 weeks of execution, the simulation had not converged to a solution. By contrast, executing in the IBM CoD environment, the model completed executing in 88 minutes. Including data transfer time (219 minutes, as in the transient analysis) brings it to a total of about five hours.

Advantages of a High-Speed Interconnect

Earlier, we mentioned the advantage of having a high-speed interconnect for the cluster for engineering simulation problems. We ran the thermal analysis on different numbers of cores on the IBM cluster, as well as a test cluster that we had at ISI. On our ISI cluster, we had gigabit Ethernet network interfaces, which have lower bandwidth and higher latency than Infiniband. Figure 7 shows the execution times of the thermal analysis for different scenarios. We can see that we get reductions in execution time when running up to 16 cores on the ISI cluster, after which the performance degrades. By contrast, the IBM performance improves up to the maximum 56 cores, although performance improvements are moderate.

Estimated Economic Benefits

The scope of the study was too small to directly measure economic benefits. Instead, the computational engineers estimated the potential benefits

Table 5. Harmonic analysis simulation times

	Desktop (Company X)	Cluster (IBM CoD)
Harmonic analysis	>4 weeks (did not finish)	88 m (1h 28m)
Data transfer	0	219 m (3h 39m)
Total	>4 weeks (did not finish)	307 m (5h 7m)
Time saved	> 4 weeks	

Figure 7. Thermal analysis execution time vs. number of cores

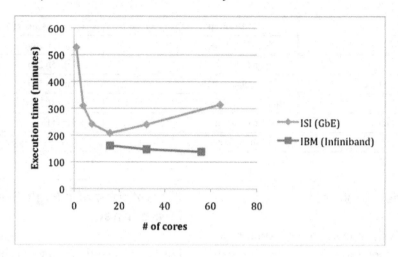

of access to larger computational resources. We considered three classes of benefits, in order of increasing payoff but also increasing uncertainty in estimates:

- Engineering productivity increase due to more efficient workflow
- Risk reduction by reducing probability of failures
- Wider adoption of HPC within the organization (manufacturing & quality)

Engineering Productivity Increase

The first estimate assumed a productivity increase by doing the same amount of work in less time. Engineers at Company X worked with one of their customers, a large aerospace company, to estimate the amount of time they would save through the use of value stream maps (Rother et al., 1999). They estimated the time savings of using HPC for their current finite element analysis, which typically involves 8 iterations, at about 43%. In addition, they estimated the time savings by applying parametric tools to set up multi-node cases and automate the post processing of results to be about 76%.

Risk Reduction

One of the primary goals of the computational engineers is to detect and prevent "design escapes": defects in the design of the nozzle that lead to problems when the nozzle is in operation. In mechanical engineering, as in software engineering, defects become exponentially more expensive to repair the later in the life cycle they are detected (Boehm, 1981). The aim of performing these simulations is to identify design defects earlier in the life cycle so that they are less expensive to fix.

Because the computational engineers have a fixed amount of time that they can devote to analysis, they must make simplifications to their models when running on the desktop so that the analysis will complete in the time available. One of the potential advantages of HPC is the ability to run higher-fidelity simulations in the same amount of time, increasing the likelihood of identifying any design defects.

While we feel that risk reduction is a significant potential economic benefit, because of data confidentiality issues, we cannot present the estimates of savings due to risk reduction in this paper.

Wider Adoption: Manufacturing and Quality

This study focuses on the computational engineers who work on identifying design defects early in the lifecycle. However, many of the development costs associated with Company X's products are due to issues in the manufacturing process, rather than the design process. Thermal analysis of the heat treatment process used in manufacturing could allow Company X to optimize the process time, increasing throughput and decreasing scrap and rework rates. The engineers estimate that productivity gains of 33% could be achievable if the manufacturing process was optimized.

Because of the uncertainties associated with the manufacturing process, it is impossible to completely eliminate manufacturing defects. In the face of manufacturing variations, the engineers must make an assessment about whether a field recall is required. This requires extensive analysis to evaluate risks, and the use of higher fidelity simulations would be directly applicable in this case, with potential effort savings on the order of 80% expected.

LESSONS LEARNED

There Are Significant Potential Benefits From Running Simulations on HPC

Modern engineering software packages are now mature enough to run well in an HPC environment provided a high-speed network interconnect is available. Our study revealed that the ANSYS software package showed performance improvements up to 56 cores, the maximum size of the cluster we had access to, when we ran a harmonic analysis on a real problem. Therefore, access to HPC resources has the potential to provide real value to solving engineering problems.

Preparing to Run on the Cloud Will Take Longer Than You Expect

It takes time to get to the point where you can run engineering simulation jobs in a cloud-computing environment. Legal issues related to contracts and software licenses can add substantial delays. The local IT department may need to make network configuration changes to allow access to external resources. Several different VPN solutions may need to be evaluated to determine which one works best with the software and network configuration. Software must be installed and configured for running parallel jobs. Doing an install on an HPC system takes longer than one on a desktop system, because of the extra configuration steps required to run in parallel. In particular, when using a high-speed interconnect like Infiniband, some additional configuration may be necessary to make sure the drivers are working and are compatible with the installed software.

Substantial IT Expertise is Required for Engineers to Leverage the Cloud Today

To assist the computational engineers, we had to install and configure ANSYS for running parallel jobs, install and configure resource management software for running many smaller, independent ANSYS jobs, and we had to write several scripts to generate the ANSYS command files and shell script files to run the simulations. In addition, we had to overcome technical issues related to network configuration for remote access to the cloud and software configuration for optimized performance within an HPC system. All of these require a level of IT knowledge that a typical computational engineer would not necessarily possess. It is particularly a problem if the computational engineer has experience primarily with Microsoft Windows-based environments, as all of these tools required substantial familiarity with Linux-based environments.

Poor Network Connectivity is a Major Obstacle for Doing Post-Processing

From a purely technical point of view, the networking "last mile" problem is the most fundamental obstacle to running engineering simulation jobs on the cloud. On this project, we developed a custom web-based front-end to work around the inadequate network bandwidth required for fully interactive remote post-processing. However, most small-to-medium engineering companies probably do not have the resources to develop this type of custom software to use the cloud.

It would be advantageous for companies if they could temporarily obtain high-speed Internet access for running these simulations, but we are not aware of any such resources currently being offered commercially.

CONCLUSION AND FUTURE WORK

The results of our feasibility study suggests that cloud computing environments with HPC-like resources are potentially useful for performing computational engineering tasks, but substantial IT expertise is required to use them effectively today. In addition, if the engineering organization lacks a high-speed Internet connection, then alternative post-processing strategies need to be pursued.

This study focused solely on estimating the benefits of using a cloud computing environment for computational engineering. For future work, we plan to measure the associated costs to provide more complete information for estimating return-on-investment. We are also interested in exploring the execution of multiphysics simulations in cloud environments, where simulations from different software packages are combined (e.g., finite-element analysis with ANSYS Mechanical, computational fluid dynamics with CD-Adapco STAR-CD). In addition, RemoteSimPortal is still under active development, and we plan to conduct usability studies with the engineers to determine

how well it can substitute for local post-processing and to tailor the interface to better fit into their workflow.

This study focused on simulations being performed in the context of a single organization. While each individual organization can potentially benefit from higher fidelity simulations, we hypothesize that significant benefits in overall engineering productivity and product quality can be best achieved through simulations that go across organizations in the supply chain. Consider the case of an automobile, where the automobile manufacturer purchases an engine from a supplier, and the engine manufacturer purchases the nozzles for the fuel injection system from a supplier. Currently, the automobile manufacturer is not able to leverage the computer-based models employed by the engine manufacturer and the fuel nozzle manufacturer, because organizations are not necessarily willing to share details of their simulations with customers or suppliers. However, if it were possible to integrate these models, then the automobile manufacturer would be able to run higher-fidelity simulations than is currently possible. Future work will explore the technical, social, and business challenges to identify how to facilitate such integrated simulations across a supply chain.

ACKNOWLEDGMENT

This research was performed under the DARPA HPC-ISP-PILOTS project under Air Force award #FA8750-08-C-0184 to the University of Southern California. We wish to acknowledge the computational engineers who contributed substantially to the content of this paper, but who must remain anonymous. We also wish to acknowledge AN-SYS, Inc., for providing us with the software licenses used in this project. We also would like to thank Jeff Carver for his feedback on an earlier draft of this paper.

REFERENCES

Amazon. (2009). *Amazon elastic compute cloud user guide* (API Version 2009-11-30). Retrieved June 22, 2010, from http://docs.amazonwebservices.com/AWSEC2/latest/UserGuide/

Armbrust, M., Fox, A., Griffith, R., Joseph, A. D., Katz, R. H., Konwinski, A., et al. (2009). *Above the clouds: A Berkeley view of cloud computing* (Tech. Rep. No. UCB/EECS-2009-28). Berkeley, CA: University of California, Berkeley.

Basili, V. R., Cruzes, D., Carver, J. C., Hochstein, L. M., Hollingsworth, J. K., Zelkowitz, M. V., & Shull, F. (2008). Understanding the high-performance-computing community: A software engineer's perspective. *IEEE Software*, *25*(4), 29–36. doi:10.1109/MS.2008.103

Bialecki, A., Cafarella, M., Cutting, D., & O'Malley, O. (2007) *Hadoop: A framework for running applications on large clusters built of commodity hardware*. Retrieved January 14, 2010, from http://hadoop.apache.org

Boehm, B. (1981). *Software engineering economics*. Upper Saddle River, NJ: Prentice Hall.

Callan, J., & Kulkarni, A. (2009). *ClueWeb09 Wiki*. Retrieved January 14, 2010, from http://boston.lti.cs.cmu.edu/clueweb09/wiki

Carver, J. C., Kendall, R. P., Squires, S. E., & Post, D. E. (2007). Software development environments for scientific and engineering software: A series of case studies. In *Proceedings of the 29th International Conference on Software Engineering*.

Cary, A., Sun, Z., Hristidis, V., & Naphtali, R. (2009). Experiences on processing spatial data with mapreduce. In *Proceedings of the 21st International Conference on Scientific and Statistical Database Management* (pp. 302-319).

Cohen, J. (2009). Graph twiddling in a MapReduce world. *Computing in Science & Engineering*, *11*(4), 29–41. doi:10.1109/MCSE.2009.120

Commander, D. (2009a). *User's guide for VirtualGL 2.1.4*. Retrieved January 14, 2010, from http://www.virtualgl.org/vgldoc/2_1/

Commander, D. (2009b). *User's guide for TurboVNC 0.6*. Retrieved January 14, 2010, from http://comments.gmane.org/gmane.comp.video.opengl.virtualgl.user/42

Council on Competitiveness. (2009) *U.S. manufacturing – Global leadership through modeling and simulation*. Retrieved January 15, 2010, from http://www.compete.org/publications/detail/652/us-manufacturingglobal-leadership-through-modeling-and-simulation

Dean, J., & Ghemawat, S. (2004). MapReduce: Simplified data processing on large clusters. In *Proceedings of the 6th Symposium on Operating Systems Design and Implementation* (pp. 107-113).

Dongarra, J. J., Otto, S. W., Snir, M., & Walker, D. (1996). A message passing standard for MPP and workstations. *Communications of the ACM*, *39*(7), 84–90. doi:10.1145/233977.234000

Easterbrook, S. M., & Johns, T. C. (2009). Engineering the software for understanding climate change. *Computing in Science & Engineering*, *11*(6), 65–74. doi:10.1109/MCSE.2009.193

Faculty Connection. (2007). *Software entrepreneurship for students curriculum: A case study*. Retrieved from http://www.microsoft.com/education/facultyconnection/articles/articledetails.aspx?cid=844&c1=en-us&c2=0

Glotzer, S. C., Kim, S., Cummings, P. T., Deshmukh, A., Head-Gordon, M., & Karniadakis, G. (2009). *WTEF panel report on international assessment of research and development in simulation-based engineering and science*. Baltimore, MD: World Technology Evaluation Center.

Heavy, A., Lauret, J., & Keahy, K. (2009, April 8). Clouds make way for STAR to shine. *International Science Grid This Week*.

Helper, S. (2009). The high road for U.S. manufacturing. *Issues in Science and Technology, 25*(2), 39–45.

Hochstein, L., & Basili, V. R. (2008). The ASC-alliance projects: A case study of large-scale parallel scientific code development. *IEEE Computer, 41*(3), 50–58.

Hoffa, C., Mehta, G., Freeman, T., Deelman, E., Keahey, K., Berriman, B., & Good, J. (2008). On the use of cloud computing for scientific workflow. In *Proceedings of the 3rd International Workshop on Scientific Workflows and Business Workflow Standards in e-Science* (pp. 640-645).

Joseph, E., Snell, A., & Willard, C. G. (2004). *Council on Competitiveness study of U.S. industrial HPC users*. Washington, DC: Council on Competitiveness.

Keahey, K., Tsugawa, M., Matsunaga, A., & Fortes, J. (2009). Sky computing. *IEEE Internet Computing, 13*(5). doi:10.1109/MIC.2009.94

Knight, J. Y. (2007). *F5 VPN command-line client*. Retrieved January 12, 2010, from http://fuhm.net/software/f5vpn-login

National Academies. (2007). *Rising above the gathering storm: Energizing and employing America for a brighter economic future*. Washington, DC: National Academies Press.

Nurmi, D., Wolski, R., Grzegorczyk, C., Obertelli, G., Soman, S., Youseff, L., & Zagorodnov, D. (2009). Eucalyptus: An open-source cloud computing infrastructure. *Journal of Physics: Conference Series, 180*.

Pinzari, G. F. (2003). *NX X protocol compression* (D-309/3-NXP-DOC). Retrieved January 14, 2010, from http://www.nomachine.com/documents/html/NX-XProtocolCompression.html

Post, D. E. (2009). The promise of science-based computational engineering. *Computing in Science & Engineering, 11*(3), 3–4. doi:10.1109/MCSE.2009.60

Post, D. E., & Votta, L. G. (2005). Computational science demands a new paradigm. *Physics Today, 58*(1), 35–41. doi:10.1063/1.1881898

Quom, S., & Bhaskaran, R. (2002). *ANSYS short course: Three-dimensional bicycle crank*. Retrieved January 14, 2010, from http://courses.cit.cornell.edu/ansys/crank

Richardson, T. (2009). *The RFB protocol, version 3.8*. Cambridge, UK: RealVNC Ltd.

Segal, J. (2005). When software engineers met research scientists: A case study. *Empirical Software Engineering, 10*(4), 517–536. doi:10.1007/s10664-005-3865-y

Shainer, G., Sparks, B., Schultz, S., Lantz, E., Liu, W., Liu, T., & Misra, G. (2010, January 26). Cloud computing will usher in a new era of science discover. *HPCwire*.

Sotomayor, B., Montero, R. S., Llorente, I. M., & Foster, I. (2008). Capacity leasing in cloud systems using the OpenNebula engine. In *Proceedings of the Cloud Computing and Applications Conference*.

Wang, L., Tao, J., Kunze, M., Castellanos, A. C., Kramer, D., & Karl, W. (2008) Scientific cloud computing: Early definition and experience. In *Proceedings of the 10th IEEE International Conference on High Performance Computing and Communications* (pp. 825-830).

Zhao, B. Y., Yan, K., Agrawal, D., & El Abbadi, A. (2009). *Massive graphics in clusters (MAGIC) project*. Retrieved January 14, 2010, from http://graphs.cs.ucsb.edu

This work was previously published in the Journal of Organizational and End User Computing, Volume 23, Issue 4, edited by Mo Adam Mahmood, pp. 31-50, copyright 2011 by IGI Publishing (an imprint of IGI Global).

Chapter 18
Scientific End–User Developers and Barriers to User/ Customer Engagement

Judith Segal
The Open University, UK

Chris Morris
STFC Daresbury Laboratory, UK

ABSTRACT

When software supports the complex and poorly understood application domain of cutting-edge science, effective engagement between its users/customers and developers is crucial. Drawing on recent literature, the authors examine barriers to such engagement. Significant among these barriers is the effects of the experience that many research scientists have of local scientific end-user development. Through a case study, the authors demonstrate that involving such scientists in a team developing software for a widely distributed group of scientists can have a positive impact on establishing requirements and promoting adoption of the software. However, barriers to effective engagement exist, which scientific end-user developers can do little to address. Such barriers stem from the essential nature of scientific practice.

INTRODUCTION

The aim of this paper is to shed some light onto the problems of user/customer engagement in scientific software development. We define scientific software to be software specifically designed either to advance science directly, for example, by providing models and simulations to investigate problems where the science is too fast or too slow or too large or too small or too complex or too dangerous to investigate in vivo (Wilson, 2006), or to support the practice of science, for example, by providing means by which a community of scientists might share remote instruments or data. There is (at least) one common factor underlying

DOI: 10.4018/978-1-4666-2059-9.ch018

every scientific software development and that is the necessity of effectively involving the users, the scientists who are going to be using the software in their workaday lives, in the development in some way. This statement is probably true of all software developments to a greater or lesser extent, but is especially relevant to scientific software where the application domain is both complex and only partially understood, even by experts in the domain. It is equally important to involve the customers effectively, that is, those scientists who control the purse-strings of the development and have overall high-level responsibility for the direction in which the science is going and hence for how the software might best support this direction. As we shall discuss later, the users and the customers may or may not be the same people.

In this paper, we suggest that the deployment of scientific end-user developers on the development team might go some way towards addressing the problem of how to optimize user/customer engagement with developers. We base our suggestion on evidence from a case study carried out by the first author over a period of about three years, tracking the development of a Laboratory Information Management System (LIMS). This LIMS is intended to support a community of biologists, that is, biologists who are not necessarily co-located but who are working on different problems within the same sub-discipline. The LIMS development team consisted of software engineers with no prior knowledge of biology, software engineers with some biological background and scientific end-user developers. The term 'scientific end-user developer' used in the context of this development team is somewhat ambiguous. The people referred to *are* scientific end-user developers in the sense that they think of themselves primarily as scientists, have little or no formal education in software development, and have considerable experience of developing software for their own use and for the use of other scientists working closely with them. They are *not*, however, representative of the potential user

group of the LIMS, though they do work alongside representatives of that group.

We begin this paper by discussing barriers to user engagement in general, and then in the particular context of software development for a distributed community of research scientists. We focus specifically on those barriers arising from the experience of many research scientists of developing their own software in a very local context, that is, in order to address a particular scientific problem at a particular point in time at a particular laboratory.

We then describe our case study and its findings. The overall finding is that including scientists with scientific end-user development experience on the development team of a significant piece of community software can have a very positive effect on the development. We are aware that some readers may baulk at the idea of taking seriously findings based on a single case study. But sometimes a case study grounded in real-world practice with all its richness of context is all that is needed to convince a reader that 'yes, this must be true'. And sometimes the findings of a case study are strengthened by their being consistent with findings in other disciplines, as is the case here with an aspect of diffusion theory (Rogers, 2003).

We continue with a discussion on the scope of our findings, both as regards the extent to which scientific end-user developers can improve user/developer relations and the extent to which our results can be generalized. As is traditional, we conclude with a summary and conclusions.

BARRIERS TO USER ENGAGEMENT IN COMMUNITY SOFTWARE DEVELOPMENT PROJECTS

We begin by discussing our use of two terms: 'community software' and 'user'. By the term 'community software', we mean software which, like the LIMS of our case study, is intended to support the activities of a distributed group of scientists working on similar scientific problems in the same scientific sub-discipline.

The term 'user' is often used to encompass the roles of both user and customer (as is the case in the heading of this section). However, sometimes the roles need to be distinguished. We consider users to be those scientists who interact with the software as part of their workaday world; the customers are those people who commission the software and hold the purse strings. The user and the customer may be the same person as is often the case in end-user development. But in the context of community software, they are likely to be different. For example, the customers might be a group of senior scientists heading a group of laboratories; the users might be the post-doctoral fellows and PhD students who do the hands-on work in the laboratories. Effective engagement of the development team with both customers and users is equally important. In the case of the customers, such engagement involves their overseeing the overall direction of the development to ensure that the software meets the aims of the science, and their providing resources (including access to users) where required. In the case of the users, their engagement is vital to ensure that the software really does meet their needs at a workaday level. If it does not, then there is every possibility that they will not use it, despite encouragement from the senior scientists.

In what follows, we shall use the term 'user' to encompass both user and customer unless we specifically state otherwise.

In the rest of this section we briefly discuss the existing literature on barriers to user (as opposed to customer) engagement in general. We then turn our attention to the context of community-based scientific software development and the barriers to both user and customer engagement arising from the cultures of scientific endeavor and of scientific end-user development.

General Barriers to User Engagement

Even when software developers use methodologies which are specifically user-centered, such as participatory or user-centered design, ensuring effective user engagement is difficult and it is not always clear as to how it might best be done (Kujala, 2003; Wagner & Piccoli, 2007). Reasons for the reluctance of potential users (as opposed to customers) to engage with software developers in general may include:

- A reluctance to interrupt their current work for the sake of a promise of future benefit
- Fear of having to learn new software and, even more importantly, new work practices
- Fear that the advent of new software might lead to a de-skilling of their jobs
- Fear that new software might put their jobs in jeopardy.

Barriers to User and Customer Engagement Arising from the Culture of Scientific Endeavor

Reward and recognition within the sciences has traditionally been based on individual achievements. Scientists have competed to be the first to make a discovery, answer some scientific question or author the first publication on a topic. Working cooperatively in a community supported by software entails a significant change in attitude on behalf of both users and customers (Star & Ruhleder, 1996; Finton, 2003). The developer has thus both to take cognizance, and to be seen by scientists to take cognizance, of issues such as motivating scientist engagement, protecting data access and providing data provenance (Hine, 2006).

Figure 1. A model of scientific end-user software development (adapted from Segal & Morris, 2008)

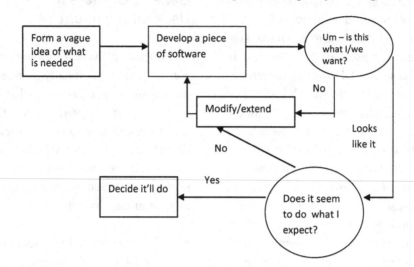

Barriers to User and Customer Engagement Arising from the Culture of Scientific End-User Development

Many users and many customers of community-based software developments have experience of scientific end-user development, and this experience colors their views of software development and professional software developers as we shall now discuss.

Scientists in general have a long tradition, dating back to the 1940s, of developing their own software to address a particular scientific problem that has just arisen for them and/or for their close colleagues. In our previous field studies of the software development activities of financial mathematicians, earth scientists, space scientists and structural biologists (Segal, 2005, 2007, 2009a, 2009b), we have articulated the culture of this scientific end-user development (also called professional end-user development) in terms of a process model of development (Figure 1), together with the values and assumptions which underlie this model. As we shall now describe, expectations raised by this culture create barriers to effective engagement between those scientists

(both users and customers) who are steeped in it, and software developers.

In this model, a scientific problem emerges for which a software solution seems feasible; the scientist writes that software and applies it to the problem. If the software does not appear to address the problem adequately, then the scientist tinkers with it until satisfied. During this tinkering, further requirements typically emerge ('it would be really nice if the software were to do *that*'). Because the software is typically intended to address a very particular scientific problem which has arisen within a very small co-located community (consisting, perhaps, only of the developer), there is no point in spending time making the software maintainable (the expectation generally is that the software will be discarded once the problem is solved) nor in testing it rigorously provided the scientist perceives that it has addressed the problem intended.

Scientists steeped in this culture are thus not used to articulating requirements fully before implementation begins. With respect to scientific software intended for a distributed community of scientists, it may or may not be the case that requirements are largely known before implementation. For example, the software development may consist of implementing a model of an aspect of natural science which is reasonably well understood, for example, stresses within beams (Smith et al., 2007). In such cases, the requirements for the software are reasonably clear. (though as Smith et al. point out, aspects such as the assumptions underlying the modeling and the accepted level of tolerances should be clearly articulated). In other cases, the gap between the commissioning of the software and its implementation – a gap in which the science and scientific practice of the application domain might evolve in unpredictable ways – can cause major problems. This is an especially relevant issue in the case of big cross-disciplinary software systems, such as e-science and cyberinfrastructure systems. These systems specifically aim to change radically the practice of scientists from working individually in a lab or with a group of co-located close colleagues to working within large, multi-site, cross-disciplinary communities (Zimmerman & Nardi, 2006). It is clearly very difficult for the scientist in such a situation to envisage the potential of such software and thus articulate their requirements for it. Indeed, even when such software exists, efforts may need to be expended to enable scientists to recognize its potential in supporting their science (e.g., the description of growing a community of users for TeraGrid in Zimmerman & Finholt, 2007).

When a requirement has been identified, the culture of scientific end-user development leads scientists to expect it to be met almost immediately. However, in the case of the development of community software, this requirement has to be considered in the light of many others from other members of the community and its priority judged before it is implemented (if it is implemented at all). And since this implementation cannot then be subject to almost instant fixes as in the scientific end-user developer model, it has to be methodically tested. And all this takes time, much to the frustration of the scientist who wants the software *now*.

A further cause of tension between scientists and software developers arising from the culture of scientific end-user development is the different perceptions of the value of software development knowledge and skill (Segal, 2007, 2008, 2009a). A pervasive assumption among the scientists in our field studies is that 'anybody can develop software'. It should be noted, however, that these studies did not include computational scientists who devote their careers to using high performance computing to develop models and simulations. The first author has been told that such 'in silica' scientists (as opposed to experimental or theoretical scientists) consider themselves an elite. Whether software development knowledge and skill is considered to be the preserve of an elite group of scientists or accessible to all, the fact remains that the questions that scientists discuss publicly are scientific ones, and not the sort of questions that software engineering attempts to answer. The scientist is thus likely to view software development knowledge and skill as having less value than scientific knowledge and skill, and so a scientist's frustration when a software developer challenges his/her opinions on aspects of the development (including that of the actual implementation, as seen in Segal, 2009a) is understandable (if not always reasonable).

All these characteristics of scientific end-user development – the general perception that requirements do not need to be specified fully upfront but rather emerge during a very iterative implementation; the expectation that scientific software can be developed quickly as part of the whole process of addressing a particular scientific problem; the perception that issues of testing rigorously and writing for maintainability are not important; the perception that scientific knowledge and skills are more important than software development knowledge and skills – pose challenges to the effective communication between professional software developers and scientist user/customers.

We now describe our case study and the light it throws on the role of scientific end-user developers in addressing these challenges.

OUR CASE STUDY

This paper is based on one aspect of a longitudinal case study of scientific software development conducted by the first author over a period of three years. The second author was appointed as project manager to the software development at the start and later came to function as software team leader. At the time of his appointment, he was a software engineer with considerable commercial experience but none of either interacting with users or of scientific software development. He is referred to in the narrative below as "the lead developer".

We begin by discussing the methodology of the case study and then consider the software at the heart of the case study and the context in which it was to be used. We go on to describe how the development was almost stymied by problems with user and customer engagement and then rescued by scientific end-user developers.

The Methodology

The data for the case study is, in general, very strongly triangulated. Over the three years of the study, the first author conducted:

- Fourteen separate observations of day-long meetings of the development team;
- Ten individual interviews with developers and customers of duration ranging from half an hour to two hours. These were audio recorded and then transcribed;
- Many phone and email communications.

In addition, the first author was given access to all the relevant project documentation. All field notes, interview transcripts and other relevant documents (e-mails, etc.) were analyzed in the same way. Semantic units were coded using an inductive iterative coding scheme and the codes were then grouped to form themes. Validation of the researcher's interpretation of the data was sought and provided by those people who had contributed most of it. More details of the case study are in Segal (2009a). That paper does not, however, focus on user engagement as this one does.

The narrative reported in this paper is based both on the triangulated case study data and on the perceptions of the second author. Where the only evidence for some phenomenon reported here is the latter, we signal this by specifically citing 'the lead developer' as the source.

The Software

The case study focuses on the grant-funded development of a laboratory information management system (LIMS) intended primarily for two research groups, both funded under another grant, and for a research laboratory doing similar work to these two, and also for the wider scientific community researching in this area. The original overall aim of the LIMS was to support a change of laboratory practice from low to high throughput. This involved a change in practice from performing experiments sequentially on a few biochemical objects identified as being important, to performing many similar experiments on many biochemical objects in parallel (Segal, 2009a). The original aim evolved over the years in ways which do not concern us now.

The product had to meet several scientific goals. Scientists wanted to record standard operating procedures, such as their experimental procedures, and often wanted to share this information. They wanted to plan their laboratory work, and then make records of what they actually did. They wanted these records to support the preparation of papers and the reproduction of experiments. They wanted to manage laboratory resources including instrument time and stocks of reagents. They needed a variety of technical calculations supported by the software. They wanted control of how they shared experimental records with colleagues.

These needs were not all clearly articulated at the beginning of the development. In particular, they were not articulated enough to be implementable. In part, this was because the LIMS was intended to support laboratory practice at a time when that practice was in a state of continual flux. But there were other problems in establishing detailed requirements as we shall now discuss.

Initial Problems with User and Customer Engagement

The development was planned to follow an iterative, incremental, feedback model. Of course, this model relies on there being user engagement in the shape of feedback and there was initially very little – the first release simply was not perceived as being usable and so was not used. One factor in the lack of usability of this first release lay in the lack of recognition of certain issues concerning the establishment of requirements as we shall now describe (other factors are discussed in Segal, 2009a).

The lead customer told the lead developer at the beginning of the development that the requirements for the system were well-known and partially met by an existing model intended to be at the heart of the software, with further requirements enumerated in a feature list. The lead customer thus saw no need to engage the users and the other customers in any further activity to establish requirements. However, the model was of the science and not of the scientific practice, that is, the workflows which the software was intended to support. And the feature list was at far too high a level to be implemented directly. For example, one of the features requested by the customers was of a library of experimental protocols. The developers needed much more information from the users – the people who conducted the experiments and recorded the actual experimental procedure and the results – before they could begin implementing the feature. Such information included answers to questions such

as: how frequently do you adapt protocols? What sort of parameterization is helpful?

Thus user-engagement activities for establishing detailed requirements were in practice very necessary. This necessity went unrecognized by the lead customer (or, indeed, by any of the other customers), so that the necessary resources, such as ensuring the availability of potential users, were not forthcoming. And the lead developer, having little previous experience of user interaction and being under great pressure to produce a timely first iteration, acquiesced with this. The result was a first version of the software which users were very reluctant to deploy.

In order to explain why the necessity to establish detailed requirements went unrecognized, we have to appeal to the culture of scientific end-user development. The lead customer had been himself a scientific end-user developer and both he and the other customers were experienced managers of end-user developers. As discussed above, establishing requirements is not a big issue in end-user development. The end-user developer *is* a potential user, and/or is embedded in a group of potential users, and so has a good initial instinct as to what the requirements are and refines them as the development iterates. Given his experience of end-user development, it was thus entirely reasonable that the lead customer believed that handing over a feature list to the lead developer sufficed to establish requirements.

Other characteristics of the culture of scientific end-user development created further challenges to effective engagement between the developers and the customers.

One such challenge was the incongruence in expectations of the time taken by software development. The lead developer reports that his time estimates for particular parts of the development were in the order of three times longer than those of the lead customer, due presumably to the latter's lack of cognizance of issues such as articulating requirements, testing, and structuring the development with a view to maintainability, such issues being largely irrelevant in the context of scientific end-user development described in an earlier section.

Another more important challenge was posed by an apparent lack of recognition on the part of the customers of the necessity of their engagement in order to negotiate and prioritize requirements. This might be traced back to the fact that a scientific end-user developer is normally developing software to meet a small number of requirements for a small and cohesive user community (consisting possibly only of the developer) and so, in the context of scientific end-user development, requirements negotiation is not a big issue.

Further factors inhibiting effective engagement between users/customers and developers in our case study are rooted in the specific context of the development. Similar software projects in the past had delivered little, and thus customers might have been loath to engage with the development themselves or to encourage the potential users to engage, seeing such engagement as a potential waste of time. In addition, potential users of the software had already started on the scientific work which it was intended to support and had established ad-hoc measures (for example, the use of Microsoft Office software such as Excel and Word) as work-arounds. They were thus understandably reluctant to take time off from their science to assist in the development of software for which work-arounds already existed.

With a first software release which was perceived as being virtually unusable and so generated little or no user feedback, with the barriers to both user and customer engagement remaining largely unrecognized and certainly unaddressed, and with the continuing problems of establishing, prioritizing and negotiating requirements, the development now appeared to be stymied.

Addressing These Problems by Means of Scientific End-User Developers

At this point of crisis, the customer at the research laboratory made available a senior scientist, with a warranted reputation as a scientific end-user developer, to act as "scientific sponsor" of the project. Another scientist in this organization, with a long and successful history of end-user development, had been involved with the development since its inception on a part-time basis. In what follows, we shall describe the important contributions that these two made to turning the development around, by acting as surrogate users, mediating communications between the customers and the developers, and reaching out to potential customer/ user communities. We should reiterate here that these people were scientific end-user developers in the sense of being primarily scientists and having considerable experience in developing software for themselves or their colleagues. They were not, in fact, potential users of the LIMS, though they did work alongside such users and had a deep understanding of both the scientific practices of such users and the underlying science.

We shall firstly consider these scientific end-user developers as surrogate users. They were especially well qualified to undertake this role. This is not only because both of them were very close to users as noted above, but also because their laboratory was unique in already using software which, while not adequately meeting their needs, was of the same type that the project was designed to supply. They were thus able to reflect on the benefits afforded by, and limitations of, this software. In the words of Kent Beck (with 'first release' in this case being the commercial software, and 'second release' being the software under development):

'The development of a piece of software changes its own requirements. As soon as the customers see the first release, they learn what they want in the second release... or what they really wanted in the first. And its valuable learning... that can only come from experience' (Beck, 2000, p. 19) (We should note here that Beck tends to be somewhat fuzzy in his writing on any distinction between the terms 'customer' and 'user'.)

Given their own deep knowledge of the science and the scientific processes used in the research laboratory, the two scientific end-user developers were able to provide important insights into the broad requirements of the software. One insight in particular became central to the product. This was that the scientific process that the software supports varies not only from laboratory to laboratory, but also over time in a given laboratory, so attempts to capture it as a somewhat static database schema could not succeed. Rather, the scientific end-user developer suggested capturing current practice as a set of templates that are used for today's records, and allowing users to add extra templates as the practice evolves. In the lead developer's view, this became a unique feature distinguishing the product from alternative solutions. Another important insight came from the scientific sponsor. He proposed adding a graphical representation of workflows derived from the actual records of experiments. The perception of the lead developer is that this graphical representation proved very popular, with some scientists using it as their principal means of navigation. The lead developer's opinion is that these two ideas together made the product uniquely suitable for recording ad hoc workflows, a key characteristic of the scientific activity it supports.

As to mediating customer interaction with the developers, the scientific sponsor took on a grueling schedule of travelling to visit all the current customer locations and agree requirements with them. This activity was made easier by his being respected by the scientific community both as a scientist and as a successful scientific end-user developer in that community. Barriers to trust between users/customers and developers, as discussed in an earlier section of this paper, were thus lowered, since the scientific sponsor spoke the same language and espoused the same values as both the scientists and the developers and was trusted by both. After this consultation, he prioritized the features requested. Some of the features were delivered quickly, and after a year's development the product was reliable and had all of the key features required. There were still problems with usability, but the product was used and user feedback began to flow in so that these problems could be addressed in an informed manner.

Though still evolving, the LIMS now very largely met the needs of the research laboratory. They adopted it, and ceased entering data into their previous LIMS. With the research project of the two primary customers now completed, attention turned to reaching the users/customers in the final population targeted by the development, those laboratories doing similar work in the wider scientific community. The scientific sponsor had considerable success in visiting these laboratories and persuading them of the potential of the LIMS to support their work, to the extent that there are currently 15 research groups in the UK and elsewhere making use of the LIMS.

So far, we have painted a very rosy picture of the role of scientific end-user developers within the development of the LIMS. There was one tension, however. Given that both scientific end-user developers came from the same organization, the lead developer perceived a feeling among the customer base that this organization was unduly favored when it came to prioritizing and implementing requirements. This perception was not, however, matched by reality, as is illustrated by the following. Towards the end of the development, the lead developer had become convinced that the greatest risk to take-up of the LIMS by new users was lack of usability. (Incidentally, this is in sharp contrast with the disregard for usability at the beginning of the development and speaks volumes for the efficacy of the development team as a learning organization). The development team invested thirty person months' development resources in improving usability. But usability was never a requirement of the organization in which the scientific end-user developers were based, as it depended heavily on robots for data entry and retrieval. In fact, the prime requirement for this organization was performance.

DISCUSSION

In the section above, we described a case study in which scientists with experience of scientific end-user development on the LIMS development team were able to improve both communications between the scientist users/customers and the development team, and the diffusion of the LIMS through the wider user/customer community. In this section, we reflect on our case study and what it tells us about how scientific end-user developers can lower the barriers to effective engagement between scientists and developers. We also discuss the scope and validity of our findings.

To the best of our knowledge, there is no large body of literature on effective user/customer engagement in the context of scientific software developments. A few recently published case studies contain some suggestions and descriptions of how the effective engagement of scientists in software development might be managed (De Roure & Goble, 2009; Macaulay et al., 2009; Thew et al., 2009). However, in each of these, the developers were aware of the importance of user-centeredness from the beginning of the development and drove the user engagement. In our case study the situation was somewhat different. The necessity of user engagement was not recognized at the beginning of the development project and this lack of recognition caused huge difficulties as described above.

The Role of Scientific End-User Developers in Lowering the Barriers to User/Customer Engagement

Our case study suggests that scientists who have previous experience as scientific end-user developers in small local projects can be effective in larger projects aimed at their wider scientific community both as surrogate users and as enablers of better communication between a distributed community of end users and professional software developers.

In having a deep understanding of the underlying science and the scientific practice, they can, like actual end-users, identify those requirements which can make the difference between a community software product being a success or a failure. Unlike other end-users, their experience in software development gives them some intuition as to which requirements are implementable. Lack of knowledge of the limits of what can and cannot be easily implemented is identified in the literature as being one of the main challenges of end-user involvement in software development (Kujala, 2003). As to their mediating communication, the fact that they have a foot in both camps means that they can lower the barriers to effective communication between scientists and developers raised by those mismatches of values and expectations described above.

The argument that they can also act as effective proselytizers for potential new user communities is supported by recent work on diffusion theory. Arguably, the most influential work synthesizing current thinking on the diffusion of innovations is the book by Rogers (2003). Rogers (2003) discusses a five stage process by which an innovation is adopted by an organization: knowledge (of the innovation); persuasion; decision (whether or not to adopt the innovation); implementation and confirmation. It is at the second stage of this process, persuasion, where we believe that scientific end-user developers can play the greatest part. Rogers is quite clear that what persuades people to adopt an innovation is not any objective evaluation but rather success stories from their near peers.

Most individuals evaluate an innovation not on the basis of scientific research by experts but through the subjective evaluations of near peers who have adopted the innovation (Rogers, 2003, p. 36).

In our case study, given his reputation as both a scientist and a scientific end-user developer, the scientific sponsor was able to play a leading role in promoting the LIMS and in persuading potential customers that it might be of use to them, by describing his organization's positive experience of its use. In the same vein, potential users were persuaded to adopt the LIMS and engage in providing feedback by a user describing his positive experiences of its use.

The Scope of Our Study

In this section we discuss two aspects regarding the scope of the study. The first is the extent to which scientific end-user developers can mediate user/developer engagement and influence the diffusion/adoption of the software through a particular research community. The second is the extent to which the findings of our case study can be generalized to other contexts of scientific software development.

With respect to the first aspect, we do not suggest that involving people with scientific end-user development experience in a software development team will solve all the challenges associated with user/developer engagement such as those concerned with the establishment of requirements and the adoption of software. This is because there are some problems which are so deeply embedded in the culture of the scientific community that they are extremely difficult to address by purely technical means, as we shall now describe.

In our case study, one such problem was due to the nature of the science. Perhaps because of the fragmented history of this particular scientific community, the science does not have a clearly defined terminology. Different terms might refer to the same concept and identical terms might refer to different concepts according to the context in which they are used (Morris & Segal, 2009). This posed obvious problems in our case study given that the back-end of the LIMS is a database. A field with a particular name might be filled with data of one type by one laboratory and with data of another type by another. Another problem of establishing requirements was due to the competitive nature of scientific endeavor as noted above in the section on barriers. The lead developer found that both users and customers were extremely uncomfortable about discussing their requirements for data access: for example, should all the data recorded in a laboratory be visible to all scientists in that laboratory? Or should data items only be visible to the scientist who deposited them? He also found that it was common for scientists to change their minds on this issue. This finding points to a tension within the research community, the well known "collaboration versus competition" dilemma: on the one hand, scientists want to be open about their data in order for the community to progress the science; on the other, they worry about premature revelations leading to other researchers exploiting their data before they have had the opportunity to do so.

As to adoption/diffusion of the software within the research community, this again was influenced by the competitive nature of science. The LIMS in our case study is designed to manage experimental data. It thus facilitates scrutiny by, and accountability to, funding sources. This facilitation might be deeply unwelcome, especially to scientists engaged in long term or risky research. In this situation, the researcher might be very loath to make interim data public in any way, and hence very loath to adopt the LIMS. The reluctance of scientists to adopt such software for this or other reasons which they are reluctant (or unable) to articulate, might manifest itself in seemingly very odd behavior. In our case study, there were two separate instances of the following unexpected occurrence. Potential users told developers they couldn't use the software because it was lacking feature A. The developers implemented feature A. The scientists then told the developers they couldn't use the software because it was lacking feature B. The developers then implemented feature B. The scientists then told the developers they couldn't use the software because it was lacking feature C. And so on, ad infinitum. The developers eventually concluded that rejection of this software had nothing to do with the presence/absence of particular features but rather was due to some other cause which the users were reluctant (or unable) to articulate to the developers. The scientific end-user developers on the development team, being part of the scientific community and having knowledge of the values and assumptions of that community, were able to offer some plausible reasons for the root cause of system rejection (perhaps rejection was due to a reluctance to adapt working practices in order to use software whose worth was not yet proven), but addressing the root cause could not be done by purely technical means. Rather, the worth of the software had to be established in one laboratory and awareness of this worth had to diffuse throughout the community, as discussed above.

The second aspect of scope is the extent to which our findings can be generalized. Given that the case study data is mostly heavily triangulated and corroborated by the feedback received when we met back with the participants in the study (Segal, 2009a), we are convinced that our findings are valid within the context in which the case study was conducted. But whether or not they are valid beyond this context remains an open question. One important characteristic of our context is that the requirements for the software were in constant flux. This is generally true of contexts where the software is designed either to support cutting-edge science or to support the practices of scientists at the cutting edge. It is probably not

so true of contexts where the science is relatively well understood, has matured enough to be able to support applications, and the problem is that of modeling the science in the software (as is the case of the software for modeling stresses in beams mentioned earlier). In addition, we have grounded our case study in the previous work of the first author in articulating a culture of scientific end-user development. This articulation has been the result of synthesizing a variety of case studies (of modelers of financial mathematics; earth scientists; space scientists; structural biologists), but none of these has involved high performance computing. It may well be that the culture of scientific high performance computing poses entirely different challenges when scientists who work within such contexts collaborate with professional software developers. The writers know of some work on the culture of scientific high performance computing in the literature (Basili, 2008; Easterbrook & Johns, 2009) but have no direct firsthand experience.

SUMMARY AND CONCLUSION

In this paper, we take for granted the importance of effective engagement between the users/customers and developers of scientific software. We examine some barriers to this engagement in the light of recent literature and previous work by the first author on the culture of scientific end-user development. We then draw on a longitudinal case study of the development of software for a community of research biologists to suggest that involving scientists with experience of scientific end-user development on a development team can have a markedly positive effect. Such end-user developers can act as surrogate users in order to help establish requirements; lower the barriers to engagement caused by mistrust and mismatching values and expectations; and support the diffusion/adoption of the software throughout the relevant scientific community. However, we also argue that there are some barriers to effective engagement which scientific end-user developers can do little or nothing to lower. These barriers are due to the essential nature of science and scientific practice, for example, the tension between competitiveness and collaboration.

Of course data from a single case study, however rich, has to be treated with a certain degree of caution. We suggest that our findings hold in contexts where the requirements for the scientific software are in a state of flux, where there is a strong tradition within the scientific community of end-user development and where this development does not involve high performance computing. The extent to which this suggestion is valid needs to be tested by further studies.

REFERENCES

Basili, V. R., Carver, J., Cruzes, D., Hochstein, L., Hollingsworth, J. K., Shull, F., & Zelkowitz, M. V. (2008). Understanding the high performance computing community: A software engineers' perspective. *IEEE Software*, *25*(4), 29–36. doi:10.1109/MS.2008.103

Beck, K. (2000). *eXtreme programming explained: Embrace change*. Reading, MA: Addison-Wesley.

De Roure, D., & Goble, C. (2009). Software design for empowering scientists. *IEEE Software*, *26*(1), 88–95. doi:10.1109/MS.2009.22

Easterbrook, S. M., & Johns, T. C. (2009). Engineering the software for understanding climate change. *Computing in Science & Engineering*, *11*(6), 65–74. doi:10.1109/MCSE.2009.193

Finton, T. A. (2003). Collaboratories as a new form of scientific organization. *Economics of Innovation and New Technology*, *12*(1), 5–25. doi:10.1080/10438590303119

Hine, C. (2006). Databases as scientific instruments and their role in the ordering of scientific work. *Social Studies of Science*, *36*(2), 269–298. doi:10.1177/0306312706054047

Kujala, S. (2003). User involvement: A review of the benefits and challenges. *Behaviour & Information Technology*, *22*(1), 1–16. doi:10.1080/01449290301782

Macaulay, C., Sloan, D., Jian, X., Forbes, P., Loynton, S., Swedlow, J. R., & Gregor, P. (2009). Usability and user-centered design in scientific development. *IEEE Software*, *26*(1), 96–102. doi:10.1109/MS.2009.27

Morris, C., & Segal, J. (2009). Some challenges facing scientific software developers: The case of molecular biology. In *Proceedings of the 5th International IEEE Conference on E-Science* (pp. 216-222).

Rogers, E. M. (2003). *Diffusion of innovations* (5th ed.). New York, NY: Free Press.

Segal, J. (2005). When software engineers met research scientists: A case study. *Empirical Software Engineering*, *10*, 517–536. doi:10.1007/s10664-005-3865-y

Segal, J. (2007). Some problems of professional end user developers. In *Proceedings of the IEEE Symposium on Visual Languages and Human-Centric Computing* (pp. 111-118).

Segal, J. (2008). Scientists and software engineers: A tale of two cultures. In *Proceedings of the Psychology of Programming Interest Group Conference*. Retrieved from http://www.ppig.org/papers/20th-segal.pdf

Segal, J. (2009a). Software development cultures and cooperation problems: A field study of the early stages of development of software for a scientific community. *Computer Supported Cooperative Work*, *18*(5-6), 581–606. doi:10.1007/s10606-009-9096-9

Segal, J. (2009b). Some challenges facing software engineers developing software for scientists. In *Proceedings of the 2nd International Software Engineering for Computational Scientists and Engineers Workshop* (pp. 9-14).

Segal, J., & Morris, C. (2008). Developing scientific software. *IEEE Software, 25*(4), 18–20. doi:10.1109/MS.2008.85

Smith, S., Lai, L., & Khedri, R. (2007). Requirements analysis for engineering computation: A systematic approach for improving reliability. *Reliable Computing, 13*, 83–107. doi:10.1007/s11155-006-9020-7

Star, S., & Ruhleder, K. (1996). Steps towards an ecology of infrastructure design and access for large information spaces. *Information Systems Research, 7*(1), 111–134. doi:10.1287/isre.7.1.111

Thew, S., Sutcliffe, A., Procter, R., De Bruijn, O., McNaught, J., Venters, C., & Buchan, I. (2009). Requirements engineering for e-science: Experiences in epidemiology. *IEEE Software, 26*(1), 80–87. doi:10.1109/MS.2009.19

Wagner, E. L., & Piccoli, G. (2007). Moving beyond user participation to achieve successful IS design. *Communications of the ACM, 50*(12), 51–55. doi:10.1145/1323688.1323694

Wilson, G. V. (2006). Where's the real bottleneck in scientific computing? *American Scientist, 94*(1), 5–6.

Zimmerman, A., & Finholt, T. A. (2007). Growing an infrastructure: The role of gateway organizations in cultivating new communities of users. In *Proceedings of the International ACM SIGGROUP Conference on Supporting Group Work* (pp. 239-248).

Zimmerman, A., & Nardi, B. (2006). Whither or whether HCI: Requirements analysis for multi-sited, multi-user cyberinfrastructures. In *Proceedings of CHI '06: Extended Abstracts on Human Factors in Computing Systems* (pp. 1601-1606).

This work was previously published in the Journal of Organizational and End User Computing, Volume 23, Issue 4, edited by Mo Adam Mahmood, pp. 51-63, copyright 2011 by IGI Publishing (an imprint of IGI Global).

Chapter 19
An Analysis of Process Characteristics for Developing Scientific Software

Diane Kelly
Royal Military College of Canada, Canada

ABSTRACT

The development of scientific software is usually carried out by a scientist who has little professional training as a software developer. Concerns exist that such development produces low-quality products, leading to low-quality science. These concerns have led to recommendations and the imposition of software engineering development processes and standards on the scientists. This paper utilizes different frameworks to investigate and map characteristics of the scientific software development environment to the assumptions made in plan-driven software development methods and agile software development methods. This mapping exposes a mismatch between the needs and goals of scientific software development and the assumptions and goals of well-known software engineering development processes.

INTRODUCTION

The methods for developing scientific software have recently garnered the attention of the software engineering community, resulting in several studies of scientists as "end-user developers". Segal (2004) differentiated scientists as "professional" end-user developers, pointing out their extensive knowledge in a technical domain and their high comfort level in writing code. There have been suggestions (Ackroyd et al., 2008; Pitt-Francis et al., 2008; Segal, 2005; Wood & Kleb, 2003) that agile software development methodologies (Highsmith & Cockburn, 2001) rather than "plan-

DOI: 10.4018/978-1-4666-2059-9.ch019

driven" methodologies (Royce, 1970) are better suited to the development of scientific software.

This paper examines the characteristics of the environment in which scientists develop their own software and compares these to the assumptions and goals of plan-driven and agile software development methodologies. The conclusion is that there is a mismatch in both types of methodologies when applied to the development of scientific software.

We first define what we mean by scientific software and discuss characteristics of its development. In Section 3, we make initial observations about these characteristics and the assumptions inherent in plan-driven and agile software development methods. Section 4 uses a variety of models to further explore the characteristics of scientific software development as opposed to inherent assumptions that underlie plan-driven and agile development methods. Section 5 summarizes our findings and Section 6 concludes.

SCIENTIFIC SOFTWARE

Definition of Scientific Software Development

A project that we categorize as scientific software development has the following three characteristics. First, the software is written to answer a scientific question. The question can be general, such as, "Is it safe to operate this nuclear generating station?" or specific, such as, "Can I track satellites with this telescope?" Second, the writing of the software necessitates the close involvement of someone with deep domain knowledge in the application area related to answering the question. The person with the deep domain knowledge, the scientist, usually writes the software. This is Segal's professional end-user developer. We have observed cases where the scientist has not written the software, but the software developer has had to become enough of an expert in the domain to understand and implement what the scientist is describing. Third, the software provides output data to support the scientific initiative. A human is in the system loop to examine the data and make observations and ultimately, answer the scientific question. The expectation is that the data provided by the computer solution is correct and will not misguide the scientist answering the question.

The science application can be solving large systems of differential equations, analyzing immense amounts of data such as imagery or bioinformatics, and computing using analytical and empirical models. Our definition of scientific software, however, excludes the following: control software whose main functioning involves the interaction with other software and hardware; user interface software that may provide the input for and report of scientific calculations; and any generalized tool that scientists may use in support of developing and executing their software, but does not of itself answer a scientific question.

Scientific software may have an extensive graphical user interface or interact with other software or hardware to obtain data. It may run on complex multi-processors and require middleware support to make use of hardware capabilities. But in the following discussion, we are talking exclusively about the code that implements the science application whether it is designed as a separate module or inserted into a product that includes these other parts. We contend that this core scientific part of the product requires different considerations when it is being developed.

Characteristics of Scientific Software

Several studies have looked at the characteristics that differentiate the environment and activities of scientific software development from the development of other types of software (Basili et al., 2008; Boisvert & Tang, 2001; Matthews et al., 2008; Post & Kendall, 2004; Sanders & Kelly, 2008; Segal, 2003, 2004). Some characteristics are invariant whether the software is developed by a scientist working alone (Sanders & Kelly, 2008) or a large distributed team (Matthew et al., 2008), or whether the code is a thousand lines of code or over a million.

The most noteworthy of the characteristics is the focus on the need for correctness, or trust, in the software. Some scientists stated baldly, "the software must never lie to me" (Sanders & Kelly, 2008). As explained by Kelly (2008) "This [trustworthiness] takes priority over efficiency, reliability, user-friendliness, cost-effectiveness, and portability … If the software gives the wrong answer, all other qualities become irrelevant." Hatton et al. (1994, 2007) and Miller (2006) both give examples of cases where correctness of the software was the sole factor in success or failure. Hatton describes an example of industrial software where the success of oil-exploration businesses depended on the accuracy of the software output. Miller describes a case where a mistake in an example of research software forced the retraction of several research papers from well-regarded journals.

Although the word "correctness" is often used, we prefer the word "trustworthiness". Hook and Kelly (2009) discuss correctness versus trustworthiness as follows. "Correctness is essentially a Boolean decision, either true or false. On the other hand, trust is on a graduated scale, ultimately determined by judgment. The level of trust can be different for every example of scientific software, and possibly different for each individual output data. When a scientist says that a computer output is acceptable within forty percent of benchmark data, he/she is talking about trust. Nothing can be said about the Boolean concept of correctness." Granted, in Hatton's (1994) and Miller's (2006) examples, there was misplaced trust in the software. This leads into a long discussion of the difficulty of testing scientific software. Weyuker's paper (1982) on testing non-testable software is a good start to that discussion.

Output from scientific software is the culmination of a series of interactions amongst scientific models, numerical techniques, data, source code, and hardware (Boisvert & Tang, 2001). Change of hardware platform or compiler option can alter the values of output. Similarly, different expressions in source code that mathematically are equivalent, can give different values of output, due to the fact that computer computations are neither associative nor distributive. Because of this, scientists are reluctant to make changes to source code or execution environments for any other reason than to improve the trust in their scientific output.

The complex interaction of scientific model, numerical technique, source code, data, and hardware affects the development environment. First, development of scientific software is a discovery process, requiring frequent iterations between coding and testing to discover the correct combination of the interacting parts. Second, correct functioning of parts of the whole, although certainly desirable, in no way guarantees the correct functioning of the whole. Third, problems identified in the output of the software almost invariably require the scientist to participate in debugging. We've found that the scientist must be involved in any meaningful assessment of the software (Kelly et al., 2010).

The scientist is often both the developer and the user of the software, obviating the need for extensive documentation at any point during the development. At the other end of the spectrum, Sanders and Kelly (2008) describe how a customer can provide the scientific question to be answered, then relinquish almost all control of the project to the scientist who is in possession of the expertise best suited to make subsequent decisions. Again, our discussion here excludes the user interface; we have found the most successful scientific software projects assiduously include the customer during the phase of user interface design.

The goal of scientific software is to answer a technical question that generally cannot be answered otherwise. As a result, there is no half-measure in the development in the software. Either the software answers the question within acceptable error tolerances – and the software is trusted to do this – or it doesn't. As well, the scope is indivisible in another way. For example, the scientific model cannot be implemented in one release and the solution technique saved for the next release.

Testing scientific software is almost wholly focused on "validation" testing (Sanders & Kelly, 2008; Boisvert & Tang, 2001). This is defined by the scientists as testing against real-world expectations, such as experimental results and monitored physical phenomena.

Characteristics of scientific software development are discussed further in the following sections, where the juxtaposition of the characteristic with assumptions associated with software development methods better illustrate the problems of mismatch.

INITIAL OBSERVATIONS ON SCIENTIFIC SOFTWARE AND DEVELOPMENT PROCESSES

Software Development Processes

Boehm and Turner (2005) divide software development processes roughly into two categories, plan-driven approaches and agile approaches. However, as noted by Segal (2005), software professionals generally tailor a methodology to their specific context, but need a base methodology that is a close fit to their situation. We provide a brief description of the two categories of development process, assuming these provide the base methodologies used by software professionals.

The original plan-driven approach, commonly known as the waterfall model, was first described by Royce (1970). There are several variations on this model (Pfleeger, 1998; Van Vliet, 2000). Royce's premise was to provide a mechanism of discussion, documentation, review, reflection, and testing that guaranteed that the code, when written, matched its requirements. The discovery

process happens mostly when the requirements are elicited and specified. The subsequent steps in the process become more and more mechanical, such that the writing of the code should involve little or no discovery. This would otherwise involve an expensive (in both time and cost) backtracking, rewriting documents, and redoing reviews. The assumption is that requirements and design can be specified in detail and that knowledge of the system and its interactions is relatively complete before implementation. Quality assessment is focused on consistency of products against specifications. The scientist's focus on validation testing, the assessment of the finished product against real-world expectations seems to be in step with plan-driven practices.

Agile methodologies (Highsmith & Cockburn, 2001) were developed as an alternative to complete up-front planning. In agile methodologies, iterations are formulated into steps that reduce planning and documentation to the minimum needed to deliver partial products in short time intervals. Agile methods are accompanied by sets of practices designed to fill the communication gap created by the reduction of documentation and up-front planning.

The two most popular agile development methodologies are Extreme Programming (XP) (Beck, 2003) and Scrum (Rawsthorne & Shimp, 2009).

The main focus of both Scrum and XP is a short time box, usually less than a month, in which a task is completed. The goal is to deliver a product on time. The accompanying assumption is that the product is readily divisible and testable.

One aspect of XP is the relationship of customer to developer. Jeffries et al. (2001) explain a key role of the customer where "the customer controls scope, deciding what to do and what to defer, to provide the best possible release by the due date" (p. 55). Often, the scientist is his/her own customer. On the other extreme, the customer provides the scientific question then defers to the scientist/developer to deal with details of the project. In between, there is a range of customers whose involvement depends on their understanding of the science. The assumption in XP is that the customer drives the development process. When applied to scientific software, the advantage of this is not always clear.

Beck (2000) discusses cost of change using XP, stating that "the cost of change [rises] slowly over time … You [can] make big decisions as late in the process as possible" (p. 23). This type of thinking allows the extensive use throughout the project of refactoring, a key practice in XP. Our experience with developing scientific software tells us that some big decisions have to be made early. The choice of modeling technique, representation of data, discretization of equations, and so on, are key to the solution being pursued. In this, the development of scientific software takes on a flavour of plan-driven development.

Jackson's Problem Domain Versus Solution Domain

Jackson (1994, 2001) emphasizes the difference between problem domain and solution domain in developing software. He points out the need for the software developer to become "... expert in those aspects of the application [problem] domain that affect the design and construction of the software" (p. 251). The purpose of requirements engineering is to encourage the software developer to explore the problem domain. This has spawned considerable work in trying to find effective means of modeling and expression in requirements documents (Pfleeger, 1998; Wiegers, 1999). The emphasis in plan-driven development models is on the need to explore the problem domain with the aim of reducing the amount of exploration when dealing with the solution domain and thus reducing costly rework. The approach is based on the statement that, "Many of the problems encountered in software development are attributable to shortcomings in the processes and practices used to gather, document, agree on,

and alter the product's requirements" (Wiegers, 1999, p. 4).

One of the realities of scientific software development is the need to explore the solution domain because of the complex interactions amongst the different contributors to the computational solution. The exploration of the solution domain is far more extensive and critical to the project in relative terms than the exploration of the problem domain. In addition, the scientist is Jackson's "expert" in the problem domain. A sufficient understanding of the problem domain is already present in the scientist/developer. As far as expressing their understanding of the problem domain, scientists have a long history of precise vocabulary and expressions for documenting their needs and ideas. In the development of scientific software, the problem domain can be precisely captured by the scientist-expert in the domain, allowing attention to be turned relatively quickly to the solution domain where extensive time and attention are required.

USING MODELS TO EXPAND OUR UNDERSTANDING

To further our understanding of both the characteristics of scientific software development and appropriateness of existing software development approaches, we looked for frameworks that could be used for our comparisons. A requirement for these frameworks was to allow us to map the characteristics of scientific software development to assumptions underlying the different methodologies. The only fully suitable framework available was that of Boehm and Turner (2005). However, we made use of various other lists of items associated with development processes to perform a partial mapping and make additional observations.

Table 1. Five knowledge domains that contribute to the development and use of a software product

Knowledge Domain (KD)	Description
Physical world knowledge	Knowledge of physical world phenomena pertinent to the problem being solved.
Theory-based knowledge	Theoretical models that provide (usually) mathematical understanding and advancement of the problem towards a solution.
Software knowledge	Representations, conventions, and practices used to construct the solution.
Execution knowledge	Knowledge of the software and hardware tools used to create, maintain, control, and run the computer solution.
Operational knowledge	Knowledge related to the use of the computer solution.

Method Engineering

Literature on comparing software development methods, such as Song and Osterweil (1992) is heavily focused on the methods themselves and provides no characterization of the environment in which the method is to be used, nor the features of methods that would allow a mapping from environment to method.

Work in the area of method engineering, such as Rolland (1997) and Henderson-Sellers (2003) aims mostly at facilitating the construction of "situation-specific methods" (Rolland, 1997). Both the works cited are heavily involved in illustrating the ability and flexibility of their approaches so that all possible development methods can be modeled. Rolland (1997) describes how to evaluate *method chunks* as to their appropriateness for a given environment. Her description characterizes method chunks by intention (e.g., reengineer high-risk project), target (e.g., business process), and approach (e.g., domain-based approach). She characterizes environment by problem domain, complexity, and risk. The intention is to map entries in a library of method chunks to an environment. It's not clear how this mapping is to be done. Rolland also describes other work where projects are characterized by seventeen contingency factors that are rated between high and low. Unfortunately, there is no description of analogous contingency factors for methods.

Henderson-Sellers (2003) describes how a selection of techniques can be mapped onto a method. However, the technique selection already presumes we have a mapping from the application environment to appropriate techniques and the paper is silent on how to get there.

Knowledge Domain Models

One point of interest in Rolland's paper (1997) is her reference to the "four worlds of IS engineering". She uses this as a framework to present the state-of-the-art in methods engineering, unfortunately with an inward focus on the methods themselves.

The idea of the "four worlds" framework can be extended to the five "knowledge domains" introduced by Kelly (2009). The five knowledge domains are summarized in Table 1 (adapted from Kelly, 2009).

To increase our understanding of the development of scientific software, we need to observe who has knowledge in these different domains. A further consideration is the impediments in transferring this knowledge across domains. Impediments include cultural, informational, and cognitive differences such as those described by Segal (2005). Impediments also include the "time at which knowledge is available", which is alluded to by the inclusion of incremental software development in agile methodologies.

Because of the tight coupling of knowledge from the five domains in the final scientific software product, transfer of knowledge between domains is necessary to affect a working solution.

Segal's discussion (2005) mainly deals with the boundary between theory-based knowledge and software knowledge. However, those designing the software still need to understand how physical world knowledge will be represented by the software solution and how the operation of the software will reflect on theory-based knowledge and possibly on interactions with decisions in the physical world. This points to a requirement for those building the software to be deeply knowledgeable about all five knowledge domains.

The acquisition of knowledge for a particular problem and its software solution can be considered on a time scale for all five knowledge domains, as progressing from general to specific knowledge. Depending on the purpose of the software solution, theory-based knowledge and physical world knowledge may be advancing as the operational knowledge of the final software product increases. This takes place on a relatively long time scale. On an intermediate time scale, theory-based knowledge, software knowledge, and execution knowledge increase as the developer/user better understands the vagaries of the new computer solution and its interplay with its underlying mathematics. The shortest time scale is the increase of software knowledge related to the execution environment in order to get a minimally working product. It is this short time scale that is most aptly addressed by the agile development processes. The long time scale associated with operating the final product to understand the world it models, is more aptly associated with plan-driven processes. Both categories of software development processes miss the intermediate time scale activities, where the scientist is grappling with code, machines, and theory.

Iron Triangle Model for Software Project Management

The "iron triangle" of project management, first described by M. Barnes in 1969, provides a simple model to understand the parameters that affect the success or failure of a project. Even though the iron triangle model is now considered too simplistic (Pearson, 2009), nothing has yet replaced it and it corresponds well with the assumptions addressed by software development processes described in software engineering literature. Indeed, Beck (2000) begins his description of the use of Extreme Programming (XP), with a commonly enhanced version of the iron triangle. He adds "quality" as a fourth corner, or variable, to accompany the usual variables of scope (functionality), schedule, and cost.

Our observations on scientific software can be mapped onto this enhanced project management model.

Quality for scientific software is primarily "correctness", or alternatively, trustworthiness. This immutable goal must be achieved. This variable is fixed in the project management plan for scientific software.

Scope is the scientific question that must be answered. This question is often difficult to divide, and so the scope is fixed.

We can argue that cost and schedule are trumped by scope and quality. Lateness of reports, theses, or decisions must be tolerated in order to achieve trustworthy and complete output from the software. The output data must be trusted to answer the scientific question. If the scientific question cannot be answered by the software, then the journal paper cannot be written and the project not completed. A fixed-schedule-fixed-cost environment is very high-risk for the scientist since the only variable left to manage is scope, and this may be difficult to sub-divide.

We can observe that the almost overwhelming assumption of existing software development methodologies is that the fixed variables are cost and schedule, and that quality and scope are mutable. We contend that the fixed variables for scientific software are quality and scope, and that schedule must be mutable.

Boehm's and Turner's Model

Boehm's and Turner's Model (2005) for comparing and blending agile and plan-driven development methods appears to be the only such model described in the literature. They present "five critical factors involved in determining the relative suitability of agile or plan driven methods in a particular project" (p. 54) and observe that "... a project that is a good fit to agile or plan-driven for four of the factors, but not for the fifth, is a project in need of ... some mix of agile and plan-driven methods" (p. 54).

Segal (2005) discusses Boehm's and Turner's advice on melding agile with plan-driven methods in the context of developing scientific instrument software. Although this is not strictly within our definition of scientific software, their need to produce significant documentation for their long-lived software along with improved and more immediate communication suggests a blending of methods is necessary.

In order to improve our understanding why this is so, we discuss each of Boehm's and Turner's factors in terms of scientific software development.

Size

Scientific software development teams are typically small (Boisvert & Tang, 2001). Boehm and Turner (2005) characterize this as a good match to agile methods. They also comment that the reliance on tacit knowledge also matches better to agile methods. Certainly, the difficulty in sharing scientific knowledge emphasizes the role of sharing tacit knowledge amongst a small team. Small teams of scientists often share understanding of existing theoretical background material for the software project, reducing the need for additional project-specific documentation.

Criticality

Scientific software is used to generate data for human decision-making. The human is in the system loop and is responsible for decisions, unlike embedded software where there is no intermediary between computational results and action taken. The scientist is an experimentalist, making changes to the environment (in this case software) and contemplating the results. This fits with Boehm and Turner's (2005) profile of an agile project.

Personnel

Boehm and Turner (2005) describe the agile development personnel as "requir[ing] continuous presence of a critical mass of ... experts" (p. 55). The development of scientific software requires the involvement of the scientist as the "critical mass". Again, this fits well with agile methods.

Culture

Cultural characteristics of agile methods are described by Boehm and Turner (2005) as "thriv[ing] in a culture where people feel comfortable and empowered by having many degrees of freedom available for them to define and work problems. This is the classic craftsman environment, where each person is expected and trusted to do whatever work is necessary ..." (p. 49). This aptly describes the scientist who is highly trained in his/her field and is comfortable with modeling and coding. The customer who poses the scientific question then leaves the subsequent project decisions to the scientist certainly expects the scientist to fulfill the role of empowerment.

Dynamism

Boehm and Turner (2005) do not define dynamism but measure it by "%requirements change per month" (p. 56). They suggest that "simple design and continuous refactoring [agile practices] are excellent for highly dynamic environments" (p. 55). This is the factor where development of scientific software fits neither agile nor plan-driven methods fully.

For scientific software, discussing requirements provides little help in the area where considerable time must be spent: exploring the solution space. The measure "%requirements change" is too difficult or impossible to assess. Yet, certain decisions must be made early in the project and get "locked in".

Development is certainly dynamic as scientists make changes to the software to explore and refine the software to better represent their science. In this way, scientists are risk-takers, but the risk-taking is limited and controlled. Changes are made as necessary to explore and improve the answers to the scientific question. To reduce collateral damage to the science, software changes for any other reason are discouraged (Boisvert & Tang, 2001). Refactoring, a defining XP activity, would be considered counter-productive to scientific development since it introduces unnecessary changes to the environment the scientist is studying. Refactoring software for a scientist would be equivalent to rearranging a bench-scientist's experimental set-up because it's ugly looking. This is not to say that simple, uncluttered code is not a good idea. However, working code for the scientist may mean leaving code in a state that would be considered less than ideal for the purpose of maintainability, say.

We can observe that plan-driven processes seek to augment exploration of the problem domain by extensive requirements and design planning and documentation. The assumption is that the complexities of the implementation step are thus reduced, possibly to the point of being pro forma.

Before developing scientific software, the scientist usually relies on theory and mathematics that have been through significant development and are well documented. This is part of the scientist's requirements and appears like plan-driven development. However, the scientist now changes mode and must spend significant time exploring the solution space of the software because of its inherent complexities and interactions. It is in this mode that the scientist appears to enter an agile development phase.

One of the goals of agile methods is to encourage exploration of the problem space, but with direct customer involvement as opposed to extensive documentation. When the scientist enters the agile-like part of the development phase, it is to explore the solution space. The customer may not be a useful participant at this point.

Another goal of agile methods is to address schedule pressures by dividing the problem into small pieces and delivering working code for these pieces. The solution space exploration phase of the scientist often requires a complete solution, obviating the division of the product. This exploration unfortunately can take considerable time and obtaining a correct answer will not yield to schedule pressures.

Even though, by Boehm's and Turner's model, scientific software development appears to be compatible with agile development, there are significant differences in underlying needs and assumptions. These are explored further by considering specific practices associated with one of the agile methods, XP.

Using Beck's Twelve Practices of XP as an Analysis Model

The popularity of agile approaches leads us to consider one of the approaches, Kent Beck's XP, in more detail. We consider the twelve practices that Beck (2000) uses to define XP to compare to scientific software development.

Table 2. Practices that define XP versus their use with scientific software development

XP Practice defined by Beck (2000, p. 54)	Scientific Software Development
The Planning Game – quickly determine the scope of the next release …	Major decisions are made up-front, more characteristic of a plan-driven approach.
Small releases – Put a simple system into production quickly, then release short versions on a short cycle.	Small releases generally are an antithesis to the indivisibility of the scientific problem being solved.
Metaphor – Guide all development with a simple shared story.	The simple shared story is the problem statement. The difficulty is the complexity of the solution domain.
Testing – Programmers continually write unit tests … Customers write tests demonstrating that features are finished.	Good testing practices are always desirable. Customers usually cannot test the software without being familiar with the scientific domain.
Refactoring – Programmers restructure the system without changing its behaviour …	Can be problematic with the complex interactions of techniques, compiler, hardware, and data. Best practices that govern the restructuring may not translate over to scientific software.
Pair programming – All production code is written by two programmers at one machine.	It may be that only one scientist has the knowledge necessary to address the scientific question. Alternate ideas to pair programming such as testing buddies and peer review may work in special cases.
Collective ownership – anyone can change code anywhere in the system at any time.	The combination of the above comments listed under "refactoring" and "pair programming" make this definition of collective ownership untenable.
Continuous integration – integrate and build the system many times a day, every time a task is completed.	Development of scientific code often involves long periods of careful consideration – thought time – and changes may not be rapid. The idea of tracking code changes in some manner is non-refutable.
On-site customer – include a real, live user on the team, available full-time to answer questions.	Customers can cover a spectrum of involvement: the scientist is his/her own customer; the customer provides the scientific question and empowers the scientist to pursue it; the customer is himself a scientist and may participate in the development and/or testing of the software.

To start, we remove any practices whose statements are not refutable. For example, maintaining that a design should be as simple as possible is not refutable. Its opposite is that the design should be as complicated as possible. As a design philosophy, the statement about simplicity is not useful for comparisons. Similarly, Beck's statement about limiting work to forty hours per week is desirable management of knowledge workers in any situation. This statement is also removed from our comparison.

The statement about coding standards is also problematic for comparisons. The reason cited for having coding standards is increased communication through the code base. Increased communication is always desirable. Using the code base as a means of communication also has a number of desirable side effects. The choice of software development method has no impact on whether this practice is used or not, and the practice itself is essentially not refutable. This is removed from our comparison.

From the remaining nine XP practices, the three overriding considerations appear to be speed of delivery of the product, communication during development, and flexibility for making possibly sweeping changes. We consider assumptions behind each of the nine practices of XP against the overriding considerations and environment characteristics of scientific software development.

We use three sets of experiences to comment on the nine remaining XP practices. Table 2 summarizes our comparisons.

The first set of experiences is our own through personal industrial work and interviews with other scientists (Kelly et al., 2004, 2006, 2008, 2010; Sanders & Kelly, 2008).

The second is an experience report (Ackroyd et al., 2008) following the adoption of XP practices at a research facility that developed data acquisition software for their own use. This software does not entirely fit our definition of scientific software since it includes a graphical user interface, and interfaces to hardware devices. However, there are some similarities that can be extracted. We designate this case study as *Tools*.

The third is a description of the introduction of agile methods into an academic environment where computational software is developed to model the electrical activity of the heart. The software uses large meshes and small time steps to solve systems of ordinary differential equations. The software is run on parallel computers with the complications of efficiency and complex hardware architectures. The authors state that the overriding concern still is correctness of their code. They are dealing with large turnover in personnel and changes in technology and underlying science. They introduced agile techniques into a four-week course, and then expanded their experience with agile methods by developing a set of maintainable libraries over a three year period. Their experience is included in the discussions below, with the designation *HeartSim*.

- **The Planning Game – quickly determine the scope of the next release:** An assumption is made here that the problem being solved is highly divisible. We have found this is often not the case with scientific software. Generally, considerable planning has to be done to determine theoretical approaches, solution techniques, data requirements, and so on, long before any code is written. This is more characteristic of plan-driven approaches. The case study for *Tools* described the difficulty in

getting users to formalize requirements and the need for developers to group requirements into regular releases. Even then, they had problems sticking to release dates. *HeartSim* did not discuss planning specifically.

- **Small releases – Put a simple system into production quickly, then release short versions on a short cycle:** This practice makes the same assumption as the previous practice and suffers from the same misalignment with scientific software development. The description from *Tools* stated that evolutionary development fits the scientific research environment without further comment or details. We agree that evolutionary development fits the need to explore the solution space and iteratively refine the interactions of different parts of that space. However, this refinement is driven by correctness and trust, not by delivery time. The difficulties described by the *Tools* case study may have come from the failure to see this mismatch.

HeartSim also agreed that the iterative approach to development seemed to fit that of scientific software, but they categorize releases into external and internal. External releases were not tied to time boxing for the customer as assumed by XP, but to publishing an academic paper. These are large indivisible efforts. Internal releases to close collaborators allow more of an XP-like approach, but *HeartSim* did not discuss time frames for releases, but mentioned the need to have working code.

- **Metaphor – Guide all development with a simple shared story:** There is some alignment with scientific software development, where the simple shared story is the scientific question to be answered. However, the story becomes very complex as the details are worked out. Who these details are shared with depends on the players in the development. But it is these details that are very important and the simple shared story provides little help with the details. If we consider the simple shared story as a management device for focusing the project, then this again becomes a non-refutable statement. Neither the *Tools* case study nor *HeartSim* mention metaphors.

- **Testing – Programmers continually write unit tests:** *Customers write tests demonstrating that features are finished.* Good testing practices are another non-refutable statement. But the interesting assumption here is the role of the customer. In our experience, unless the customer is intimately familiar with the scientific domain of the software, the customer cannot write and evaluate acceptance tests. The *Tools* case study simply said that they needed to improve their testing practices. *HeartSim* discussed the different types of modules typically found in scientific software, where some are easy to test and others extremely difficult. They obviously agreed that good testing techniques and practices are beneficial, and it was also obvious that a great deal of technical knowledge was needed to do their testing, not a role assumed by a non-technical customer.

- **Refactoring – Programmers restructure the system without changing its behavior:** Refactoring is defined as restructuring the system without changing its behaviour in order to increase desirable properties of the code. Again, this appears like a non-refutable statement: a globally good thing to do. The assumption behind this is that anyone on the team can change the code anywhere at any time. Discussions with some scientists have raised some very strong reactions against this practice. The interactions of code with compilers, solution

techniques, hardware, and data can create a complex situation where the scientist becomes very conservative in their desire to change the code. Anyone who makes a code change needs to have a full understanding of these interactions. In addition, "best practices" have been usually defined for software other than scientific software. In one case study, the practice of breaking modules up to reduce their size actually reduced the potential comprehensibility of the software for the scientific programmer (Kelly, 2006). The *Tools* case study only commented that the automatic refactoring provided by an integrated development environment was helpful. *HeartSim* did not specifically discuss refactoring.

- **Pair programming – All production code is written by two programmers at one machine:** The assumption is that programmers have overlapping knowledge, though possibly differences in experience. The specialized knowledge of scientists writing scientific software is often spread very thinly and would make the idea of two highly specialized scientists sitting together to work on one code a rarity. The only situation where we have seen something similar to this idea at work is with a scientific specialist who did not write code and a scientist non-specialist who did. The non-specialist had become extremely familiar with the specialized field, worked closely with the specialist, and wrote the code in such a way that it was "readable" by the specialist. Pair programming in this case transformed into pair peer reviews, sitting together at the terminal and reading the code. The *Tools* case study found it was difficult to enforce pair programming, for various reasons. *HeartSim* enforced pair programming in group development sessions held one day a week. Outside those sessions, all development was solo. Almost all code developed by students and researchers for their own projects was done solo. However, some projects were peer-reviewed, adding back some of the benefits of pair programming.

- **Collective ownership – anyone can change code anywhere in the system at any time:** This practice makes the same assumptions as refactoring and suffers from the same limitations when applied to scientific software development. The *Tools* case study made attempts at enforcing collective ownership, but particular project pressures resulted in individuals taking ownership of specific parts of the code. As discussed by *HeartSim*, collective ownership in an academic environment is problematic in many cases. For the core libraries, they discussed mechanisms such as peer review and change-set tracking to maintain some on-going expertise in key areas of code in the face of large personnel turnover. The concern is not ubiquitous change by anyone at any time, but the maintenance of corporate memory to do changes over long time periods. This speaks to the much longer time scale that is typically of concern with scientific software.

- **Continuous integration – integrate and build the system many times a day, every time a task is completed.** The underlying philosophy of being organized, frequently checking code changes, testing frequently, and maintaining reproducibility is, again, non-refutable. The problem in the statement of this practice is "many times a day". The focus is on speed of delivery rather than careful consideration of problems, alternatives, and next steps. This practice also suggests a much less complex solution space than the scientists are dealing with. The *Tools* case study agreed that continuous integration was helpful though

sometimes difficult to enforce. Their comment on time came from an attempt to implement daily meetings. Their change to weekly meetings worked better and also fits with our observations that a longer time box is required for scientific software. Continuous integration is not discussed by *HeartSim*.

- **On-site customer – include a real, live user on the team, available full-time to answer questions:** The customer is often the scientist who is writing the software. In other cases, the customer may have devolved all responsibility for making key decisions to the scientist, so the customer's involvement is minimal. In other cases that we have seen, the customer is another scientific group co-located with the scientists who are writing the software. Their interactions are typically informal, frequent, and very detailed, often sharing both code writing and testing.

The underlying philosophy of this practice is on-going, meaningful communication with the customer, which is again non-refutable. The means of promoting this practice is highly dependent on the situation. The *Tools* case study mentions team collaborations but doesn't specifically say if they had on-site customers. *HeartSim* discusses the wide variation in roles for the customers and agrees with our experience that how this is handled is highly dependent on the situation.

SUMMARY

The discussions in this paper have exposed a number of mismatches between the characteristics of scientific software and the assumptions behind both plan-driven and agile software development methodologies.

Both categories of methodologies make assumptions about time required to explore problem space as opposed to solution space. These assumptions are at odds with scientific software development. For scientific software, the theory exploration as a part of the problem domain is usually previously established and documented. The overwhelming need is to dynamically explore the solution space because of complex interactions of elements of the solution. This often requires large portions of the code to be assembled before this exploration can take place. This is also is a process that is not amenable to the participation of a non-specialist customer.

The highest priority concern with scientific software is trustworthiness. The focus of agile methods is mainly speed of delivery of the product. It can be argued that plan-driven methods are more aligned with trust in the product rather than speed of delivery.

Time factors coupled with knowledge acquisition play a role in the mismatch. There are three cycles of knowledge acquisition in scientific software development that evolve at different rates. There is a short cycle of learning about the software and execution environment that lends itself to agile timelines. There is a long cycle of learning about the final product and its relation to the exterior environment that could line up with plan-driven timelines. However, there is an intermediate cycle of learning about the interactions of software, hardware, and theory that is not aligned with either agile timelines or plan-driven timelines. This is an important cycle of learning and plays havoc with the application of either category of methodology.

Both categories of methodologies assume the customer drives out the detailed requirements of the project. With scientific software, the customer may simply pose the problem, that is the "metaphor", and leave details to the scientist.

Scientific software is positioned between plan-driven and agile methodologies in respect to changes to the code. Before implementation begins the scientist invariable starts with a developed theory and plan. There will be parts of these plans that will be immutable, such as using particular equations of state and a matrix solution technique. As the intermediate cycle of learning begins, changes will be made to the code. At this point, development appears to be iterative as in agile approaches. However, it is not schedule-based, nor is it divisible. As well, changes are contained and focused on the exploration. This "phase "of scientific software development fits neither plan-driven nor agile approaches.

The goal of XP's refactoring is to allow anyone anytime to improve the quality of the code base. Unfortunately, quality is in the eyes of the beholder, and many qualities considered "best practices" for other types of software have not been validated for scientific software. Qualities important to the scientist are different from those typically considered important to the software engineer.

The time constant for scientific software is much longer than that assumed by agile methodologies. Understanding a change may take weeks rather than hours. As well, the lifetime of scientific software is often decades. This requires that learning be preserved over a number of years and over a large number of people disassociated by time. The focus of agile is sharing knowledge within the initial development of the product with temporally coincident participants. Unfortunately, plan-driven methods also fail to address this problem, with documents focused on describing the problem space of the product. Scientists already have a long and well-established history of documentation of their science. This is essentially their systems requirements. Their need is for documentation that supports long-term evolution of their products. All this points to a need for research into viable methods of preserving and transferring knowledge over time for scientific software products and for extending documentation scientists already produce as part of this knowledge transfer.

CONCLUSION

We define scientific software as the core part of the scientific product exclusive of user and hardware interfaces and of support libraries. The characteristics of the development of scientific software as we defined it are such that they do not fully mesh with the assumptions of either plan-driven or agile software development methodologies. These fundamental mismatches demonstrate the unsuitability of whole-heartedly using either type of process. There is a danger of imposing activities that detract from fundamental exploration that must be done and substituting values that do not fit with the goals of scientific software. Problems are already arising where quality standards wholly based on plan-driven development methodologies have been enforced on the development of scientific software in particular application areas.

This mismatch would also strongly argue for the careful partitioning of projects so that user interfaces, for example, are developed separately from the scientific part of the product. Development of other parts of the product more likely matches the assumptions inherent in agile and plan-driven methodologies. Certainly, the time lines, quality goals, and knowledge domains are very different when we compare the scientific core of the product to the creation of the user interface. This supports the argument for careful project partitioning.

There is considerable interest in research into the application of agile methodologies to scientific software development. Understanding the dynamics of developing scientific software will help to profitably move this research forward. Dynamics include the interaction of time, knowledge, software quality, and goals. Once that understanding is sufficient, we can assess the value of particular activities that comprise a suitable software development methodology.

Knowledge transfer over an extended period of time and across possibly temporally disassociated contributors to the software product is a major problem with the long-term evolution and use of scientific software. This problem is in addition to the knowledge transfer amongst co-temporal contributors as the product is developed. This is an important research area that would contribute to the quality of the scientific code bases.

There are irrefutable good software goals that all software developers should consider. However, successfully weaving activities to achieve these goals into a process that supports the needs of scientists writing software is not easy. We need to consider factors not traditionally included in software engineering processes, factors that, so far, seem to be unique to the development of scientific software.

REFERENCES

Ackroyd, K. S., Kinder, S. H., Mant, G. R., Miller, M. C., Ramsdale, C. A., & Stephenson, P. C. (2008). Scientific software development at a research facility. *IEEE Software, 25*(4), 44–51. doi:10.1109/MS.2008.93

Basili, V., Cruzes, D., Carver, J., Hochstein, L., Hollingsworth, J., Zelkowitz, M., & Shull, F. (2008). Understanding the high-performance-computing community: A software engineer's perspective. *IEEE Software, 25*(4), 29–36. doi:10.1109/MS.2008.103

Beck, K. (2000). *Extreme programming explained.* Reading, MA: Addison-Wesley.

Beck, K. (2003). *Test-driven development by example.* Reading, MA: Addison-Wesley.

Boehm, B., & Turner, R. (2005). *Balancing agility and discipline.* Reading, MA: Addison-Wesley.

Boisvert, R. F., & Tang, P. T. P. (Eds.). (2001). *The architecture of scientific software.* Boston, MA: Kluwer Academic.

Hatton, L. (2007). The chimera of software quality. *IEEE Computer, 40*(8), 102–104.

Hatton, L., & Roberts, A. (1994). How accurate is scientific software? *IEEE Transactions on Software Engineering, 20*(10), 786–797. doi:10.1109/32.328993

Henderson-Sellers, B. (2003). Method engineering for OO system development. *Communications of the ACM, 46*(10), 73–78. doi:10.1145/944217.944242

Highsmith, J., & Cockburn, A. (2001). Agile software development: The business of innovation. *IEEE Computer, 34*(9), 120–122.

Hook, D., & Kelly, D. (2009). Testing for trustworthiness in scientific software. In *Proceedings of the 2nd International Workshop on Software Engineering for Computational Science and Engineering*, Vancouver, BC, Canada.

Jackson, M. (1994). Problems, methods, and specialisation. *IEEE Software Engineering Journal, 9*(6), 249–255.

Jackson, M. (2001). *Problem frames*. Reading, MA: Addison-Wesley.

Jeffries, R., Anderson, A., & Hendrickson, C. (2001). *Extreme programming installed*. Reading, MA: Addison-Wesley.

Kelly, D. (2006). A study of design characteristics in evolving software using stability as a criterion. *IEEE Transactions on Software Engineering, 32*(5), 315–329. doi:10.1109/TSE.2006.42

Kelly, D. (2008). Innovative standards for innovative software. *IEEE Computer, 41*(7), 88–89.

Kelly, D. (2009). Determining factors that affect long-term evolution in scientific application software. *Journal of Systems and Software, 82*, 851–861. doi:10.1016/j.jss.2008.11.846

Kelly, D., & Shepard, T. (2004). Task-directed inspection. *Journal of Systems and Software, 73*(2), 361–368. doi:10.1016/j.jss.2003.10.027

Kelly, D., Thorsteinson, S., & Hook, D. (2010). Scientific software testing, analysis with four dimensions. *IEEE Software, 28*(3), 84–90. doi:10.1109/MS.2010.88

Matthews, D., Wilson, G., & Easterbrook, S. (2008). Configuration management for large-scale scientific computing at the UK Met office. *IEEE Computational Science & Engineering, 10*(6), 56–64.

Miller, G. (2006). Scientific publishing: A scientist's nightmare: Software problem leads to five retractions. *Science, 314*(12), 1856–1857. doi:10.1126/science.314.5807.1856

Pearson, C. (2009). *The demise of the iron triangle*. Retrieved February 19, 2010, from http://mosaicprojects.wordpress.com/2009/11/19/the-demise-of-the-iron-triangle/

Pfleeger, S. L. (1998). *Software engineering theory and practice*. Upper Saddle River, NJ: Prentice Hall.

Pitt-Francis, J., Bernabeu, M. O., Cooper, J., Garny, A., Momtahan, L., & Osborne, J. (2008). Chaste: Using agile programming techniques to develop computational biology software. *Philosophical Transactions of the Royal Society, 366*, 3111–3136. doi:10.1098/rsta.2008.0096

Post, D., & Kendall, R. (2004). Software project management and quality engineering practices for complex, coupled, multiphysics, massively parallel computational simulations: Lessons learned from ASCI. *International Journal of High Performance Computing Applications, 18*(4), 399–416. doi:10.1177/1094342004048534

Rawsthorne, D., & Shimp, D. (2009). *Scrum in a nutshell*. Retrieved February 19, 2010, from http://www.scrumalliance.org/articles/151-scrum-in-a-nutshell

Rolland, C. (1997). A primer for method engineering. In *Proceedings of the IFIP TC8, WG8.1/8.2 Working Conference on Method Engineering: Principles of Method Construction and Tool Support*, Toulouse, France (pp. 1-7).

Royce, W. (1970, August). Managing the development of large software systems. In *Proceedings of the 9th International Conference on Software Engineering* (pp. 328-338).

Sanders, R., & Kelly, D. (2008). Scientific software: Where's the risk and how do scientists deal with it? *IEEE Software, 25*(4), 21–28. doi:10.1109/MS.2008.84

Segal, J. (2004). *Professional end user developers and software development knowledge* (Tech. Rep. No. 2004/25). Milton Keynes, UK: Open University.

Segal, J. (2005). When software engineers met research scientists: A case study. *Empirical Software Engineering, 10*(4), 517–536. doi:10.1007/s10664-005-3865-y

Song, X., & Osterweil, L. J. (1992). Toward objective systematic design-method comparisons. *IEEE Software, 9*(3), 43–53. doi:10.1109/52.136166

Van Vliet, H. (2000). *Software engineering principles and practices*. Chichester, UK: John Wiley & Sons.

Weyuker, E. (1982). On testing non-testable programs. *The Computer Journal, 25*(4), 465–470.

Wiegers, K. E. (1999). *Software requirements*. Redmond, WA: Microsoft Press.

Wood, W. A., & Kleb, W. L. (2003). Exploring XP for scientific research. *IEEE Software, 20*(3), 30–36. doi:10.1109/MS.2003.1196317

This work was previously published in the Journal of Organizational and End User Computing, Volume 23, Issue 4, edited by Mo Adam Mahmood, pp. 64-79, copyright 2011 by IGI Publishing (an imprint of IGI Global).

Compilation of References

Abai, M. (2007). *Your organization needs an information strategy.* Retrieved July 8, 2009, from http://www.cioupdate.com/trends/article.php/3667451/Your-Organization-Needs-an-Information-Strategy

Abdul-Gader, A. H., & Kozar, K. A. (1995). The Impact of Computer Alienation of Information Technology Investment Decisions: An Exploratory Cross-National Analysis. *Management Information Systems Quarterly*, *19*(4), 535–559. doi:10.2307/249632

Ackoff, R. L. (1981). *Creating the corporate future.* New York: Wiley.

Ackroyd, K. S., Kinder, S. H., Mant, G. R., Miller, M. C., Ramsdale, C. A., & Stephenson, P. C. (2008). Scientific software development at a research facility. *IEEE Software*, *25*(4), 44–51. doi:10.1109/MS.2008.93

Adams, J. A. (1987). Historical review and appraisal of research on the learning, retention, and transfer of human motor skills. *Psychological Bulletin*, *101*(1), 41–74. doi:10.1037/0033-2909.101.1.41

Adams, J. S. (1963). Toward an Understanding of Inequity. *Journal of Abnormal Psychology*, *67*(5), 422–436. doi:10.1037/h0040968

Adkins, M., Burgoon, M., & Nunamaker, J. F. Jr. (2003). Using group support systems for strategic planning with the United States Air Force. *Decision Support Systems*, *34*(3), 315–337. doi:10.1016/S0167-9236(02)00124-0

Adler, P. S., & Kwon, S.-W. (2002). Social capital: prospects for a new concept. *Academy of Management Review*, *27*(1), 17–40. doi:10.2307/4134367

Agarwal, R., Brown, C., Ferratt, T., & Moore, J. E. (2006). Five mindsets for retaining IT staff. *MIS Quarterly Executive*, *5*(3), 137–150.

Agarwal, R., & Ferratt, T. W. (2001). Crafting an HR strategy to meet the need for IT workers. *Communications of the ACM*, *44*(7), 58–64. doi:10.1145/379300.379314

Agarwal, R., & Ferratt, T. W. (2002). Enduring practices for managing IT professionals. *Communications of the ACM*, *45*(9), 73–79. doi:10.1145/567498.567502

Agarwal, R., & Karahanna, E. (2000). Time Flies When You're Having Fun: Cognitive Absorption and Beliefs about Information Technology Usage. *Management Information Systems Quarterly*, *24*(4), 665–694. doi:10.2307/3250951

Agarwal, R., & Prasad, J. (1998). Conceptual and Operational Definition of Personal Innovativeness in the Domain of Information Technology. *Information Systems Research*, *9*(2), 204–215. doi:10.1287/isre.9.2.204

Agarwal, R., Sambamurthy, V., & Stair, R. M. (2000). The Evolving relationship between general and specific computer self-efficacy – An empirical assessment. *Information Systems Research*, *11*(4), 418–430. doi:10.1287/isre.11.4.418.11876

Agarwal, R., & Venkatesh, V. (2002). Assessing a Firm's Web Presence: A Heuristic Evaluation Procedure for the Measurement of Usability. *Information Systems Research*, *13*(2), 168–186. doi:10.1287/isre.13.2.168.84

Ahuja, M., & Carley, K. (1999). Network Structure in Virtual Organizations. *Organization Science*, *10*(6), 741–757. doi:10.1287/orsc.10.6.741

Ajzen, I. (1991). The theory of planned behavior. *Organizational Behavior and Human Decision Processes*, *50*, 179–211. doi:10.1016/0749-5978(91)90020-T

Ajzen, I., & Fishbein, M. (2005). The influence of attitudes on behavior. In Albarracin, D., Johnson, B. T., & Zanna, M. P. (Eds.), *The handbook of attitudes* (pp. 173–221). Mahwah, NJ: Lawrence Erlbaum Associates.

Akkermans, H., & Van Helden, K. (2002). Vicious and Virtuous Cycles in ERP Implementation: A Case Study of Interrelations between Critical Success Factors. *European Journal of Information Systems, 11*, 35–46. doi:10.1057/palgrave/ejis/3000418

Aladwani, A. M. (2002). Organizational actions, computer attitudes, and end-user satisfaction in public organizations: An empirical study. *Journal of Organizational and End User Computing, 14*(1), 42–49. doi:10.4018/joeuc.2002010104

Aladwani, A. M., & Palvia, P. C. (2002). Developing and validating an instrument for measuring user-perceived Web quality. *Information & Management, 39*, 467–476. doi:10.1016/S0378-7206(01)00113-6

Alavi, M. (1985). End-user computing: The MIS manager's perspective. *Information & Management, 8*(3), 171–178. doi:10.1016/0378-7206(85)90046-1

Alavi, M. (1994). Computer-mediated collaborative learning: an empirical evaluation. *Management Information Systems Quarterly, 18*(2), 159–174. doi:10.2307/249763

Alavi, M., & Leidner, D. (2001). Review: Knowledge Management and Knowledge Management Systems: Conceptual Foundations and Research Issues. *Management Information Systems Quarterly, 25*(1), 107–136. doi:10.2307/3250961

Alavi, M., & Leidner, D. E. (2001). Review: Knowledge management and knowledge management systems: Conceptual foundations and research issues. *Management Information Systems Quarterly, 25*(1), 107–136. doi:10.2307/3250961

Alavi, M., Marakas, G. M., & Yoo, Y. (2002). A comparative study of distributed learning environments on learning outcomes. *Information Systems Research, 13*(4), 404–415. doi:10.1287/isre.13.4.404.72

Alavi, M., Yoo, Y., & Vogel, D. R. (1997). Using information technology to add value to management education. *Academy of Management Journal, 40*(6), 1310–1333. doi:10.2307/257035

Al-Gahtani, S. S. (2001). The applicability of TAM outside North America: An empirical test in the United Kingdom. *Information Resources Management Journal, 14*(3), 37.

Al-Gahtani, S. S. (2003). Computer technology adoption in Saudi Arabia: Correlates of perceived innovation attributes. *Information Technology for Development, 10*(1), 57–69. doi:10.1002/itdj.1590100106

Al-Gahtani, S. S. (2004). Computer technology acceptance success factors in Saudi Arabia: An exploratory study. *Journal of Global Information Technology Management, 7*(1), 5–29.

Al-Gahtani, S. S. (2008). Testing for the Applicability of the TAM Model in the Arabic Context: Exploring an Extended TAM with Three Moderating Factors. *Information Resources Management Journal, 21*(4), 1.

Al-Gahtani, S. S., Hubona, G. S., & Wang, J. (2007). Information technology (IT) in Saudi Arabia: Culture and the acceptance and use of IT. *Information & Management, 44*(8), 681–691. doi:10.1016/j.im.2007.09.002

Al-Jabri, I. M., & Al-Khaldi, M. A. (1997). Effects of user characteristics on computer attitudes among undergraduate business students. *Journal of End User Computing, 9*(2), 16–22.

Al-Khaldi, M. A., & Al-Jabri, I. M. (1998). The relationship of attitudes to computer utilization: New evidence from a developing nation. *Computers in Human Behavior, 14*(1), 23–42. doi:10.1016/S0747-5632(97)00030-7

Alliger, G. M., Tannenbaum, S. I., Bennett, W. Jr, Traver, H., & Shotland, A. (1997). A meta analysis of the relations among training criteria. *Personnel Psychology, 50*, 341–358. doi:10.1111/j.1744-6570.1997.tb00911.x

Alvarez, R., & Urla, J. (2002). Tell me a Good Story: Using Narrative Analysis to Examine Information Requirements Interviews during an ERP Implementation. *The Data Base for Advances in Information Systems, 33*(1), 38–52.

Alvesson, M. (1985). A critical framework for organizational analysis. *Organization Studies, 6*(2), 117–138. doi:10.1177/017084068500600202

Amabile, T. M. (1988). A Model of Creativity and Innovation in Organizations. In Straw, B. M., & Cummings, L. L. (Eds.), *Research in Organizational Behavior* (*Vol. 10*, pp. 123–167). Greenwich, CT: JAI Press.

Amazon. (2009). *Amazon elastic compute cloud user guide* (API Version 2009-11-30). Retrieved June 22, 2010, from http://docs.amazonwebservices.com/AWSEC2/latest/UserGuide/

American Society for Training and Development. (2007). *2007 ASTD State of the Industry Report.*

Amoroso, D. L. (1988). Organizational Issues of End-user Computing. *ACM SIGMIS Database, 19*(3-4), 49–58. doi:10.1145/65766.65773

Anandarajan, M., Igbaria, M., & Anakwe, U. P. (2002). IT acceptance in a less-developed country: A motivational factor perspective. *International Journal of Information Management, 22*(1), 47–65. doi:10.1016/S0268-4012(01)00040-8

Anastasi, A. (1968). *Psychological Testing* (3rd ed.). New York: Macmillan.

Ancona, D. G., Goodman, P. S., Lawrence, B. S., & Tushman, M. L. (2001). Time: A new research lens. *Academy of Management Review, 26*(4), 645–663. doi:10.2307/3560246

Anderson, C., Al-Gahtani, S. S., & Hubona, G. S. (2007). *Evaluating the Antecedents of the Technology Acceptance Model in Saudi Arabia.* Paper presented at the Sixth Annual Workshop on HCI Research in MIS.

Anderson, C., Al-Gahtani, S. S., & Hubona, G. S. (2008). *Evaluating TAM Antecedents in Saudi Arabia.* Paper presented at the Southern Association of Information Systems Conference.

Anderson, N. R., Lee, E. S., Brockenbrough, J. S., Minie, M. E., Fuller, S., Brinkley, J., & Tarczy-Hornoch, P. (2007). Issues in biomedical research data management and analysis: Needs and barriers. *Journal of the American Medical Informatics Association, 14*(4), 478–488. doi:10.1197/jamia.M2114

Ang, S., & Slaughter, S. A. (2000). The missing context of information technology personnel: a review and future directions for research. In Zmud, R. W. (Ed.), *Framing the Domains of IT Management* (pp. 305–327). Cincinnati, OH: Pinnaflex Educational Resources.

Anonymous,. (2008). 2008 Industry Report. *Training (New York, N.Y.), 45*(9), 16–34.

Ansoff, H. I. (1964). A quasi-analytical approach of the business strategy problem. *Management Technology, IV,* 67–77.

Ansoff, H. I. (1965). *Corporate strategy: An analytic approach to business policy for growth and expansion.* New York: McGraw Hill.

Arbaugh, J. B. (2000). An exploratory study of the effects of gender on student learning and class participation in an Internet-based MBA course. *Management Learning, 31*(4), 503–519. doi:10.1177/1350507600314006

Arbaugh, J. B. (2001). How instructor immediacy behaviors affect student satisfaction and learning in web-based courses. *Business Communication Quarterly, 64*(4), 42–54. doi:10.1177/108056990106400405

Argote, L. (1999). *Organizational learning: Creating, retailing and transferring knowledge.* Norwell, MA: Kluwer Academic Publishers.

Argyris, C. (1960). *Understanding Organizational Behavior.* Homewood, IL: Dorsey Press.

Armbrust, M., Fox, A., Griffith, R., Joseph, A. D., Katz, R. H., Konwinski, A., et al. (2009). *Above the clouds: A Berkeley view of cloud computing* (Tech. Rep. No. UCB/EECS-2009-28). Berkeley, CA: University of California, Berkeley.

Arndt, S., Clevenger, J., & Meiskey, L. (1985). Students' attitudes toward computers. *Computers and the Social Sciences, 14*(3-4), 181–190.

Aryee, S., & Chay, Y. W. (1994). An examination of the impact of career-oriented mentoring on work commitment attitudes and career satisfaction among professional and managerial employees. *British Journal of Management, 5*(4), 241–249. doi:10.1111/j.1467-8551.1994.tb00076.x

Ash, J. S., Anderson, N., & Tarczy-Hornoch, P. (2008). People and organizational issues in research systems implementation. *Journal of the American Medical Informatics Association, 15*(3), 283–289. doi:10.1197/jamia.M2582

Ash, J. S., Berg, M., & Coiera, E. (2004). Some unintended consequences of information technology in health care: The nature of patient care information system-related errors. *Journal of the American Medical Informatics Association, 11*(2), 104–112. doi:10.1197/jamia.M1471

Astin, A. W., Vogelgesang, L. J., Ikeda, E. K., & Yee, J. A. (2000). *How service learning affects students*. Los Angeles, CA: Higher Education Research Institute.

Au, N., Ngai, E. W. T., & Cheng, T. C. E. (2008). Extending the Understanding of End User Information Systems Satisfaction Formation: An Equitable Needs Fulfillment Model Approach. *Management Information Systems Quarterly*, *32*(1), 43–66.

Avison, D. E., & Wood-Harper, A. T. (1990). *Multiview: An exploration in information systems development*. London: McGraw-Hill.

Avison, D., & Fitzgerald, G. (2006). *Information systems development methodologies, techniques and tools* (4th ed.). London: McGraw-Hill International.

Avison, D., Lau, F., Myers, M., & Nielsen, P. A. (1999). Action research. *Communications of the ACM*, *42*(1), 94–97. doi:10.1145/291469.291479

Bagby, R. M., Taylor, G. J., & Parker, J. D. A. (1994). The twenty-item Toronto alexithymia scale-II: convergent, discriminant, and concurrent validity. *Journal of Psychosomatic Research*, *38*(1), 33–40. doi:10.1016/0022-3999(94)90006-X

Bagozzi, P. (1980). *Causal methods in marketing*. New York: John Wiley & Sons.

Bagozzi, R. P. (1980). Evaluating structural equation models with unobservable variables and measurement error: a comment. *Journal of Marketing Research*, *13*, 375–381.

Bagozzi, R. P. (1994). Measurement in marketing research: basic principles of questionnaire design. In Bagozzi, R. P. (Ed.), *Principles of Marketing Research*. Oxford, UK: Blackwell Publishers.

Bagozzi, R. P., & Phillips, L. W. (1982). Representing and testing organizational theories: a holistic construal. *Administrative Science Quarterly*, *27*(3), 459–489. doi:10.2307/2392322

Bailey, J. E., & Pearson, S. W. (1983). Development of a tool for measurement and analysing computer user satisfaction. *Management Science*, *29*(5), 530–545. doi:10.1287/mnsc.29.5.530

Bailey, K. D. (1978). *Methods of Social Research*. New York, NY: Free Press.

Baker, K. R., Powell, S. G., Lawson, B., & Foster-Johnson, L. (2006). Comparison of Characteristics and Practices among Spreadsheet Users with Different Levels of Experience. In *Proceedings of the European Spreadsheet Risks Interest Group Conference*.

Baker, T., Miner, A. S., & Eesley, D. T. (2003). Improvising firms: Bricolage, account giving and improvisational competencies in the founding process. *Research Policy*, *31*(2), 255–276. doi:10.1016/S0048-7333(02)00099-9

Balasubramanian, S., Konana, P., & Menon, N. M. (2003). Customer Satisfaction in Virtual Environments: A Study of Online Investing. *Management Science*, *49*(7), 871–889. doi:10.1287/mnsc.49.7.871.16385

Bandura, A. (1997). *Self-efficacy: The Exercise of Control*. New York, NY: W.H. Freeman and Company.

Barclay, D., Higgins, C., & Thompson, R. (1995). The partial least squares (PLS) approach to causal modeling: Personal computer adoption and use as an illustration. *Technology Studies*, *2*(2), 285–309.

Barker, S. K. (2007). End-user Computing and End-user Development: Exploring Definitions for the 21st century. In *Proceedings of the 2007 Information Resources Management Association International Conference*, Vancouver, BC, Canada (pp. 249-252). Hershey, PA: IGI Global.

Barker, T., & Frolick, M. N. (2003). ERP Implementation Failure: A Case Study. *Information Systems Management*, *20*(4), 43–49. doi:10.1201/1078/43647.20.4.200 30901/77292.7

Barki, H., & Hartwick, J. (1994). Measuring User Participation, User Involvement, and User Attitude. *Management Information Systems Quarterly*, *18*(1), 59–82. doi:10.2307/249610

Barki, H., & Hartwick, J. (2001). Interpersonal Conflict and its Management in Information System Development. *Management Information Systems Quarterly*, *25*(2), 195–228. doi:10.2307/3250929

Barki, H., Rivard, S., & Talbot, J. (1993). Toward an Assessment of Software Development Risk. *Journal of Management Information Systems*, *10*(2), 203–225.

Barnard, J., & Price, A. (1994). Managing Code Inspection Information. *IEEE Software*, *11*(2), 59–69. doi:10.1109/52.268958

Barnes, S., & Vidgen, R. (2000). WebQual: an exploration of website quality. In *Proceedings of the 8th European Conference on Information Systems*, Vienna, Austria.

Barrett, E., & Lally, V. (1999). Gender differences in an on-line learning environment. *Journal of Computer Assisted Learning*, *15*, 48–60. doi:10.1046/j.1365-2729.1999.151075.x

Bartlett, J. C., & Toms, E. G. (2005). Developing a protocol for bioinformatics analysis: An integrated information behavior and task analysis approach. *Journal of the American Society for Information Science and Technology*, *56*(5), 469–482. doi:10.1002/asi.20136

Basili, V. R., Carver, J., Cruzes, D., Hochstein, L., Hollingsworth, J. K., Shull, F., & Zelkowitz, M. V. (2008). Understanding the high performance computing community: A software engineers' perspective. *IEEE Software*, *25*(4), 29–36. doi:10.1109/MS.2008.103

Basili, V., Cruzes, D., Carver, J., Hochstein, L., Hollingsworth, J., Zelkowitz, M., & Shull, F. (2008). Understanding the high-performance-computing community: A software engineer's perspective. *IEEE Software*, *25*(4), 29–36. doi:10.1109/MS.2008.103

Bassellier, G., & Benbasat, I. (2004). Business Competence of Information Technology Professionals: Conceptual Development and Influence on IT-Business Partnership. *Management Information Systems Quarterly*, *28*(4), 673–694.

Beath, C. M., & Orlikowski, W. J. (1994). The contradictory structure of systems development methodologies: Deconstructing the IS-user relationship in information engineering. *Information Systems Research*, *5*(4), 350–377. doi:10.1287/isre.5.4.350

Becerra-Fernandez, I., & Sabherwal, R. (2001). Organizational knowledge management: A contingency perspective. *Journal of Management Information Systems*, *18*(1), 23–55.

Beck, K. (2000). *Extreme programming explained: Embrace change*. Reading, MA: Addison-Wesley.

Beck, K. (2003). *Test-driven development by example*. Reading, MA: Addison-Wesley.

Beckwith, L., Sorte, S., Burnett, M., Wiedenbeck, S., Chintakovid, T., & Cook, C. (2005, September). Designing Features for Both Genders in End-User Software Engineering Environments. In *Proceedings of the IEEE Symposium on Visual Languages and Human-Centric Computing* (pp. 153-160).

Bentley, Y. (2005). *A critical approach to the development of a framework to support the evaluation of information strategies in UK higher education institutions.* Unpublished doctoral dissertation.

Bentley, Y., & Clarke, S. A. (2006a). Using Ethnographic research to investigate the development process of a university's information strategy. In *Proceedings of ECRM 2006: The 5th European Conference on Research Methods in Business and Management Studies*, Dublin, Ireland.

Bentley, Y., & Clarke, S. A. (2006b). Reflections from an information management project development. In *Proceedings of EMCIS 2006: The Third European and Mediterranean Conference on Information Systems*, Alicante, Spain.

Bernstam, E. V., Hersh, W. R., Johnson, S. B., Chute, C. G., Nguyen, H., & Nagarajan, R. (2009). Synergies and distinctions between computational disciplines in biomedical research: Perspective from the Clinical and Translational Science Award programs. *Academic Medicine*, *84*(7), 964–970. doi:10.1097/ACM.0b013e3181a8144d

Besnard, D., & Cacitti, L. (2005). Interface changes causing accidents - An empirical study of negative transfer. *International Journal of Human-Computer Studies*, *62*(1), 105–125. doi:10.1016/j.ijhcs.2004.08.002

Bhattacherjee, A. (2001). Understanding information systems continuance: An expectation-confirmation model. *Management Information Systems Quarterly*, *25*(3), 351–370. doi:10.2307/3250921

Bhattacherjee, A. (2002). Individual Trust in Online Firms: Scale Development and Initial Test. *Journal of Management Information Systems*, *19*(1), 211–241.

Bialecki, A., Cafarella, M., Cutting, D., & O'Malley, O. (2007) *Hadoop: A framework for running applications on large clusters built of commodity hardware*. Retrieved January 14, 2010, from http://hadoop.apache.org

Bies, R. J., & Tripp, T. M. (1996). Beyond Distrust: 'Getting Even' and the Need for Revenge. In Kramer, R. M., & Tyler, T. R. (Eds.), *Trust in Organizations: Frontiers of Theory and Research* (pp. 246–260). Thousand Oaks, CA: Sage.

Billig, S. H. (2000). Research on K-12 school-based service-learning: The evidence builds. *Phi Delta Kappan, 81*, 658–664.

Biner, P. M. (1993). The development of an instrument to measure student attitudes toward televised courses. *American Journal of Distance Education, 7*(1), 63–73. doi:10.1080/08923649309526811

Biner, P. M., Welsh, K. D., Barone, N. M., Summers, M., & Dean, R. S. (1997). The impact of remote-site group size on student satisfaction and relative performance in interactive telecourses. *American Journal of Distance Education, 11*(1), 23–33. doi:10.1080/08923649709526949

Bingi, P., Sharma, M. K., & Godla, J. K. (1999). Critical Issues Affecting an ERP Implementation. *Information Systems Management, 16*(3), 7–14. doi:10.1201/1078/43197.16.3.19990601/31310.2

Biocca, F., Harms, C., & Burgoon, J. K. (2003). Toward a more robust theory and measure of social presence: Review and suggested criteria. *Presence (Cambridge, Mass.), 12*(5). doi:10.1162/105474603322761270

Bishop, B., & McDaid, K. (2007). An Empirical Study of End-User Behaviour in Spreadsheet Error Detection & Correction. In *Proceedings of the European Spreadsheet Risk Interest Group Conference*, Greenwich, UK.

Bjerke, B., & Al-Meer, A. (1993). Culture's consequences: Management in Saudi Arabia. *Leadership and Organization Development Journal, 14*(2), 30–35. doi:10.1108/01437739310032700

Blum, K. (1998). Gender differences in CMC-based distance education. *Feminista, 2*(5).

Blum, K. (1999). Gender differences in asynchronous learning in higher education. *Journal of Asynchronous Learning Networks, 3*(1), 46–47.

Boehm, B. (1981). *Software engineering economics*. Upper Saddle River, NJ: Prentice Hall.

Boehm, B., Grunbacher, P., & Briggs, R. (2007). Developing groupware for requirements negotiation: Lessons learned. In Selby, R. (Ed.), *Software Engineering* (pp. 301–314). Hoboken, NJ: Wiley.

Boehm, B., & Turner, R. (2004). *Balancing agility and discipline: A guide for the perplexed*. Reading, MA: Addison-Wesley.

Bohlmann, M. (2007). *Managing relationships with end-users*. Retrieved March 22, 2010, from http://bmighty.informationweek.com/network/showArticle.jhtml?articleID=202200206

Boisvert, R. F., & Tang, P. T. P. (Eds.). (2001). *The architecture of scientific software*. Boston, MA: Kluwer Academic.

Boland, R. J. Jr, & Tenkasi, R. V. (1995). Perspective making and perspective taking in communities of knowing. *Organization Science, 6*(4), 350–372. doi:10.1287/orsc.6.4.350

Bollen, K. A. (1984). Multiple indicators: Internal consistency or no necessary relationship? *Quality & Quantity, 18*(4), 377–385. doi:10.1007/BF00227593

Boneva, B., Kraut, R., & Frohlich, D. (2001). Using e-mail for personal relationships. *The American Behavioral Scientist, 45*(3), 530–549. doi:10.1177/00027640121957204

Borgatti, S. P., & Cross, R. (2003). A relational view of information seeking and learning in social networks. *Management Science, 49*(4), 432–445. doi:10.1287/mnsc.49.4.432.14428

Bostrom, R. P., Olfman, L., & Sein, M. K. (1990). The importance of learning style in end-user training. *Management Information Systems Quarterly, 14*(1), 101–119. doi:10.2307/249313

Boudreau, M. C., Gefen, D., & Straub, D. W. (2001). Validation in information systems research: a state-of-the-art assessment. *Management Information Systems Quarterly, 25*(1), 1–16. doi:10.2307/3250956

Boudreau, M. C., & Seligman, L. (2005). Quality of use of a complex technology: a learning-based model. *Journal of Organizational and End User Computing, 17*(4), 1–22. doi:10.4018/joeuc.2005100101

Bourdieu, P. (1986). The forms of capital. In Richardson, J. G. (Ed.), *Handbook of Theory and Research for the Sociology of Education* (pp. 241–258). New York: Greenwood.

BPR Online Learning Center. (2010). *Module 2: The importance of scope and objectives.* Retrieved March 11, 2010, from http://www.prosci.com/tutorial-project-plan-mod2.htm

Brancheau, J. C., & Brown, C. V. (1993). The Management of End-user Computing: Status and Directions. *ACM Computing Surveys, 26*(4), 437–482. doi:10.1145/162124.162138

Brennan, L. L. (2009). A strategic approach to IT-enabled access and immediacy. *Journal of Organizational and End User Computing, 21*(4), 63–72. doi:10.4018/joeuc.2009062604

Briggs, R. O., Adkins, M., Mittleman, D., Kruse, J., Miller, S., & Nunamaker, J. F. Jr. (1999). A technology transition model derived from field investigation of GSS use aboard the U.S.S. Coronado. *Journal of Management Information Systems, 15*(3), 151–195.

Briggs, R. O., Vreede, G.-J. D., & Nunamaker, J. F. Jr. (2003). Collaboration Engineering with ThinkLets to Pursue Sustained Success with Group Support Systems. *Journal of Management Information Systems, 19*(4), 31–64.

Brislin, R. W. (1986). *The Wording and Translation of Research Instrument.* Beverly Hills, CA: Sage.

Brock, D. B., & Sulsky, L. M. (1994). Attitudes toward computers: construct validation and relations to computer use. *Journal of Organizational Behavior, 15*, 17–35. doi:10.1002/job.4030150104

Brody, N. (1992). *Intelligence.* San Diego, CA: Academic Press.

Brown, J. D. (1995). *The elements of language curriculum: A systematic approach to program development.* Boston, MA: Heinle & Heinle.

Brown, J. S., & Duguid, P. (1991). Organizational Learning and Communities-Of-Practice: Toward a Unified View of Working, Learning, and Innovation. *Organization Science, 2*(1), 40–57. doi:10.1287/orsc.2.1.40

Brown, S. P., & Chin, W. W. (2004). Satisfying and Retaining Customers through Independent Service Representative. *Decision Sciences, 35*(1), 527–550. doi:10.1111/j.0011-7315.2004.02534.x

Brown, S., Rust, C., & Gibbs, G. (1994). *Strategies for Diversifying Assessments in Higher Education.* Oxford, UK: Oxford Centre for Staff Development.

Brüderl, J., & Preisendörfer, P. (1998). Network support and the success of newly founded business. *Small Business Economics, 10*(3), 213–225. doi:10.1023/A:1007997102930

Burke, K., & Chidambaram, L. (1999). How much bandwidth is enough? A longitudinal examination of media characteristics and group outcomes. *Management Information Systems Quarterly, 23*(4), 557–579. doi:10.2307/249489

Burkhardt, M. E., & Brass, D. J. (1990). Changing patterns or patterns of change: the effects of a change in technology on social network structure and power. *Administrative Science Quarterly, 35*(1), 104–127. doi:10.2307/2393552

Burnett, M., Cook, C., & Rothermel, G. (2004). End User Software Engineering. *Communications of the ACM, 47*(9), 53–58. doi:10.1145/1015864.1015889

Burns, J. R., Jung, D. G., & Hoffman, J. J. (2009). Capturing and Comprehending the Behavioral/Dynamical Interactions within an ERP Implementation. *Journal of Organizational and End User Computing, 21*(2), 67–90. doi:10.4018/joeuc.2009040104

Byrd, T. A., & Turner, D. E. (2000). Measuring the Flexibility of Information Technology Infrastructure: Exploratory Analysis of a Construct. *Journal of Management Information Systems, 17*(1), 167–208.

Callan, J., & Kulkarni, A. (2009). *ClueWeb09 Wiki.* Retrieved January 14, 2010, from http://boston.lti.cs.cmu.edu/clueweb09/wiki

Caloghirou, Y., Kastelli, I., & Tsakanikas, A. (2004). Internal capabilities and external knowledge sources: Complements or substitutes for innovative performance? *Technovation, 24*(1), 29–39. doi:10.1016/S0166-4972(02)00051-2

Campbell, D. T. (1969). Variation and Selective Retention in Socio-cultural Evolution. *General Systems, 16*, 69–85.

Campbell, D. T., & Fiske, D. W. (1959). Convergent and discriminant validity by the multitrait-multimethod matrix. *Psychological Bulletin, 56*(2), 81–105. doi:10.1037/h0046016

Campbell, D., & Stonehouse, G. (2002). *Business strategy: An introduction* (2nd ed.). London: Bill Houston.

Campus Compact. (2008). *Service Statistics 2008*. Retrieved March 22, 2010, from http://www.compact.org/wp-content/uploads/2009/10/2008-statistics1.pdf

Cappelli, M. (2000). A market-driven approach to retaining talent. *Harvard Business Review*, 103–111.

Carlile, P. R. (2002). A pragmatic view of knowledge and boundaries: Boundary objects in new product development. *Organization Science, 13*(4), 442–455. doi:10.1287/orsc.13.4.442.2953

Carlson, R., & Grabowski, B. (1992). The effects of computer self-efficacy on direction-following behavior in computer-assisted instruction. *Journal of Computer-Based Instruction, 19*(1), 6–11.

Carol, C. (2006). End user computing ergonomics: Facts or Fads? *Journal of Organizational and End User Computing, 18*(3), 66–76. doi:10.4018/joeuc.2006070104

Carr, C. L. (2007). The FAIRSERV Model: Consumer Reactions to Services Based on a Multidimensional Evaluation of Service Fairness. *Decision Sciences, 38*(1), 107–130. doi:10.1111/j.1540-5915.2007.00150.x

Carroll, J. M. (1997). Human-computer interaction: psychology as a science of design. *Annual Review of Psychology, 48*(1), 61–83. doi:10.1146/annurev.psych.48.1.61

Carswell, A. D., & Venkatesh, V. (2002). Learner outcomes in a distance education environment. *International Journal of Human-Computer Studies, 56*(5), 475–494. doi:10.1006/ijhc.2002.1004

Carver, J. C., Kendall, R. P., Squires, S. E., & Post, D. E. (2007). Software development environments for scientific and engineering software: A series of case studies. In *Proceedings of the International Conference on Software Engineering* (pp. 550-559).

Carver, J. C. (2009). First international workshop on software engineering for computational science & engineering. *Computing in Science & Engineering, 11*(2), 7–11. doi:10.1109/MCSE.2009.30

Cary, A., Sun, Z., Hristidis, V., & Naphtali, R. (2009). Experiences on processing spatial data with mapreduce. In *Proceedings of the 21st International Conference on Scientific and Statistical Database Management* (pp. 302-319).

Cassel, C. M., Hackl, P., & Westlund, A. H. (2000). On Measurement of Intangible Assets: A Study of Robustness of Partial Least Squares. *Total Quality Management, 11*(7), 897–907. doi:10.1080/09544120050135443

Cassiman, B., & Veugelers, R. (2006). In search of complementarity in innovation strategy: Internal R&D and external knowledge acquisition. *Management Science, 52*(1), 68–82. doi:10.1287/mnsc.1050.0470

Chang, J. C.-J., & King, W. R. (2005). Measuring the Performance of Information Systems: A Functional Scorecard. *Journal of Management Information Systems, 22*(1), 85–115.

Chang, P.-L., & Shen, P.-D. (1997). A conceptual framework for managing end-user computing by the total quality management strategy. *Total Quality Management, 8*(1), 91–102. doi:10.1080/09544129710477

Chan, Y. E., & Storey, V. C. (1996). The use of spreadsheet in organizations: determinants and consequences. *Information & Management, 31*, 119–134. doi:10.1016/S0378-7206(96)00008-0

Chatterji, M. (2003). *Designing and using tools for educational assessment*. Upper Saddle River, NJ: Pearson Education.

Chau, P. Y. (2001). Influence of computer attitude and self-efficacy on IT usage behavior. *Journal of Organizational and End User Computing, 13*(1), 26–33. doi:10.4018/joeuc.2001010103

Chau, P. Y. K., & Hu, P. J.-H. (2002). Investigating healthcare professionals' decisions to accept telemedicine technology: An empirical test of competing theories. *Information & Management, 39*(4), 297–311. doi:10.1016/S0378-7206(01)00098-2

Checkland, P. B. (1983). OR and the System Movement: Mappings and Conflicts. *The Journal of the Operational Research Society, 34*(8), 661–675.

Checkland, P. B. (1999). *Systems Thinking, Systems Practice: includes a 30-year retrospective*. Chichester, UK: John Wiley and Sons.

Checkland, P. B., & Haynes, M. G. (1994). Varieties of systems thinking: the case of soft systems methodology. *System Dynamics, 10*(2-3), 189–197. doi:10.1002/sdr.4260100207

Chelariu, C., Johnston, W. J., & Young, L. (2002). Learning to Improvise, Improvising to Learn: A Process of Responding to Complex Environments. *Journal of Business Research, 55*(2), 141–147. doi:10.1016/S0148-2963(00)00149-1

Chen, Y., & Chan, H. (2000). An Exploratory Study of Spreadsheet Debugging Processes. In *Proceedings of the 4th Pacific Asia Conference on Information Systems* (pp. 143-155).

Chen, L., Soliman, K. S., Mao, E., & Frolick, M. N. (2000). Measuring user satisfaction with data warehouses: an exploratory study. *Information & Management, 37*(3), 103–110. doi:10.1016/S0378-7206(99)00042-7

Chickering, A. W., & Ehrmann, S. C. (1996). Implementing the Seven Principles: Technology as Lever. *AAHE Bulletin,* 3-6.

Chilana, P. K., Palmer, C. L., & Ko, A. J. (2009). Comparing bioinformatics software development by computer scientists and biologists: An exploratory study. In *Proceedings of the ICSE Workshop on Software Engineering for Computational Science and Engineering* (pp. 72-79).

Chin, W. W. (1998). Issues and opinion on structural equation modeling. *Management Information Systems Quarterly, 22*(1), 7–16.

Chin, W. W. (1998). The Partial Least Squares Approach to Structural Equation Modeling. In Marcoulides, G. A. (Ed.), *Modern Methods for Business Research* (pp. 295–336). Mahwah, NJ: Lawrence Erlbaum Associates.

Chin, W. W., & Gopal, A. (1995). Adoption Intention in GSS: Relative Importance of Beliefs. *The Data Base for Advances in Information Systems, 26*(2-3), 42–64.

Chin, W. W., Gopal, A., & Salisbury, W. D. (1997). Advancing the Theory of Adaptive Structuration: The Development of a Scale to Measure Faithfulness of Appropriation. *Information Systems Research, 8*(4), 342–366. doi:10.1287/isre.8.4.342

Chin, W. W., & Newsted, P. R. (1999). Structural equation modeling analysis with small samples using partial least squares. In Hoyle, R. (Ed.), *Statistical strategies for small sample research* (pp. 307–341). Thousand Oaks, CA: Sage.

Chiou, J.-S. (2004). The antecedents of consumers' loyalty toward Internet Service Providers. *Information & Management, 41*, 685–695. doi:10.1016/j.im.2003.08.006

Chudoba, K. M., Watson-Manheim, M. B., Lee, C. S., & Crowston, K. (2005). *Meet Me in Cyberspace: Meetings in the Distributed Work Environment*. Paper presented at the Academy of Management Conference.

Chung, R. H. G., Kim, B. S., & Abreu, J. M. (2004). Asian American Multidimensional Acculturation Scale: Development, Factor Analysis, Reliability, and Validity. *Cultural Diversity & Ethnic Minority Psychology, 10*(1), 66–80. doi:10.1037/1099-9809.10.1.66

Ciborra, C. U. (1999). Notes on Improvisation and Time in Organizations. *Accounting Management and Information Technologies, 9*(2), 77–94. doi:10.1016/S0959-8022(99)00002-8

Clarke, S. A. (2001). Mixing methods for organisational intervention: Background and current status. In M. G. Nicholls, S. Clarke, & B. Lehaney (Eds.), *Mixed-mode modelling: Mixing methodologies for organisational intervention* (1-18). Dordrecht, The Netherlands: Kluwer Academic.

Clarke, S. A. (2000). From socio-technical to critical complementarist: A new direction for information systems development. In Coakes, E., Lloyd-Jones, R., & Willis, D. (Eds.), *The New Socio Tech: Graffiti on the Long Wall* (pp. 61–72). London: Springer.

Clarke, S. A. (2004). A new direction for organizational end-user computing. *Journal of Organizational and End User Computing, 16*(2), i–viii.

Clarke, S. A. (2007). *Information systems strategic management: an integrated approach* (2nd ed.). London: Routledge.

Clarke, S. A., & Lehaney, B. (1998). A theoretical framework for facilitating methodological choice. *Systemic Practice and Action Research*, *11*(3), 295–318. doi:10.1023/A:1022952114289

Clarke, S. A., & Lehaney, B. (2002). Human-centered methods in information systems: Boundary setting and methodological choice. In Szewczak, E., & Snodgrass, C. (Eds.), *Human Factors in Information Systems* (pp. 20–30). Hershey, PA: IRM Press.

Clark, G. L., & Kaminski, P. (1986). Bringing a reality into the classroom: The team approach to a client-financed marketing research project. *Journal of Education for Business*, *62*(2), 61–65.

Cobb, A. T., Wooten, K. C., & Folger, R. (1995). Justice in the Making: Toward Understanding the Theory and Practice of Justice in Organizational Change and Development. In Pasmore, W. A., & Woodman, R. W. (Eds.), *Research in Organizational Change and Development* (pp. 243–295). London, UK: Jai Press.

Cockburn, A. (2002). *Agile software development*. Reading, MA: Addison-Wesley.

Cohen-Charash, Y., & Spector, P. E. (2001). The Role of Justice in Organizations: A Meta-Analysis. *Organizational Behavior and Human Decision Processes*, *86*(2), 278–321. doi:10.1006/obhd.2001.2958

Cohen, J. (1988). *Statistical power analysis for the behavioral sciences* (2nd ed.). Philadelphia, PA: Lawrence Erlbaum Associates.

Cohen, J. (2009). Graph twiddling in a MapReduce world. *Computing in Science & Engineering*, *11*(4), 29–41. doi:10.1109/MCSE.2009.120

Cohen, J., & Cohen, P. (1983). *Applied Multiple Regression/Correlation Analysis for the Behavioral Sciences*. Hillsdale, NJ: Lawrence Erlbaum Associates.

Cohen, P., Cohen, J., Teresi, J., Marchi, M., & Velez, C. N. (1990). Problems in the measurement of latent variables in structural equations causal models. *Applied Psychological Measurement*, *14*(2), 183–196. doi:10.1177/014662169001400207

Commander, D. (2009a). *User's guide for VirtualGL 2.1.4*. Retrieved January 14, 2010, from http://www.virtualgl.org/vgldoc/2_1/

Commander, D. (2009b). *User's guide for TurboVNC 0.6*. Retrieved January 14, 2010, from http://comments.gmane.org/gmane.comp.video.opengl.virtualgl.user/42

Compeau, D. R. (2002). The role of trainer behavior in end user software training. *Journal of End User Computing*, *14*(1), 23–32. doi:10.4018/joeuc.2002010102

Compeau, D. R., & Higgins, C. A. (1995a). Application of social cognitive theory to training for computer skills. *Information Systems Research*, *6*(2), 118–143. doi:10.1287/isre.6.2.118

Compeau, D. R., & Higgins, C. A. (1995b). Computer self-efficacy: development of a measure and initial test. *Management Information Systems Quarterly*, *19*(2), 189–202. doi:10.2307/249688

Constant, D., Sproull, L., & Kiesler, S. (1996). The kindness of strangers: The usefulness of electronic weak ties for technical advice. *Organization Science*, *7*(2), 119–135. doi:10.1287/orsc.7.2.119

Conway, N., & Briner, R. B. (2002). A Daily Diary Study of Affective Responses to Psychological Contract Breach and Exceeded Promises. *Journal of Organizational Behavior*, *23*(3), 287–302. doi:10.1002/job.139

Cougar, J. D. (1995). Implied creativity no longer appropriate for IS curriculum. *Journal of IS Education*, *7*(1), 12–13.

Couger, J. D., Oppermann, E. B., & Amoroso, D. L. (1994). Changes in motivation of IS managers – Comparison over a Decade. *Information Resources Management Journal*, *7*(2), 5–13.

Coulson, T., Zhu, J., Stewart, W., & Rohm, T. (2004). The importance of database application knowledge in successful ERP system training. *Communications of the IIMA*, *4*(3), 95–122.

Council on Competitiveness. (2009) *U.S. manufacturing – Global leadership through modeling and simulation*. Retrieved January 15, 2010, from http://www.compete.org/publications/detail/652/us-manufacturingglobal-leadership-through-modeling-and-simulation

Coyle-Shapiro, J. A.-M. (2002). A Psychological Contract Perspective on Organizational Citizenship Behavior. *Journal of Organizational Behavior*, *23*(8), 927–946. doi:10.1002/job.173

Cramton, C. D. (2001). The mutual knowledge problem and its consequences for dispersed collaboration. *Organization Science*, *12*(3), 346–371. doi:10.1287/orsc.12.3.346.10098

Cranny, C. J., Smith, P. C., & Stone, E. F. (1992). *Job satisfaction: how people feel about their jobs and how it affects their performance*. New York: Lexington Press.

Croll, G. (2005). The importance and criticality of spreadsheets in the city of London. In *Proceedings of the European Spreadsheet Risks Interest Group Conference*.

Cronbach, L. J. (1951). Coefficient Alpha and the Internal Structure of Tests. *Psychometrika*, *16*, 297–334. doi:10.1007/BF02310555

Cronbach, L. J. (1990). *Essentials of Psychological Testing* (5th ed.). New York: Harper-Row.

Cronbach, L. J., & Meehl, P. E. (1955). Construct validity in psychological tests. *Psychological Bulletin*, *55*(4), 281–302. doi:10.1037/h0040957

Cronin, J. J., Brady, M. K., & Hult, G. T. M. (2000). Assessing the effects of quality, value, and customer satisfaction on consumer behavioral intentions in service environments. *Journal of Retailing*, *76*(2), 193–218. doi:10.1016/S0022-4359(00)00028-2

Cropanzano, R., & Folger, R. (1989). Referent Cognitions and Task Decision Autonomy: Beyond Equity Theory. *The Journal of Applied Psychology*, *74*(2), 293–299. doi:10.1037/0021-9010.74.2.293

Crossan, M., Cunha, M. P. E., Vera, D., & Cunha, J. (2005). Time and Organizational Improvisation. *Academy of Management Review*, *30*(1), 129–145. doi:10.5465/AMR.2005.15281441

Cross, R., & Cummings, J. N. (2004). Tie and network correlates of individual performance in knowledge-intensive work. *Academy of Management Journal*, *47*(6), 928–937. doi:10.2307/20159632

Crowston, K., Howison, J., Masango, C., & Eseryel, U. Y. (2005). *Face-to-face interactions in self-organizing distributed teams*. Paper presented at the Academy of Management Conference.

Cummings, T. G., & Worley, C. G. (2005). *Organization Development and Change* (8th ed.). Mason, OH: Thomson South-Western.

Cunha, M. P. E., Kamoche, K., & Cunha, R. C. E. (2003). Organizational improvisation and leadership: A field study in two computer-mediated settings. *International Studies of Management & Organization*, *33*(1), 34–57.

Daft, R. L., & Lengel, R. H. (1986). Organizational information requirements: Media richness and structural design. *Management Science*, *32*(5), 554–571. doi:10.1287/mnsc.32.5.554

D'Ambra, J., & Rice, R. E. (2001). Emerging factors in user evaluation of the World Wide Web. *Information & Management*, *38*, 373–384. doi:10.1016/S0378-7206(00)00077-X

Darr, E. D., Argote, L., & Epple, D. (1995). The acquisition, transfer, and depreciation of knowledge in service organizations: Productivity in franchises. *Management Science*, *41*(11), 1750–1762. doi:10.1287/mnsc.41.11.1750

Davenport, T., & Volpel, S. (2001). The rise of knowledge towards attention management. *Journal of Knowledge Management*, *5*(3), 212–221. doi:10.1108/13673270110400816

Davidson, J. O. C. (1994). The Sources and Limits of Resistance in a Privatized Utility. In Jermier, J. M., Knights, D., & Nord, W. R. (Eds.), *Resistance and Power in Organizations* (pp. 69–101). London, UK: Routledge.

Davis, F. D. (1989). Perceived usefulness, perceived ease of use, and user acceptance of information technology. *Management Information Systems Quarterly*, *13*(3), 318–340. doi:10.2307/249008

Davis, F. D., Bagozzi, R. P., & Warshaw, P. R. (1992). Extrinsic and intrinsic motivation to use computers in the workplace. *Journal of Applied Social Psychology*, *22*(14), 1111–1132. doi:10.1111/j.1559-1816.1992.tb00945.x

Davis, G. B., & Olson, M. H. (1985). *Management Information Systems: Conceptual Foundations, Structure, and Development* (2nd ed.). New York, NY: McGraw-Hill.

Davison, R. (1997). An instrument for measuring meeting success. *Information & Management*, *32*, 163–176. doi:10.1016/S0378-7206(97)00020-7

De Cremer, D. (2005). Procedural and Distributive Justice Effects Moderated by Organizational Identification. *Journal of Managerial Psychology*, *20*(1), 4–13. doi:10.1108/02683940510571603

de los Santos, G., & Jensen, T. D. (1985). Client-sponsored projects: Bridging the gap between theory and practice. *Journal of Marketing Education*, *7*(2), 45–50. doi:10.1177/027347538500700207

De Roure, D., & Goble, C. (2009). Software design for empowering scientists. *IEEE Software*, *26*(1), 88–95. doi:10.1109/MS.2009.22

De Vreede, G.-J., Jones, N., & Mgaya, R. J. (1998). Exploring the application and acceptance of group support systems in Africa. *Journal of Management Information Systems*, *15*(3), 197–234.

Dean, J., & Ghemawat, S. (2004). MapReduce: Simplified data processing on large clusters. In *Proceedings of the 6th Symposium on Operating Systems Design and Implementation* (pp. 107-113).

DeSanctis, G., & Poole, M. S. (1994). Capturing the complexity in advanced technology use: Adaptive structuration theory. *Organization Science*, *5*(2), 121–147. doi:10.1287/orsc.5.2.121

Devaraj, S., Fan, M., & Kohli, R. (2002). Antecedents of B2C channel satisfaction and preference: Validation e-commerce metrics. *Information Systems Research*, *13*(3), 316–333. doi:10.1287/isre.13.3.316.77

Diamantopoulos, A., & Winklhofer, H. (2001). Index construction with formative indicators: An alternative to scale development. *JMR, Journal of Marketing Research*, *38*(2), 269–277. doi:10.1509/jmkr.38.2.269.18845

Dillman, D. A. (2000). *Mail and Internet Surveys: the Tailored Design*. New York: John Wiley & Sons.

Doll, W. J., & Torkzadeh, G. (1988). The measurement of end user computing satisfaction. *Management Information Systems Quarterly*, *12*(2), 259–275. doi:10.2307/248851

Doll, W. J., & Torkzadeh, G. (1991). The measurement of end-user computing satisfaction: Theoretical and methodological issues. *Management Information Systems Quarterly*, *15*(1), 5–10. doi:10.2307/249429

Doll, W. J., & Torkzadeh, G. (1998). Developing a Multi-dimensional Measure of System-Use in an Organizational Context. *Information & Management*, *33*(4), 171–185. doi:10.1016/S0378-7206(98)00028-7

Dongarra, J. J., Otto, S. W., Snir, M., & Walker, D. (1996). A message passing standard for MPP and workstations. *Communications of the ACM*, *39*(7), 84–90. doi:10.1145/233977.234000

Downey, J. (2004). Toward a comprehensive framework: EUC research issues and trends (1990-2000). *Journal of Organizational and End User Computing*, *16*(4), 1–16. doi:10.4018/joeuc.2004100101

Downey, J. P., & Bartczak, C. A. (2005). End-user Computing Research Issues and Trends (1990-2000). In Mahmood, M. A. (Ed.), *Advanced Topics in End-user Computing* (*Vol. 4*, pp. 1–20). Hershey, PA: Idea Group.

Doyle, M., & Strauss, D. (1976). *How to make meetings work*. Chicago, IL: Playboy Press.

Dubie, D. (2007). Can IT and end users get along? *Network World*. Retrieved March 11, 2010, from http://www.networkworld.com/news/2007/121207-it-end users-relationships.html

Ducheneaut, N., Yee, N., Nickell, E., & Moore, R. J. (2006). *"Alone together?" Exploring the social dynamics of massively multiplayer online games*. Paper presented at the ACM CHI 2006 Conference on Human Factors in Computing Systems, Montreal, Quebec, Canada.

Dukes, R. L., Discenza, R., & Couger, J. D. (1989). Convergent Validity of Four Computer Anxiety Scales. *Educational and Psychological Measurement*, *49*, 195–203. doi:10.1177/0013164489491021

Dyrud, M. A. (2001). Group projects and peer review. *Business Communication Quarterly*, *64*(4), 106–112. doi:10.1177/108056990106400412

Eagly, A. H. (1987). *Sex differences in social behavior: A social role interpretation*. Hillsdale, NJ: Earlbaum.

Eagly, A. H., & Chaiken, S. (1995). Attitude Strength, Attitude Structure, and Resistance to Change. In Petty, R. E., & Krosnick, J. A. (Eds.), *Attitude Strength: Antecedents and Consequences* (pp. 413–432). Mahwah, NJ: Lawrence Erlbaum Associates.

Earl, M. J. (1989). *Management strategies for information technology*. London: Prentice Hall.

Easterbrook, S. M., & Johns, T. C. (2009). Engineering the software for understanding climate change. *Computing in Science & Engineering, 11*(6), 65–74. doi:10.1109/MCSE.2009.193

Edelsky, C. (1981). Who's got the floor? *Language in Society, 10*.

Eden, D., & Kinnar, J. (1991). Modeling Galatea: boosting self-efficacy to increase volunteering. *The Journal of Applied Psychology, 76*(6), 770–780. doi:10.1037/0021-9010.76.6.770

Ehn, P. (1988). *Work-oriented design of computer artifacts*. Falköping, Sweden: Pelle Ehn and Arbetslivscentrum.

Einhorn, H., & Hogarth, R. (1981). Behavioral decision theory: Processes of judgment and choice. *Annual Review of Psychology, 32*(1), 53–88. doi:10.1146/annurev.ps.32.020181.000413

Eisenhardt, K. M. (1989). Building theory from case study research. *Academy of Management Review, 14*(4), 532–550. doi:10.2307/258557

Eisenhardt, K. M., & Tabrizi, B. N. (1995). Accelerating Adaptive Processes: Product Innovation in the Global Computer Industry. *Administrative Science Quarterly, 40*(1), 84–110. doi:10.2307/2393701

Elbeltagi, I., McBride, N., & Hardaker, G. (2005). Evaluating the factors affecting DSS usage by senior managers in local authorities in Egypt. *Journal of Global Information Management, 13*(2), 42.

Elsner, P. (2000). *Civic responsibility and higher education*. Phoenix, AZ: Oryx Press.

Espinosa, J. A., Cummings, J. N., Wilson, J. M., & Pearce, B. M. (2003). Team boundary issues across multiple global firms. *Journal of Management Information Systems, 19*(4), 157–190.

Etezadi-Amoli, J., & Farhoomand, A. (1991). On End-User Computing Satisfaction. *Management Information Systems Quarterly, 15*(1), 1–4. doi:10.2307/249428

Evans, G. E., & Simkin, M. G. (1989). What best predicts computer proficiency? *Communications of the ACM, 32*(11), 1322–1327. doi:10.1145/68814.68817

Faculty Connection. (2007). *Software entrepreneurship for students curriculum: A case study.* Retrieved from http://www.microsoft.com/education/facultyconnection/articles/articledetails.aspx?cid=844&c1=en-us&c2=0

Falk, R. F., & Miller, N. B. (1992). *A Primer for Soft Modeling*. Akron, OH: The University of Akron.

Feiertag, H. (2001). Front office attitude training ensures business in tough times. *Hotel and Motel Management, 216*(20), 22.

Ferratt, T. W., Agarwal, R., Brown, C., & Moore, J. E. (2005). IT Human resource management configurations and IT turnover: theoretical synthesis and empirical analysis. *Information Systems Research, 16*(3), 237–255. doi:10.1287/isre.1050.0057

Ferratt, T. W., Short, L. E., & Agarwal, R. (1993). Measuring the Information Systems Supervisor's Work-Unit Environment and Demonstrated Skill at Supervising. *Journal of Management Information Systems, 9*(4), 121–144.

Finton, T. A. (2003). Collaboratories as a new form of scientific organization. *Economics of Innovation and New Technology, 12*(1), 5–25. doi:10.1080/10438590303119

Fishbein, M., & Ajzen, I. (1975). *Belief, Attitude, Intention and Behavior: An Introduction to Theory and Research*. Reading, MA: Addison-Wesley.

Fjermestad, J., & Hiltz, S. R. (1999). An assessment of group support systems experiment research: Methodology and results. *Journal of Management Information Systems, 15*(3), 7–149.

Flap, H., & Völker, B. (2001). Goal specific social capital and job satisfaction effects of different types of networks on instrumental and social aspects of work. *Social Networks, 23*, 297–320. doi:10.1016/S0378-8733(01)00044-2

Flood, R. L., & Jackson, M. C. (1991). *Creative problem solving: Total systems intervention*. Chichester, UK: Wiley.

Folger, R. (1993). Reactions to Mistreatment at Work. In Murnighan, J. K. (Ed.), *Social Psychology in Organizations: Advances in Theory and Research* (pp. 161–183). Englewood Cliffs, NJ: Prentice-Hall.

Folger, R., & Skarlicki, D. P. (1999). Unfairness and Resistance to Change: Hardship as Mistreatment. *Journal of Organizational Change Management, 12*(1), 35–50. doi:10.1108/09534819910255306

Foltz, P., Davies, S. E., Polson, P. G., & Kieras, D. E. (1988). Transfer between menu systems. In E. Soloway, D. Frye, & S. B. Sheppard (Eds.), *Proceedings of the ACM SIGCHI Conference on Human Factors in Computing Systems*, Washington, DC (pp. 107-112).

Ford, D. P., Connelly, C. E., & Meister, D. B. (2003). Information systems research and Hofstede's culture's consequences: An uneasy and incomplete partnership. *IEEE Transactions on Engineering Management, 50*(1), 8–25. doi:10.1109/TEM.2002.808265

Ford, J. K., & Noe, R. A. (1987). Self-assessed training needs: the effects of attitudes toward training, managerial level, and function. *Personnel Psychology, 40*, 39–53. doi:10.1111/j.1744-6570.1987.tb02376.x

Fornell, C., & Larcker, D. F. (1981). Evaluating Structural Equation Models with Unobservable Variables and Measurement Error. *JMR, Journal of Marketing Research, 18*(1), 39–50. doi:10.2307/3151312

Foucault, M. (1983). The subject and power. In Dreyfus, H. L., & Rabinow, P. (Eds.), *Michel Foucault: Beyond structuralism and hermeneutics* (2nd ed.). Chicago: University of Chicago Press.

Fulk, J., & Collins-Jarvis, L. (2001). Wired meetings: Technological mediation of organizational gatherings. In Jablin, F. M., & Putnam, L. L. (Eds.), *The New Handbook of Organizational Communication: Advances in Theory* (pp. 624–663). Research, and Methods.

Galletta, D. F., Hartzel, K. S., Johnson, S., Joseph, J., & Rustagi, S. (1996). An Experimental Study of Spreadsheet Presentation and Error Detection. In *Proceedings of the 29th Annual Hawaii International Conference on System Sciences* (Vol. 2, p. 336).

Galletta, D. F., Abraham, D., El Louadi, M., Lekse, W., Pollailis, Y. A., & Sampler, J. L. (1993). An Empirical Study of Spreadsheet Error-Finding Performance. *Journal of Accounting, Management, and Information Technology, 3*(2), 79–95. doi:10.1016/0959-8022(93)90001-M

Galliers, R. D. (1991). Strategic information systems planning, myths and reality. *European Journal of Information Systems, 3*, 199–213.

Galliers, R. D. (1993). Towards a flexible information architecture: Integrating business strategies, information systems strategies and business process redesign. *Journal of Information Systems, 3*(3), 199–213. doi:10.1111/j.1365-2575.1993.tb00125.x

Galliers, R. D., & Land, F. F. (1987). Choosing Appropriate information systems research methodologies. *Communications of the ACM, 30*(11), 900–902. doi:10.1145/32206.315753

Garavan, T. N., & McCracken, C. (1993). Introducing End-user Computing: The Implications for Training and Development-Part 1. *Industrial & Commercial Training, 25*(7), 8–14. doi:10.1108/00197859310042443

Gartner Research. (2001). *U.S. Syposium/ITxpo 2001.* Retrieved May 15, 2010, from http://www.courtesy-computers.com/Networking/enduser%20computing%20best%20practices.pdf

Gatignon, H., Tushman, M. L., Smith, W., & Anderson, P. (2002). A Structural Approach to Assessing Innovation: Construct Development of Innovation Locus, Type, and Characteristics. *Management Science, 48*(9), 1103–1122. doi:10.1287/mnsc.48.9.1103.174

Gattiker, U. E. (1992). Computer skills acquisition: a review and future directions for research. *Journal of Management, 18*(3), 547–574. doi:10.1177/014920639201800307

Gattiker, U. E., & Hlavka, A. (1992). Computer attitudes and learning performance: Issues for management education and training. *Journal of Organizational Behavior, 13*(1), 89–101. doi:10.1002/job.4030130109

Gefen, D., & Keil, M. (1998). The impact of developer responsiveness on perceptions of usefulness and ease of use: An extension of the technology acceptance model. *The Data Base for Advances in Information Systems, 29*(2), 35–49.

Gerrity, T. P., & Rockart, J. F. (1986). End-user computing: Are you a leader or a laggard? *Sloan Management Review, 27*(4), 25–34.

Ghiselli, E. E., Campbell, J. P., & Zedeck, J. P. (1981). *Measurement Theory for the Behavioral Sciences.* San Francisco, CA: Freeman.

Ghorab, K. E. (1997). The impact of technology acceptance considerations on system usage, and adopted level of technological sophistication: An empirical investigation. *International Journal of Information Management, 17*(4), 249–259. doi:10.1016/S0268-4012(97)00003-0

Gilmour, D. (2003). How to fix knowledge management. *Harvard Business Review, 81*(10), 16–17.

Ginzberg, M. J. (1981). Early Diagnosis of MIS Implementation Failure: Promising Results and Unanswered Questions. *Management Science, 27*(4), 459–478. doi:10.1287/mnsc.27.4.459

Glorfeld, K. D., & Cronan, T. P. (1992). Computer information satisfaction: a longitudinal study of computing systems and EUC in a public organisation. *Journal of End User Computing, 5*(1), 27–36.

Glotzer, S. C., Kim, S., Cummings, P. T., Deshmukh, A., Head-Gordon, M., & Karniadakis, G. (2009). *WTEF panel report on international assessment of research and development in simulation-based engineering and science.* Baltimore, MD: World Technology Evaluation Center.

Goodhue, D. L. (1998). Development and measurement validity of a task-technology fit instrument for user evaluations of information systems. *Decision Sciences, 29*(1), 105–138. doi:10.1111/j.1540-5915.1998.tb01346.x

Gorriz, C. M., & Medusa, C. (2000). Engaging girls with computers through software games. *Communications of the ACM, 43*(1), 42–49. doi:10.1145/323830.323843

Govindarajulu, C., & Reithel, B. J. (1998). Beyond the Information Center: An Instrument to Measure End User Computing Support for Multiple Sources. *Information & Management, 33*, 241–250. doi:10.1016/S0378-7206(98)00030-5

Granovetter, M. S. (1973). The strength of weak ties. *American Journal of Sociology, 78*(6), 1360–1380. doi:10.1086/225469

Granovetter, M. S. (1982). The strength of weak ties: a network theory revisited. In Mardsen, P. V., & Lin, N. (Eds.), *Social Structure and Network Analysis* (pp. 105–130). Beverly Hills, CA: Sage.

Grant, R. M. (1996). Prospering in dynamically-competitive environments: Organizational capability as knowledge integration. *Organization Science, 7*(4), 375–387. doi:10.1287/orsc.7.4.375

Gray, P. H., & Meister, D. B. (2004). Knowledge sourcing effectiveness. *Management Science, 50*(6), 821–834. doi:10.1287/mnsc.1030.0192

Gray, P. H., & Meister, D. B. (2006). Knowledge sourcing methods. *Information & Management, 43*(2), 142–156. doi:10.1016/j.im.2005.03.002

Greenberg, J. (1990). Employee Theft as a Reaction to Underpayment Inequity: The Hidden Costs of Pay Cuts. *The Journal of Applied Psychology, 75*(5), 561–568. doi:10.1037/0021-9010.75.5.561

Greene, S. L. (2002). Characteristics of applications that support creativity. *Communications of the ACM, 45*(10), 100–104. doi:10.1145/570907.570941

Griffith, T. L., Sawyer, J. E., & Neale, M. A. (2003). Virtualness and knowledge in teams: Managing the love triangle of organizations, individuals and teams. *Management Information Systems Quarterly, 27*(3), 265–287.

Guimaraes, T., & Igbaria, M. (1997). Assessing user computing effectiveness: An integrated model. *Journal of End User Computing, 9*(2), 3–14.

Gunawardena, C. N. (1995). Social presence theory and implications for interaction and collaborative learning in computer conferences. *International Journal of Educational Telecommunications, 1*(2-3), 147–166.

Gunawardena, C. N., Lowe, X., Constance, A., & Anderson, T. (1997). Analysis of a global debate and the development of an interaction analysis model for examining social construction of knowledge in computer conferences. *Journal of Educational Computing Research, 17*(4), 397–431. doi:10.2190/7MQV-X9UJ-C7Q3-NRAG

Gupta, A. (2000). Enterprise Resource Planning: The Emerging Organizational Value Systems. *Industrial Management & Data Systems, 100*(3), 114–118. doi:10.1108/02635570010286131

Guzman, I. R., Stam, K. R., & Stanton, J. M. (2008). The occupational culture of IS/IT personnel within organizations. *The Data Base for Advances in Information Systems, 39*, 33–50.

Habermas, J. (1972). *Knowledge and human interests.* London: Heinemann.

Hackos, J., & Redish, J. (1998). *User and task analysis for interface design.* New York, NY: John Wiley & Sons.

Haines, R., & Cooper, R. (2008). The Influence of Workspace Awareness on Group Intellective Decision Effectiveness. *European Journal of Information Systems, 17*(6), 631–648. doi:10.1057/ejis.2008.51

Hair, J. F., Anderson, R. E., Tatham, R. L., & Grablowsky, B. J. (1998). *Multivariate Data Analysis* (5th ed.). New York, NY: Prentice Hall.

Hambrick, D. C., Geletkanycz, M. A., & Frederickson, J. W. (1993). Top executive commitment to the status quo: Some tests of its determinants. *Strategic Management Journal, 14*, 401–418. doi:10.1002/smj.4250140602

Hansen, F. (1972). *Consumer choice behaviour.* New York, NY: Free Press.

Hansen, F. (1976). Psychological theories of consumer choice. *The Journal of Consumer Research, 3*(3), 117–142. doi:10.1086/208660

Hansen, M. T., & Haas, M. R. (2001). Competing for attention in knowledge markets: Electronic document dissemination in a management consulting company. *Administrative Science Quarterly, 46*(1), 1–28. doi:10.2307/2667123

Harris, A. L. (1994). Developing the systems project course. *Journal of Information Systems Education, 6*(4), 192–197.

Harrison, A. W., & Rainer, R. K. (1992). The influence of individual differences on skill in end-user computing. *Journal of Management Information Systems, 9*(1), 93–111.

Harvey, L., & Rousseau, R. (1995). Development of text-editing skills: from semantic and syntactic mappings to procedures. *Human-Computer Interaction, 10*(4), 345–400. doi:10.1207/s15327051hci1004_1

Hatch, M. J. (1997). Jazzing up the theory of organizational improvisatione. In Walsh, J. P., & Huff, A. S. (Eds.), *Advances in Strategic Management* (pp. 181–191). Greenwich, CT: JAI Press.

Hatton, L. (2007). The chimera of software quality. *IEEE Computer, 40*(8), 102–104.

Hatton, L., & Roberts, A. (1994). How accurate is scientific software? *IEEE Transactions on Software Engineering, 20*(10), 786–797. doi:10.1109/32.328993

Haughey, D. (2009). *Stop scope creep running away with your project.* Retrieved October 15, 2008, from http://www.projectsmart.co.uk

Heavy, A., Lauret, J., & Keahy, K. (2009, April 8). Clouds make way for STAR to shine. *International Science Grid This Week.*

HEFCE. (1998). Information systems and technology management: Value for money study. In *Management Review Guide.* London: VFM Steering Group.

Heinssen, R. K., Glass, C. R., & Knight, L. A. Jr. (1987). Assessing Computer Anxiety: Development and Validation of the Computer Anxiety Rating Scale. *Computers in Human Behavior, 3*(1), 49–59. doi:10.1016/0747-5632(87)90010-0

Helliwell, J., & Putnam, R. (2007). Education and social capital. *Eastern Economic Journal, 33*, 1–19. doi:10.1057/eej.2007.1

Helper, S. (2009). The high road for U.S. manufacturing. *Issues in Science and Technology, 25*(2), 39–45.

Henderson, J. C., & Venkatraman, N. (1993). Strategic alignment: Leveraging information technology for transforming organisations. *IBM Systems Journal, 32*(1), 4–16. doi:10.1147/sj.382.0472

Henderson, R., & Divett, M. J. (2003). Perceived usefulness, ease of use and electronic supermarket use. *International Journal of Human-Computer Studies, 59*(3), 383–395. doi:10.1016/S1071-5819(03)00079-X

Henderson-Sellers, B. (2003). Method engineering for OO system development. *Communications of the ACM, 46*(10), 73–78. doi:10.1145/944217.944242

Highsmith, J., & Cockburn, A. (2001). Agile software development: The business of innovation. *IEEE Computer, 34*(9), 120–122.

Hill, C. E., Loch, K. D., Straub, D. W., & El-Sheshai, K. (1998). A qualitative assessment of Arab culture and information technology transfer. *Journal of Global Information Technology Management*, 29–38.

Hill, K. (2003). *System Designs that Start at the End (User)*. CRM Daily.

Hillman, D. C. A., Willis, D. J., & Gunawardena, C. N. (1994). Learner-interface interaction in distance education: An extension of contemporary models and strategies for practitioners. *American Journal of Distance Education*, *8*(2), 30–42. doi:10.1080/08923649409526853

Hiltz, S. R. (1994). *The Virtual Classroom: Learning Without Limits via Computer Networks*. Norwood, NJ: Ablex Publishing Company.

Hiltz, S. R., Zhang, Y., & Turoff, M. (2002). Studies of effectiveness of learning networks. In *Elements of Quality Online Education*. Needham, MA: SCOLE.

Hine, C. (2006). Databases as scientific instruments and their role in the ordering of scientific work. *Social Studies of Science*, *36*(2), 269–298. doi:10.1177/0306312706054047

Hirschheim, R. A. (1986). Understanding the office: A social- analytic perspective. *ACM Transactions on Office Information Systems*, *4*(4), 331–344. doi:10.1145/9760.9763

Hirschheim, R. A., & Klein, H. K. (1989). Four paradigms of information systems development. *Communications of the ACM*, *32*(10), 1199–1216. doi:10.1145/67933.67937

Hirschheim, R., & Newman, M. (1988). Information Systems and User Resistance: Theory and Practice. *The Computer Journal*, *31*(5), 398–408. doi:10.1093/comjnl/31.5.398

Hochstein, L., & Basili, V. R. (2008). The ASC-alliance projects: A case study of large-scale parallel scientific code development. *IEEE Computer*, *41*(3), 50–58.

Hoffa, C., Mehta, G., Freeman, T., Deelman, E., Keahey, K., Berriman, B., & Good, J. (2008). On the use of cloud computing for scientific workflow. In *Proceedings of the 3rd International Workshop on Scientific Workflows and Business Workflow Standards in e-Science* (pp. 640-645).

Hofstede, G. (1980). *Culture's Consequences: International Differences in Work-Related Values*. Beverly Hills, CA: Sage.

Hofstede, G. (2001). *Culture's Consequences: Comparing Values, Behaviors, Institutions and Organizations Across Nations*. Newbury Park, CA: Sage.

Hook, D., & Kelly, D. (2009). Testing for trustworthiness in scientific software. In *Proceedings of the 2nd International Workshop on Software Engineering for Computational Science and Engineering*, Vancouver, BC, Canada.

Hornik, S. R., Johnson, R. D., & Wu, Y. (2007). When technology does not support learning: The negative consequences of dissonance of individual epistemic beliefs in technology mediated learning. *Journal of Organizational and End User Computing*, *19*(2), 23–46.

Hornik, S. R., Saunders, C. S., Li, Y., Moskal, P. D., & Dzuiban, C. D. (2008). The impact of paradigm development and course level on performance in technology-mediated learning environments. *Informing Science*, *11*, 35–58.

Horton, W. K. (2000). *Designing web-based training: How to teach anyone anything anywhere anytime*. Hoboken, NJ: John Wiley & Sons.

Howe, H., & Simkin, M. G. (2006). Factors Affecting the Ability to Detect Spreadsheet Errors. *Decision Sciences Journal of Innovative Education*, *4*(1), 101–122. doi:10.1111/j.1540-4609.2006.00104.x

Hoxmeier, J. A., Nie, W., & Purvis, G. T. (2000). The impact of gender and experience on user confidence in electronic mail. *Journal of Organizational and End User Computing*, *12*(4), 11–20. doi:10.4018/joeuc.2000100102

Huang, M.-H. (2005). Web performance scale. *Information & Management*, *42*, 841–852. doi:10.1016/j.im.2004.06.003

Hunton, J. E. (1996). Involving Information System Users in Defining System Requirements: The Influence of Procedural Justice Perceptions on User Attitudes and Performance. *Decision Sciences*, *27*(4), 647–671. doi:10.1111/j.1540-5915.1996.tb01830.x

Hu, P. J., Chau, P. Y. K., Sheng, O. R. L., & Tam, K. Y. (1999). Examining the Technology Acceptance Model Using Physician Acceptance of Telemedicine Technology. *Journal of Management Information Systems*, *16*(2), 91–112.

Hussain, D., & Hussain, K. M. (1984). *Information Resource Management*. Homewood, IL: Richard D. Irwin.

Hutchins, E. (1991). Organizing Work by Adaptation. *Organization Science*, *2*(1), 14–39. doi:10.1287/orsc.2.1.14

Ibarra, H. (1995). Race, Opportunities, and diversity of social circles in managerial networks. *Academy of Management Journal*, *38*, 673–703. doi:10.2307/256742

Igbaria, M., & Baroudi, J. J. (1993). A Short-Form Measure of Career Orientations: A Psychometric Evaluation. *Journal of Management Information Systems*, *10*(2), 131–154.

Igbaria, M., Iivari, J., & Maragahh, H. (1995). Why do individuals use computer technology? A Finnish case study. *Information & Management*, *29*(5), 227–238. doi:10.1016/0378-7206(95)00031-0

Igbaria, M., & Nachman, S. A. (1990). Correlates of user satisfaction with end user computing: an exploratory study. *Information & Management*, *19*(2), 73–82. doi:10.1016/0378-7206(90)90017-C

Igbaria, M., Zinatelli, N., Cragg, P., & Cavaye, A. L. M. (1997). Personal computing acceptance factors in small firms: A structural equation model. *Management Information Systems Quarterly*, *21*(3), 279–305. doi:10.2307/249498

Invest in Developing Employee Attitude. (1998, August 14). *Business Line*, p. 1.

Ives, B., Valacich, J., Watson, R. T., Zmud, R., Alavi, M., & Baskerville, R. (2002). What every business student needs to know about information systems. *Communications of the AIS*, *9*, 467–477.

Jackson, M. (1994). Problems, methods, and specialisation. *IEEE Software Engineering Journal*, *9*(6), 249–255.

Jackson, M. (2001). *Problem frames*. Reading, MA: Addison-Wesley.

Jackson, M. C. (1987). Present positions and future prospects in management science. *Omega*, *15*, 455. doi:10.1016/0305-0483(87)90003-X

Jackson, M. C. (2000). *Systems approaches to management*. New York: Kluwer Academic/Plenum Publishers.

Jackson, M. C. (2003). *Systems thinking: Creative holism for managers*. Chichester, UK: Wiley.

Jackson, M. C., & Keys, P. (1984). Towards a system of systems methodologies. *The Journal of the Operational Research Society*, *35*, 473–486.

Jackson, M., & Poole, M. S. (2003). Idea generation in naturally-occurring contexts: Complex appropriation of a simple procedure. *Human Communication Research*, *29*, 560–591. doi:10.1093/hcr/29.4.560

Jacobson, S., Shepherd, J., D'Aquila, M., & Carter, K. (2007). *The ERP Market Sizing Report, 2006-2011*. Stamford, CT: AMR Research.

Jarvis, C. B., Mackenzie, S. B., & Podsakoff, P. M. (2003). A critical review of construct indicators and measurement model misspecification in marketing and consumer research. *The Journal of Consumer Research*, *30*(2), 199–218. doi:10.1086/376806

Jawahar, I. M., & Elango, B. (2001). The effect of attitudes, goal setting, and self-efficacy on end-user performance. *Journal of Organizational and End User Computing*, *13*(2), 40–45. doi:10.4018/joeuc.2001040104

Jeffries, R., Anderson, A., & Hendrickson, C. (2001). *Extreme programming installed*. Reading, MA: Addison-Wesley.

Jenkins, R., Deshpande, Y., & Davison, G. (1998). Verification and validation and complex environments: a study in service sector. In *Proceedings of the 1998 Winter Simulation Conference* (Vol. 2, pp. 1433-1440).

Jessup, H. (1991). A model for workteam success. *Journal for Quality and Participation*, *14*(3), 70–74.

Jessup, H. (1992). The road to results for teams. *Training & Development*, *46*(9), 65–68.

Jex, S. M. (2002). *Organizational Psychology*. New York, NY: John Wiley & Sons.

JISC. (1998a). *Guidelines for developing an information strategy*. London: Author.

JISC. (1998b). *Case studies on developing information strategies*. London: Author.

JISC. (2004). Renewal and growth: Using technology to support learning, teaching, management and research. *Inform (Silver Spring, Md.)*, *5*.

Johnson, L. (2008). *NMC Virtual Worlds Announces Plans for 2008*. Retrieved June 8, 2009, from http://virtualworlds.nmc.org/wp-content/uploads/2008/01/press-release-nmc-virtual-worlds-2008-plans.pdf

Johnson, G., Scholes, K., & Whittington, R. (2007). *Exploring corporate strategy*. London: Prentice Hall.

Johnson, R. D., Hornik, S. R., & Salas, E. (2008). An empirical examination of factors contributing to the creation of successful e-learning environments. *International Journal of Human-Computer Studies, 66*, 356–369. doi:10.1016/j.ijhcs.2007.11.003

Johnson, R. D., & Marakas, G. M. (2000). Research report: the role of behavioral modeling and computer skills acquisition-toward refinement of the model. *Information Systems Research, 11*(4), 402–417. doi:10.1287/isre.11.4.402.11869

Johnson, R. D., & Marakas, G. M. (2000). The Role of Behavioral Modeling in Computer Skills Acquisition: Toward Refinement of the Model. *Information Systems Research, 11*(4), 402–417. doi:10.1287/isre.11.4.402.11869

Johnson, R. D., Marakas, G. M., & Palmer, J. W. (2006). Differential social attributions toward computing technology: An empirical examination. *International Journal of Human-Computer Studies, 64*(5), 446–460. doi:10.1016/j.ijhcs.2005.09.002

Jones, M. C., Cline, M., & Ryan, S. (2006). Exploring knowledge sharing in ERP implementation: an organizational culture framework. *Decision Support Systems, 41*(2), 411–434. doi:10.1016/j.dss.2004.06.017

Jones, M. C., & Harrison, A. W. (1996). Is project team performance: an empirical assessment. *Information & Management, 31*(2), 57–65. doi:10.1016/S0378-7206(96)01068-3

Jones, M. C., & Young, R. (2006). ERP usage in practice: An empirical investigation. *Information Resources Management Journal, 19*(1), 23–42.

Joreskog, K. G., & Wold, H. (1982). *Systems under Indirect Observation*. Amsterdam, The Netherlands: North Holland.

Joseph, E., Snell, A., & Willard, C. G. (2004). *Council on Competitiveness study of U.S. industrial HPC users*. Washington, DC: Council on Competitiveness.

Joshi, K. (1989). The Measurement of Fairness or Equity Perceptions of Management Information Systems Users. *Management Information Systems Quarterly, 13*(3), 343–358. doi:10.2307/249010

Joshi, K. (1990). An Investigation of Equity as a Determinant of User Information Satisfaction. *Decision Sciences, 21*(4), 786–807. doi:10.1111/j.1540-5915.1990.tb01250.x

Joshi, K. (1991). A Model of Users' Perspective on Change: The Case of Information Systems Technology Implementation. *Management Information Systems Quarterly, 15*(2), 229–242. doi:10.2307/249384

Joshi, K. (1992). A Causal Path Model of the Overall User Attitudes Toward the MIS Function: The Case of User Information Satisfaction. *Information & Management, 22*, 77–88. doi:10.1016/0378-7206(92)90063-L

Kanellis, P., & Paul, R. J. (2005). User Behaving Badly: Phenomena and Paradoxes from an Investigation into Information Systems Misfit. *Journal of Organizational and End User Computing, 17*(2), 64–91. doi:10.4018/joeuc.2005040104

Kanfer, R. (1990). Motivation theory in I/O psychology. In Dunnette, M. D., & Hough, L. M. (Eds.), *Handbook of industrial and organizational psychology* (pp. 75–170). Palo Alto, CA: Consulting Psychological Press.

Kanfer, R., & Ackerman, P. L. (1989). Motivation and cognitive abilities: An integrative/aptitude-treatment interaction approach to skill acquisition. *The Journal of Applied Psychology, 74*(4), 657–690. doi:10.1037/0021-9010.74.4.657

Kankanhalli, A., Tan, B. C. Y., & Wei, K. K. (2005). Contributing Knowledge to Electronic Knowledge Repositories: An Empirical Investigation. *Management Information Systems Quarterly, 29*(1), 113–143.

Kappel, T. A., & Rubenstein, A. H. (1999). Creativity in Design: The Contribution of Information Technology. *IEEE Transactions on Engineering Management, 46*(2), 132–143. doi:10.1109/17.759140

Karahanna, E., Agarwal, R., & Angst, C. M. (2006). Reconceptualizing compatibility beliefs in technology acceptance research. *Management Information Systems Quarterly, 30*(4), 781–804.

Karahanna, E., Straub, D. W., & Chervany, N. L. (1999). Information technology adoption across time: a cross-sectional comparison of pre-adoption and post-adoption beliefs. *Management Information Systems Quarterly*, *23*(2), 183–213. doi:10.2307/249751

Keahey, K., Tsugawa, M., Matsunaga, A., & Fortes, J. (2009). Sky computing. *IEEE Internet Computing*, *13*(5). doi:10.1109/MIC.2009.94

Kelly, D. (2006). A study of design characteristics in evolving software using stability as a criterion. *IEEE Transactions on Software Engineering*, *32*(5), 315–329. doi:10.1109/TSE.2006.42

Kelly, D. (2008). Innovative standards for innovative software. *IEEE Computer*, *41*(7), 88–89.

Kelly, D. (2009). Determining factors that affect long-term evolution in scientific application software. *Journal of Systems and Software*, *82*, 851–861. doi:10.1016/j.jss.2008.11.846

Kelly, D. F. (2007). A software chasm: Software engineering and scientific computing. *IEEE Software*, *24*(6), 120–119. doi:10.1109/MS.2007.155

Kelly, D., & Shepard, T. (2004). Task-directed inspection. *Journal of Systems and Software*, *73*(2), 361–368. doi:10.1016/j.jss.2003.10.027

Kelly, D., Thorsteinson, S., & Hook, D. (2010). Scientific software testing, analysis with four dimensions. *IEEE Software*, *28*(3), 84–90. doi:10.1109/MS.2010.88

Kelly, R., & Caplan, J. (1993). How Bell Labs creates star performers. *Harvard Business Review*, *71*(4), 128–139.

Kendall, R. P., Carver, J., Mark, A., Post, D., Squires, S., & Shaffer, D. (2005). *Case study of the hawk code project* (Tech. Rep. No. LA-UR-05-9011). Los Alamos, CA: Los Alamos National Laboratories.

Kendall, R. P., Mark, A., Post, D., Squires, S., & Halverson, C. (2005). *Case study of the condor code project* (Tech. Rep. No. LA-UR-05-9291). Los Alamos, CA: Los Alamos National Laboratories.

Kendall, R. P., Post, D., & Mark, A. (2007). *Case study of the NENE code project* (Tech. Rep. No. CMUI/SEI-2006-TN-044). Pittsburgh, PA: Carnegie Mellon University.

Kendall, R. P., Post, D., Squires, S., & Carver, J. (2006). *Case study of the eagle code project* (Tech. Rep. No. LA-UR-06-1092). Los Alamos, CA: Los Alamos National Laboratories.

Kendall, R. P., Carver, J. C., Fisher, D., Henderson, D., Mark, A., & Post, D. (2008). Development of a weather forecasting code: A case study. *IEEE Software*, *25*(4), 59–65. doi:10.1109/MS.2008.86

Kernan, M., & Howard, G. S. (1990). Computer anxiety and computer attitudes: An investigation of construct and predictive validity issues. *Educational and Psychological Measurement*, *50*, 681–690. doi:10.1177/0013164490503026

Kernfeld, B. (1995). *What to Listen for in Jazz*. New Haven, CT: Yale University Press.

Kesling, G. (1989). A community project approach to teaching management information systems. *Journal of Education for Business*, 341–344. doi:10.1080/08832323.1989.10117386

Kettinger, W. J., & Lee, C. C. (1994). Perceived service quality and user satisfaction with the information services functions. *Decision Sciences*, *25*(5-6), 737–766. doi:10.1111/j.1540-5915.1994.tb01868.x

Kickul, J. R., Neuman, G., Parker, C., & Finkl, J. (2001). Settling the Score: The Role of Organizational Justice in the Relationship between Psychological Contract Breach and Anticitizenship Behavior. *Employee Responsibilities and Rights Journal*, *13*(2), 77–93. doi:10.1023/A:1014586225406

Kielsmeier, J. C. (2000). A time to serve, a time to learn: Service learning and the promise of democracy. *Phi Delta Kappan*, *81*, 652–657.

Kim, D. H. (1993). The Link between Individual and Organizational Learning. *Sloan Management Review*, *35*(1), 37–50.

Kim, G., Shin, B., & Lee, H. G. (2006). A study of factors that affect user intentions toward email service switching. *Information & Management*, *43*(7), 884–893. doi:10.1016/j.im.2006.08.004

Kim, K. K., & Umanath, N. S. (2005). Information transfer in B2B procurement: an empirical analysis and measurement. *Information & Management, 42*, 813–828. doi:10.1016/j.im.2004.08.004

King, W. R. (1988). How effective is your information systems planning? *Long Range Planning, 21*(5), 103–112. doi:10.1016/0024-6301(88)90111-2

King, W. R., & He, J. (2006). A meta-analysis of the technology acceptance model. *Information & Management, 43*(6), 740–755. doi:10.1016/j.im.2006.05.003

King, W. R., & Lekse, W. J. (2006). Deriving managerial benefit from knowledge search: A paradigm shift? *Information & Management, 43*(7), 874–883. doi:10.1016/j.im.2006.08.005

Kirkman, B. L., Shapiro, D. L., Novelli, L. Jr, & Brett, J. M. (1996). Employee Concerns Regarding Self-Managing Work Teams: A Multidimensional Perspective. *Social Justice Research, 9*(1), 47–67. doi:10.1007/BF02197656

Kirkpatrick, D. L. (1976). Evaluation of training. In Craig, R. L. (Ed.), *Training and Development Handbook: A Guide to Human Resource Development*. New York: McGraw-Hill.

Klaus, T., & Blanton, J. E. (2010). User Resistance Determinants and the Psychological Contract in Enterprise System Implementations. *European Journal of Information Systems, 19*, 625–636.

Klaus, T., Wingreen, S. C., & Blanton, J. E. (2010). Resistant Groups in Enterprise System Implementations: A Q-methodology Examination. *Journal of Information Technology, 25*, 91–106.

Knight, J. Y. (2007). *F5 VPN command-line client*. Retrieved January 12, 2010, from http://fuhm.net/software/f5vpn-login

Knights, D., & Vurdubakis, T. (1994). Foucault, Power, Resistance, and All That. In Jermier, J. M., Knights, D., & Nord, W. R. (Eds.), *Resistance and Power in Organizations* (pp. 167–198). London, UK: Routledge.

Ko, D.-G., Kirsch, L. J., & King, W. R. (2005). Antecedents of Knowledge Transfer from Consultants to Clients in Enterprise System Implementations. *Management Information Systems Quarterly, 29*(1), 59–85.

Ko, D., Kirsch, L. J., & King, W. R. (2005). Antecedents of knowledge transfer from consultants to clients in Enterprise Systems implementations. *Management Information Systems Quarterly, 29*(1), 59–85.

Koeszegi, S., Vetschera, R., & Kersten, G. (2004). National cultural differences in the use and perception of Internet-based NSS: Does high or low context matter? *International Negotiation, 9*(1), 79. doi:10.1163/1571806041262070

Koh, C., Ang, S., & Straub, D. W. (2004). IT Outsourcing Success: A Psychological Contract Perspective. *Information Systems Research, 15*(4), 356–373. doi:10.1287/isre.1040.0035

Kolb, D. A. (1984). *Experiential Learning: Experience as the Source of Learning and Deployment*. Englewood Cliffs, NJ: Prentice-Hall.

Komischke, T., & Burmester, M. (2000). User-centered standardization of industrial process control user interface. *International Journal of Human-Computer Interaction, 12*(3-4), 375–386. doi:10.1207/S15327590IJHC1203&4_8

Kontogiannis, T., & Shepherd, A. (1999). Training conditions and strategic aspects of skill transfer in a simulated process control task. *Human-Computer Interaction, 14*(4), 355–393. doi:10.1207/S15327051HCI1404_1

Krackhardt, D. (1992). The strength of strong ties: the importance of philos in Organizations. In Nohria, N., & Eccles, R. (Eds.), *Networks and Organizations: Structures, Form and Action* (pp. 216–239). Boston: Harvard Business School Press.

Kraemmergaard, P., & Rose, J. (2002). Managerial Competences for ERP Journeys. *Information Systems Frontiers, 4*(2), 199–211. doi:10.1023/A:1016054904008

Kramarae, C. (2001). *The third shift: Women learning online*. Washington, DC: Association of American University Women.

Krasner, H. (2000). Ensuring E-Business Success by Learning from ERP Failures. *IT Professional, 2*(1), 22–27. doi:10.1109/6294.819935

Kruck, S. E., Maher, J. J., & Barkhi, R. (2003). Framework for cognitive skill acquisition and spreadsheet training. *Journal of End User Computing, 15*(1), 20–37. doi:10.4018/joeuc.2003010102

Kuhl, P. K., Tsao, F.-M., & Liu, H.-M. (2003). Foreign-language experience in infancy: Effects of short-term exposure and social interaction on phonetic learning. *Proceedings of the National Academy of Sciences of the United States of America*, *100*(15), 9096–9101. doi:10.1073/pnas.1532872100

Kujala, S. (2003). User involvement: A review of the benefits and challenges. *Behaviour & Information Technology*, *22*(1), 1–16. doi:10.1080/01449290301782

Kulkarni, U. R., Ravindran, S., & Freeze, R. (2006). A Knowledge Management Success Model: Theoretical Development and Empirical Validation. *Journal of Management Information Systems*, *23*(3), 309–347. doi:10.2753/MIS0742-1222230311

Kumar, B. P., Selvam, J., Meenakshi, V. S., Kanthi, K., Suseels, A. L., & Kumar, L. K. (2007). Business Decision Making Management and Information Technology. *Ubiquity*, *8*(8), 1. doi:10.1145/1226690.1232401

Lamb, C. H., Swinth, R. L., Vinton, K. L., & Lee, J. B. (1998). Integrating service learning into a business school curriculum. *Journal of Management Education*, *22*(5), 637–655. doi:10.1177/105256299802200506

Larsen, T. J., & Sørebø, Ø. (2005). Impact of Personal Innovativeness on the Use of the Internet among Employees at Work. *Journal of Organizational and End User Computing*, *17*(2), 43–63. doi:10.4018/joeuc.2005040103

Latham, G. P., & Saari, L. M. (1979). Application of social-learning theory to training supervisors through behavioral modeling. *The Journal of Applied Psychology*, *64*(3), 239–246. doi:10.1037/0021-9010.64.3.239

Laughlin, R. C. (1987). Accounting systems in organizational contexts: A case for critical theory. *Accounting, Organizations and Society*, *12*(5), 479–502. doi:10.1016/0361-3682(87)90032-8

Lazar, J., & Preece, J. (1999). Implementing service learning in an online communities course. In *Proceedings of the International Academy for Information Management 1999 Conference* (pp. 22-27).

Lazar, J., & Norcio, A. (2000). Service-research: Community partnerships for research and training. *Journal of Informatics Education and Research*, *2*(3), 21–25.

Lee, D. M. S. (1994). Social ties, task-related communication and first job performance of young engineers. *Journal of Engineering and Technology Management*, *11*(3-4), 203–229. doi:10.1016/0923-4748(94)90010-8

Leedy, P. D., & Ormrod, J. E. (2005). *Practical research: Planning and Design* (8th ed.). Upper Saddle River, NJ: Prentice Hall.

Lee, H., & Liebenau, J. (1999). Time in organizational studies: Towards a new research direction. *Organization Studies*, *20*(6), 1035–1058. doi:10.1177/0170840699206006

Lee, H.-Y., Ahn, H., & Han, I. (2007). VCR: Virtual community recommender using the technology acceptance model and the user's needs type. *Expert Systems with Applications*, *33*(4), 984–995. doi:10.1016/j.eswa.2006.07.012

Lee, K. C., Lee, S., & Kang, I. W. (2005). KMPI: measuring knowledge management performance. *Information & Management*, *42*, 469–482. doi:10.1016/j.im.2005.10.003

Lee, P. C. B. (2004). Social support and leaving intention among computer professionals. *Information & Management*, *41*(3), 323–334. doi:10.1016/S0378-7206(03)00077-6

Lee, Y. W., Strong, D. M., Kahn, B. K., & Wang, R. Y. (2002). AIMQ: a methodology for information quality assessment. *Information & Management*, *40*, 133–146. doi:10.1016/S0378-7206(02)00043-5

Lee, Y., Kozar, K. A., & Larsen, K. R. T. (2003). The technology acceptance model: Past, present, and future. *Communications of the Association for Information Systems*, *12*(50), 752–780.

Legris, P., Ingham, J., & Collerette, P. (2003). Why do people use information technology? A critical review of the technology acceptance model. *Information & Management*, *40*(3), 191–204. doi:10.1016/S0378-7206(01)00143-4

Leonard, A. C. (2000). The importance of having a multidimensional view of IT-end user relationships for the successful restructuring of IT departments. In Khosrow-Pour, M. (Ed.), *Challenges of Information Technology Managers in the 21st Century* (pp. 492–495). Hershey, PA: Idea Group.

Letondal, C. (2005). Participatory programming: Developing programmable bioinformatics tools for end users. In Lieberman, H., Paterno, F., & Wulf, V. (Eds.), *End-user development* (pp. 207–242). Dordrecht, The Netherlands: Springer-Verlag.

Lewis, B. R., Snyder, C. A., & Rainer, R. K. (1995). An Empirical Assessment of the Information Resource Management Construct. *Journal of Management Information Systems*, *12*(1), 199–223.

Lewis, W., Agarwal, R., & Sambamurthy, V. (2003). Sources of influence on beliefs about information technology use: An empirical study of knowledge workers. *Management Information Systems Quarterly*, *27*(4), 657–678.

Liaw, S.-S., Chang, W.-C., Hung, W.-H., & Huang, H.-M. (2006). Attitudes toward search engines as a learning assisted tool: Approach of Liaw and Huang's research model. *Computers in Human Behavior*, *22*(2), 177–190. doi:10.1016/j.chb.2004.09.003

Liaw, S.-S., & Huang, H.-M. (2003). An investigation of user attitudes toward search engines as an information retrieval tool. *Computers in Human Behavior*, *19*(6), 751–765. doi:10.1016/S0747-5632(03)00009-8

Li, D., Browne, G. J., & Chau, P. Y. K. (2006). An empirical investigation of web site use using a commitment-based model. *Decision Sciences*, *37*(3), 427–444. doi:10.1111/j.1540-5414.2006.00133.x

Li, D., Browne, G. J., & Wetherbe, J. C. (2007). Online consumers' switching behavior: A buyer-seller relationship perspective. *Journal of Electronic Commerce in Organizations*, *5*(1), 30–42. doi:10.4018/jeco.2007010102

Lim, C. K. (2001). Computer self-efficacy, academic self-concept, and other predictors of satisfaction and future participation of adult distance learners. *American Journal of Distance Education*, *15*(2), 41–51. doi:10.1080/08923640109527083

Lin, C. Y., Kuo, T. H., Kuo, Y. K., Ho, L. A., & Kuo, Y. L. (2007). The KM chain - Empirical study of the vital knowledge sourcing links. *Journal of Computer Information Systems*, *48*(2), 91–99.

Lin, J. C.-C., & Lu, H. (2000). Towards an understanding of the behavioural intention to use a web site. *International Journal of Information Management*, *20*(3), 197–208. doi:10.1016/S0268-4012(00)00005-0

Lin, N. (2000). Inequality in social Capital. *Contemporary Sociology*, *29*(6), 785–795. doi:10.2307/2654086

Linne, J., & Plers, L. (2002). The DNA of an e-tutor. *ITTraining*, 26-28.

Lin, T.-C., & Huang, C.-C. (2008). Understanding knowledge management system usage antecedents: An integration of social cognitive theory and task technology fit. *Information & Management*, *45*(6), 410–417. doi:10.1016/j.im.2008.06.004

Lipscomb, T., Hotard, D., Shelley, K., & Baldwin, Y. (2002). Business students' attitudes toward statistics: A preliminary investigation. *Proceedings of the Academy of Educational Leadership*, *7*(1), 47–50.

Lohmoller, J.-B. (1988). The PLS Program System: Latent Variables Path Analysis with Partial Least Squares Estimation. *Multivariate Behavioral Research*, *23*(1), 125–127. doi:10.1207/s15327906mbr2301_7

Loyd, B. H., & Gressard, C. (1984). Reliability and factoral validity of computer attitude scales. *Educational and Psychological Measurement*, *44*(2), 501–555. doi:10.1177/0013164484442033

Loyd, B. H., & Gressard, C. (1984). The effects of sex, age, and computer experience on computer attitudes. *AEDS Journal*, 67–77.

Lu, J., Liu, C., Yu, C.-S., & Yao, J. E. (2003). Exploring factors associated with wireless Internet via mobile technology acceptance in mainland China. *Communications of the International Information Management Association*, *3*(1), 101–120.

Luong, A., & Rogelberg, S. G. (2005). Meetings and more meetings: The relationship between meeting load and the daily well-being of employees. *Group Dynamics*, *9*(1), 58–67. doi:10.1037/1089-2699.9.1.58

Lyytinen, K. (1992). Information systems and critical theory. In Alvesson, M., & Willmott, H. (Eds.), *Critical Management Studies* (pp. 159–180). London: Sage.

Lyytinen, K., & Hirschheim, R. (1989). Information systems and emancipation: Promise or threat? In Klein, H. K., & Kumar, K. (Eds.), *Systems Development for Human Progress* (pp. 115–139). Amsterdam, The Netherlands: North Holland.

Mabert, V. A., Soni, A., & Venkatramanan, M. A. (2000). Enterprise resource planning survey of US manufacturing firms. *Production and Inventory Management, 41*(2), 52–58.

Macaulay, C., Sloan, D., Jian, X., Forbes, P., Loynton, S., Swedlow, J. R., & Gregor, P. (2009). Usability and user-centered design in scientific development. *IEEE Software, 26*(1), 96–102. doi:10.1109/MS.2009.27

Mackie, D. M., Worth, L. T., & Asuncion, A. G. (1990). Processing of persuasive in-group messages. *Journal of Personality and Social Psychology, 58*, 812–822. doi:10.1037/0022-3514.58.5.812

MacMullen, W. J., & Denn, S. O. (2005). Information problems in molecular biology and bioinformatics. *Journal of the American Society for Information Science and Technology, 56*(5), 447–456. doi:10.1002/asi.20134

Majchrzak, A., & Malhotra, A. (2004). *Virtual Workspace Technology Use and Knowledge-Sharing Effectiveness in Distributed Teams: The Influence of a Team's Transactive Memory*. Los Angeles, CA: Marshall School of Business, University of Southern California.

Majchrzak, A., Rice, R. E., Malhotra, A., King, N., & Ba, S. (2000). Technology Adaptation: The Case of a Computer-Supported Inter-Organizational Virtual Team. *Management Information Systems Quarterly, 24*(4), 569–600. doi:10.2307/3250948

Mak, B. L., & Sockel, H. (2001). A confirmatory factor analysis of IS employee motivation and retention. *Information & Management, 38*, 265–276. doi:10.1016/S0378-7206(00)00055-0

Makin, P. J., Cooper, C. L., & Cox, C. J. (1996). *Organizations and the Psychological Contract: Managing People at Work*. London, UK: Wiley-Blackwell.

Malone, T. W. (1997). Is Empowerment Just a Fad? Control, Decision Making, and IT. *Sloan Management Review, 38*(2), 23–35.

Mankins, M. (2004). Stop wasting valuable time. *Harvard Business Review, 82*(9), 58.

Mao, E., & Palvia, P. (2006). Testing an extended model of IT acceptance in the Chinese cultural context. *The Data Base for Advances in Information Systems, 37*(2-3), 20–32.

Mao, E., Srite, M., Thatcher, J. B., & Yaprak, O. (2005). A research model for mobile phone service behaviors: Empirical validation in the U.S. and Turkey. *Journal of Global Information Technology Management, 8*(4), 7.

Mao, J., & Brown, B. R. (2005). The effectiveness of online task support vs. instructor-led training. *Journal of Organizational and End User Computing, 17*(3), 27–46. doi:10.4018/joeuc.2005070102

Marakas, G. M., Johnson, R. D., & Clay, P. F. (2007). The evolving nature of the computer self-efficacy construct: An empirical investigation of measurement construction, validity, reliability, and stability over time. *Journal of the Association for Information Systems, 8*(1), 16–46.

Marakas, G. M., Yi, M. Y., & Johnson, R. D. (1998). The multilevel and multifaceted character of computer self-efficacy: Toward clarification of the construct and an integrative framework for research. *Information Systems Research, 9*(2), 126–163. doi:10.1287/isre.9.2.126

Marakas, G., & Hornik, S. (1996). Passive Resistance Misuse: Overt Support and Covert Recalcitrance in IS Implementation. *European Journal of Information Systems, 5*, 208–219. doi:10.1057/ejis.1996.26

Marcolin, B. L., Compeau, D. R., Munro, M. C., & Huff, S. L. (2000). Assessing user competence: Conceptualization and measurement. *Information Systems Research, 11*(1), 37–60. doi:10.1287/isre.11.1.37.11782

Margevicius, M. (2001). *End user computing best practices*. Retrieved January 5, 2010, from http://www.courtesycomputers.com/Networking/enduser%20computing%20best%20practices.pdf

Markus, M. L. (2001). Toward a theory of knowledge reuse: Types of knowledge reuse situations and factors in reuse success. *Journal of Management Information Systems, 18*(1), 57–93.

Markus, M. L., Axline, S., Petrie, D., & Tanis, C. (2003). Learning from Adopters' Experiences with ERP: Problems Encountered and Success Achieved. In Shanks, G., Seddon, P. B., & Willcocks, L. P. (Eds.), *Second-Wave Enterprise Resource Planning Systems*. Cambridge, UK: Cambridge University Press.

Markus, M. L., Majchrzak, A., & Gasser, L. (2002). A Design Theory for Systems that Support Emergent Knowledge Processes. *Management Information Systems Quarterly*, *26*(3), 179–212.

Markus, M. L., & Tanis, C. (2000). The Enterprise System Experience - From Adoption to Success. In Zmud, R. W. (Ed.), *Framing the Domains of IT Management* (pp. 173–208). Cincinnati, OH: Pinnaflex Educational Resources.

Marsden, P. V. (1988). Homogeneity in confiding networks. *Social Networks*, *10*, 57–76. doi:10.1016/0378-8733(88)90010-X

Marsden, P. V., & Campbell, K. E. (1984). Measuring tie strength. *Social Forces*, *63*(2), 482–501. doi:10.2307/2579058

Martocchio, J. J., & Webster, J. (1992). Effects of feedback and cognitive playfulness on performance in microcomputer software training. *Personnel Psychology*, *45*, 553–578. doi:10.1111/j.1744-6570.1992.tb00860.x

Mason, J. (1996). *Qualitative Researching*. London, UK: Sage.

Massar, J., Travers, M., Elhai, J., & Shrager, J. (2005). BioLingua: A programmable knowledge environment for biologists. *Bioinformatics (Oxford, England)*, *21*(2), 199–207. doi:10.1093/bioinformatics/bth465

Massa, S., & Testa, S. (2005). Data warehouse-in-practice: Exploring the function of expectations in organizational outcomes. *Information & Management*, *42*(5), 709–718. doi:10.1016/j.im.2004.06.002

Mathe, J., Duncavage, S., Werner, J., Malin, B., Ledeczi, A., & Sztipanovits, J. (2007). Implementing a model-based design environment for clinical information systems. In *Proceedings of the ACM/IEEE International Workshop on Model-Based Trustworthy Health Information Systems* (pp. 399-408).

Mathieson, K., Peacock, E., & Chin, W. W. (2001). Extending the Technology Acceptance Model: The Influence of Perceived User Resources. *The Data Base for Advances in Information Systems*, *32*(3), 86–112.

Matthews, D., Wilson, G., & Easterbrook, S. (2008). Configuration management for large-scale scientific computing at the UK Met office. *IEEE Computational Science & Engineering*, *10*(6), 56–64.

Maurer, M. (1983). *Development and validation of a measure of computer anxiety*. Unpublished master's thesis, Iowa State University.

Maurer, R. (2002). *Plan for the Human Part of ERP*. Workforce Online.

Mayers, M. D. (1999). Investigating information systems with ethnographic research. *Communications of the Association for Information Systems*, *2*, 23.

Maznevski, M. L., & Chudoba, K. M. (2000). Bridging space over time: Global virtual team dynamics and effectiveness. *Organization Science*, *11*(5), 473–492. doi:10.1287/orsc.11.5.473.15200

McCoy, S., Everard, A., & Jones, B. M. (2005). An examination of the technology acceptance model in Uruguay and the US: A focus on culture. *Journal of Global Information Technology Management*, 27–45.

McCoy, S., Galletta, D. F., & King, W. R. (2005). Integrating national culture into IS research: The need for current individual level measures. *Communications of the Association for Information Systems*, *15*, 1.

McCoy, S., Galletta, D. F., & King, W. R. (2007). Applying TAM across cultures: the need for caution. *European Journal of Information Systems*, *16*(1), 81. doi:10.1057/palgrave.ejis.3000659

McDonald, D. J., & Makin, P. J. (2000). The Psychological Contract, Organisational Commitment and Job Satisfaction of Temporary Staff. *Leadership and Organization Development Journal*, *21*, 84–91. doi:10.1108/01437730010318174

McDonough, E. F. III. (2000). Investigation of factors contributing to the success of cross-functional teams. *Journal of Product Innovation Management*, *17*(3), 221–235. doi:10.1016/S0737-6782(00)00041-2

McGrath, J. E. (1991). Time, interaction, and performance (TIP): A theory of groups. *Small Group Research, 22*(2), 147–174. doi:10.1177/1046496491222001

McGrath, J. E., Arrow, H., Gruenfeld, D. H., Hollingshead, A. B., & O'Connor, K. M. (1993). Groups, tasks, and technology: The effects of experience and change. *Small Group Research, 24*, 406–420. doi:10.1177/1046496493243007

McHaney, R., & Cronan, T. P. (1998). Computer Simulation Success: On the Use of the End-User Computing Satisfaction Instrument. *Decision Sciences, 29*(2), 525–536. doi:10.1111/j.1540-5915.1998.tb01589.x

McKinney, V., Yoon, K., & Zahedi, F. (2002). The Measurement of Web-Customer Satisfaction: An Expectation and Disconfirmation Approach. *Information Systems Research, 13*(3), 296–315. doi:10.1287/isre.13.3.296.76

McKnight, D. H., Choudhury, V., & Kacmar, C. (2002). Developing and Validating Trust Measures for e-Commerce: An Integrative Typology. *Information Systems Research, 13*(3), 334–359. doi:10.1287/isre.13.3.334.81

McLeod, R. Jr, & Schell, G. P. (2000). *Management information systems* (8th ed.). Upper Saddle River, NJ: Pearson-Prentice Hall.

McMurtrey, M., McGaughey, R., & Downey, J. (2008). Seniors and information technology: Are we shrinking the digital divide? *Journal of International Technology and Information Management, 17*(2), 121–136.

McQuitty, S., Finn, A., & Wiley, J. B. (2000). Systematically varying consumer satisfaction and its implications for product choice. *Academy of Marketing Science Review, 10*, 1–16.

Meckler, M., Drake, B. H., & Levinson, H. (2003). Putting Psychology Back into Psychological Contracts. *Journal of Management Inquiry, 12*(3), 217–228. doi:10.1177/1056492603256338

Menon, T., & Pfeffer, J. (2003). Valuing internal vs. External knowledge: Explaining the preference for outsiders. *Management Science, 49*(4), 497–513. doi:10.1287/mnsc.49.4.497.14422

Menon, T., Thompson, L., & Choi, H.-S. (2006). Tainted knowledge vs. Tempting knowledge: People avoid knowledge from internal rivals and seek knowledge from external rivals. *Management Science, 52*(8), 1129–1144. doi:10.1287/mnsc.1060.0525

Microsoft. (2006). *Certifications*. Retrieved April 12, 2006, from http://office.microsoft.com/ marketplace/ default.aspx?

Miles, M. B., & Huberman, A. M. (1994). *Qualitative Data Analysis: An Expanded Sourcebook* (2nd ed.). Thousand Oaks, CA: Sage.

Miller, G. (1956). The magic number seven, plus or minus two: Some limits on our capacity for processing information. *Psychological Review, 63*(2), 81–97. doi:10.1037/h0043158

Miller, G. (2006). Scientific publishing: A scientist's nightmare: Software problem leads to five retractions. *Science, 314*(12), 1856–1857. doi:10.1126/science.314.5807.1856

Miller, R. H., & Sim, I. (2004). Physicians' use of electronic medical records: barriers and solutions. *Health Affairs, 23*(2), 116–126. doi:10.1377/hlthaff.23.2.116

Miner, A. S., Bassoff, P., & Moorman, C. (2001). Organizational Improvisation and Learning: A Field Study. *Administrative Science Quarterly, 46*(2), 304–337. doi:10.2307/2667089

Mingers, J., & Gill, A. (1997). Commentary. In Mingers, J., & Gill, A. (Eds.), *Multimethodology: The theory and practice of combining management science methodologies*. Chichester, UK: Wiley.

Mintzberg, H., & Westley, F. (1992). Cycles of organizational change. *Strategic Management Journal*.

Mintzberg, H., Quinn, J. B., & Ghoshal, S. (1998). The strategy process (Revised European edition). Hemel Hempstead, UK: Prentice Hall.

Mintzberg, H. (1973). *The Nature of Managerial Work*. New York, NY: Harper & Row.

Mintzberg, H. (1987). Crafting strategy. *Harvard Business Review, 65*(4), 66–75.

Mintzberg, H. (1990). The design school: Reconsidering the basic premises of strategic management. *Strategic Management Journal*, *11*(3), 171–195. doi:10.1002/smj.4250110302

Mintzberg, H. (1994). Rethinking strategic planning. Part I: Pitfalls and fallacies. *Long Range Planning*, *27*(3), 12–21. doi:10.1016/0024-6301(94)90185-6

Mitchell, V. L. (2006). Knowledge integration and information technology project performance. *Management Information Systems Quarterly*, *30*(4), 919–939.

Molla, A., & Licker, P. S. (2005). eCommerce adoption in developing countries: a model and instrument. *Information & Management*, *42*, 877–899. doi:10.1016/j.im.2004.09.002

Moore, G. C., & Benbasat, I. (1991). Development of an Instrument to Measure the Perceptions of Adopting an Information Technology Innovation. *Information Systems Research*, *2*(3), 192–222. doi:10.1287/isre.2.3.192

Moore, J. E., & Burke, L. S. (2002). How to turn around 'turnover culture' in IT. *Communications of the ACM*, *45*(2), 73–78. doi:10.1145/503124.503126

Moore, M. G. (2002). Editorial: What does research say about the learners using computer-mediated communication in distance learning? *American Journal of Distance Education*, *16*(2), 61–64. doi:10.1207/S15389286AJDE1602_1

Moorman, C., & Miner, A. S. (1998a). Organizational Improvisation and Organizational Memory. *Academy of Management Review*, *23*(4), 698–723.

Moorman, C., & Miner, A. S. (1998b). The Convergence of Planning and Execution: Improvisation in New Product Development. *Journal of Marketing*, *62*(3), 1–20. doi:10.2307/1251740

Moorman, R. H. (1991). Relationship between Organizational Justice and Organizational Citizenship Behaviors: Do Fairness Perceptions Influence Employee Citizenship? *The Journal of Applied Psychology*, *76*(6), 845–855. doi:10.1037/0021-9010.76.6.845

Moran, P. (2005). Structural vs. relational embeddedness: social capital and managerial performance. *Strategic Management Journal*, *26*, 1129–1151. doi:10.1002/smj.486

Morgan, G. (1986). *Images of Organisation*. Beverley Hills, CA: Sage.

Morris, C., & Segal, J. (2009). Some challenges facing scientific software developers: The case of molecular biology. In *Proceedings of the 5th International IEEE Conference on E-Science* (pp. 216-222).

Morrison, E. W. (1993). Newcomer information seeking: Exploring types, modes, sources, and outcomes. *Academy of Management Journal*, *36*(3), 557–589. doi:10.2307/256592

Morrison, E. W. (2002). Newcomers' relationships: the role of social network ties during socialization. *Academy of Management Journal*, *45*(6), 1149–1160. doi:10.2307/3069430

Morrison, E. W., & Robinson, S. L. (1997). When Employees Feel Betrayed: A Model of How Psychological Contract Violation Develops. *Academy of Management Review*, *22*(1), 226–256.

Morrison, E. W., & Vancouver, J. B. (2000). Within-person analysis of information seeking: The effects of perceived costs and benefits. *Journal of Management*, *26*(1), 119–137. doi:10.1016/S0149-2063(99)00040-9

Mortensen, M., Woolley, A. W., & O'Leary, M. B. (2007). Conditions Enabling Effective Multiple Team Membership. In K. Crowston, S. Sieber, & E. Wynn (Eds.), *Proceedings of the IFIP Working Group 8.2 Working Conference on Virtuality and Virtualization*, Portland OR (Vol. 236, pp. 215–228). Berlin, Germany: Springer.

Mullins, L. S., & Kopelman, R. E. (1988). Toward an Assessment of the Construct Validity of Four Measures of Narcissism. *Journal of Personality Assessment*, *52*(4), 610–625. doi:10.1207/s15327752jpa5204_2

Mumford, E. (1995). *Effective requirements analysis and systems design: The ETHICS Method*. Basingstoke, UK: Macmillan.

Munro, M. C., Huff, S. L., Marcolin, B. L., & Compeau, D. R. (1997). Understanding and measuring user competence. *Information & Management*, *33*, 45–57. doi:10.1016/S0378-7206(97)00035-9

Murphy, S., Gainer, V., & Chueh, H. (2003). A visual interface designed for novice users to find research patient cohorts in a large biomedical database. In *Proceedings of the AMIA Annual Symposium* (pp. 489-493).

Murphy, C. A., Coover, D., & Owen, S. V. (1989). Development and validation of the computer self-efficacy scale. *Educational and Psychological Measurement, 49*(2), 893–899. doi:10.1177/001316448904900412

Muylle, S., Moenaert, R., & Despontin, M. (2004). The conceptualization and empirical validation of Web site user satisfaction. *Information & Management, 41*, 543–560. doi:10.1016/S0378-7206(03)00089-2

Myers, R. H. (1990). *Classical and modern regression with applications.* Boston, MA: PWS and Kent Publishing.

Mykytyn, P. (2007). Educating our students in computer application concepts: a case for problem-based learning. *Journal of Organizational and End User Computing, 19*(1), 51–61. doi:10.4018/joeuc.2007010103

Nahapiet, J., & Ghoshal, S. (1998). Social capital, intellectual capital and the organizational advantage. *Academy of Management Review, 23*(2), 242–266. doi:10.2307/259373

Nah, F. F.-H., & Lau, J. L.-S. (2001). Critical Factors for Successful Implementation of Enterprise Systems. *Business Process Management Journal, 7*(3), 285–296. doi:10.1108/14637150110392782

National Academies. (2007). *Rising above the gathering storm: Energizing and employing America for a brighter economic future.* Washington, DC: National Academies Press.

National Center for Education Statistics. (1999). *Community Service in K-12 Public Schools.* Washington, DC: U.S. Department of Education.

National Research Council. (1999). *The Changing Nature of Work: Implications for Occupational Analysis.* Washington, DC: National Academy Press.

National Service Learning Clearinghouse. (2008). *What is service-learning?* Retrieved February 10, 2010, from http://www.servicelearning.org/what-service-learning

Neiberg, M. S. (2000). *Making Citizen Soldiers: ROTC.* Cambridge, MA: Harvard University Press.

Netemeyer, R. G., Bearden, W. O., & Sharma, S. (2003). *Scaling Procedures Issues and Applications.* Thousand Oaks, CA: Sage.

Neter, J., Kutner, M., Nachtsheim, C., & Wasserman, W. (1996). *Applied Linear Statistical Models* (4th ed.). Chicago, IL: Irwin.

Neuendorf, K. A. (2002). *The content analysis guidebook.* Beverley Hills, CA: Sage.

Nickell, G. S., & Pinto, J. N. (1986). The computer attitude scale. *Computers in Human Behavior, 2*, 301–306. doi:10.1016/0747-5632(86)90010-5

Niederman, F., Boggs, D. J., & Kundu, S. (2002). International business and global information managment research: Toward a cumulative tradition. *Journal of Global Information Management, 10*(1), 33–47.

Niehoff, B. P., & Moorman, R. H. (1993). Justice as a Mediator of the Relationship between Methods of Monitoring and Organizational Citizenship Behavior. *Academy of Management Journal, 36*(3), 527–556. doi:10.2307/256591

Noe, R. A. (1986). Trainees' attributes and attitudes: neglected influences on training effectiveness. *Academy of Management Review, 11*(4), 736–749.

Nonaka, I. (1994). A Dynamic Theory of Organizational Knowledge Creation. *Organization Science, 5*(1), 14–38. doi:10.1287/orsc.5.1.14

Norman, D. A. (1999). *The Invisible Computer.* Cambridge, MA: MIT Press.

Nunamaker, J. F., Dennis, A. R., Valacich, J. S., Vogel, D. R., & George, J. F. (1991). Electronic meeting systems to support group work. *Communications of the ACM, 34*(7), 40–61. doi:10.1145/105783.105793

Nunnally, J. (1978). *Psychometric theory.* New York, NY: McGraw-Hill.

Nunnally, J. C., & Bernstein, I. (1994). *Psychometric Theory.* New York, NY: McGraw-Hill.

Nurmi, D., Wolski, R., Grzegorczyk, C., Obertelli, G., Soman, S., Youseff, L., & Zagorodnov, D. (2009). Eucalyptus: An open-source cloud computing infrastructure. *Journal of Physics: Conference Series, 180*.

Nygaard, K. (1986). *Program development as social activity in information processing*. Amsterdam, The Netherlands: Elsevier Science Publishers.

O'Leary, M. B., & Mortensen, M. (2010). Go (Con) figure: Subgroups, Imbalance, and Isolates in Geographically Dispersed Teams. *Organization Science, 21*(1), 115–131.

Okoli, C., & Pawlowski, S. D. (1995). The Delphi method as a research tool: An example, design considerations and application. *Information & Management, 42*(1), 15–29.

O'Leary, D. E. (2000). *Enterprise Resource Planning Systems: Systems, Life Cycle, Electronic Commerce, and Risk*. Cambridge, UK: Cambridge University Press.

Olson, G. M., & Olson, J. S. (2003). Human-computer interaction: psychological aspects of the human use of computing. *Annual Review of Psychology, 54*(1), 491–516. doi:10.1146/annurev.psych.54.101601.145044

Ong, C. S., & Lai, J. Y. (2006). Gender differences in perceptions and relationships among dominants of e-learning acceptance. *Computers in Human Behavior, 22*, 816–829. doi:10.1016/j.chb.2004.03.006

O'Reilly, C. A., & Pfeffer, J. (2000). *Hidden Value: How Great Companies Achieve Extraordinary Results with Ordinary People*. Cambridge, MA: Harvard Business School Press.

Orlikowski, W. J. (1992). The Duality of Technology: Rethinking the Concept of Technology in Organizations. *Organization Science, 3*(3), 398–427. doi:10.1287/orsc.3.3.398

Orlikowski, W. J. (2000). Using Technology and Constituting Structures: A Practice Lens for Studying Technology in Organizations. *Organization Science, 11*(4), 404–428. doi:10.1287/orsc.11.4.404.14600

Orlikowski, W. J. (2002). Knowing in practice: Enacting a collective capability in distributed organizing. *Organization Science, 13*(3), 249–273. doi:10.1287/orsc.13.3.249.2776

Orlikowski, W., & Robey, D. (1991). Information technology and the structuring of organizations. *Information Systems Research, 2*(2), 143–169. doi:10.1287/isre.2.2.143

Orr, J. (1987a). Narratives at Work: Story Telling as Cooperative Diagnostic Activity. *Field Service Manager*, 47-60.

Orr, J. (1990a). *Talking About Machines: An Ethnography of a Modern Job*. Unpublished doctoral dissertation, Cornell University, Ithaca, NY.

Orr, J. (1987b). *Talking About Machines: Social Aspects of Expertise*. Palo Alto, CA: Xerox, Palo Alto Research Center.

Orr, J. (1990b). Sharing Knowledge, Celebrating Identity: War Stories and Community Memory in a Service Culture. In Middleton, D. S., & Edwards, D. (Eds.), *Collective Remembering: Memory in Society* (pp. 169–189). Beverley Hills, CA: Sage.

Osmundson, J. S., Michael, J. B., Machniak, M. J., & Grossman, M. A. (2003). Quality management metrics for software development. *Information & Management, 40*, 799–812. doi:10.1016/S0378-7206(02)00114-3

Palmer, J. W. (2002). Web Site Usability, Design, and Performance Metrics. *Information Systems Research, 13*(2), 151–167. doi:10.1287/isre.13.2.151.88

Palvia, P. C. (1996). A model and instrument for measuring small business user satisfaction with information technology. *Information & Management, 31*, 151–163. doi:10.1016/S0378-7206(96)01069-5

Palvia, P. C. (1997). Developing a model of the global and strategic impact of information technology. *Information & Management, 32*, 229–244. doi:10.1016/S0378-7206(97)00023-2

Panko, R. (2000, July 17-18). What We Know About Spreadsheet Errors. In *Proceedings of the Spreadsheet Risk Symposium,* Greenwich, UK.

Panko, R. (1998). What We Know About Spreadsheet Errors. *Journal of End User Computing, 10*(2), 15–21.

Panko, R. (1999). Applying Code Inspection to Spreadsheet Testing. *Journal of Management Information Systems, 16*(2), 159–176.

Parboteeah, D. V., Parboteeah, K. P., Cullen, J. B., & Basu, C. (2005). Perceived Usefulness Of Information Technology: A Cross-National Model. *Journal of Global Information Technology Management, 8*(4), 29.

Paulk, M. C., Weber, C., Curtis, B., & Chrissis, M. (1995). *The capability maturity model: Guidelines for improving the software process*. Reading, MA: Addison-Wesley.

Peace, A. G., Galletta, D. F., & Thong, J. Y. L. (2003). Software Piracy in the Workplace: A Model and Empirical Test. *Journal of Management Information Systems*, *20*(1), 153–177.

Pearson, C. (2009). *The demise of the iron triangle*. Retrieved February 19, 2010, from http://mosaicprojects.wordpress.com/2009/11/19/the-demise-of-the-iron-triangle/

Perrin, C. (2007). Work with end users -- Not against them -- To improve security. *TechRepublic*. Retrieved March 11, 2010, from http://blogs.techrepublic.com.com/security/?p=290

Pett, M. A., Lackey, N. R., & Sullivan, J. J. (2003). *Making sense of factor analysis: the use of factor analysis for instrument development in health care research*. Beverly Hills, CA: Sage.

Pfleeger, S. L. (1998). *Software engineering theory and practice*. Upper Saddle River, NJ: Prentice Hall.

Phillips, L. A., Calantone, R., & Lee, M.-T. (1994). International technology adoption: Behavior structure, demand certainty and culture. *Journal of Business and Industrial Marketing*, *9*(2), 16. doi:10.1108/08858629410059762

Piccoli, G., Ahmad, R., & Ives, B. (2001). Web-based virtual learning environments: a research framework and a preliminary assessment of effectiveness in basic IT skills training. *Management Information Systems Quarterly*, *25*(4), 401–426. doi:10.2307/3250989

Piccoli, G., & Ives, B. (2003). Trust and the Unintended Effects of Behavior Control in Virtual Teams. *Management Information Systems Quarterly*, *27*(3), 365–395.

Pinzari, G. F. (2003). *NX X protocol compression* (D-309/3-NXP-DOC). Retrieved January 14, 2010, from http://www.nomachine.com/documents/html/NX-XProtocolCompression.html

Pitt-Francis, J., Bernabeu, M. O., Cooper, J., Garny, A., Momtahan, L., & Osborne, J. (2008). Chaste: Using agile programming techniques to develop computational biology software. *Philosophical Transactions of the Royal Society*, *366*, 3111–3136. doi:10.1098/rsta.2008.0096

Plaisant, C., Lam, S., Shneiderman, B., Smith, M., Roseman, D., Marchand, G., et al. (2008). Searching electronic health records for temporal patterns in patient histories: A case study with microsoft Amalga. In *Proceedings of the AMIA Annual Symposium* (pp. 601-605).

Png, I. P. L., Tan, B. C. Y., & Khai-Ling, W. (2001). Dimensions of national culture and corporate adoption of IT infrastructure. *IEEE Transactions on Engineering Management*, *48*(1), 36–45. doi:10.1109/17.913164

Podsakoff, P. M., MacKenzie, S. B., Lee, J. Y., & Podsakoff, N. P. (2003). Common method biases in behavioral research: A critical review of the literature and recommended remedies. *The Journal of Applied Psychology*, *88*(5), 879–903. doi:10.1037/0021-9010.88.5.879

Podsakoff, P. M., MacKenzie, S. B., Lee, J., & Podsakoff, N. (1986). Common method biases in behavioral research: a critical review of the literature and recommended remedies. *The Journal of Applied Psychology*, *99*(5), 879–903.

Podsakoff, P. M., & Organ, D. W. (1986). Self-Reports in Organizational Research: Problems and Prospects. *Journal of Management*, *12*(4), 531–544. doi:10.1177/014920638601200408

Polson, P. G., Muncher, E., & Engelbeck, G. (1986). A test of a common elements theory of transfer. In M. Mantei & P. Orbeton (Eds.), *Proceedings of the ACM SIGCHI Conference on Human Factors in Computing Systems*, Boston, MA (pp. 78-83).

Polson, P. G., Bovair, S., & Kieras, D. (1987). Transfer between text editors. *ACM SIGCHI Bulletin*, *18*, 27–32. doi:10.1145/1165387.30856

Popovich, P. M., Hyde, K. R., & Zakrajsek, T. (1987). The Development of the Attitudes Toward Computer Usage Scale. *Educational and Psychological Measurement*, *47*(1), 261–269. doi:10.1177/0013164487471035

Porter, L. W., Pearce, J. L., Tripoli, A. M., & Lewis, K. M. (1998). Differential Perceptions of Employers' Inducements: Implications for Psychological Contracts. *Journal of Organizational Behavior*, *19*(S1), 769–782. doi:10.1002/(SICI)1099-1379(1998)19:1+<769::AID-JOB968>3.0.CO;2-1

Porter, M. E. (1980). *Competitive strategy: Techniques for analyzing industries and competitors*. New York: Free Press.

Porter, M. E. (1990). *The competitive advantage of nations*. London: Macmillan.

Post, D. E., Kendall, R. P., & Whitney, E. (2005). Case study of the falcon project. In *Proceedings of the 2nd International Workshop on Software Engineering for High Performance Computing Systems Applications*, St. Louis, MO (pp. 22-26).

Post, D. E. (2009). The promise of science-based computational engineering. *Computing in Science & Engineering, 11*(3), 3–4. doi:10.1109/MCSE.2009.60

Post, D. E., & Votta, L. G. (2005). Computational science demands a new paradigm. *Physics Today, 58*(1), 35–41. doi:10.1063/1.1881898

Post, D., & Kendall, R. (2004). Software project management and quality engineering practices for complex, coupled, multiphysics, massively parallel computational simulations: Lessons learned from ASCI. *International Journal of High Performance Computing Applications, 18*(4), 399–416. doi:10.1177/1094342004048534

Poston, R. S., & Royne, M. B. (2008). Rating scheme bias in e-commerce: preliminary insights. *Journal of Organizational and End User Computing, 20*(4), 45–73. doi:10.4018/joeuc.2008100103

Potosky, D., & Bobko, P. (2001). A model for predicting computer experience from attitudes toward computers. *Journal of Business and Psychology, 15*(3), 391–403. doi:10.1023/A:1007866532318

Powell, S. G., Baker, K. R., & Lawson, B. (2007). *Errors in Operational Spreadsheets*. Retrieved November 1, 2008, from http://mba.tuck.dartmouth.edu/spreadsheet/product_pubs.html

Powell, A., & Moore, J. E. (2002). The focus of research in end user computing: where have we come since the 1980s? *Journal of End User Computing, 14*(1), 3–22. doi:10.4018/joeuc.2002010101

Powell, A., Piccoli, G., & Ives, B. (2004). Virtual teams: A review of current literature and directions for future research. *The Data Base for Advances in Information Systems, 35*(1), 6–36.

Powell, S. G., Baker, K. R., & Lawson, B. (2008). A critical review of the literature on spreadsheet errors. *Decision Support Systems, 46*(1), 128–138. doi:10.1016/j.dss.2008.06.001

Prabhakararao, S., Cook, C., Ruthruff, J., Creswick, E., Main, M., Durham, M., & Burnett, M. (2003, October). Strategies and Behaviors of End-User Programmers with Interactive Fault Localization. In *Proceedings of the IEEE Symposium on Human-Centric Computing Languages and Environments*, Auckland, New Zealand (pp. 15-22).

Pressing, J. (1984). Cognitive processes in improvisation. In Crozier, W. R., & Chapman, A. J. (Eds.), *Cognitive Processes in the Perception of Art* (pp. 345–363). Amsterdam, The Netherlands: North-Holland. doi:10.1016/S0166-4115(08)62358-4

Preston, A. (1987). Improvising Order. In Mangham, I. L. (Ed.), *Organization Analysis and Development* (pp. 81–102). New York, NY: John Wiley & Sons.

Probert, S. K. (1993). Interpretive analytics and critical information systems: A framework for analysis. In *Systems Science*. Addressing Global Issues.

Purser, M., & Chadwick, D. (2006). Does an awareness of different types of spreadsheet errors aid end-users in identifying spreadsheet errors? In *Proceedings of the European Spreadsheet Risks Interest Group Conference*, London, UK.

Putnam, R. D. (1993). The Prosperous community: social capital and public life. *The American Prospect, 13*, 35–42.

Quinn, J. B. (1980). Formulating strategy one step at a time. *The Journal of Business Strategy*, 42–63.

Quom, S., & Bhaskaran, R. (2002). *ANSYS short course: Three-dimensional bicycle crank*. Retrieved January 14, 2010, from http://courses.cit.cornell.edu/ansys/crank

Raghunathan, B., Raghunatan, T. S., & Tu, Q. (1999). Dimensionality of the Strategic Grid Framework: The Construct and its Measurement. *Information Systems Research, 10*(4), 343–355. doi:10.1287/isre.10.4.343

Rainer, R. K., & Harrison, A. W. (1993). Toward development of the end-user computing construct in a university setting. *Decision Sciences, 24*(6), 1187–1202. doi:10.1111/j.1540-5915.1993.tb00510.x

Rajlich, V., Syed, W. A., & Martinez, J. (2000). Perceptions of contribution in software teams. *Journal of Systems and Software*, *54*(1), 61–63. doi:10.1016/S0164-1212(00)00026-1

Rao, S. S. (2000). Enterprise Resource Planning: Business Needs and Technologies. *Industrial Management & Data Systems*, *100*(2), 81–88. doi:10.1108/02635570010286078

Rasch, R. H., & Tosi, H. L. (1992). Factors affecting software developers' performance: an integrated approach. *Management Information Systems Quarterly*, *16*(3), 395–413. doi:10.2307/249535

Raskin, R., & Hall, C. S. (1981). The Narcissistic Personality Inventory: Alternate Form Reliability and Further Evidence of Construct Validity. *Journal of Personality Assessment*, *45*(2), 159–162. doi:10.1207/s15327752jpa4502_10

Rawsthorne, D., & Shimp, D. (2009). *Scrum in a nutshell.* Retrieved February 19, 2010, from http://www.scrumalliance.org/articles/151-scrum-in-a-nutshell

References

Richardson, T. (2009). *The RFB protocol, version 3.8.* Cambridge, UK: RealVNC Ltd.

Ringle, C. M., Wende, S., & Will, A. (2005). *SmartPLS 2.0 (beta),* www.smartpls.de. Hamburg, Germany: University of Hamburg.

Robey, D., Ross, J. W., & Boudreau, M.-C. (2002). Learning to Implement Enterprise Systems: An Exploratory Study of the Dialectics of Change. *Journal of Management Information Systems*, *19*(1), 17–46.

Robinson, S. L. (1996). Trust and Breach of the Psychological Contract. *Administrative Science Quarterly*, *41*(4), 574–599. doi:10.2307/2393868

Robinson, S. L., & Morrison, E. W. (1995). Psychological Contracts and OCB: The Effect of Unfulfilled Obligations on Civic Virtue Behavior. *Journal of Organizational Behavior*, *16*, 289–298. doi:10.1002/job.4030160309

Robinson, S. L., & Rousseau, D. M. (1994). Violating the Psychological Contract: Not the Exception but the Norm. *Journal of Organizational Behavior*, *15*, 245–259. doi:10.1002/job.4030150306

Rockart, J. F., & Flannery, L. S. (1983). The management of end user computing. *Communications of the ACM*, *26*(10), 776–784. doi:10.1145/358413.358429

Rodan, S., & Galunic, C. (2004). More than network structure: how knowledge heterogeneity influences managerial performance and innovation. *Strategic Management Information*, *25*, 541–562.

Rogelberg, S. G., Leach, D. J., Warr, P. B., & Burnfield, J. L. (2006). "Not another meeting!"? Are meeting time demands related to employee well-being? *The Journal of Applied Psychology*, *1*, 86–96.

Rogers, E. M. (2003). *Diffusion of innovations* (5th ed.). New York, NY: Free Press.

Rogers, T. B. (1995). *The Psychological Testing Enterprise*. Belmont, CA: Brooks/Cole.

Rolland, C. (1997). A primer for method engineering. In *Proceedings of the IFIP TC8, WG8.1/8.2 Working Conference on Method Engineering: Principles of Method Construction and Tool Support*, Toulouse, France (pp. 1-7).

Rolland, C., & Pernici, C. T. (1998). A comprehensive view of process engineering. In *Proceedings of the 10th International Conference CAiSE'98* (LNCS 1413, pp. 1-24).

Romano, N. C., Jr., & Nunamaker, J. F., Jr. (2001). Meeting Analysis: Findings from Research and Practice. In *Proceedings of the 34th Hawaii International Conference on System Sciences.*

Rooney, P. (2000). Constructive Controversy: A new approach to designing team projects. *Business Communication Quarterly*, *63*(1), 53–61. doi:10.1177/108056990006300106

Rose, A. F., Schnipper, J. L., Park, E. R., Poon, E. G., Li, Q., & Middleton, B. (2005). Using qualitative studies to improve the usability of an EMR. *Journal of Biomedical Informatics*, *38*(1), 51–60. doi:10.1016/j.jbi.2004.11.006

Rose, G., & Straub, D. W. (1998). Predicting general IT use: Applying TAM to the Arab world. *Journal of Global Information Management*, *6*(3), 39–46.

Rosenberg, M. (1968). *Society and the adolescent self-image*. Princeton, NJ: Princeton University Press.

Rosenthal, E. A. (1996). *Social Network and Team Performance*. Unpublished doctoral dissertation, University of Chicago.

Rosenzweig, M. R., & Bennett, E. L. (1978). Experiential influences on brain anatomy and brain chemistry in rodents. In Gottlieb, G. (Ed.), *Studies on the Development of Behavior and the Nervous System* (pp. 289–387). New York: Academic Press.

Rothermel, G., Burnett, M., Li, L., Dupuis, C., & Sheretov, A. (2001). A Methodology for Testing Spreadsheets. *ACM Transactions on Software Engineering and Methodology, 10*(1), 110–147. doi:10.1145/366378.366385

Rousseau, D. M. (1995). *Psychological Contracts in Organizations: Understanding Written and Unwritten Agreements*. Thousand Oaks, CA: Sage.

Rousseau, D. M. (1996). Changing the Deal While Keeping the People. *The Academy of Management Executive, 10*(1), 51–61. doi:10.5465/AME.1996.9603293198

Royce, W. (1970, August). Managing the development of large software systems. In *Proceedings of the 9th International Conference on Software Engineering* (pp. 328-338).

Rumelt, R. (1995). The Evaluation of Business Strategy. In Mintzberg, H., Quinn, J. B., & Voyer, J. (Eds.), *The Strategy Process* (pp. 73–79). Englewood Cliffs, NJ: Prentice Hall.

Rusbult, C. E., & Farrell, D. (1983). A longitudinal test of the investment model: The impact on job satisfaction, job commitment, and turnover of variations in rewards, costs, alternatives, and investments. *The Journal of Applied Psychology, 68*(3), 429–438. doi:10.1037/0021-9010.68.3.429

Rutner, P. S., Hardgrave, B. C., & McKnight, D. H. (2008). Emotional dissonance and the information technology professionals. *Management Information Systems Quarterly, 32*(3), 635–652.

Ryan, K., & Cooper, J. (1998). *Those who can, Teach*. Boston, MA: Houghton Mifflin.

Ryan, S. D., Harrison, D. W., & Schkade, L. L. (2002). Information-technology investment decisions: when do costs and benefits in the social subsystem matter? *Journal of Management Information Systems, 19*(2), 85–127.

Saarinen, T. (1996). An expanded instrument for evaluation information system success. *Information & Management, 31*, 103–118. doi:10.1016/S0378-7206(96)01075-0

Sabherwal, R., & King, W. R. (1995). An Empirical Taxonomy of the Decision-Making Processes Concerning Strategic Applications of Information Systems. *Journal of Management Information Systems, 11*(4), 177–214.

Sahay, S. (2004). Beyond utopian and nostalgic views of information technology and education: Implications for research and practice. *Journal of the Association for Information Systems, 5*(7), 282–313.

Salas, E., & Cannon-Bowers, J. A. (2001). The science of training: A decade of progress. *Annual Review of Psychology, 52*(1), 471–499. doi:10.1146/annurev.psych.52.1.471

Salas, E., DeRouin, R., & Littrell, L. (2005). Research-based guidelines for designing distance learning: What we know so far. In Gueutal, H. G., & Stone, D. L. (Eds.), *The Brave New World of e-HR: Human Resources Management in the Digital Age*. San Francisco: Jossey-Bass.

Salisbury, W. D., Chin, W. W., Gopal, A., & Newsted, P. R. (2002). Research Report: Better Theory Through Measurement-Developing a Scale to Capture Consensus on Appropriation. *Information Systems Research, 13*(1), 91–103. doi:10.1287/isre.13.1.91.93

Sanderson, P., & Vollmar, K. (2000). A primer for applying service learning to computer science. In *Proceedings of the ACM Conference on Computer Science Education (SIGCSE)* (pp. 222-226).

Sanders, R., & Kelly, D. (2008). Scientific software: Where's the risk and how do scientists deal with it? *IEEE Software, 25*(4), 21–28. doi:10.1109/MS.2008.84

Sarker, S., & Sahay, S. (2002). Information Systems Development by US-Norwegian Virtual Teams: Implications of Time and Space. In *Proceedings of the 35th Annual Hawaii International Conference on System Sciences* (pp. 1–10).

Saunders, C. S., & Jones, W. J. (1992). Measuring Performance of the Information Systems Function. *Journal of Management Information Systems, 8*(4), 63–82.

Savage, A., & Mingers, J. (1996). A framework for linking soft systems methodology (SSM) and Jackson system development (JSD). *Information Systems Journal, 6*, 109–129. doi:10.1111/j.1365-2575.1996.tb00008.x

Scaffidi, C., Shaw, M., & Myers, B. (2005). *The 55M End-User Programmers Estimate Revisited* (Tech. Rep. No. CMU-ISRI-05-100). Pittsburgh, PA: Institute for Software Research, Carnegie Mellon University.

Scales, P. C., & Roehlkepartain, E. C. (2004). *Community service and service-learning in U.S. public schools, 2004: Findings from a National Survey*. St. Paul, MN: National Youth Leadership Council. Retrieved March 11, 2010, from http://www.nylc.org/objects/inaction/initiatives/2 004G2G/2004G2GCompleteSurvey.pdf

Schein, E. H. (1996). Three cultures of management: The key to organizational learning. *Sloan Management Review*.

Schein, E. H. (1980). *Organizational Psychology* (3rd ed.). Englewood Cliffs, NJ: Prentice-Hall.

Schepers, J., & Wetzels, M. (2007). A meta-analysis of the technology acceptance model: Investigating subjective norm and moderation effects. *Information & Management*, *44*(1), 90–103. doi:10.1016/j.im.2006.10.007

Schmidt, A. M., & Ford, J. K. (2003). Learning within a learner control training environment: The interactive effects of goal orientation and metacognitive instruction on learning outcomes. *Personnel Psychology*, *56*(2), 405–429. doi:10.1111/j.1744-6570.2003.tb00156.x

Schmidt, R. C. (1997). Managing Delphi surveys using nonparametric statistical techniques. *Decision Sciences*, *28*(3), 763–774. doi:10.1111/j.1540-5915.1997.tb01330.x

Scholtz, J., & Wiedenbeck, S. (1990). Learning second and subsequent programming languages: A problem of transfer. *International Journal of Man-Computer Interaction*, *2*(1), 51–72. doi:10.1080/10447319009525970

Schriesheim, C. A. (1979). The Similarity of Individual Directed and Group Directed Leader Behavior Descriptions. *Academy of Management Journal*, *22*(2), 345–355. doi:10.2307/255594

Schuelke, M. J., Day, E. A., McEntire, L. E., Boatman, P. R., Boatman, J. E., Kowollik, V., & Wang, X. (2009). Relating indices of knowledge structure coherence and accuracy to skill-based performance: Is there utility in using a combination of indices? *The Journal of Applied Psychology*, *94*(4), 1076–1085. doi:10.1037/a0015113

Schuldt, B. A. (1991). 'Real-world' versus 'simulated' projects in database instruction. *Journal of Education for Business*, *67*(1), 35–39. doi:10.1080/08832323.199 1.10117514

Schultze, U., & Orlikowski, W. J. (2001). Metaphors of virtuality: Shaping an emergent reality. *Information and Organization*, *11*, 45–77. doi:10.1016/S1471-7727(00)00003-8

Schwartzman, H. (1986). The meeting as a neglected social form in organizational studies. In Staw, B., & Cummings, L. (Eds.), *Research in organizational behavior* (Vol. 9, pp. 233–258). Greenwich, CT: JAI.

Schwartzman, H. (1989). *The meeting: Gatherings in organizations and communities*. New York, NY: Plenum.

Schweiger, D. M., & Denisi, A. S. (1991). Communication with Employees following a Merger: A Longitudinal Field Experiment. *Academy of Management Journal*, *34*(1), 110–135. doi:10.2307/256304

Segal, J. (2004). *Professional end user developers and software development knowledge* (Tech. Rep. No. 2004/25). Milton Keynes, UK: Open University.

Segal, J. (2007). Some problems of professional end user developers. In *Proceedings of the IEEE Symposium on Visual Languages and Human-Centric Computing* (pp. 111-118).

Segal, J. (2008). Scientists and software engineers: A tale of two cultures. In *Proceedings of the Psychology of Programming Interest Group Conference*. Retrieved from http://www.ppig.org/papers/20th-segal.pdf

Segal, J. (2009b). Some challenges facing software engineers developing software for scientists. In *Proceedings of the 2nd International Software Engineering for Computational Scientists and Engineers Workshop* (pp. 9-14).

Segal, J. A. (2008). Models of scientific software. In *Proceedings of the 1st International Workshop on Software Engineering for Computational Science and Engineering*.

Segal, J. (2005). When software engineers met research scientists: A case study. *Empirical Software Engineering*, *10*, 517–536. doi:10.1007/s10664-005-3865-y

Segal, J. (2009). Software development cultures and cooperation problems: A field study of the early stages of development of software for a scientific community. *Computer Supported Cooperative Work*, *18*(5-6), 581–606. doi:10.1007/s10606-009-9096-9

Segal, J., & Morris, C. (2008). Developing scientific software. *IEEE Software*, *25*(4), 18–20. doi:10.1109/MS.2008.85

Segars, A. H. (1997). Assessing the Unidimensionality of Measurement: A Paradigm and Illustration within the Context of Information Systems Research. *Omega*, *25*(1), 107–121. doi:10.1016/S0305-0483(96)00051-5

Segars, A. H., & Grover, V. (1998). Strategic Information Systems Planning Success: An Investigation of the Construct and Its Measurement. *Management Information Systems Quarterly*, *22*(2), 139–163. doi:10.2307/249393

Sehoole, C. T., & Moja, T. (2003). Pedagogical issues and gender in cyberspace education: Distance education in South Africa. *African and Asian Studies*, *2*(4), 475–496. doi:10.1163/156920903773004022

Seibert, S., Kraimer, M. L., & Liden, R. (2001). A social capital theory of career success. *Academy of Management Journal*, *44*(2), 219–237. doi:10.2307/3069452

Sein, M. K., Bostrom, R. P., & Olfman, L. (1999). Rethinking end-user training strategy: applying a hierarchical knowledge-level model. *Journal of End User Computing*, *11*(1), 32–39.

Sekaran, U. (2003). *Research Methods for Business: A Skill Building Approach*. New York: John Wiley & Sons.

Selim, H. M. (2003). An empirical investigation of student acceptance of course websites. *Computers & Education*, *40*(4), 343–360. doi:10.1016/S0360-1315(02)00142-2

Sethi, R., Smith, D. C., & Whan, C. (2000). Cross-functional product development teams, creativity, and the innovativeness of new consumer products. *JMR, Journal of Marketing Research*, *38*(1), 73–85. doi:10.1509/jmkr.38.1.73.18833

Sethi, V., & King, W. R. (1994). Development of Measures to Assess the Extent to Which an Information Technology Application Provides Competitive Advantage. *Management Science*, *40*(12), 1601–1627. doi:10.1287/mnsc.40.12.1601

Seyal, A. H., Rahim, M. M., & Rahman, M. N. A. (2000). Computer attitudes of non-computing academics: a study of technical colleges in Brunei Darussalam. *Information & Management*, *37*, 169–180. doi:10.1016/S0378-7206(99)00045-2

Shah, P. P. (1998). Who are employees' social referents: using a network perspective to determine referent others. *Academy of Management Journal*, *41*(3), 249–268. doi:10.2307/256906

Shainer, G., Sparks, B., Schultz, S., Lantz, E., Liu, W., Liu, T., & Misra, G. (2010, January 26). Cloud computing will usher in a new era of science discover. *HPCwire*.

Shang, S. S. C., & Su, T. C. C. (2004, August). *Managing User Resistance in Enterprise Systems Implementation*. Paper presented at the Americas Conference on Information Systems, New York.

Shapiro, D. L., Buttner, E. H., & Barry, B. (1994). Explanations: What Factors Enhance Their Perceived Adequacy. *Organizational Behavior and Human Decision Processes*, *58*, 346–368. doi:10.1006/obhd.1994.1041

Shapiro, D. L., & Kirkman, B. L. (1999). Employees' Reaction to the Change to Work Teams: The Influence of "Anticipatory" Injustice. *Journal of Organizational Change Management*, *12*, 51–66. doi:10.1108/09534819910255315

Sharma, R., & Yetton, P. (2003). The contingent effects of management support and task interdependence on successful information systems implementation. *Management Information Systems Quarterly*, *27*(4), 533–555.

Shaw, N., Lee-Partidge, J., & Ang, J. S. K. (2003). Understanding the hidden dissatisfaction of users toward end-user computing. *Journal of End User Computing*, *15*(2), 1–22. doi:10.4018/joeuc.2003040101

Shayo, C., & Olfman, L. (1998). The role of conceptual models in formal software training. In S. Poltrock & J. Grudin (Eds.), *Proceedings of the 1998 ACM Conference on Computer Personal Research (SIGCPR)*, Seattle, WA (pp. 242-253).

Sheppard, B. H., Lewicki, R. J., & Minton, J. W. (1992). *Organizational Justice: The Search for Fairness in the Workplace*. New York, NY: Lexington Books.

Shih, M. H., Yu, T. Y., & Soni, B. K. (1994). Interactive grid generation and NURBS applications. *Journal of Applied Mathematics and Computation, 65*(1-3), 279–294. doi:10.1016/0096-3003(94)90183-X

Shin, S. K., Ishman, M., & Sanders, G. L. (2007). An empirical investigation of socio-cultural factors of information sharing in China. *Information & Management, 44*(2), 165–174. doi:10.1016/j.im.2006.11.004

Shi, Z., Kunnathur, A. S., & Ragu-Nathan, T. S. (2005). IS outsourcing management competence dimensions: instrument development and relationship exploration. *Information & Management, 42*, 901–919. doi:10.1016/j.im.2004.10.001

Shneiderman, B. (1998). Relate-Create-Donate: a teaching/learning philosophy for the cyber-generation. *Computers & Education, 31*, 25–39. doi:10.1016/S0360-1315(98)00014-1

Short, J., Williams, E., & Christie, B. (1976). *The Social Psychology of Telecommunications*. New York: John Wiley & Sons.

Simonson, M. R., Maurer, M., Montag-Torardi, M., & Whitaker, M. (1987). Development of a standardized test of computer literacy and a computer anxiety index. *Journal of Educational Computing Research, 3*(2), 231–247. doi:10.2190/7CHY-5CM0-4D00-6JCG

Singley, M. K., & Anderson, J. R. (1985). The transfer of text-editing skill. *International Journal of Man-Machine Studies, 22*(4), 403–423. doi:10.1016/S0020-7373(85)80047-X

Singley, M. K., & Anderson, J. R. (1988). A keystroke analysis of learning and transfer in text editing. *Human-Computer Interaction, 3*(3), 223–274. doi:10.1207/s15327051hci0303_2

Skarlicki, D. P., & Folger, R. (1997). Retaliation in the Workplace: The Roles of Distributive, Procedural, and Interactional Justice. *The Journal of Applied Psychology, 82*(3), 434–443. doi:10.1037/0021-9010.82.3.434

Smelcer, J. B., & Walker, N. (1993). Transfer of knowledge across computer command menus. *International Journal of Human-Computer Interaction, 5*(2), 147–165. doi:10.1080/10447319309526062

Sminia, S., & Van Nistelrooij, A. (2006). Strategic Management and Organization Development: Planned Change in a Public Sector Organization. *Journal of Change Management, 6*(1), 99–113. doi:10.1080/14697010500523392

Smith, B. L., & MacGregor, J. T. (1992). *What is Collaborative Learning?* University Park, PA: National Center on Postsecondary Teaching, Learning, and Assessment, Pennsylvania State University.

Smith, H. J., Milberg, S. J., & Burke, S. J. (1996). Information Privacy: Measuring Individual's Concerns about Organisational Practices. *Management Information Systems Quarterly, 20*(2), 167–196. doi:10.2307/249477

Smith, S., Lai, L., & Khedri, R. (2007). Requirements analysis for engineering computation: A systematic approach for improving reliability. *Reliable Computing, 13*, 83–107. doi:10.1007/s11155-006-9020-7

Smits, M. T., van der Poel, K. G., & Ribbers, P. M. A. (1997). Assessment of information strategies in insurance companies in the Netherlands. *The Journal of Strategic Information Systems, 6*, 129–148. doi:10.1016/S0963-8687(97)00004-8

Song, X., & Osterweil, L. J. (1992). Toward objective systematic design-method comparisons. *IEEE Software, 9*(3), 43–53. doi:10.1109/52.136166

Soni, B. K., Thompson, J., Stokes, M. L., & Shih, M. H. (1992). GENIE++, EAGLE View and TIGER: General and special purpose graphically interactive grid systems. In *Proceedings of the 30th AIAA Aerospace Science Meeting*, Reno, NV.

Sonnentag, S. (2001). High performance and meeting participation: An observational study in software design teams. *Group Dynamics, 5*(1), 3–18. doi:10.1037/1089-2699.5.1.3

Sotomayor, B., Montero, R. S., Llorente, I. M., & Foster, I. (2008). Capacity leasing in cloud systems using the OpenNebula engine. In *Proceedings of the Cloud Computing and Applications Conference*.

Spannagel, C., Gläser-Zikuda, M., & Schroeder, U. (2005). Application of Qualitative Content Analysis in User-Program Interaction Research. *Forum Qualitative Sozial Forschung, 6*(2).

Spitler, V. (2005). Learning to use IT in the workplace: mechanisms and masters. *Journal of Organizational and End User Computing, 17*(2), 1–25. doi:10.4018/joeuc.2005040101

Spitler, V. K. (2005). Learning to Use IT in the Workplace: Mechanisms and Masters. *Journal of Organizational and End User Computing, 17*(2), 1–25. doi:10.4018/joeuc.2005040101

Spreitzer, G. M. (1995). Psychological Empowerment in the Workplace: Dimensions, Measurement, and Validation. *Academy of Management Journal, 38*(5), 1442–1465. doi:10.2307/256865

Spreitzer, G. M. (1996). Social Structural Characteristics of Psychological Empowerment. *Academy of Management Journal, 39*(2), 483–504. doi:10.2307/256789

Spreitzer, G. M., De Janasz, S. C., & Quinn, R. E. (1999). Empowered to lead: the role of psychological empowerment in leadership. *Journal of Organizational Behavior, 20*(4), 511–526. doi:10.1002/(SICI)1099-1379(199907)20:4<511::AID-JOB900>3.0.CO;2-L

Srite, M., & Karahanna, E. (2006). The role of espoused national cultural values in technology acceptance. *Management Information Systems Quarterly, 30*(3), 679–704.

Stanton, J. M. (1998). An empirical assessment of data collection using the Internet. *Personnel Psychology, 51*(3), 709–725. doi:10.1111/j.1744-6570.1998.tb00259.x

Star, S. L. (1989). The structure of ill-structured solutions: Boundary objects and heterogeneous distributed problem solving. In Gasser, L., & Huhns, M. N. (Eds.), *Distributed Artificial Intelligence* (*Vol. 2*, pp. 37–54). San Francisco, CA: Morgan Kaufmann.

Star, S., & Ruhleder, K. (1996). Steps towards an ecology of infrastructure design and access for large information spaces. *Information Systems Research, 7*(1), 111–134. doi:10.1287/isre.7.1.111

Stein, T. (1999). Making ERP Add Up. *InformationWeek*, 59-68.

Stephens, K., & Davis, J. (2009). The social influences on electronic multitasking in organizational meetings. *Management Communication Quarterly, 23*(1), 63–83. doi:10.1177/0893318909335417

Sternad, S., Bobek, S., Dezelak, Z., & Lampret, A. (2009). Critical Success Factors (CSFs) for Enterprise Resource Planning (ERP) Solution Implementation in SMEs: What Does Matter for Business Integration. *International Journal of Enterprise Information Systems, 5*(3), 27–46. doi:10.4018/jeis.2009070103

Stowell, F., & West, D. (1994). *Client-led design – A systemic approach to information systems definition.* Maidenhead, UK: McGraw-Hill.

Strand, K., Marullo, S., Cutforth, N., Stoecker, R., & Donohue, P. (2003). *Community-based research and higher education: Principles and practices.* San Francisco, CA: Jossey-Bass.

Stratman, J. K., & Roth, A. V. (2002). Enterprise Resource Planning (ERP) Competence Constructs: Two-Stage Multi-Item Scale Development. *Decision Sciences, 33*(4), 601–628. doi:10.1111/j.1540-5915.2002.tb01658.x

Straub, D. W. (1989). Validating instruments in MIS research. *Management Information Systems Quarterly, 13*(2), 147–169. doi:10.2307/248922

Straub, D. W. (1994). The effect of culture on IT diffusion: E-mail and fax in Japan and the United States. *Information Systems Research, 5*(1), 23–47. doi:10.1287/isre.5.1.23

Straub, D. W., Boudreau, M. C., & Gefen, D. (2004). Validation guidelines in IS positivist research. *Communications of the Association for Information Systems, 14*, 380–426.

Straub, D. W., Keil, M., & Brenner, W. (1997). Testing the technology acceptance model across cultures: A three country study. *Information & Management, 33*(1), 1–11. doi:10.1016/S0378-7206(97)00026-8

Straub, D. W., Loch, K. D., & Hill, C. (2001). Transfer of information technology to the Arab world: A test of cultural influence modeling. *Journal of Global Information Management, 9*, 6–28.

Straub, D., Limayem, M., & Karahanna-Evaristo, E. (1995). Measuring system usage: Implications for IS theory testing. *Management Science, 41*(8), 1328–1342. doi:10.1287/mnsc.41.8.1328

Strauss, A., & Corbin, J. (1990). *Basics of qualitative research: Grounded theory procedures and techniques.* Newbury Park, CA: Sage.

Strauss, A., & Corbin, J. (1998). *Basics of qualitative research: Techniques and procedures for developing grounded theory*. Newbury Park, CA: Sage.

Stryker, S. (1980). *Symbolic Interactionism: A Social Structural View*. Reading, MA: Benjamin Cummings.

Stylianou, A. C., Robbins, S. S., & Jackson, P. (2003). Perceptions and attitudes about eCommerce development in China: An exploratory study. *Journal of Global Information Management*, *11*(2), 31.

Sugianto, L. F., Tojib, D. R., & Burstein, F. (2007). A practical measure of employee satisfaction with B2E portals. In *Proceedings of the 28th International Conference on Information Systems*, Montreal, Quebec, Canada.

Sugianto, L. F., & Tojib, D. R. (2007). Portal Power. *Monash Business Review*, *3*(1), 25. doi:10.2104/mbr07016

Suh, B., & Han, I. (2002). Effect of trust on customer acceptance of Internet banking. *Electronic Commerce Research and Applications*, *1*(3-4), 247–263. doi:10.1016/S1567-4223(02)00017-0

Sukkar, A. A., & Hasan, H. (2005). Toward a model for the acceptance of Internet banking in developing countries. *Information Technology for Development*, *11*(4), 381–398. doi:10.1002/itdj.20026

Summerfield, B. (2006). Working with end users toward an effective solution. *Certification Magazine*. Retrieved March 11, 2010, from http://www.certmag.com/read.php?in=1869# Tanniru, M. R., & Agarwal, R. (2002). Applied technology in business program. *e-Service Journal*, 5-23.

Sun, H., & Zhang, P. (2006). The role of moderating factors in user technology acceptance. *International Journal of Human-Computer Studies*, *64*(2), 53–78. doi:10.1016/j.ijhcs.2005.04.013

Swift, J., & Smith, A. (1992). Attitudes to language learning. *Journal of European Industrial Training*, *17*(7), 7–15.

Szajna, B., & Mackay, J. M. (1995). Predictors of learning performance in a computer-user training environment: A path analytic study. *International Journal of Human-Computer Interaction*, *7*(2), 167–185. doi:10.1080/10447319509526118

Szalma, J. L., Hancock, P. A., Dember, W. N., & Warm, J. S. (2006). Training for vigilance: The effect of knowledge of results format and dispositional optimism and pessimism on performance and stress. *The British Journal of Psychology*, *97*, 115–135. doi:10.1348/000712605X62768

Tan, H. H., & Zhao, B. (2003). Individual and contextual level antecedents of individual technical information inquiry in organizations. *The Journal of Psychology*, *137*(6), 579–621. doi:10.1080/00223980309600637

Tannen, D. (1990). *You Just Don't Understand: Women and Men in Conversation*. New York: Ballantine.

Tannen, D. (1993). The relativity of linguistic strategies: Rethinking power and solidarity in gender and dominance. In Tannen, D. (Ed.), *Gender and Conversational Interaction* (pp. 165–188). New York: Oxford University Press.

Teasley, B. E. (1994). The effects of naming style and expertise on program comprehension. *International Journal of Human-Computer Studies*, *40*(5), 757–770. doi:10.1006/ijhc.1994.1036

Teigland, R., & Wasko, M. M. (2003). Integrating knowledge through information trading: Examining the relationship between boundary spanning communication and individual performance. *Decision Sciences*, *34*(2), 261–286. doi:10.1111/1540-5915.02341

Templeton, G. F., Lewis, B. R., & Snyder, C. A. (2002). Development of a Measure for the Organisational Learning Construct. *Journal of Management Information Systems*, *19*(2), 175–218.

Teo, H. H., Wei, K. K., & Benbasat, I. (2003). Predicting Intention to Adopt Interorganisational Linkages: An Institutional Perspective. *Management Information Systems Quarterly*, *27*(1), 19–49.

Teo, H.-H., Chan, H.-C., Wei, K.-K., & Zhang, Z. (2003). Evaluating information accessibility and community adaptivity features for sustaining virtual learning communities. *International Journal of Human-Computer Studies*, *59*(5), 671–697. doi:10.1016/S1071-5819(03)00087-9

Teo, T., & Lee-Partridge, J. (2001). Effects on error factors and prior incremental practice on spreadsheet error detection: An experimental study. *Omega*, *29*, 445–456. doi:10.1016/S0305-0483(01)00037-8

Terry, M., & Mynatt, E. D. (2002). Supporting experimentation with side-views. *Communications of the ACM*, *45*(10), 106–108. doi:10.1145/570907.570942

Tesser, A., Millar, M., & Moore, J. (1988). Some affective consequences of social comparison and reflection processes: The pain and pleasure of being close. *Journal of Personality and Social Psychology*, *54*(1), 49–61. doi:10.1037/0022-3514.54.1.49

Thaler, R. (1985). Mental accounting and consumer choice. *Marketing Science*, *4*(3), 199–214. doi:10.1287/mksc.4.3.199

Thatcher, J. B., & Perrewe, P. L. (2002). An empirical examination of individual traits as antecedents to computer anxiety and computer self-efficacy. *Management Information Systems Quarterly*, *26*(4), 381–396. doi:10.2307/4132314

Thew, S., Sutcliffe, A., Procter, R., De Bruijn, O., McNaught, J., Venters, C., & Buchan, I. (2009). Requirements engineering for e-science: Experiences in epidemiology. *IEEE Software*, *26*(1), 80–87. doi:10.1109/MS.2009.19

Thibodeaux, M. S., & Favilla, E. (1995). Strategic management and organizational effectiveness in colleges of business. *Journal of Education for Business*, *70*(4), 189–196. doi:10.1080/08832323.1995.10117748

Thomas, C. D., & Freeman, R. J. (1990). The Body Esteem Scale: Construct Validity of the Female Subscales. *Journal of Personality Assessment*, *54*(1-2), 204–212. doi:10.1207/s15327752jpa5401&2_20

Thompson, J. F., Warsi, Z. U. A., & Mastin, C. W. (1985). *Numerical grid generation: Foundations and applications*. New York, NY: Elsevier North-Holland.

Thompson, L. F., & Lynch, B. J. (2003). Web-based instruction: Who is inclined to resist and why? *Journal of Educational Computing Research*, *29*(3), 375–385. doi:10.2190/3VQ2-XTRH-08QV-CAEL

Thompson, R. L., Higgins, C. A., & Howell, J. M. (1991). Personal computing: Toward a conceptual model of utilization. *Management Information Systems Quarterly*, *15*(1), 125–143. doi:10.2307/249443

Tojib, D. R. (2007). *Development and Validation of the Business-to-Employee Portal User Satisfaction (B2E-PUS) Scale*. Unpublished doctoral dissertation, Monash University, Australia.

Tojib, D. R., & Sugianto, L. F. (2007). The Development and Empirical Validation of B2E Portal User Satisfaction (B2EPUS) Scale. *Journal of End User Computing*, *19*(3), 1–18.

Tojib, D. R., Sugianto, L. F., & Sendjaya, S. (2008). User Satisfaction with Business-to Employee (B2E) Portals: Conceptualization and Scale Development. *European Journal of Information Systems*, *17*(6), 649–667. doi:10.1057/ejis.2008.55

Torkzadeh, G., & Dhillon, G. (2002). Measuring factors that influence the success of internet commerce. *Information Systems Research*, *13*(2), 187–204. doi:10.1287/isre.13.2.187.87

Torkzadeh, G., & Doll, W. J. (1993). The place and value of documentation in end-user computing. *Information & Management*, *24*(3), 147–158. doi:10.1016/0378-7206(93)90063-Y

Torkzadeh, G., & Doll, W. J. (1999). The Development of a Tool for Measuring the Perceived Impact of Information Technology on Work. *Omega*, *27*(7), 327–339. doi:10.1016/S0305-0483(98)00049-8

Torkzadeh, G., Koufteros, X., & Doll, W. J. (2005). Confirmatory factor analysis and factorial invariance of the impact of information technology instrument. *Omega*, *33*(2), 107–118. doi:10.1016/j.omega.2004.03.009

Torkzadeh, G., & Lee, J. (2003). Measures of perceived end-user computing skills. *Information & Management*, *40*, 607–615. doi:10.1016/S0378-7206(02)00090-3

Totterdell, P., Wall, T., Holman, D., & Epitropaki, O. (2004). Affect network: A structural analysis of the relationship between work ties and job –related affect. *The Journal of Applied Psychology*, *89*(5), 854–867. doi:10.1037/0021-9010.89.5.854

Touliatos, J., Bedeian, A. G., Mossholder, K. W., & Barkman, A. I. (1984). Job-related perceptions of male and female government, industrial and public accountants. *Social Behavior and Personality*, *12*, 61–68. doi:10.2224/sbp.1984.12.1.61

Tran, D., Dubay, C., Gorman, P., & Hersh, W. (2004). Applying task analysis to describe and facilitate bioinformatics tasks. *Medinfo, 11*(2), 818–822.

Trochim, W., & Donnelly, J. (2007). *Research Methods Knowledge Base* (3rd ed.). Mason, OH: Atomic Dog Publishing.

Tropman, J. (1996). *Effective meetings: Improving group decision making*. Thousand Oaks, CA: Sage.

Tu, C.-H., & McIsaac, M. (2002). The Relationship of Social Presence and Interaction in Online Classes. *American Journal of Distance Education, 16*(3), 131–150. doi:10.1207/S15389286AJDE1603_2

Ulrich, W. (1983). *Critical heuristics of social planning: A new approach to practical philosophy*. Chichester, UK: Berne, Haupt, and J. Wiley.

Umarji, M., & Seaman, C. (2008). Informing design of a search tool for bioinformatics. In *Proceedings of the ICSE Workshop on Software Engineering for Computational Science and Engineering.*

Umble, E. J., & Umble, M. M. (2002). Avoiding ERP Implementation Failure. *Industrial Management (Des Plaines), 44*(1), 25–33.

University of North Carolina. (2009). *Service-learning overview*. Retrieved March 22, 2010, from http://olsl.uncg.edu/svl/about/

University of Southern California. (n.d.). *Service learning theory and practice*. Retrieved March 11, 2010, from http://college.usc.edu/service-learning-theory-practice/eval.htm

University of York. (2005). *Information strategy 2004-9*. Retrieved August 8, 2009, from http://www.york.ac.uk/admin/po/infostrat/informationstrategy

Urbaczewski, A., & Wheeler, B. C. (2001). Do sequence and concurrency matter? An investigation of order and timing effects on student learning of programming languages. *Communications of the AIS, 5*(1).

van Birgelen, M., de Jong, A., & de Ruyter, K. (2006). Multi-channel service retailing: The effects of channel performance satisfaction on behavioral intentions. *Journal of Retailing, 82*(4), 367–377. doi:10.1016/j.jretai.2006.08.010

Van der Heijden, H., & Verhagen, T. (2004). Online store image: conceptual foundations and empirical measurement. *Information & Management, 41*, 609–617. doi:10.1016/S0378-7206(03)00100-9

van Raaij, E. M., & Schepers, J. J. L. (2008). The acceptance and use of a virtual learning environment in China. *Computers & Education, 50*(3), 838–852. doi:10.1016/j.compedu.2006.09.001

Van Vliet, H. (2000). *Software engineering principles and practices*. Chichester, UK: John Wiley & Sons.

Vandenbosch, B., & Higgins, C. (1996). Information acquisition and mental modes: An investigation into the relationship between behavior and learning. *Information Systems Research, 7*(2), 198–214. doi:10.1287/isre.7.2.198

Veiga, J. F., Floyd, S., & Dechant, K. (2001). Towards modelling the effects of national culture on IT implementation and acceptance. *Journal of Information Technology, 16*(3), 145–158. doi:10.1080/02683960110063654

Venkatesh, V. (2000). Determinants of perceived ease of use: Integrating control, intrinsic motivation, and emotion into the technology acceptance model. *Information Systems Research, 11*(4), 342. doi:10.1287/isre.11.4.342.11872

Venkatesh, V., & Davis, F. D. (2000). A theoretical extension of the technology acceptance model: Four longitudinal field studies. *Management Science, 46*(2), 186–204. doi:10.1287/mnsc.46.2.186.11926

Venkatesh, V., & Morris, M. G. (2000). Why don't men ever stop to ask for directions? Gender, social influence, and their role in technology acceptance and usage behavior. *Management Information Systems Quarterly, 24*(1), 115–139. doi:10.2307/3250981

Vetschera, R., Kersten, G., & Koeszegi, S. (2006). User assessment of Internet-based negotiation support systems: An exploratory study. *Journal of Organizational Computing and Electronic Commerce, 16*(2), 123. doi:10.1207/s15327744joce1602_3

Vygotsky, L. S. (1978). *Mind in Society*. Cambridge, MA: Harvard University Press.

Wade, M. R., & Parent, M. (2001). Relationships between job skills and performance: a study of webmasters. *Journal of Management Information Systems*, *18*(3), 71–76.

Wagner, E. L., & Piccoli, G. (2007). Moving beyond user participation to achieve successful IS design. *Communications of the ACM*, *50*(12), 51–55. doi:10.1145/1323688.1323694

Wang, L., Tao, J., Kunze, M., Castellanos, A. C., Kramer, D., & Karl, W. (2008) Scientific cloud computing: Early definition and experience. In *Proceedings of the 10th IEEE International Conference on High Performance Computing and Communications* (pp. 825-830).

Wang, T. D., Plaisant, C., Quinn, A. J., Stanchak, R., Murphy, S., & Shneiderman, B. (2008). Aligning temporal data by sentinel events: discovering patterns in electronic health records. In *Proceeding of the SIGCHI Conference on Human Factors in Computing Systems* (pp. 457-466).

Wang, S. (2005). Business software specifications for consumers: towards a standard format. *Journal of Organizational and End User Computing*, *17*(1), 23–37. doi:10.4018/joeuc.2005010102

Wang, Y.-S. (2003). Assessment of learner satisfaction with asynchronous electronic learning systems. *Information & Management*, *41*, 75–86. doi:10.1016/S0378-7206(03)00028-4

Watson-Manheim, M. B., & Belanger, F. (2002). Exploring Communication-Based Work Processes in Virtual Work Environments. In *Proceedings of the 35th Hawaii International Conference on System Sciences (HICSS-35)*.

Watson-Manheim, M. B., Chudoba, K. M., & Crowston, K. (2002). Discontinuities and continuities: A new way to understand virtual work. *Information Technology & People*, *15*(3), 191–209. doi:10.1108/09593840210444746

Watson, W. E., Johnson, L., & Merritt, D. (1998). Team orientation, self orientation, and diversity in task groups. *Group & Organization Management*, *23*(2), 161–188. doi:10.1177/1059601198232005

Webster, J., & Martocchio, J. J. (1992). Microcomputer Playfulness: Development of a Measure with Workplace Implications. *Management Information Systems Quarterly*, *16*(2), 201–226. doi:10.2307/249576

Webster, J., & Martocchio, J. J. (1993). Microcomputer playfulness: Development of a measure with workplace implications. *Management Information Systems Quarterly*, *16*(2), 201–226. doi:10.2307/249576

Weick, K. E. (1993a). Organizational Redesign as Improvisation. In Huber, G. P., & Glick, W. H. (Eds.), *Organizational Change and Redesign* (pp. 346–379). New York, NY: Oxford University Press.

Weick, K. E. (1993b). The Collapse of Sensemaking in Organizations: The Mann Gulch Disaster. *Administrative Science Quarterly*, *38*(4), 628–652. doi:10.2307/2393339

Weick, K. E. (1998). Improvisation as a Mindset for Organizational Analysis. *Organization Science*, *9*(5), 543–555. doi:10.1287/orsc.9.5.543

Weick, K. E., & Meader, D. K. (1993). Sensemaking and group support systems. In Jessup, L., & Valacich, J. (Eds.), *Group Support Systems: New Perspectives* (pp. 230–252). New York, NY: Macmillan.

Werner, J. M., & Lester, S. W. (2001). Applying a team effectiveness framework to the performance of student case teams. *Human Resource Development Quarterly*, *12*(4), 385–402. doi:10.1002/hrdq.1004

Wetzels, M., Odekerken-Schroder, G., & van Oppen, C. (2009). Using PLS path modeling for assessing hierarchical construct models: guidelines and empirical illustration. *Management Information Systems Quarterly*, *33*(1), 177–195.

Weyuker, E. (1982). On testing non-testable programs. *The Computer Journal*, *25*(4), 465–470.

Whitley, B. E. (1997). Gender Differences in Computer-Related Attitudes and Behavior: A Meta-Analysis. *Computers in Human Behavior*, *13*(1), 1–22. doi:10.1016/S0747-5632(96)00026-X

Wiegers, K. E. (1999). *Software requirements*. Redmond, WA: Microsoft Press.

Willcocks, L. P., & Sykes, R. (2000). The Role of the CIO and IT Function in ERP. *Communications of the ACM*, *43*(4), 32–38. doi:10.1145/332051.332065

Wilson, B., & Henseler, J. (2007). Modeling reflective higher-order constructs using three approaches with PLS path modeling: a Monte Carlo comparison. In M. Thyne, K. R. Deans, & J. Gnoth (Eds.), *Proceedings of the Australian and New Zealand Marketing Academy Conference* (pp. 791-800).

Wilson, G. V. (2006). Where's the real bottleneck in scientific computing? *American Scientist, 94*(1), 5–6.

Winston, E. R., & Dologite, D. (2002). How does attitude impact IT implementation: A study of small business owners. *Journal of Organizational and End User Computing, 14*(2), 16–29. doi:10.4018/joeuc.2002040102

Winter, S. J., Chudoba, K. M., & Gutek, B. A. (1997). Misplaced resources? Factors associated with computer literacy among end-users. *Information & Management, 32*, 29–42. doi:10.1016/S0378-7206(96)01086-5

Wirt, J., & Livingston, A. (2004). *Condition of Education 2002 in Brief (Rep. No. 2002011)*. Washington, DC: National Center for Education Statistics.

Wit, B., & Meyer, R. (2005). *Strategy synthesis: Resolving strategy paradoxes to create competitive advantage* (2nd ed.). London: International Thomson Business.

Wixom, B. H., & Todd, P. A. (2005). A theoretical integration of user satisfaction and technology acceptance. *Information Systems Research, 16*(1), 85–102. doi:10.1287/isre.1050.0042

Woodroof, J., & Burg, W. (2003). Satisfaction/Dissatisfaction: Are Users Predisposed? *Information & Management, 40*, 317–324. doi:10.1016/S0378-7206(02)00013-7

Woodrow, J. E. (1991). A comparison of four computer attitude scales. *Journal of Educational Computing Research, 7*(2), 165–187. doi:10.2190/WLAM-P42V-12A3-4LLQ

Wood, W. A., & Kleb, W. L. (2003). Exploring XP for scientific research. *IEEE Software, 20*(3), 30–36. doi:10.1109/MS.2003.1196317

Wood, W., & Rhodes, N. D. (1992). Sex differences in interaction style in task groups. In Ridgeway, C. (Ed.), *Gender, Interaction, and Inequality* (pp. 97–121). New York: Springer-Verlag.

Wrzesniewski, A., & Dutton, J. (2001). Crafting A Job: Revisioning Employees as Active Crafters of Their Work. *Academy of Management Review, 28*(2), 179–201.

Wynekoop, J. L., & Walz, D. B. (1998). Revisiting the perennial question: are IT people different? *The Data Base for Advances in Information Systems, 29*(2), 62–72.

Xia, W., & Lee, G. (2005). Complexity of Information Systems Development Projects: Conceptualization and Measurement Development. *Journal of Management Information Systems, 22*(1), 45–83.

Yamnill, S., & McLean, G. N. (2001). Theories supporting transfer of training. *Human Resource Development Quarterly, 12*(2), 195–208. doi:10.1002/hrdq.7

Yang, Z., Cai, S., Zhou, Z., & Zhou, N. (2005). Development and validation of an instrument to measure user perceived service quality of information presenting Web portals. *Information & Management, 42*, 575–589. doi:10.1016/S0378-7206(04)00073-4

Ye, N. (1991). *Development and Validation of a Cognitive Model of Human Knowledge System: Toward an Effective Adaptation to Differences in Cognitive Skills*. Unpublished doctoral dissertation, Purdue University, Lafayette, IN.

Yi, M. Y., & Im, K. S. (2004). Predicting computer task performance: personal goal and self-efficacy. *Journal of Organizational and End User Computing, 16*(2), 20–37. doi:10.4018/joeuc.2004040102

Yin, R. K. (2002). *Case study research: Design and methods (applied social research methods)* (3rd ed.). Thousand Oaks, CA: Sage.

Yoo, Y., & Alavi, M. (2001). Media and group cohesion: Relative influences on social presence, task participation, and group consensus. *Management Information Systems Quarterly, 25*, 371–390. doi:10.2307/3250922

Zaltman, G., Duncan, R., & Holbek, J. (1973). *Innovations and Organizations*. New York: Wiley.

Zerhouni, E. (2007). Translational research: Moving discovery to practice. *Clinical Pharmacology and Therapeutics, 81*, 126–128. doi:10.1038/sj.clpt.6100029

Zhao, B. Y., Yan, K., Agrawal, D., & El Abbadi, A. (2009). *Massive graphics in clusters (MAGIC) project*. Retrieved January 14, 2010, from http://graphs.cs.ucsb.edu

Zhu, K., & Kraemer, K. L. (2002). E-commerce Metrics for Net-Enhanced Organisations: Assessing the Value of e-Commerce to Firm Performance in the Manufacturing Sector. *Information Systems Research, 13*(3), 275–295. doi:10.1287/isre.13.3.275.82

Zimmerman, A., & Finholt, T. A. (2007). Growing an infrastructure: The role of gateway organizations in cultivating new communities of users. In *Proceedings of the International ACM SIGGROUP Conference on Supporting Group Work* (pp. 239-248).

Zimmerman, A., & Nardi, B. (2006). Whither or whether HCI: Requirements analysis for multi-sited, multi-user cyberinfrastructures. In *Proceedings of CHI '06: Extended Abstracts on Human Factors in Computing Systems* (pp. 1601-1606).

Zviran, M., Pliskin, N., & Levin, R. (2005). Measuring user satisfaction and perceived usefulness in the ERP context. *Journal of Computer Information Systems, 45*(3), 43–52.

About the Contributors

Ashish Dwivedi is the Deputy Director, Centre for Systems Studies at Hull University Business School, UK. Previously, he was the Deputy Graduate Research Director at Hull University Business School, and was also associated with the management of the high-tech Management Learning Laboratory). His primary research interests are in knowledge management (in which he obtained his PhD), supply chain management healthcare management and information and communication technologies. He has published 4 books and over 50 journal and conference papers. He has served as an Invited reviewer and Guest Editor for several journals, including the IEEE Transactions on Information Technology in Biomedicine.

Steve Clarke, Ph.D.received a BSc in Economics from The University of Kingston Upon Hull, an MBA from the University of Luton and a PhD in human centred approaches to information systems development from Brunel University (UK). He is a professor of Information Systems in the University of Hull Business School (UK). He has extensive experience in management systems and information systems consultancy and research, focusing primarily on the identification and satisfaction of user needs and issues connected with knowledge management. His research interests include: social theory and information systems practice; strategic planning; and the impact of user involvement in the development of management systems. Professor Clarke is the co-editor of two books, Socio-Technical and Human Cognition Elements of Information Systems, 2003 published by Idea Group Publishing and Beyond Knowledge Management, 2004 published by Idea Group Publishing.

* * *

Yongmei Bentley has a PhD in information systems management and is now Senior Lecturer of Business Systems at the University of Bedfordshire. She has taught a range of business and management related subjects at both undergraduate and postgraduate levels. She has also taken an active part in a number of EC-funded research projects working with partners across Europe and from China in areas such as the applications of ICTs, supply-chain management, and the development of e-learning courses for SMEs. Her current research interests include information systems management, applications of e-learning technologies, the impact of climate change on logistics management, and supply chain management in recessionary times. She has published a number of papers related to her teaching and research. She has also been a referee, session chair and committee member at a number of international conferences, and has refereed papers for a variety of academic journals.

Chad Anderson is a Ph.D. Candidate in the Computer Information Systems Department of Georgia State University. He holds an MBA from Eastern Kentucky University and Bachelor of Science degrees in Occupational Therapy and Psychology from Eastern Kentucky University and Business Administration from North Dakota State University. His work has been presented at the Workshop on HCI Research in MIS and the Southern Association for Information Systems Conference. Chad is currently the Associate Managing Editor for MIS Quarterly.

Said S. Al-Gahtani is an associate professor of computer information systems in the Department of Administrative Sciences at King Khalid University, Abha, Saudi Arabia. He has a B.Sc. in Systems Engineering from King Fahad University of Petroleum & Minerals, M.Sc. in Computer Sciences from Atlanta University, Atlanta, Georgia, and a PhD in Computer-Based Information Systems from Lougborough University, Loughborough, UK. His research interests include the user acceptance of information technologies, the modeling of IT acceptance, end-user computing, and organizational cross-cultural research. He has published journal research articles in Information & Management, Information Technology & People, Journal of Global Information Management, Journal of Global Information Technology Management, Information Technology for Development, Information Resources Management Journal, and the Behaviour & Information Technology.

Geoffrey S. Hubona currently holds a faculty appointment as an Associate Professor of computer information systems in the J. Mack Robinson College of Business at Georgia State University in Atlanta, Georgia. He received a BA in Psychology from the University of Virginia (1972), an MBA from George Mason University (1980), and a Ph.D. in MIS from the University of South Florida (1993). His research interests include the user acceptance of information technologies, the human perception of computer visualizations, and methodological issues relating to quantitative research in MIS. In addition to the Journal of Organizational and End User Computing, he has published journal articles in MIS Quarterly, ACM Transactions on Computer-Human Interaction, IEEE Transactions on Systems, Man and Cybernetics, Part A: Systems and Humans, DATA BASE for Advances in Information Systems, Information & Management, Information Technology & People, International Journal of Human Computer Studies, International Journal of Technology and Human Interaction, and the Journal of Information Technology Management.

Yinglei Wang is an Assistant Professor at the Fred C. Manning School of Business at Acadia University. His research interests include virtual organizations, knowledge management, e-learning, and information technology adoption. His work has been published in scholarly journals such as Information & Management and Information Systems Journal, as well as various conference proceedings. He received his PhD from the Richard Ivey School of Business at the University of Western Ontario.

Darren Meister is the Faculty Director of the HBA and MSc Programs and an Associate Professor of Information Systems at the Richard Ivey School of Business. His interests focus on the role of technology in enhancing organizational effectiveness, specifically as it concerns innovation processes. He investigates this question primarily within three settings: technology adoption, knowledge management and interorganizational systems. His work has appeared in Management Science, MIS Quarterly and other leading journals and conferences. He was a Rotary International Foundation scholar, attending the University of Cambridge. Subsequently, he earned his PhD at the University of Waterloo, Canada.

Peter Gray is an Assistant Professor at the McIntire School of Commerce, University of Virginia. He has a background in electronic commerce and online information services, and has worked in a variety of information technology and management consulting positions. Peter's research interests include knowledge management and knowledge management systems, virtual teams, computer-supported cooperative work, computer-mediated communication, social technologies, and online communities. He has published in the MIS Quarterly, Management Science, Information Systems Research, the Journal of Management Information Systems, the Journal of Strategic Information Systems, Information Technology & People, Information & Management, Decision Support Systems, and the Communications of the AIS, as well as various national and international conference proceedings. His PhD in Management is from Queen's University (Canada).

Sandra Barker B. Sc. Chem, is a Lecturer in the School of Management at the University of South Australia. She is involved in teaching Information and Resource Management as well as Organisational Learning and Leadership to undergraduate and postgraduate students locally and overseas. Her research interests include management of end users, end-user computing, distance education and the use of technology in tertiary education. She is currently completing a PhD thesis investigating the management of end-users and end-user development of applications.

Brenton Fiedler B.A.Acc, M.Bus, F.C.A., is an Associate Professor and the Associate Head of the School of Commerce at the University of South Australia. He is involved in teaching Auditing at undergraduate and postgraduate levels (both local and international) and his research interests include auditing, accounting education, end-user computing, and resident funded retirement village accounting.

Brian Bishop is a researcher and software developer in the Software Technology Research Centre at Dundalk Institute of Technology. His areas of interest are spreadsheet development and testing tools, and the interaction between end-users and spreadsheet applications. After receiving a BSc in Commercial Computing, he completed an MSc in the area of end-user spreadsheet engineering and reliability.

Kevin McDaid is a lecturer in Computing at Dundalk Institute of Technology. He leads research in the area of Spreadsheet Engineering in the Software Technology Research Centre. His background is in Mathematics and Statistics and his research focuses on the application of statistical methods to decision problems in software development. Before working at Dundalk, he lectured at Dun Laoghaire Institute of Art, Design and Technology and was employed as a Statistician with the Central Statistics Office in Dublin, Ireland.

Yuan Li is an assistant professor in the Division of Business, Mathematics and Sciences at Columbia College in Columbia, South Carolina, U.S.A. He received his Ph.D. in Management Information Systems from the University of South Carolina. His current research focuses on knowledge management at the organizational and individual levels, and knowledge and skills transfer in end user computer training. His research appeared in the Journal of the Association for Information Systems and the European Journal of Information Systems.

Kuo-Chung Chang is an assistant professor of the Department of Information Management at Yuan Ze University, Taiwan, R.O.C. He received his Ph.D. from the University of South Carolina, U.S.A. His current research focuses on IS project management, information security, and knowledge management. His work has been published in journals such as Information and Management, Industrial Management and Data Systems, Information and Software Technology, and Journal of Systems and Software.

Laura L. Hall is an Associate Professor in the Department of Information and Decision Sciences, in the College of Business at the University of Texas at El Paso. Dr. Hall's major research interests are in e-Commerce with emphasis on supply chain management, distance education development, and minority workforce training. She has published in the Journal of Organizational Computing and Electronic Commerce, the Journal of Behavior and Information Technology, Information Resource Management, the International Online Conference on Teaching Online in Higher Education, the Proceedings of the Decision Sciences Institute, and the Proceedings on Human Computer Interaction.

Roy D. Johnson is currently an Extraordinary Professor of Informatics at the University of Pretoria where he previously served as a Fulbright Scholar. His research focuses on Systems Analysis, Project Management and the Transfer of Knowledge. His current work involves the study of critical and reflective thinking. He is involved in organizing and promoting Active Learning in the classroom and has published in numerous journals. He currently serves as Association for Information Systems (AIS) VP of Accreditation and was previously elected AIS VP of Education. He was the founder in 1986 of the Academy of Information Systems (IAIM), which is now the special interest group of AIS for education (SIGED: IAIM).

Lixuan Zhang is Assistant Professor in the Hull College of Business at Augusta State University, Augusta, GA. She holds a MS and MBA in MIS from the University of Oklahoma and a Ph.D. from the University of North Texas. Her papers have appeared, or are forthcoming, in the Journal of Information Technology and Management, Journal of Organizational and End User Computing, Information Resources Management Journal and Information Systems Management.

Mary C. Jones is Professor of Information Systems and Chair of the Information Technology and Decision Sciences Department at the University of North Texas. She received her doctorate from the University of Oklahoma in 1990. Her research interests are in the organizational impacts of emerging and large scale information technologies. Her work appears in numerous journals including MIS Quarterly, European Journal of Information Systems, Behavioral Science, Decision Support Systems, Communications of the Association for Information Systems, and Information and Management.

Dewi Rooslani Tojib is a Research Fellow in the Department of Marketing at Monash University. She holds Bachelor of Business Systems (H1) and Doctor of Philosophy (PhD) from Monash University. Her research has been published in a number of academic and practitioner journals as well as presented in numerous international conferences. Her research interests include web-based Information Systems, mobile technology, consumer behavior, services marketing, experimental designs and choice modeling.

Ly-Fie Sugianto is Senior Lecturer in the Faculty of Business and Economics, Monash University. She has coauthored 70+ refereed articles published in journals and conference proceedings. She has also received a number of awards and grants for her work and contribution in Optimization and IS research. Her research reflects her on-going interests in the support tools and techniques for intelligent decision making. Other positions she has held include serving as program and international board committee for international conferences and journal editorial. Ly-Fie has been appointed as an Expert of International Standing by the Australian Research Council College of Expert.

Richard D. Johnson (Ph.D. University of Maryland) is an Assistant Professor of Management at the University at Albany, State University of New York. His primary interests focus on HRIS, psychological and sociological impacts of computing technology, e-learning, and issues surrounding the digital divide. Dr. Johnson's research has appeared in journals such as *Information Systems Research, Journal of the Association for Information Systems, Journal of Applied Social Psychology,* and *International Journal of Human Computer Studies*. He teaches courses on Human Resources and human resource information systems.

Katherine M. Chudoba is Associate Professor of MIS in the Jon M. Huntsman School of Business at Utah State University. Her research focuses on the nature of work in distributed environments, and how Information and Communication Technologies (ICTs) are used and integrated into work practices. She has published in journals such as *MIS Quarterly, Organization Science,* and *Information Systems Journal*. She earned her Ph.D. at the University of Arizona, and her bachelor's degree and MBA at the College of William and Mary. Before joining academe, she worked as an analyst and manager with IBM.

Mary Beth Watson-Manheim is an Associate Professor in the Information Decision Sciences Department in the College of Business Administration at the University of Illinois, Chicago. She obtained her Ph.D. in Information Technology Management from Georgia Institute of Technology. She is actively involved in research on issues related to the use of information and communication technologies (ICT) in distributed work environments. Her work has been published in *MIS Quarterly, Journal of Management Information Systems, Information Systems Journal, Information Technology and People, MIS Quarterly Executive, IEEE Transactions on Professional Communications, Group Decision and Negotiation* and others. Her research has been funded by Intel Corporation, IBM Corporation, Lotus Development Corporation, and various university centers. She was awarded a Fulbright-Nehru Senior Research Scholarship to India for 2009-2010. Prior to obtaining her PhD, she worked in the telecommunications industry.

Kevin Crowston joined the School of Information Studies at Syracuse University in 1996. He received his Ph.D. in Information Technologies from the Sloan School of Management, Massachusetts Institute of Technology (MIT) in 1991. Before moving to Syracuse he was a founding member of the Collaboratory for Research on Electronic Work at the University of Michigan and of the Centre for Coordination Science at MIT. His current research focuses on new ways of organizing made possible by the extensive use of information technology.

Chei Sian Lee obtained her PhD in Management Information Systems from the University of Illinois at Chicago and is currently an Assistant Professor of the Division of Information Studies, Wee Kim Wee School of Communication and Information at Nanyang Technological University. Her research interests include computer-mediated communication, distributed work environment, and social computing. Her work has been published in international journals and conference proceedings.

Tim Klaus is an Associate Professor of Management Information Systems at Texas A&M University – Corpus Christi. He earned his PhD (Management Information Systems) from University of South Florida. His primary research interests are User Resistance, ERP implementations, IT personnel, and Web Usage. He has published papers in journals such as Communications of the ACM (CACM), Journal of International Technology (JIT), and European Journal of Information Systems (EJIS). He also is a consultant in the area of IT-enabled change, helping organizations better understand the process of change as well as the impact of user attitudes and behaviors.

William J. Doll is a Professor of MIS and Strategic Management at the University of Toledo. Dr. Doll holds a doctoral degree in Business Administration from Kent State University and has published extensively on information system and manufacturing issues in academic and professional journals including *Management Science, Communications of the ACM, MIS Quarterly, Academy of Management Journal, Journal of Management Information Systems, and Information Systems Research*. Dr. Doll has published on a variety of topics including system success measures, computer integrated manufacturing, executive steering committees, top management involvement in MIS development, strategic information systems, information systems downsizing, and end-user computing.

Xiaodong Deng is an Associate Professor of Management Information Systems at Oakland University. He received his Ph.D. in Manufacturing Management from The University of Toledo. His research has been appeared in *Journal of Management Information Systems, Information and Management, Decision Sciences, Information Resources Management Journal, and Journal of Intelligent Manufacturing*. His research interests are in post-implementation information technology learning, information systems benchmarking, and information technology acceptance and diffusion.

James P. Downey is an associate professor in the MIS Department in the College of Business at the University of Central Arkansas. He received his Ph.D. in Management Information Systems from Auburn University. He spent 25 years as a Naval officer, including a tour at the U.S. Naval Academy, before leaving the Navy in November 2004. His current research interests include project management, database management, and individual differences in behavior in human-computer interactions and end-user computing. He has published articles in *Journal of Organizational and End User Computing, International Journal of Training & Development, Journal of Information Systems Education, Interacting with Computers, Journal of Information Technology Education*, among others.

Lloyd A. Smith is Professor and Chair of the Computer Science Department in the College of Natural and Applied Science, in Springfield, MO. He received a bachelor's degree in Music, a master's degree in Speech Communication, and a Ph.D. in Computer Science from the University of North Texas. His research interests include applications of speech recognition, computer analysis of music and data and text mining. He was a member of the New Zealand Digital Library project and the Waikato Environment for Knowledge Analysis (WEKA) data mining project at the University of Waikato, in Hamilton, New Zealand. He has published in journals ranging from the *Journal of the Acoustical Society of America* to the *Journal of Quantitative Analysis in Sports*.

Jeffrey C. Carver received the PhD degree in Computer Science from the University of Maryland. He is an assistant professor in the Department of Computer Science at the University of Alabama. His main research interests include software engineering for computational science and engineering, empirical software engineering, software quality, software architecture, human factors in software engineering and software process improvement. Carver has a PhD in computer science from the University of Maryland. He is a member of the IEEE Computer Society and the ACM.

Parmit K. Chilana is a PhD student at the Information School of the University of Washington (UW), specializing in human-computer interaction (HCI). A research interest of Parmit has been the application of usability and user-centered design principles for improving software design in biomedical and health informatics domains and she has collaborated on various projects within UW Medicine. Based on this experience, she has done further investigation into the challenges of usability practices in other highly complex domains and has synthesized pedagogical implications for HCI. Parmit's current research focus is on better understanding post-deployment software usability and redesigning help interfaces. Her recent projects have looked at how users express unwanted software behaviors and how support professionals and developers respond to user-reported software issues in open source and commercial software development contexts. She is currently exploring the design of contextual-help tools for web applications which leverage crowdsourced solutions from users. Parmit received her MS in Library and Information Science from the University of Illinois at Urbana-Champaign and BSc in Computing Science from Simon Fraser University, Canada.

Elishema Fishman is a recent graduate of the MS in Information Management program at the University of Washington's Information School. She holds a BA in English from UCLA. Elishema's research interests lie in user-centered design and content management, with a goal of effectively looking at and understanding user needs. Throughout her studies at the University of Washington, Elishema worked as a Research Assistant at the Institute of Translational Health Sciences, part of UW Medicine. Elishema spent several years working as a public relations specialist, prior to obtaining her Masters degree where she gained experience in writing and editing. Elishema's long-term career interests lie in information architecture with a focus on User Experience design.

Estella Geraghty is an Assistant Professor of Clinical Internal Medicine at the University of California, Davis. She earned her MD from UC Davis in 2002 and also holds both Masters of Medical Informatics and Masters in Public Health degrees from UC Davis. Dr. Geraghty is board certified in Internal Medicine and is also among the charter class of Certified in Public Health professionals. Her research interests revolve around spatial epidemiology and geographic information systems (GIS) as methodologies for understanding the interplay between health and the environment. One current project investigates the relationship between aerial pesticide spraying for West Nile virus and health effects. She is also working on a multi-disciplinary, multi-scalar, mixed-method approach to understanding youth outcomes in a

9-county Sacramento region. Outcomes, including health, are analyzed by their geographies to under-stand disparities and vulnerabilities in the population. For the last two years she has been involved in the evaluation component of the Cohort Discovery Tool, powered by i2b2 (informatics for integrating the bench and the bedside). This is a multi-institution project leveraging EMR data to improve cohort discovery among NIH designated CTSAs.

Peter Tarczy-Hornoch is an elected Fellow of the American College of Medical Informatics and an elected member of the Society for Pediatric Research. He serves as the Head of the Division of Bio-medical and Health Informatics. He also serves in a variety of leadership roles throughout the School of Medicine including leading the research and service activities of the Biomedical Informatics Core of the Institute of Translational Health Sciences (the regional CTSA award) and serving as the Director of Research and Data Integration for UW Medicine Information Technology Services (the operational clinical computing group). His current research focuses on data integration of biomedical and health data including looking at ways of handling semi structured data, representing uncertainty at various levels in the system, and doing computerized reasoning over integrated data. His research builds on collaborations with biologists and clinical and translational researchers looking at: a) large scale func-tional gene annotation, b) SNPs for elucidation of disease mechanisms, and c) as part of the Institute of Translational Health Sciences and the Northwest Institute of Genetic Medicine research in the area of collaborative integrated analysis of a combination of clinical data, experimental biological data, and clinical/translational research study data.

Fredric M. (Fred) Wolf is Professor and Chair of the Department of Medical Education and Bio-medical and Health Informatics in the School of Medicine and Adjunct Professor of Health Services and Epidemiology in the School of Public Health at the University of Washington. He was formerly Professor of Medical Education and Director of the Learning Resource Center and the Laboratory for Computing and Cognition at the University of Michigan Medical School. He has many years of experi-ence in educational psychology/evaluation and measurement, medical education, and health services research. He is a member of the international Cochrane Collaboration and former Visiting Scholar at UK Cochrane Centre and Green College, University of Oxford. His research has focused on a) dissemina-tion and evaluation of new technology, including decision support systems, b) clinical decision making and judgment under uncertainty, c) evidence based medicine, systematic reviews and meta-analysis of educational and healthcare interventions, and d) evaluation of clinical and translational research inter-ventions and training.

Nick Anderson's academic research areas include user needs analysis of information management issues faced by small research laboratories, clinical decision support, knowledge transformation and de-livery, and the study of how information tools support policy development for biospecimen data sharing. He is the Principal Investigator of the Cross-Institutional Clinical Translational Research project(CICTR), which is building a collaboration to provide query discovery across de-identified clinical data from 3 academic research hospitals for cohort recruitment, and Co-Investigator on the Cancer Biospecimen Portal project, a collaboration between the University of Washington, Fred Hutchinson Cancer Research Center and Seattle Children's Research Institute. He also leads a range of collaborative large-scale clinical data projects that are exploring the boundaries of sharing of sensitive clinical information under evolving regulatory policies. He has faculty appointments as Assistant Professor in the Division of Biomedical Health Informatics and Adjunct Assistant Professor in the Department of Bioethics and Humanities at the University of Washington and is the Associate Director of the Biomedical Informatics Core for the Institute of Translational Health Sciences (ITHS).

Lorin Hochstein is a computer scientist at the University of Southern California's Information Sciences Institute (ISI). His current research focuses on improving the productivity of computational science and engineering software developers and end-users, in particular through the application of high-performance computing and cloud computing. Prior to ISI, he was an assistant professor in the Computer Science & Engineering Department at the University of Nebraska-Lincoln. He received a PhD in computer science from the University of Maryland, an M.S. in electrical engineering from Boston University, and a B.Eng. in computer engineering from McGill University.

Brian Schott is a Project Leader at the University of Southern California's Information Sciences Institute (ISI). He has over 15 years experience in the development of high performance computing systems, embedded computing systems, and computer networks. Presently, he leads the Dynamic On-Demand Computing Systems (DODCS) project, which is extending existing cloud computing technologies to support heterogeneous architectures and dynamic network topologies for technical computing applications. Since starting at ISI in 1997, Mr. Schott has managed DARPA, NASA, and DoD-funded research projects involving large multidisciplinary teams of researchers from leading academic institutions, national laboratories, and industry. Mr. Schott has two patents. He received an MS in computer science from George Washington University and a BS in computer science from the University of Maryland.

Robert B. Graybill is CEO and President of Nimbis Services. He has more than 35 years of HPC-related senior-level experience as a business leader, government program manager and technology researcher. Prior to founding Nimbis, he was Director of National Innovation Initiatives at the University of Southern California's Information Sciences Institute (ISI), where he fostered development of advanced national high performance computing (HPC) collaborative environments to help companies, universities and national laboratories share high performance computing systems and computational science expertise. Before joining ISI, he spent six years at DARPA, where he designed, developed and implemented six new transformational programs in high-end computing architectures and responsive embedded computing hardware, software and network systems. He was a member of the Senior Science Team for government HPC studies conducted by the Defense Science Board task force on DoD Supercomputing Needs and the High-End Computing Revitalization Task Force. He has an MS in computer science from Johns Hopkins University and a BS in electrical engineering from Pennsylvania State University.

Judith Segal works in the Department of Computing at the Open University in the UK and is a member of the Empirical Studies of Software Development research group. She has published widely on the practice of scientific software development. She has a PhD in Algebra and her current research originated in her investigations as to how mathematicians use and develop software.

Chris Morris is a software developer and project manager at Daresbury Laboratory UK, part of the Science and Technology Facilities Council. He has been developing software for over twenty years, including roles at an internet service provider, and in the electricity supply industry. Eventually, he realised that the coding is not the hardest part of the job. He has a degree in pure mathematics from the Queen's College, Oxford.

Diane Kelly is an associate professor at the Royal Military College of Canada (RMC). She has over twenty years of industrial experience in scientific software development and has spent the past ten years combining academic research with what she's learned from her industrial experience. Dr. Kelly has a B.Sc in pure mathematics and B.Ed. in mathematics and computer science, both from the University of Toronto. Her M.Eng and PhD are in software engineering, both from RMC. She is a member of ACM and a senior member of IEEE.

Index